Space, Time and Language in Plutarch

Edited by
Aristoula Georgiadou and Katerina Oikonomopoulou

DE GRUYTER

ISBN 978-3-11-053771-0
e-ISBN (PDF) 978-3-11-053947-9
e-ISBN (EPUB) 978-3-11-053811-3
ISSN 1862-1139

Library of Congress Cataloging-in-Publication Data
A CIP catalog record for this book has been applied for at the Library of Congress.

Bibliographic information published by the Deutsche Nationalbibliothek
The Deutsche Nationalbibliothek lists this publication in the Deutsche Nationalbibliografie; detailed bibliographic data are available on the Internet at http://dnb.dnb.de.

Printing and binding: CPI books GmbH, Leck
♾ Printed on acid-free paper
Printed in Germany

MIX
Papier aus verantwortungsvollen Quellen
FSC® C083411
www.fsc.org

www.degruyter.com

To the memory of
Françoise Frazier

Contents

9 Space, place, landscape: symbolic and metaphorical aspects

List of contributors

Alcalde Martín Carlos, Universidad de Málaga

Alexiou Evangelos, Aristotle University of Thessaloniki

Almagor Eran, Independent Scholar

Aloumpi Myrto, University of Oxford

Beck Mark, University of South Carolina

Berardi Elisabetta, Università degli Studi di Torino

Brenk Frederick E., Pontificio Istituto Biblico

Catanzaro Andrea, Università degli Studi di Genova

Demulder Bram, Katholieke Universiteit Leuven

Desideri Paolo, Università degli Studi di Firenze

Driscoll David F., University of Miami

Duff Timothy E., University of Reading

Fernández Delgado José Antonio, Universidad de Salamanca

Fletcher Lucy, University of Reading

†Frazier Françoise, Université Paris Ouest-Nanterre La Défense

Geiger Joseph, The Hebrew University of Jerusalem

Goeken Johann, Université de Strasbourg

Jacobs Susan, Independent Scholar

Kim Lawrence, Trinity University, TX

Lipka Michael, University of Patras

Lucchesi Michele A., Facoltà Teologica dell'Italia Settentrionale, Torino

Meeusen Michiel, King's College London

Nikolaidis Anastasios G., University of Crete

Oikonomopoulou Katerina, University of Patras

Pelling Christopher B. R., University of Oxford

Pordomingo Francisca, Universidad de Salamanca

Pugh Ginn Joshua, University of Cambridge

Roskam Geert, Katholieke Universiteit Leuven

Vamvouri Ruffy Maria, Université de Lausannne

Volpe Cacciatore Paola, Università degli Studi di Salerno

Xenophontos Sophia, University of Glasgow

Preface

The present volume derives its inspiration from the papers presented at the 10th Conference of the International Plutarch Society, titled *Space, Time and Language in Plutarch's Visions of Greek Culture: Introversion, Imperial Cosmopolitanism and Other Forms of Interaction with the Past and Present*, which was held at the European Cultural Center at Delphi, 16 – 18 May 2014. Our choice of Delphi as a venue for the meeting was closely connected with Plutarch's long-standing and deep ties with the city and its sanctuary: ties which extended from his visit to Delphi with his teacher Ammonius on the occasion of Nero's tour of Greece in 67 (*The E at Delphi* 385B) to his election as one of the two permanent priests of the shrine (perhaps in Trajan's reign),[1] down to the reign of Hadrian, when as *epimelete* of the Delphic Amphictyony he supervised the erection of a statue for the emperor (Dittenberger, *Syll.*^{3} 829A). The wide range of participants included faculty from North American Universities, the United Kingdom, Austria, France, Italy, Germany, Belgium, Switzerland and Spain, Portugal, Greece, Israel, Poland, as well as researchers and graduate students in Plutarchan studies.

The papers that were delivered at the Conference aimed to demonstrate how in Plutarch's works spaces, geographical sites, topographical landmarks, historical locations and locales, religious and mythological landscapes (real or imagined) can prompt reflection on a variety of issues: these include the relationship between local culture (in the Greek cities) and the Roman Empire (an inclusive, cosmopolitan space); the nature of the different kinds of interactions (cultural, military, linguistic, mythological and other) among Greeks, Romans and others at different moments in history (thus opening an avenue for understanding Plutarch's perception and construction of time); and the uses of spatial and temporal concepts and terminology in Plutarch's works.

The present volume includes revised and expanded versions of some of the papers presented at the Conference, with an additional contribution by Mark Beck. It addresses not only Plutarch scholars and Classicists, but anyone in the Humanities and Social Sciences interested in the concepts of space and time, and their codification through literary discourse.

Naturally, this volume does not exhaust all research avenues into the topics of space, time and language, as far as Plutarch is concerned. A next step would involve exploring Plutarch's handling of time and space in relation to other imperial authors, Greek or Latin. Moreover, it would be beneficial to enquire whether there are divergences in the concepts of time and space (and their linguistic representation) between the *Moralia* and the *Lives*, or across the different genres in which Plutarch writes. But we believe that what emerges clearly from all contributions is both con-

1 Jones (1971) 26, 31.

cepts' unquestionable value for gaining a richer understanding of Plutarch's engagement with the past, his versions of Greek *paideia*, his philosophical layers, and his biographical techniques and moralism.

* * *

The 10th Conference of the International Plutarch Society would not have been feasible without the financial assistance of the International Plutarch Society (I.P.S.), the Municipality of Delphi, the Archdiocese of Thebes and Lebadeia, and the University of Patras Special Account for Research Funds (ΕΛΚΕ). We gratefully acknowledge their support.

Warm thanks go to Anastatios Nikolaidis, Christopher Pelling and Frances Titchener, who provided valuable advice on organisational matters as well as on the preparation and publication of this volume. We are especially grateful to Vasiliki Maria Vlachaki, who at the time was a postgraduate student at the University of Patras, for the zeal and efficiency with which she assisted us at all phases of the Conference's preparation. We are also indebted to Angeliki Tzanetou, who offered stimulating insights and sharp observations on the notions of space and time at critical moments of the project. Last but not least, we wish to express our gratitude to Peter von Möllendorff, the editor of the Millenium-Studien, for his advice and guidance during the preparation of this volume, as well as to the editorial team of the series for overseeing this book's passage into print.

Aristoula Georgiadou and Katerina Oikonomopoulou

Introduction: Reading Plutarch through space, time and language

The confluence of space, time and language: Plutarch's Delphi

Delphi is a place that 'is essential to understanding Plutarch in his historical and social context'.[1] In Plutarch's Pythian dialogues (*The E at Delphi*, *On the Oracles of the Pythia*, *The Obsolescence of Oracles*) the sanctuary is described in a manner that is reminiscent of Pausanias' descriptions of Greek religious sites (especially Delphi and Olympia) and their monuments. According to Jaś Elsner, 'Together, woven as a web of interconnected cross-references, the places and objects (that part of the Pausanian project which actually is a descriptive topography of Greece) constitute much more than a material account: they evoke, they *are* an imaginative geography in which each site and all the sites together are infused with the myth-historical essence of Greekness'.[2] For Plutarch, too, Delphi was a sanctuary of panhellenic significance, and a place whose monuments and dedications evoked manifold episodes of Greek myth and history. Precisely because of this, Delphi provides an apt introduction into the ways the three concepts (space, time, language) that form the main focus of this volume interact and fertilise each other (Pelling in this volume). In the Pythian dialogues, the sanctuary of Delphi is not simply the backdrop to the recorded conversations, but also functions as a place of pilgrimage for people travelling from distant parts of the world, as a repository of valuable objects,[3] and as a sacred space that triggers reflection on the past and present and prompts enquiry into oracular language and metre (Brenk, Kim and Lucchesi in this volume). Pythia, the priestess of Delphi, herself represents a confluence of the three concepts, through her ability to travel across space and time, in all directions simultaneously, and her divine way of prophesying.[4]

The frame dialogue between Basilocles and Philinus in *On the Oracles of the Pythia*[5] provides a most vivid illustration of how the three concepts are intimately linked (394E, transl. F.C. Babbitt, Loeb):

1 Stadter (2004) 19.
2 Elsner (2001) 6.
3 Bal (2009) 138: 'objects have a spatial status'.
4 Purves (2010) 154.
5 Widely known as *De Pythiae Oraculis*. See Brenk and Kim in this volume.

https://doi.org/10.1515/9783110539479-001

> *Basilocles.* You people have kept it up till well into the evening, Philinus, escorting the foreign visitor around among the statues and votive offerings. For my part, I had almost given up waiting for you.
>
> *Philinus.* The fact is, Basilocles, that we went slowly, sowing words, and reaping them straightway with strife, like the men sprung from the Dragon's teeth, words with meanings behind them of the contentious sort, which sprang up and flourished along our way.

The space that the sanctuary of Delphi and its monuments and treasures occupy offers visitors who travel to the site a profound religious and cultural experience. Philinus' reference to the men sprung from the Dragon's teeth (Σπαρτοί) on one level evokes the foundation myth of Thebes according to which Cadmus was given instructions by the Delphic oracle to found his city. Thus, the myth accentuates the oracle's omniscient command of Greek history and its diachronic involvement in the shaping of Greek identity. On another level, Philinus' comparison of the Spartoi to the 'warlike' conversations (λόγους ... πολεμικούς) that sprouted and grew, as it were (βλαστάνοντας ... καὶ ὑποφυομένους), from the occasion of the interlocutors' tour of the sanctuary underscores the site's capacity to engender combative discourse.[6] Space, time and language are inextricably woven and as such decisively shape the texture of the ensuing dialogue. As Frederick Brenk states in this volume, 'Though highly engaged with the past, this is a dialogue which also points to the future, both of the Roman Empire and civilisation to come. In his desire for one world and universal peace, the real prophetic voice is no longer that of the Pythia, but of Theon'. Thus, '[t]he new space spoken of at the end [of the dialogue] is that of the Roman Empire, the new time, the present, and the new language, the prose of the Pythia' (pp. 86 and 85, respectively).

Methodological approaches to space and time

The so-called 'spatial turn' in the study of ancient Greek and Latin literature[7] has helped spur a new understanding of the role descriptions of space play across different genres (such as epic, historiography, novel, biography). As scholarship has repeatedly shown, geographical locations and locales in ancient texts are not merely background settings for action or discussion, nor are they always portrayed in terms that we associate with 'scientific' geography: rather, ancient authors represent or imagine spaces in ways that are suggestive of how those spaces were experienced by human agents, and invested with emotions and ideas by them. In this context, scholars often discuss 'space and time' as constituting 'a fundamental unity',[8]

6 On this topic, see Brenk in this volume.

7 Gilhuly and Worman (2014) 1: "spatial turn', a term used to describe the confluence of interests across many disciplines regarding what it means to be situated in space'. See also Warf and Arias (2009).

8 Bemong et al. (2010) 3. See also Bridgeman (2007).

since constructions of time, themselves relative,[9] are essential to how space is perceived and constructed in turn, and vice versa (Beck in this volume). 'Space and time' thus yield a richly interdisciplinary field, as they allow for the methods of linguistic analysis and narratological theory to engage in dialogue with novel approaches from the sciences of geography, sociology and anthropology. These sciences also stress the importance of making a distinction between the concepts of place and space: the former should be understood as a site of human beings' interaction with their natural and social environment, and charged with feelings from the uses people make of it; the latter is to be thought of in less concrete terms as 'the area defined by a network of places'.[10]

In recent years there has been a considerable upsurge in scholarly publications in the field of Classics and Ancient History which explore space, place, landscape and territory, and time and temporality from the vantage points of philosophy, archaeology and social anthropology, landscape studies, memory studies, linguistics, gender studies and narrative theory. Particularly notable are Irene de Jong and René Nünlist's *Time in Ancient Greek Literature* (Leiden and Boston 2007), and de Jong's *Space in Ancient Greek Literature* (Leiden and Boston 2012), as they showcase the application of narratology through spatial and temporal descriptions in a wide array of texts, from epic to the Greek novel. Alex Purves' *Space and Time in Ancient Greek Narrative* (Cambridge 2010) draws further attention to the temporal and spatial relations depicted in poetic and prose works, and, from a Bakhtinian perspective, demonstrates the impact of time on the perception and representation of space in narrative. Kate Gilhuly and Nancy Worman's *Space, Place and Landscape in Ancient Greek Literature and Culture* (Cambridge 2014) shifts the focus toward the cultural, social and political projections and representations of places in literature. In addition, there are specialised studies on the concepts of space and place in ancient Greek philosophy, such as Keimpe Algra's *Concepts of Space in Greek Thought* (Leiden and New York 1995), and Benjamin Morrison's *On Location. Aristotle's Concept of Place* (Oxford 2002).

An increasing output of scholarship offers examinations of space and time in specific ancient genres. Richard Seaford's *Cosmology and the Polis: the Social Construction of Space and Time in the Tragedies of Aeschylus* (Cambridge 2012) analyses critical themes such as reciprocity, ritual and money through Bakhtin's concept of the chronotope.[11] The study of space in relation to theatrical space has obviously been the object of focus study in drama scholarship, such as David Wiles' *Tragedy in Athens: Performance, Space and Theatrical Meaning* (Cambridge 1997) and Rush Rehm's *The Play of Space: Spatial Transformation in Greek Tragedy* (Princeton 2002). Moving on to epic, Christos Tsagalis' *From Listeners to Viewers: Space in the*

9 See Clarke (2008) 7–46. See also Gawlinski (2015); Hannah (2015a) and (2015b).

10 Tuan (1977) 12. See also Pelling in this volume.

11 Bakhtin (1981) 84–258. See also Beck in this volume.

Iliad (Cambridge, Mass. and London 2012) offers close readings of the *Iliad*'s spatio-temporal framework; also Marios Skempis and Ioannis Ziogas' *Geography, Topography, Landscape: Configurations of Space in Greek and Roman Epic* (Berlin 2013) delves into the rich territory of the configurations of Greek and Roman epic space with attention to ethnography, power, alterity, real and fictional landscapes. Most importantly, William Thalmann's *Apollonius of Rhodes and the Spaces of Hellenism* (Oxford 2011) focuses on the *Argonautica*'s so-called 'production of space'. As he explains, the voyage of the Argonauts serves to signify, organise and order space on the basis of human (and especially Greek) cultural activity and relations. As Thalmann notes, the production of space within the text is inextricable from the cultural politics of Apollonius' era.

Ancient history and the study of ancient religion have also concerned themselves with space and its political, social and ritual functions. Irad Malkin's *Myth and Territory in the Spartan Mediterranean* (Cambridge 1994) is a notable contribution to the exploration of colonisation, ethnicity and cult viewed through a whole spectrum of attitudes to territories and settlement in the Greek world. Lisa Nevett's study of the physical organization of domestic space (i.e. the *oikos*, which encompasses both the physical house and its occupants) in *House and Society in the Ancient Greek World* (Cambridge 1999) sheds light on the relationship between material culture and social behavior. Susan Guettel Cole's *Landscapes, Gender, and Ritual Space: The Ancient Greek Experience* (Berkeley 2004) shifts the focus to the relationship between different types of landscapes (natural space, community space and sacred/ritual space) in order to uncover the role of gender in them. Natural landscapes (the mountains, the sea and its shore, the caves, the springs) are tackled by Richard Buxton in his *Imaginary Greece: The Contexts of Mythology*, Cambridge 1994, 81–96), in the context of the Greek *imaginaire*.

Roman imperial space in its relationship to imperial structures of power is an area of investigation where interdisciplinary approaches to the study of space have yielded particularly rich insights. Thus, Claude Nicolet's *L'inventaire du monde. Géographie et politique aux origines de l'Empire romain* (Paris 1988), and Richard Talbert and Kai Brodersen's collection of essays in *Space in the Roman World: Its Perception and Presentation* (Münster 2004) view ancient geographical texts as systems of knowledge which organised space and served the ideological and cultural interests of the Graeco-Roman world. Similarly, in *Frontiers of the Roman Empire. A Social and Economic Study* (Baltimore 1994), C.R. Whittaker studies imperial borders and frontiers primarily as cartographic icons of the Roman Empire's power. Looking at specific imperial Graeco-Roman genres and authors, space both in a broad sense (cities, travels, roads, place-names) and in connection to specific themes (*locus amoenus*, *ekphrasis* or single-action space) is explored in Michael Paschalis' and Stavros Frangoulidis' *Space in the Ancient Novel* (Groningen 2002). The essays in Susan Alcock, John Cherry and Jaś Elsner (eds.), *Pausanias: Travel and Memory in Roman Greece* (Oxford 2001) link the theme of travel and tourism in Pausanias' *Periegesis* with Roman power, cultural memory and religious pilgrimage. Last but not least, William

Hutton's *Describing Greece: Landscape and Literature in the Periegesis of Pausanias* (Cambridge 2005) examines the topographical principles that underpin the city and territorial descriptions of the *Periegesis*. Notably, it links these principles with Pausanias' linguistic choices, which, as Hutton finds, deliberately evoke Herodotus.

The conceptions and representations of space in antiquity have also been the theme of digital humanities projects in such as the *Hestia* project on Herodotus' representation of the ancient world (http://hestia.open.ac.uk), and the TOPOI project in Berlin (https://www.topoi.org/). Furthermore, the concept of space has been the central theme in recent international and multidisciplinary conferences and colloquia.[12]

Situating space, time and language in Plutarch

Despite the richness of scholarship on space and time in ancient Greek and Roman culture as well as across different genres of Greek and Latin literature, there is to date no extensive study devoted to representations of space and time in Plutarch. Similarly, as we have shown, representations of space and time in other imperial Greek authors have been studied on the basis of a very limited sample (mainly the novels and Pausanias). Neither has special attention been paid to the significance of language as a means of portraying space and time or reflecting on them.

Plutarch's significance for such a line of study lies in the fact that his diverse and wide-ranging oeuvre offers a much more extensive set of case studies on how space and time are conceived, portrayed, or interlinked across different genres, when compared with that of other ancient Greek authors. The de Jong and the de Jong and Nünlist volumes on *Space in Ancient Greek Literature*, and *Time in Ancient Greek Literature* respectively, include chapters on Plutarch by Mark Beck, both of which focus principally on select *Lives*. In this project, we have undertaken an investigation of these two concepts across a much broader range of Plutarchan writings (both the *Lives* and the *Moralia*). While narratological and Bakhtinian approaches do have a presence in the volume (see especially Beck, Duff and Fletcher), as do theoretical concepts utilised by traditional and 'new' cultural geographers (see Pelling and Oikonomopoulou), these methodologies are not restrictive of the volume's scope. In fact, one of the volume's aims is to show how philological approaches (close reading and intertextual reading) in their own right can shed light on Plutarch's spatial terminology or linguistic choice, when it comes to the representation of space, concrete or metaphorical notions of space in his writings (e.g., Frazier, Alcalde- Martín, Ca-

12 For example, the conference on *Psychogeographies: Space and Place in Latin Literature*, organised by King's College London and Royal Holloway, in July 2013, on places as products of the interrelationship of humans and their natural environment; and the more recent conference *Re-visioning Space(s), Time and Bodies* (Sydney, April 2015), whose main aim was to 'open up new insights and conversation between the arts, humanities, business studies and natural/social sciences' (http://www.iiinz.org/call-for-papers.html).

tanzaro), and the ways in which space can illuminate aspects of his biographical, philosophical, religious and political thought. Similarly, philological approaches, in conjunction with socio-cultural readings of Plutarch's writings, can clarify his conceptions of time, especially in terms of the ways in which he situates himself in the Second Sophistic's fascination with the past. Thus, some chapters discuss time in terms of how Plutarch's works initiate a dialogue between past and present, or in terms of how the past is received in Plutarch's writings and defines his thought (e. g., Geiger, Goeken, Roskam and Driscoll).

Accordingly, in the volume we seek to explore how space is depicted and described within certain types of narrative settings (such as in the context of religious pilgrimage or the symposium: see e. g., Brenk, Kim, Fernández Delgado and Pordomingo, Driscoll, Nikolaidis), as well as to chart various types of space and their historical, philosophical, religious or political dimensions (e. g., Alexiou, Demulder, Meeusen, Lipka, Vamvouri Ruffy and Volpe). We further investigate time as a concept that is intrinsically linked to that of space, as its perception is often shaped by spatial representations, and as a concept in its own right, which is central to Plutarch's thought (see especially the contributions in parts 4 and 5). In this way, we revisit some key themes in Plutarch scholarship, namely, moralism, Greek and Roman identity, *paideia*, relationship to Empire. Language forms a key part of this horizon of concerns for Plutarch and his Second Sophistic contemporaries. Consequently, some chapters explore the ways in which conceptions of space and time in Plutarch's writings may interact with or influence his views on and about language as a key component of cultural identity, as well as his choice of linguistic idiom (especially Berardi, Brenk and Kim).

The volume is subdivided into thematic sections, each of which is treating a special theme. The first part—under the heading: 'Moving through space and time in Plutarch'—consists of the contributions of Christopher Pelling and Mark Beck and introduces the readers to some major aspects of the interrelationship between space and time in Plutarch's works, as well as the theoretical tools and concepts that can be used in order to analyse it.

Christopher Pelling underlines the importance of experiencing space hodologically (that is, as a journey or route travelled, as opposed to the vision of a bird's-eye map) in Plutarch's dialogue *On the Oracles of the Pythia* and in the *Life of Alexander*: in the former text, the characters' tour of the site of Delphi provides them with the opportunity to reflect on the past (marked by a long history of Greek strife) as well as the present (the ways Greek affairs have improved thanks to Rome). The pace and register of the dialogue itself, moving from the combative to the calm, mirror this transition. In the latter instance, Alexander's military journey eastwards prompts in him an intense reflection of the past (namely, the fate of the Persian kings Cyrus and Xerxes), through visits to particular *lieux de mêmoire* (Xerxes' fallen statue and Cyrus' grave).

Mark Beck, next, analyses the narrative texture of Plutarch's *Lives* in terms of the narrator's manipulation of time and depictions of space. Making systematic use of well-known narratological concepts, he discusses the role of temporal acceleration or deceleration in specific *Lives*, and the function of *ellipses*, *analepses*, foreshadowing and *prolepses*, achronic narrations or references to the narrator's own time. Secondly, using Mikhail Bakhtin's concept of the 'chronotope', he examines instances within the *Lives* where physical monumental structures (such as the Parthenon in the *Life of Pericles*) are discussed in terms of their role in re-enacting or reviving key episodes of history and collective memory, thus perfectly fusing space and time for specific biographical aims (exemplarity, characterisation, moralisation).

Part 2 ('Time manipulation and narrative signification') contains three contributions whose common theme is Plutarch's manipulation of time in his narratives for various purposes. Françoise Frazier explores Plutarch's construction of the 'monumental landscape of Athens' in his *Lives*. She carefully plots through Plutarch's use of temporal markers, tense variety (especially the delicate distinction between the present and imperfect tenses) and choice of verbs that point to the preservation, location and lore surrounding Athenian monuments, sanctuaries or dedications. In this way, she shows that Plutarch's descriptions of Athenian monuments seek to link past and present in a way that is inverse to that of Pausanias: the aim is not to treat the present as a starting-point for evoking the past, but, rather, to inscribe the past onto the present shared by Plutarch and his imperial readers, thus fashioning imperial Athens as a space of living memory.

Timothy Duff, next, demonstrates how aspectual choices in Plutarchan narrative create meaning, by distinguishing between two key functions of the imperfective aspect (conveyed in particular through the use of imperfect indicatives and present participles). In the first instance, the 'backgrounding function' of the imperfective aspect serves to present events of the wider historical context as backdrop to the principal actions of a biographical subject's life (which are usually in such cases narrated by a main verb in the aorist tense); in the second instance, the imperfective aspect slows down the narrative speed to create static 'tableaux', which the readers experience from a 'participant' perspective (that is, as if they were present themselves). As Duff demonstrates, Plutarch explicitly theorised the narrative advantages of the second function, which he and other critics associated with greater narrative vividness.

Lucy Fletcher, finally, discusses temporal foreshadowing and anticipation in Plutarch's *Life of Nicias*, which, as she argues, serve to underscore the significance of key events (most importantly, the Sicilian expedition) which unfold later in Nicias' life. Further, she notes that this process of signification extends beyond the textual space and time of the *Life of Nicias* itself, reaching the end of the *Nicias–Crassus* pair.

Part 3 ('Religious locales as places of reflection on language, discourse and time') includes three contributions on the ways in which the religious space of Delphi functions as a means of reflecting on the unity or disunity between different phases of history, as well as of providing the opportunity to ponder the element of change (especially in linguistic usage) over time.

Frederick Brenk teases out the complex interconnection between space, time and language in Plutarch's dialogue *On the Oracles of the Pythia*, discussing how the space of the Delphic sanctuary opens up a large vista of reflection on Greek history, which encompasses manifold and geographically diverse Greek communities. As he observes, the dialogue is structured upon the apparent contrast between the distant past and the present reality, with its second half praising the new space of the Roman Empire. The dialogue's attitude to prophetic language (prose as opposed to verse) follows this pattern, with the second half praising (through the character Theon) the new prose speech by the Pythia.

Lawrence Kim looks closely at *On the Oracles of the Pythia* as well, but with a focus on how Theon's positive attitude to discourse shift (pertaining to the change from poetry to prose in the style of the Pythia's oracular responses), distinct from that of his interlocutors, shades into a positive appraisal of moral and cultural change, from an extravagant past to a moderate present.

Delphi is also the focus of Michele Lucchesi's study of the *Lives of Lycurgus and Lysander*: as he shows, the oracle features in these *Lives* as a symbolic place whose oracular responses and monuments serve to associate different important phases of Spartan history.

Katerina Oikonomopoulou's contribution, finally, regards relative and relational space as key concepts through which we can interpret the way in which the aetiological enquiries contained within Plutarch's collection of *Greek Questions* attempt to link the past (meaning the mythical and pre-classical past of Greek communities) with the imperial present shared by Plutarch and his readers. After mapping out the main types of spatial experience depicted across the different aetiologies, it discusses the special role the enquiries assign to the oracle of Delphi, as the only centre to an otherwise chronically fragmented and polycentric Greek world.

In parts 4 and 5 ('Models of the past I: configurations of memory and history for Plutarch's imperial readers' and 'Models of the past II: Plutarch and the classical era', respectively) the contributions explore time in Plutarch's works in terms of his attitudes to and perceptions of the historical past. They link these attitudes to Plutarch's political, ethical, and broader ideological concerns.

Joseph Geiger argues that, despite Plutarch's long and serious engagement with Roman history and antiquities, the scarcity of references to contemporary Roman subjects and monuments in the *Parallel Lives* and the *Moralia* may be attributed to his political cautiousness.

Joshua Pugh Ginn, next, discusses Plutarch's perception of mid-republican Roman culture, at the moment of its first contact with Greek culture. As he demonstrates, this was not just a story of Greek culture migrating to Rome, but also of Roman virtues spreading to Greece.

Susan Jacobs views Plutarch's *Lives* as texts which seek to conflate past and present by incorporating contemporary political concerns into their depiction of the motives and strategies of historical figures. In this way, they aim to offer credible *exempla* for men active in imperial Graeco-Roman political life.

Eran Almagor traces parallelisms between the relative conceptions of time and space in the *Lives of Agesilaus and Pompey.* As he shows, these prompt reflection on the place of Greece and Greek civilisation in the fall of the Roman Republic.

Geert Roskam explores Plutarch's treatise *On the Malice of Herodotus* in terms of its conception of the great Greek historian of the 5th century Herodotus: as he argues, the treatise in question is not about historical exactitude, but about moralism. It thus betrays Plutarch's moral approach towards literature.

Paolo Desideri considers the theme of travel in Plutarch's *Life of Solon.* As he shows, Solon's journeys into foreign lands and the people he encountered there were valuable sources of knowledge which decisively shaped the lawgiver's political career and reforms. The wisdom Solon acquired during these trips (especially his meeting with Croesus) provides a paradoxical link with the *Life of Publicola* (with which *Solon* is paired), as the Greek lawgiver's knowledge is in a way transposed to Publicola's time and life.

Elisabetta Berardi examines the evolution of Plutarch's language between the epideictic work *On the Glory of the Athenians* and his later ethical-pedagogical treatise *On Listening to Lectures*, from moderate Atticism to a high *koine* influenced by Atticism, respectively. As she observes, this linguistic change relates to a shift in Plutarch's relationship to his classical models: the former work reveals a scholastic adherence to texts such as Thucydides, whereas the latter a more creative use of literary allusion (especially to Plato).

Myrto Aloumpi compares the connotations of the concept of *philotimia* in Plutarch's Athenian *Lives*, with the import the term *philotimia* carried in 5th and 4th century Athenian sources (such as Thucydides or Demosthenes). As she argues, the distance between Plutarch's conception of *philotimia* as a quality inherent in the individual (whose manifestations however vary depending on the context), and *philotimia* in democratic Athens (a civic virtue, whose public dimension is favoured over its private aspect) bespeaks different socio-political conditions, as well as of genre.

The two contributions of part 6 ('Philosophy and religion between past and present') examine processes of integrating non-Greek knowledge (particularly Egyptian) in Plutarch's moral-philosophical and religious writings, and discuss the dialogue between past and present that these processes generate.

Bram Demulder examines how considerations of space (meaning Greek vs. non-Greek cultural space) and time (pre-Platonic past vs. Plutarch's middle Platonic stance) interact and shape Plutarch's dualism (the idea that reality ultimately consists of two non-reducible principles) into a multi-layered, culturally and historically informed notion. After arguing for a presence of different types of dualistic worldviews in Plutarch's thought (depending on whether the subject is Platonic ontology and epistemology or ethics), he discusses the wider intellectual context in which these views are articulated, marked as it is by Plutarch's conscious attempt to integrate non-Greek and pre-Platonic (Egyptian and Zoroastrian) knowledge into his dualistic philosophy.

Michiel Meeusen, lastly, stresses that the symposia depicted in the *Table Talk* function as much more than spaces for the contemplation of Greek (or Graeco-Roman) cultural tradition. As he argues, Egyptian knowledge in the *Table Talk* has a special role to play in the forging of what he calls a 'transcultural morality'. In this construct, Greek knowledge is allied to Egyptian religion and culture in order to contribute to the sympotic speculation about philosophical truth, thus transcending issues of cultural identity.

Part 7 ('Space, time and notions of community') explores the relevance of the concepts of time and space in perceptions of community (local or cosmopolitan) in Plutarch.

Taking his cue from rhetorical *topoi*, Evangelos Alexiou reads *cultural topoi* as collective attitudes and as moral indicators of personal attributes which are in line with or in contradiction to collective attitudes. As he argues, cultural *topoi* serve to map out distinctions or continuities between the past and the present.

Maria Vamvouri Ruffy argues that Plutarch's treatise *On Exile* promotes a notion of a cosmopolitan space, which overrides that of local space. She shows how this notion is constructed within the text by means of re-interpreting Athenian myths of autochthony, and re-contextualising Athenian heroes, philosophers and poets, such as Theseus, Socrates, and Euripides, in terms of their exile, cosmopolitan outlook or migrant life. Lastly, she explores the treatise's notion of exile as a constructed condition, opposed to the natural laws of the world. Man's true homeland is the celestial landscape which envelopes that of the earth.

Paola Volpe Cacciatore traces semantic shifts in the term *xenos* (stranger/guest/exile) between the classical era and Plutarch's time. Taking her case-studies from Plutarch's *Lives* as well as from the treatise *On Exile*, she associates the term's different meanings with Plutarch and his contemporaries' multiple identities (Greek and Roman), Plutarch's relationship to the Roman Empire, and ideas of cosmopolitanism in his works.

In part 8 ('Sympotic spaces: forging links between past and present') the contributions focus on one particular type of space in Plutarch, that of the symposium, in terms of the ways in which sympotic conversation serves to link the past with the present.

Anastasios Nikolaidis discusses the ways in which the sympotic conversations in the *Table Talk* focus on the past, for the most part. However, the instances where Roman participants or specialists (such as grammarians or doctors) feature in the same sympotic space as Greek participants afford the opportunity to situate the dialogues in their contemporary cultural context.

David Driscoll explores the social and cultural dynamics of sympotic space by looking closely at Homeric quotation in *Table-Talk* 1.2, which is concerned with assigning seating at the symposium. As he observes, the social hierarchies of Plutarch's world are mapped out in the sympotic space not only physically, in the seating arrangement of the guests, but also verbally, as correct knowledge of poetry legitimises one's elite status and right to be present at the symposium.

Johann Goeken argues that, through rhetoric, which becomes the common language of the *pepaideumenoi* during the Roman Empire, Plutarch transforms the symposium into an open space of communication between Greeks and Romans. He does so by occasionally taking distance from Plato's *Symposium* in the *Table Talk* and the *Symposium of the Seven Sages,* in order to foreground the role of rhetoric as a 'champs du savoir'. José Antonio Fernández Delgado and Francisca Pordomingo further underline the importance of rhetoric for Plutarch's construction of symposium as an intellectual space, by examining the influence of the rhetorical *thesis*-theory on the structure and argument of a group of the convivial *quaestiones* ('Whether...?'–questions) debated in the *Table-Talk.*

The contributions of part 9 ('Space, place, landscape: symbolic and metaphorical aspects') discuss different types of space in Plutarch's works, including symbolic and metaphorical uses of the concept of space in different contexts.

Carlos Alcalde-Martín treats monumental space in connection to eyewitness testimony in Plutarch's *Lives*. Questioning Buckler's (1992) claim that monuments in Plutarch serve primarily to corroborate literary sources, he argues that statues and other monuments contribute also to the moral portrait of his protagonists, as well as serve to validate the link between past and present. In this way, like Françoise Frazier, he stresses the role monuments play as means of forging a link between past and present.

Michael Lipka discusses sacred space in Plutarch's works (such as holy precincts and sanctuaries) in connection to his conceptions of the divine. As he argues, when mention is made of sacred space in Plutarch, this is always in connection to the old, individuated gods of the polytheistic past. For Plutarch, the gods who actively affect human affairs appear under abstract names (God, Tyche or Daimon) and are detached from the ritual geography of the human lifeworld.

Sophia Xenophontos focuses on military space in the *Lives of Pyrrhus and Marius* as a vital sphere for the construction and interpretation of the biographical account. This is because it helps cast light on how the two heroes behave in other contexts, such as the family, politics, philosophy, and rhetoric, which in turn has implications for the heroes' morality and cultural identity.

The final paper by Andrea Catanzaro considers the way in which Plutarch and Dio Chrysostom treat the sun's course as a metaphor for the imperial ruler's space of action. At stake in both authors, he argues, is the issue of the imperial ruler's limits of power and relationship to his subjects. He carefully teases out the spatial and temporal language used in the treatment of this metaphor.

1 Moving through space and time in Plutarch

Christopher Pelling

Space travel and time travel in Plutarch

Abstract: One important insight of recent scholarship has been the importance of figuring space 'hodologically', as a lived experience as one travels through it, rather than (or, occasionally, as well as) through the vision of a bird's-eye map. Plutarch's own use of the Delphic Sacred Way in *On the Oracles of the Pythia* is a particularly clear and evocative hodological account, exploiting the suggestions of 'place' as well as 'space' (to adopt another useful modern distinction) to stimulate reflection on the entire course and rhythm of Greek history, with memories of internecine Greek conflict giving way to the calm of the Roman present: the move from combativeness to more tranquil conversation also mimics this process. The chapter then explores *Alexander* and the differences made as the narrative moves eastwards and then back towards the west. Outlandish experiences certainly cluster towards the edges of the world, as we might expect, but is there evidence that these generate any change in Alexander himself? The chapter argues that the perceptible change in Alexander's character has little to do with the east entering his soul; *lieux de mémoire* are however relevant, again prompting reflections on the whole of Greek history and provoking the sense of melancholy and even macabre that pervades the final chapters. Life as a journey: that particular cliché began its journey a long time ago.

Space travelling is all the scholarly rage. There has been a lot of interest recently in how ancient authors figure space in their narratives; or 'place' rather than 'space', in the favourite theoretical distinction. Space is a matter more of nature, place of culture: space is what is given us by geography, the facts of the physical landscape; place is what humans have done to it, building their cities and their monuments, endowing particular localities with associations and human liveliness. Spaces are covered by air, places embedded in 'atmosphere'. It is important too that ancient texts often treat place and space in a 'hodological' way: that is, a journey tends to be described by the impressions as one goes, by visualising each stage in turn, rather than with the take-it-all-in-with-a-single-view image that we get from a bird's-eye map. There were of course such bird's-eye maps in antiquity: there is the famous story of Aristagoras wielding one in front of Cleomenes in Herodotus (5.49). But Cleomenes is bewildered by it all, and it needs to be explained to him. It may be second nature to us to cry out for a bird's-eye map to go with, say, a narrative like Caesar's *Gallic Wars*, or even to start mapping one out mentally for ourselves on to that vague shape of France that we already have in our head. The ancient visualising equivalent would be more like a sat-nav reconstruction, once again seeing place as something travelled through sequentially. (Equally, one should not overstate the difference: if one is asked to describe a journey one knows well, say from one's home to one's of-

https://doi.org/10.1515/9783110539479-002

fice, one typically figures it in a hodological way, and may be quite surprised by a later bird's-eye view of the curves in a familiar road.)

Another interest has been metatextual, seeing how journeys in the text may have analogies in the way the text itself works, turning the reader into a sort of narrative journeyer. Purves (2010), in particular, took that approach a long way. The textual grounding for such an approach is of course secure, however far we decide to push it. The 'path of song' is familiar from archaic times;[1] many will think too of how Herodotus promises to 'go forward' (προβήσομαι) 'journeying through' (ἐπεξιών) cities big and small alike (1.5.14), covering them in his text as earlier he had in his travels.[2] Herodotus has indeed been the focus of a project in which I have been involved myself, the Herodotus Encoded Space-Text-Image Archive (HESTIA):[3] Among other things, that has been concerned with alternative ways of digitally 'mapping' the place-names appearing in Herodotus' text. During that project we noticed how often questions of space or place overlap with questions of time. It might be a question of distance: did things happen in the same way, following the same physical rules, in the distant past as they do today, and do they happen in the same way in distant lands in the present? (Compare Thucydides' use in the *Archaeology* of distant practices in the present to cast light on his reconstruction of practices in the distant past, 1.6.5–6.) But it is also striking how often local disputes over place —whose territory should this be?—become disputes over the past, over traditional claims and legends echoing back into time immemorial.[4]

Not that this overlapping of space-questions and time-questions is any surprise. One need only think of the way that Aeschylus' *Persians* is so unusual among Greek tragedies, but replaces distance in time with distance in space. And that same early programmatic chapter of Herodotus goes on to explain how his travels have given him an insight into human mutability, into big cities becoming small and small cities becoming big: travel through space, or rather through places (for 'cities' are quintessentially places), has given him insight into time (1.5.3–4)—just as, a little into his narrative, the much-travelled Solon will have such insight into human change and vulnerability.

And what of Plutarch? I shall take two texts, *On the Oracles of the Pythia* and the *Life of Alexander*, seeing how place works on people and does so sequentially and 'hodologically', and in particular tracing that interaction of place and the past, of space and time.

---

1 From *Od.* 8.73–74 and *h. Herm.* 451 on: further passages are collected by Lefkowitz (1991) 27 n. 44.

2 More on this at Pelling (1999b) 331–333, 356, with further bibliography.

3 http://wiki.digitalclassicist.org/HESTIA. Cf. Barker et al. (2010); Pelling (2011c) 3–4; Barker et al. (2013).

4 More on this in Barker et al. (2016), which also includes more on HESTIA. Barker and Pelling (2016) explore the link of space and time in, particularly, Herodotus 5.

On the Oracles of the Pythia

If one wants an example of hodologicality in Plutarch, Delphi, the scene of the conference from which this book springs, is the place to look. *On the Oracles of the Pythia*, particularly the dialogue's first half, describes the conversation as the group wind their way up the Sacred Way, and the climb is described in terms of what they see and the effect this has on them: 'place', indeed, and all that this very holy and very special place can suggest. As always with Plutarch conversations, it ranges widely and learnedly. The first topic centres on the rusting process: what can it be that gives those statues of the navarchs their peculiar blue-green tinge, appropriate as it seems for those old sea-dogs, 'standing there with the true complexion of the sea and its depths' (395B)? Then the conversation turns to matters of religion and history, with one prompt or another given by whatever they are passing: that statue of Hieron the tyrant—could it be coincidence that it fell down on the very day he died, any more than it was coincidence that the statue of a certain Spartan lost its eyes just before his death at Leuctra (397E–398A)? A little later we get to the treasury of Cypselus: why Cypselus, and not the Corinthians as a whole ... (400D–F)? Next, those statues of courtesans (401A): are they not shaming? Yet ponder the history of Greece: isn't it better to commemorate the odd prostitute than all those infamous battles of one Greek against one another? And so it goes on, until their guest suggests it might be time to sit down and get back to the question they had originally raised, why oracular answers are now given in prose when the famous cases of the old days were given in verse (ch. 17). Here too place matters:

> Boëthus immediately observed that the place itself helped to solve our visitor's problem. 'There used to be a shrine of the Muses here,' he said, 'near the outlet of the stream ... Simonides speaks of the place...' (*De Pyth. or.* 402C)[5]

Admittedly, *how* the place helps is not clear, as the text is defective: it is probably something about how the place used to inspire, not just because of its beauty and the presence of Apollo, but also because of that cult of the Muses that 'used to be'.[6] But, somehow, it matters.

Arguably, place matters a good deal more: this is not a dialogue that could be happening just anywhere. That atmosphere of Delphi has its effect. This is initially the case in the most literal way: the air here is particularly thick, and it has affected that rusting process (396A). But this discussion in those early chapters also introduces other themes that are going to come back in interestingly different registers: how

5 Translations from *de Pyth. or.* are adapted from Russell (1993); those from *Alexander* are my own.
6 As Ziegler assumed, (1949) 193 = (1951) 830. Schröder doubts this in his commentary, (1990) 310–312, on the grounds that in that case '[i]hre Function wäre einzig die einer szenischen Ausmalung': that 'einzig' is misjudged in view of the general importance of place in the dialogue and its capacity to give inspiration and intellectual guidance.

far, for instance, purely physical explanations are enough to explain those things that look like coincidences, like the sea-colouring of the navarchs—but are they, really, just coincidences, when there are so many of them and there is so much of a godly presence in the air? And it is not just religion that is in the air, but history, all that Greek history that is commemorated there, for good or for ill.

> Do you not feel pity for the Greeks as you read the inscriptions of shame on these beautiful dedications: 'Brasidas and the Acanthians, from Athenian spoils'; 'The Athenians from Corinthian spoils'; 'The Phocians from Thessalian spoils'; 'The Orneates from Sicyonian spoils'; 'The Amphictyons from Phocian spoils'. (*De Pyth. or.* 15.401C–D)

That is a favourite theme of the *Lives* as well, of course, where Plutarch several times dwells on the senselessness of the Greeks throughout their history in fighting one another, so that eventually it had to be left to the Roman Flamininus to give them that peace that their own bickering had denied them for so long (*Flam.* 11). (Admittedly, not all of that emphasis carries across to the dialogue: Roman memorials, including those of Flamininus, are not mentioned either.[7] There may be a reason for that as well, as we will later see.) Once more, then, though in a rather different way from Herodotus, Plutarch's hodological moving through space encourages insight into time: these *lieux de mémoire* are dripping with memory, the wrong sorts of memory. Too many battles, too many tyrants, too much Greek blood … . Notice the memories that do *not* figure here: no Marathon, no Salamis, no Plataea (though Plataea does figure in the sister dialogue *On the Decline of the Oracles*, and the climbers must have passed the Tripod of Plataea just before getting to Hieron); no, it is the Peloponnesian War and Leuctra and Lysander fighting Thebans that get the space. The Persian Wars figure only once—in the mention of the statue of Apollo carrying a spear set up by the Megarians 'in consequence of the victory *that expelled the Athenians from the city* after the Persian Wars' (402A). So even there it is Greek against Greek. The silence is echoing.

Still, times change: the second half of the dialogue is concerned with that, as Theon—a real person,[8] but still a significant name—gives his explanation of why the oracles no longer come in verse. There is a lot of insight there, including a plea against overstatement: some of them still do come in verse, and even in the past some doubtless came in prose. And religion and history are still firmly in the air, and firmly intertwined. A lot has the air of Tacitus' *Dialogus* about it, despite the difference of theme: here too there is a nuanced treatment of an issue, arguably but also questionably one of 'decline', weaving it into a broader treatment of cultural change. For it is not just about oracles: philosophy and astronomy too tended to come in metre back then; any change can be seen as part of a more general tendency to do away with flummery and concentrate on clarity. And 'decline'? By the end of

7 McInerney (2004) 49–50.
8 *RE* nr. 10, v² (1934) 2059–2066 (Ziegler); Puech (1992) 4886. Cf. Swain (1991) 326–327.

the work, religion and history are coming together in a different register. Look around you—place again—and you get a different view: are things really so bad? All this prosperity, all the healthy state of Delphi in (presumably) the early second century CE:[9]

> Our wealth has given her form and beauty and a splendour of temples and meeting-houses and water-courses such as she has never had for a thousand years. The inhabitants of Galaxion in Boeotia became aware of the epiphany of the god because of the abundant flow of milk ... But here he gives us a brighter, stronger, clearer sign. He has given us prosperity and splendour and honour, in place of the drought of our former desolation. I love myself [this is still Theon speaking] for my zeal and service in this cause, with Polycrates and Petraeus. I love also the leader[10] of this policy, who takes thought and cares for most of these achievements [then a 25-letter lacuna, which perhaps—frustratingly—might have clarified who this 'leader' was[11]]. But so great a change cannot have happened in so short a time by mere human effort, without the presence of the god among us and his divine guidance of the oracle. (*De Pyth. or.* 409B–C)

Religion and history, once again. Place—Delphi—gives you insight into both, and shows how you cannot understand the one without feeling the presence of the other; past and present drip from the monuments; and the god is everywhere. And if that 'leader of this policy' is indeed Hadrian—a big 'if'—[12] or even if he is a

9 Which is when the work has usually been dated: there is no suggestion of a dramatic date any different from that of composition. However, Müller (2014) 65–66 n. 2 prefers a late first-century date, largely because of the thematic similarities to *De def. or.*, whose most likely dramatic date is c. 83 CE according to Ogilvie (1967). If this were true, both these works would date from before the period when Plutarch became a priest at Delphi himself (c. 95 CE). The same might be true of *De E* as well, which Obsieger (2013) 19 follows Ziegler (1949) 75 in dating to the 90s. But (a) Ogilvie himself dates the composition date of *De def. or.* later, between c. 95 and 115 CE: all these dates may be less precise than Ogilvie and Müller assume, but there seems no reason to assume that the dramatic and composition dates are close to each other. Were Plutarch in fact writing *De def. or.* in, say, the 110s, he could easily have chosen to retroject the dramatic date to link it with Demetrius' voyage (410A). (b) Whatever we decide about *De def. or.* or about *De E*, there is no reason to assume that *De Pyth. or.* is particularly close in date because of its similarity of theme.

10 Russell (1993) has 'initiator'.

11 Flacelière builds boldly on his hypothesis (see next n.) and prints <αὐτοκράτορ' Ἀδριανὸν Καίσαρα> in his Budé edition, (1974) 40 and 82.

12 The matter is complicated: for Hadrian, Flacelière (1971) and, e.g., Holzhausen (1993); *contra*, Jones (1966), 63–65 = (1995) 100–104, Schröder (1990) 15–20 and the full and careful discussion of Swain (1991), though along the way Swain discredits many of the arguments used by others who reach the same conclusion. In particular, the epigraphic arguments turn out to be very indecisive. The other favoured candidate for this 'leader' has been Plutarch himself, as Hirzel (1895) ii. 205 n. 1, Ziegler (1949) 25 = (1951) 661, and others, including Swain and Sieveking and Gärtner in the Teubner text (*Plutarchus ipse videtur esse*), have thought: not impossible, but I am still not convinced that this would sit comfortably in the work of the man who also wrote *On Inofffensive Self-Praise*. Schröder (1990) 21 decides for a person unknown to us, whose name would have been specified in the lacuna: disappointing if so, but that may well be right. One point in the debate is relevant here. Jones (1966) 64 = (1995) 101–102 observed that the reference to the Pylaea at 409A points to Thermopylae rather than Delphi; but the argument has moved on since that passage, and the empha-

Roman governor, then that gets even closer to that insight of the *Lives*, that the Romans have eventually brought to Greece what Greece has been unable to provide for itself. That, indeed, may help to explain why the dialogue passes over those memorials of Roman successes: it was not the bloodshed of the Roman past that mattered for Plutarch now, it was the peace of the Roman present.[13]

Maybe indeed the place, and those insights that the place affords, drive the work in a further sense. That history of Greek strife giving way to calm; in a way, the text mimics that too. The first half of the dialogue has not been conspicuously good-tempered. There was a gibe at the expense of the guides (395A);[14] there was quite a lot of quarrelsomeness too, between Stoics and Epicureans and between physical reductionists and those that insist on the divine presence.[15] The initial description of the conversation had some edge to it:

> We were walking slowly, Basilocles, sowing and harvesting in battle festering[16] and warlike words that kept sprouting and growing under our feet on the way, like the warriors that sprang from the dragon's teeth. (*De Pyth. or.* 394E)

Admittedly, some of the *Table Talk* becomes combative too, but there the good symposiarch knows how to damp things down; here there is no symposiarch to do that. Yet it calms down anyway. That quarrelsomeness has gone by the second half, as they sit quietly in the sun and muse on time and its changes. The rhythm of the dialogue itself has mimicked the rhythm of the history on which they muse, and calm has broken out.

Alexander

Breaking out into the unknown can be expected to bring some uncanny experiences. It is noticeable that the more outlandish 'marvels' in Herodotus tend to be at the ends

---

sis throughout the dialogue on their physical surroundings—cf. 'you see', also at 409A—would be dissipated were the final focus to shift abruptly and exclusively to 50 kilometres away. Flacelière (1971) 182 gets this right, also stressing the close relationship between the two: 'Entre les deux sanctuaires, il n'y a pas concurrence, mais symbiose'. See also Brenk in this volume, and his footnote 22.

13 So this is a qualification of McInerney (2004) 51: 'The Pythian *logoi*, in fact, excise Delphi's recent past ... Plutarch is often seen as unusually accommodating towards Rome, but that accommodation stopped at the doors of Apollo's sanctuary'. Not excision, I think, but selectivity: what mattered was not whether, but how, one welcomed Roman thoughts to the sanctuary.

14 Though admittedly Theon, the one whose voice is most often equated with Plutarch's own, was polite enough to let the guides have their say first at 397D–E.

15 So perhaps not so 'exemplary' of the conversational virtues initially paraded in ch. 1 (394D–395A) as Müller (2014) 73 suggests. On *De Pyth. or.* 394E, see also the Introduction to this volume.

16 On the reading here cf. Bolkestein (1964) 367–368, defending the MSS reading ὑπούλους: 'the word indicates the evil that is festering under the surface and suits well the image of the Spartoi ...'. It is accepted by Flacelière and Sieveking-Gärtner, but not by Schröder.

of the earth: gold-digging ants, flying snakes, Ethiopians who regularly live to the age of 120, and 'dog-headed men, and headless men with eyes in their breasts, so the Libyans say, ... and other beasts in huge numbers, not at all fabulous'.[17] And as Plutarch's Alexander presses on into the unknown, he certainly comes across some outlandish things: some of them are physical—the spontaneously combusting air in Babylonia (*Alex.* 35), the well that miraculously produces oil in India (57.5–6)—and some more to do with the people, as with the bizarre practical experiment that Alexander's retinue try with that Babylonian combustability, trying to set a slave boy on fire.

As the *Life* moves eastwards, then, do we sense the world is changing? And is Alexander changing with it? That has been argued recently by Tim Whitmarsh,[18] in one of two outstanding discussions (the other is by Judith Mossman).[19] In his view, the change and decline in Alexander's behaviour are related not merely to his distance from Hellenicity but to the way in which the east enters his soul (my words, not his, and he emphasises too that the soul itself aids the process by already being 'fiery'): a challenge, he argues, to Hellenic identity itself.

My own emphasis would be different.[20] Certainly Alexander has changed by the end of the *Life*; but it is hard to see much interest in anything the East has done to him, in anything those eastern places or peoples had to offer. The points are about Alexander, not about place. When we get to Indian philosophers—the Gymnosophists, then Dandamis and Calanus—we may well feel the distance from the clear air of Aristotle and the Hellenic philosophy of his youth, but the emphasis is on what he has lost, not on anything he has gained. When Richard Stoneman tried to find genuine Indian thought in those 'naked philosophers', he did get somewhere —but not with the Plutarch versions, but with the stories or related stories in other sources.[21] Plutarch just does not seem very interested in anything that Indian thought has to offer, other than a spot of nakedness and bizarrerie. Yes, odd things continue to happen over there, none odder than when Calanus builds his own funeral pyre and self-immolates. But there does not seem much to *learn* from that, either for Alexander or for Plutarch's readers. Whitmarsh argues that Plutarch is here 'test[ing] his own conceptions of Hellenism in the crucible of narrative' ... offering 'a voyage of self-discovery (and in a sense self-destruction) for his readers as well as his subject'.[22] Yet this is not a particularly *harrowing* test, and there is not much here to make any complacent Greek lose his sleep. There is nothing wrong with Hellenicity; it is Alexander that has gone wrong.

17 Gold-digging ants: Hdt. 3.102.2. Flying snakes: 2.75.1, 3.107.2, 3.109.1. Long-lived Ethiopians: 3.23.1; dog-headed men etc., 4.191. Cf. Romm (1994), esp. 57–59, 91–92, 95–96.

18 Whitmarsh (2002), esp. 186–192.

19 Mossman (2006), esp. 287–301.

20 I discuss this more fully in Pelling (2016).

21 Stoneman (1995).

22 Whitmarsh (2002) 191–192.

Let us return to the connection of space and time. There is certainly a perspective of time in the *Alexander*, and it is connected with 480 and all that: is all this vengeance for Xerxes' invasion? Vengeance for those great battles of the past is marked by the special destination of some of the spoils sent home after Gaugamela, to Plataea and (oddly) to Croton, 34.2–4. This is also where he sees the fallen statue of Xerxes lying on the ground in Susa.

> He stood over it, and spoke to it as if it were a living thing: 'There you lie. Shall we pass you by, remembering your campaign against Greece? Or shall we raise you up because of your greatness of soul and goodness in other respects?' He spent a long time pondering in silence; then continued on his way. (*Alex.* 37.5)

But the time he takes already suggests that this is no easy decision. A chapter later, and the drink-fuelled arson of the royal palace driven by Thais gives a clearer example of how vengeance can be out of joint, and Alexander himself swiftly repents.

That Xerxes moment must be in our minds later when he returns close to the scene of Thais' disgrace, and comes to Cyrus' grave at Pasargadae. He finds it forced upon and desecrated.

> He executed the perpetrator, even though he was not the least noble of the Pellaean, a man called Poulamachus. He read the inscription, and gave orders that a version in Greek letters should be engraved below. It read as follows: 'Reader, whoever you are and wherever you come from—for this much I know, that you will come—I am Cyrus, who gained the Persians their empire. Do not, then, begrudge me this handful of earth that covers my body'. This affected Alexander very much, as he reflected on uncertainty and change.[23] (*Alex.* 69.4–5)

Once again, it requires faith to find anything distinctively eastern in the insight:[24] the stress on human mutability, on how even the greatest are brought low, is quintessentially Herodotean, even if Herodotus too uses the fate of the great eastern dynasts to make it particularly clear. The stress on the 'earth' is particularly evocative, that great swathe of earth, γῆ, that he conquered for Persia, contrasting with the handful that now he pleads to retain. (That neat touch is likely to be Plutarch's own: Arrian 6.29.8 and Strabo 15.3.7 simply have 'do not begrudge me my memorial', τοῦ μνήματος,[25] and Plutarch has a similar 'land' conceit in Cleopatra's marvellous lament at *Antony*

23 Ziegler here adds from Zonaras <τῶν πραγμάτων> before τὴν ἀδηλότητα καὶ μεταβολήν, and he is followed by Flacelière. Zonaras not infrequently imports slight alterations to Plutarch's phrasing, and this is not the only place where Ziegler is over-influenced by his choice of words. The addition does not seriously affect the sense, but without the extra words the phrasing is even more evocatively generalised.

24 Similarly Zadorojnyi (2013) 381–382, stressing the 'essentially Greek optics' of Alexander's viewing and the distinctive Greekness of the lesson that is learned.

25 So does Eustathius on Dionys. Perieget. 1069. Strabo explicitly attributes this version to Aristobulus, and Arrian too has just cited Aristobulus for a related item.

84.6.[26]) Nor does Arrian have Alexander particularly thoughtful or affected by this, though he is clearly annoyed and eager to track down the culprit.[27]

That thoughtfulness in Plutarch certainly adds to the atmosphere of those final chapters. Mossman called them 'melancholy':[28] perhaps they are more, really 'macabre', not least in that episode of Calanus' self-incineration that immediately follows. Death, most certainly, is in the air; and whatever else one says about this, it is pretty clear that place, this particular *lieu de mémoire*, is having a strong effect on the person who observes. What is more difficult is to pin down exactly what sort of effect this is. Mossman compares it to the previous Xerxes moment, and suggests that the earlier occasion showed more Hellenicity while this second one shows a worrying affinity to the Persian king.[29] I am not so sure; of course, there was more reason for vengeance against Xerxes than against Cyrus, even though Cyrus too did some enslaving of Greek cities, in his case in Asia Minor. Maybe it is rather a matter of different forms of Hellenicity, the quest for vengeance (Xerxes) or the calmer insight into the nature of the human condition (Cyrus). Notice that the instruction is to take the existing Persian inscription and provide a Greek version underneath. This is wisdom for everyone, both Greek and Persian. I am tempted to find this Alexander wiser than the earlier one, even if only momentarily before he relapses into his next drunken stupor; wiser, but also, as Mossman rightly underlines, much, much sadder.

Conclusion

So maybe, to adopt another Mossman formulation,[30] travel has indeed broadened Alexander's mind; maybe it is that Herodotean insight once again, the way that travel through space can give insight into the workings of time, especially the changes that time can bring—that 'uncertainty and change' of the Cyrus inscription. It is harder to find any particular effect on Alexander of the places themselves, of the sort we saw in the Delphic dialogue; it is more a question of the journey he has taken, the distance he has travelled, the amount he has lost as well as, perhaps, that final bit of wisdom that he has gained; and once again, this is more a point about Alexander the individual, not of any effect on him of the *specific* places or peoples that he encounters.

---

26 κινδυνεύομεν δὲ τῷ θανάτῳ διαμείψασθαι τοὺς τόπους, σὺ μὲν ὁ Ῥωμαῖος ἐνταῦθα κείμενος, ἐγὼ δ' ἡ δύστηνος ἐν Ἰταλίᾳ, τοσοῦτο τῆς σῆς μεταλαβοῦσα χώρας μόνον. ('It seems that death will force us to change places. You, the Roman, have found a grave in Egypt, and I, unhappy woman, will lie in Italy, gaining just enough of your country for that', translation adapted from Scott-Kilvert and Pelling [2011].)

27 Alexander's 'distress': 6.29.4. In his version the investigation proves fruitless, and there is no mention of 'the Pellaean'.

28 Mossman (2006) 294.

29 Mossman (2006) 293–294: 'a progression in his character, if not (I would suggest) a deterioration'.

30 Mossman (2006) 292.

Life as ‘a journey’: the cliché makes one shudder. Googling the book-title ‘My journey’ showed 2 million+ hits: the first page showed *My Journey: from Horses and Iceboxes to Aero Planes and Refrigeration* as well as *My Journey in Karate: The Sabaki Way,* and I could go no further. But there is some of that in *Alexander* too, just as after all there was in the *Odyssey* and the *Aeneid.* The journey of this particular cliché began a long, long time ago.

Mark Beck

Time and space in Plutarch's *Lives*[1]

Abstract: Plutarch is a sophisticated narrator. Many years of research have established this beyond question. This chapter focuses on his conscious manipulation of time and space in his narratives of the *Lives*. After setting forth the narratological terminology used in this analysis, the chapter goes on to analyse departures from the normal chronological sequence of events (*analepsis*, *prolepsis*, achronic narratives), variations in the narrative rhythm (acceleration, deceleration), and the reasons behind Plutarch's deployment of such techniques. The chapter then turns to an analysis of some of the key narratives involving space in the *Lives* with the application of Mikhail Bakhtin's concept of the chronotope. Plutarch's chronotopic narratives tie his protagonists' actions with places, monuments, and physical structures that serve to memorialise the superlative nature of their achievements. In particular, this chapter draws attention to Plutarch's narrative construction of time and space as it relates to the various modes of characterisation and vivid dramatic portraiture encountered in the *Lives*.

Plutarch, as narrator, finely constructs the narration of individuals' lives with multiple aims in mind. His overriding aim is the representation of exemplary individuals in a lifelike, fairly detailed way so that we, the narratees, may acquire a vivid and lasting impression that personally motivates us in our own lives to imitate or emulate what we can of the biographical subjects' great qualities. Exemplarity, *mimesis*, characterisation and moralisation figure strongly in the attainment of this complex and multifaceted end. The narratological construction of time and space thus serves these aims in the *Lives*. Before specifically focusing on these narrative techniques, we should commence our analysis with a brief description of the challenges confronting Plutarch as narrator.

In terms of time we may approach this from the perspective of 'layers'.[2] Four layers may be envisioned: material, *fabula*, story, and text. Collection of material for construction of the story and text would constitute the initial layer. The material in this instance would consist of prior narrative accounts, either historiographical or biographical texts of Plutarch's predecessors (earlier historians and biographers) or ancillary material he might gather himself or witness via, for example, autopsy (particularly important for the construction of space). The *fabula* consists of 'the aggregate of events reported in a narrative in their chronological order'.[3] The restructuring or

1 This chapter is based on my two more extensive contributions on time and space in Plutarch published in de Jong and Nünlist (2007) and de Jong (2012a), respectively. In citing Plutarch's *Lives* I follow the Loeb edition by B. Perrin.

2 de Jong and Nünlist (2007) 2–3.

3 de Jong and Nünlist (2007) 2–3.

https://doi.org/10.1515/9783110539479-003

rearrangement of this tight chronological order that the narrator undertakes in his text is termed the story.[4] The narratee may reconstruct the *fabula* from the story and the text.[5] The narrator may vary the temporal arrangement (anachrony) by altering the order of events (order), by dwelling to a greater or lesser extent on some events (rhythm), and by relating events once or more than once (frequency).[6]

In characterising the narration of space, we may also distinguish between *fabula*-space and story-space with *fabula*-space being a total depiction of the location(s) that come into play in a narrative, whereas story-space refers to the actual place depicted or referred to.[7] Detailed and rich descriptions of space or objects (*enargeia, ekphrasis*) that assume a greater importance than the simple narration of space as a backdrop to the narration of events engage our attention and usually warrant special interpretive consideration vis-à-vis the work's narrative strategy as a whole.[8] I view such detailed descriptions as the spatial correlate to a slowing of the temporal rhythm.

Time

In most cases Plutarch had access to an abundance of mostly historical sources that afforded him adequate material for the writing of the *Lives*.[9] The 'important' events that Plutarch stressed in constructing his narrative are revealing of character and were incorporated in preference to other events that his historical sources may have magnified. This means that Plutarch gives cursory attention to what he might regard as insignificant detail.[10] The well-known proem to his *Lives of Alexander and Caesar* articulates this fundamental contrast between the historian's approach and the biographer's (*Alex.* 1.2).[11] The first layer of material collection leaves therefore much on the cutting room floor. The silence of his sources on the personal lives of historical figures would also hinder a complete cradle to grave reconstruction simply because the ancients usually paid little attention to early events in the lives of great individuals prior to their becoming great. Childhood was thus usually overlooked in the chronological sequence. Nevertheless he generally appears to use whatever information he does have about his protagonists' childhood in constructing the narrative.[12] The *fabula* for biography comprises an individual's sequential lifetime experi-

4 de Jong and Nünlist (2007) 3.
5 de Jong and Nünlist (2007) 3.
6 de Jong and Nünlist (2007) 3.
7 de Jong (2012) 2–3.
8 For a different viewpoint on *ekphrasis* see de Jong (2012) 5–8.
9 On Plutarch's use of copious historical source material see now Schettino (2014) 417–436.
10 On the major characteristics of the genre of ancient historiography see Marincola (1997).
11 On biography as the *genus proximum* of ancient historiography see Geiger (1985) 22 and Burridge (1992) and (1997) 371–391.
12 Rosenmeyer (1992) 210. Plutarch frequently displays concern for chronological accuracy, e.g., *Them.* 2.5.

ences beginning at birth and ending with the inevitability of death.[13] The story is necessarily a much abbreviated account vis-à-vis the *fabula*, being on average only about 35 pages in length, in which chronologically disparate events, representative of central themes or characteristics, are frequently grouped together or receive elaboration in the narrative not unlike Suetonius's *per species* method of categorisation. Both techniques exert a significant influence on the order, rhythm, frequency of the *Lives'* narrative structure.[14] The comparative structure of the *Parallel Lives* itself imposes an organisational principle which dictates to some extent Plutarch's selection of his source material and construction of the narrative. The concluding analeptic comparative/evaluative postscript (*Synkrisis*) appended to all but four of the 22 pairs of *Lives* underscores this structural feature and revisits interpretatively the commonalities and differences.

Narrative deceleration I: The 'grand scenes'

Some key instances of narrative slowing or deceleration are induced to some degree by the comparative structure of the *Lives*, as indicated above. These 'grand scenes', to use a term coined by Françoise Frazier, appear as instances of narrative deceleration in the service of greater representational significance vis-à-vis character and personality.[15] One very famous example of these 'grand scenes' represents the breaking in of Bucephalas by the young Alexander (*Alex.* 6.1–5). This anecdote is to be contrasted with an incident in the paired *Life of Caesar*, his capture by the pirates. Plutarch's Caesar achieves a certain intellectual dominance over his captors and actually succeeds in winning them over through his insulting, jocular, even contemptuous manner of addressing them (2.3–4). The first significant detail concerns the ransom (*Caes.* 2.1). Caesar laughingly raises it from twenty to fifty talents stating that they did not know whom they had captured. The salient aspects of Caesar's personality, his indomitable pride, his ability to charm, his sanguine disposition in the face of adversity are clearly revealed in Plutarch's version and are paralleled in Alexander's breaking of Bucephalas. The intuitive psychological control he exerts over the pirates reveals, in the human sphere, what Alexander's breaking of Bucephalas does on another level.[16] It is apparent that Plutarch employs both of these key scenes of narra-

---

13 Momigliano (1993) 11: 'An account of the life of a man from birth to death is what I call biography'. Cf. Geiger (1985) 14–29 and Pelling (2002) 365–386.

14 Moles (1988) 9.

15 Beck (2007) 399; Frazier (1992) 4487–4535. Briefer, less dramatic versions of these 'grand scenes' are termed by her 'micro-scenes'. Cf. the story of Timocleia (*Alex.* 12).

16 On this anecdote and its relation to the pirate episode see Frazier (1992) 4496–4499; Stadter (1996).

tive deceleration to showcase traits associated with brilliant leadership ability. Reading these two *Lives* together reinforces this lesson.[17]

Narrative deceleration II: Key historical events

These scenes are to be distinguished from the extended narration of significant historical events which I categorise as another variety of narrative deceleration.[18] In representing these political or military events, Plutarch, as narrator, adapts narratives of these events from historical sources such as Herodotus, Thucydides, Theopompus, Ephorus, Timaeus, Polybius, Dionysius of Halicarnassus, Sallust, and Livy, by reducing them in size and occasionally supplementing them with additional information and material gleaned from his research (cf. *Nic.* 1.5).[19] These events which may take up a very large part of the biography may reflect anything from only a few months to several years of the subject's life. In the *Nicias*, for example, this major event is the Sicilian expedition. Of the thirty chapters of the *Life* this event is treated in chapters 12–30, or in other words it takes up 63% of the biography (story time), but the event itself occupies only approximately two years of Nicias' 57-year lifetime[20] (*fabula*-time). The paired *Crassus* has a correspondingly long narrative of the equally disastrous and brief Parthian expedition (16–33). Other examples of this type of narrative deceleration include the account of Themistocles' role in the second Persian War against Xerxes (*Them.* 7–18)[21] and Aristides' involvement in the war against Xerxes' forces (Salamis and Plataea) (*Arist.* 8–21).[22] In the *Lucullus* the war against Mithridates is recounted at length (*Luc.* 7–36). Many more examples of this type could be adduced.[23] It is important to note that Plutarch varies the pace and mode of

---

17 Other notable 'grand scenes' include, e.g.: *Alex.* 30 (Darius and the eunuch), 50–51 (murder of Cleitus), 60 (Alexander and Porus), *Caes.* 32 (Caesar at the Rubicon), 61 (Caesar at the Lupercalia), 66 (assassination of Caesar), *Sol.* 27 (meeting with Croesus), *Cor.* 33 (Valeria and Volumnia), *Ant.* 77 (the death of Antony and Cleopatra) and 83 (Octavian and Cleopatra), *Agis* 17 (Chilonis), *Brut.* 23 (Brutus and Porcia), *Aem.* 36 (Aemilius on *Fortuna*), *Dem.* 29 (death of Demosthenes), *Pomp.* 78–79 (murder of Pompey), *Crass.* 31 (death of Crassus), *Them.* 11 (council of war prior to Salamis), *Ca. Mi.* 68–70 (suicide at Utica), *Fab.* 13 (recognition of Fabius' prudence by Minucius), *Pyrrh.* 2–3.5 (childhood episode).

18 Beck (2007) 400–401.

19 On Plutarch's historical sources, see now Schettino (2014) 417–436.

20 c. 470–413 BCE.

21 480 BCE. The story of Themistocles' ostracism and life in the Persian Empire might also be included in this category (*Them.* 23–31).

22 His actions at Marathon receive only one chapter (*Arist.* 5).

23 See, e.g., also *Fab.* 2–27 (Second Punic War against Hannibal), *Cim.* 6–14 (leadership role in allied activities), *Cam.* 15–30 (repulsion of Gauls under Brennus), *Ca. Ma.* 12–14 (war against Antiochus), *Lys.* 3–21 (final phase of Peloponnesian War and postwar activities), *Sull.* 11–24 (Mithridatic War), *Cic.* 10–22 (conspiracy of Catiline), *Ages.* 6–19 (campaigns in Asia Minor and Greece), 27–35 (Agesilaus vs. Epaminondas, incl. Leuctra and Mantinea), 36–40 (service in Egypt), *Pomp.* 24–29

the narrative continuously within these sections as he develops some scenes and employs both singulative (an event is told once) and iterative (repeated events are told only once) narrative modes. Very often these episodes of narrative deceleration approach the 'grand scenes' in terms of their density and dramatic portraiture.

Narrative acceleration

Plutarch often consciously abbreviates his narration of some events which he judges to be inconsequential to his main purpose of depicting character and moral attributes. This renders the rhythm of his narrative much faster-paced in some places, especially when compared to historical sources. His narratives of some of the great battles in his biography of Alexander provide excellent examples of this. For example Plutarch's account of the actual battle of Issus is very brief, only one sentence (*Alex.* 20.8).[24] Instead, he focuses on relatively minor incidents that display the trust Alexander placed in Philip (*Alex.* 19.4–9), Alexander's cool response to the amenities of Darius' captured camp (*Alex.* 20.11–13), and his treatment of the captured Persian women, a demonstration of his great restraint (*Alex.* 21.1–11). It is not surprising that he transports into this narrative sequence, in the form of an internal *prolepsis*, the story of Barsine, the widow of Memnon, whom he takes as his mistress and who is 'the only woman he was intimate with prior to his marriage' (*Alex.* 21.7–9).[25]

Ellipses

Sometimes this narrative acceleration takes the form of *ellipsis*.[26] Plutarch regularly hastens through uneventful periods of his subjects' lives by mentioning only notable offices or a few major acts, while silently skipping over years at a time, a very common narrative technique in the biographies that lends them their episodic quality.[27] *Ellipsis* frequently occurs in his accounts of childhood and youth, and may be due to lack of information, but this is often difficult or impossible to verify.

(*bellum piraticum*), 30–42 (war against Mithridates), 59–79 (Civil War, incl. flight and death of Pompey), *Per.* 25–28.3 (war against the Samians), *Ant.* 33–52 (Parthian expedition), and *Brut.* 8–53 (the assassination of Caesar and Philippi).

24 Observed by Stadter (1996) 296f.

25 The battle of Issus took place in November of 333 BCE, the capture of Damascus, at which time Barsine fell into Parmenio's hands, occurred in December of that year.

26 Genette (1980) 43: '... *ellipsis* or leap forward without any return is, obviously, not an anachrony but a simple acceleration of the narrative ...'.

27 On this technique in the *Cicero*, cf. Moles (1988) 33.

In the *Crassus*, for example, we first hear a report about his father Publius Licinius Crassus who was accused of corrupting a vestal virgin out of avarice (*Crass.* 1). The theme of avarice is then continued proleptically in the account of Crassus' acquisitiveness in adulthood and the questionable ways in which he accumulated his enormous fortune (*Crass.* 2). This section, which contains some rough chronological indicators, is followed by a chronologically indefinite passage in the iterative mode that relates his adult habits and pursuits (*Crass.* 3). When Plutarch then returns to a chronologically dateable event, the seizure of power by Cinna and Marius in 87 BCE, we encounter a Crassus who is now nearly 20 years old (*Crass.* 4.1). In other words Plutarch has ushered us through nearly two decades of his subject's lifetime in three chapters without having told us really anything that Crassus said or did before the age of 19. Such cursory or lacunose treatments of childhood and youth in which the narrator accelerates through the first third or half of his subject's life abound in the *Lives*.[28]

Later on in the *Life*, when narrating the events of Crassus' and Pompey's joint consulship, Plutarch is compelled to admit that 'their contentiousness rendered their consulship barren politically and without achievement', thereby absolving himself of the need to narrate at length the events of the year 70 BCE (*Crass.* 12.3). After an *ellipsis* of 4 years Plutarch feels obliged to mention his uneventful censorship (65 BCE), which 'passed without any results or achievements whatever' (*Crass.* 13.1). Another silent *ellipsis* of one year brings us to the eventful narrative of the Conspiracy of Catiline (63 – 62 BCE) (*Crass.* 13.3). The next event mentioned, Caesar's return to Rome to canvass for the consulship, occurs in 60 BCE (*Crass.* 14.1).

Internal and external *analepses*

Usually *analepses* serve to introduce past events that have been suppressed (*paralipsis*) into a context later in the narrative which enhances their significance. The episode of Pompey's dealings with Metellus on Crete, which reflects negatively on Pompey and tarnishes somewhat his triumph over the pirates, analeptically concludes the narrative of the *bellum piraticum* (*Pomp.* 29). This is a typical example of the use of an internal completing *analepsis* to set off for thematic purposes an event from a prior sequence. We are, for example, encouraged to assess in a different light Agesilaus' desire to gratify his son's wishes in the Sphodrias affair when we learn that he was exceedingly fond of children and used to join in his childrens' play when they were very young and was seen riding 'about on a stick at home pretending that it was a horse' (*Ages.* 25.11).

28 See, e.g., the *Lives of Solon, Publicola, Lycurgus, Numa, Camillus, Aristides, Cato the Elder, Cimon, Lucullus, Fabius Maximus, Nicias, Coriolanus, Lysander, Sulla, Agesilaus, Pompey, Pelopidas, Marcellus, Dion, Timoleon, Aemilius Paulus, Sertorius, Eumenes, Phocion, Demetrius, Antony, Marius, Agis, Cleomenes, Tiberius and Gaius Gracchus, Philopoemen,* and *Flamininus.*

External *analepses* are more interesting and less common because of their potential significance. Plutarch often incorporates external *analepses* into the early sections of the *Lives* that usually recount events or family histories preceding and leading up to the birth of the biographical subject (I understand the *fabula*-time in biography to begin with the birth of the subject and to end with his or her death). Some notable examples of this practice include the story of Pittheus, the grandfather of Theseus, and Aegeus, Theseus' father, that relates how Theseus came to be conceived from Pittheus' daughter Aethra and how Aegeus left a sword and a pair of sandals tucked under a rock for his future son, if she bore a male (*Thes.* 3). In the *Life of Romulus* Plutarch recounts the diverse traditions surrounding the foundation and naming of Rome, until finally concluding with the majority view recounted by Diocles of Peparethus and Fabius Pictor that contains the story of the birth and exposure of the twins Romulus and Remus, and their suckling by a she-wolf, etc. (*Rom.* 1–3). Plutarch recounts the early history of the Thesprotians and Molossians leading up to the birth of Pyrrhus in his *Life* (1). The family of the Fabii receive brief narrative treatment in the prologue to the *Life of Fabius Maximus* (1.1–1.2), as do the Marcii in the *Life of Coriolanus* (1.1), the Aemilii in the *Life of Aemilius Paulus* (2.1–2.4), and the Antonii in the *Life of Antony* (1.1–1.3). The *Cicero* traces the origins of the family back to Tullus Attius, 'an illustrious king of the Volscians who waged war upon the Romans with great ability' (*Cic.* 1.2). In the *Agis* the king's lineage is traced back six generations to Agesilaus 'who crossed into Asia and became the most powerful Greek of his time' (*Agis* 3).

Occasionally external *analepses* are employed to foreshadow future events or inherited characteristics. The story of the father's (Marcus Antonius Creticus') cowardly submissiveness towards his wife, Julia, foreshadows Marc Antony's own fateful docile submissiveness towards Cleopatra (*Ant.* 1). In the *Life of the Elder Cato*, Plutarch cites the Censor's report (employing him as a secondary narrator) about his father and grandfather (1). His hard-charging grandfather apparently had five horses killed under him in battle. The grandson too appears to have displayed a certain callousness with regard to living creatures that had served their purpose, including the horse that he left in Spain even though it had served him well during his consular campaign (*Ca. Ma.* 5).

Foreshadowing and proleptic[29] references

In the proems or early sections of the *Lives*, Plutarch often inserts references that foreshadow future events or behavior. The *Life of Alexander* provides some excellent exam-

---

29 Genette (1980) 40 suggests this term as preferable to the more subjectively loaded term 'anticipation' and defines it as 'any narrative maneuver that consists of narrating or evoking in advance an event that will take place later'. The 'evoking in advance' part of this definition is crucial for my pur-

ples of this technique as indicated above in the external analeptic narrative that contain several references presaging future greatness which precede and accompany Alexander's birth (*Alex.* 2.2–3.9).[30] Other *Lives* exhibit this structural feature to some degree.[31]

Very often Plutarch incorporates early in the *Life* information drawn from a chronologically later period to explain or illuminate physical appearance, education, and other attributes that become fully manifest in adulthood. The *Pericles* incorporates citations of comic poets to describe humorously the physical anomaly of the mature Pericles' elongated head shape (*Per.* 3.3–7). The account of Pericles' education contains numerous chronological dislocations of a proleptic nature that are for the most part concealed, as Plutarch collapses the information he has about Pericles' association with Damon, Zeno the Eleatic, and Anaxagoras into one compact, chronologically obscure section (*Per.* 4–6). The *Life* then continues with the narration of Pericles' youth, early military service, prior to his involvement in politics, and his political ascent that is causally related to the death of Aristides (c. 467 BCE) and Cimon's removal from Athens due to his foreign campaigns (476–463 BCE) (*Per.* 7.1–5).

The *Crassus*, as noted above, incorporates information deriving from later periods in the statesman's life and transfers it to the beginning sections in discussing Crassus' avaricious behavior (*Crass.* 2–3).[32] Similarly the proem of the *Cicero* contains two anecdotes referring to the time of his entry into politics and his quaestorship in Sicily (*Cic.* 1.5–6). The prologue of the *Aristides* begins with a striking statement which cites sources who claim that the statesman was of modest means, during his lifetime, and his daughters, after his death, could not marry for a very long time because of their indigence (*Arist.* 1.1). In the ensuing discussion Plutarch examines the credibility of this assertion in light of the known details of Aristides' adult life. The proem of the *Cato Major* launches into a description of the Censor's appearance and behavior that is clearly taken from his adulthood, since it contains references to his military service and oratorical ability. The proem of the *Sulla* contains an anecdote that refers to a time after the war with Jugurtha and flashes forward from there to another anecdote concerning a freedman, and former fellow lodger, whom he had executed during the proscriptions, when he 'had at last become absolute in power' (*Sull.* 1).

Sometimes Plutarch openly acknowledges chronological displacements. For example, he concludes his discussion of the Younger Cato's divorce of Marcia, in a section devoted to Cato's relationship with women, with the remark: 'This incident occurred at a later time, it is true, but since I had taken up the topic of the women of

---

pose here since Plutarch's narrative strategy is more often allusive than explicit. Tim Rood has suggested the term 'implied *prolepsis*' for those instances in the narrative which allusively adumbrate future events.

30 See Stadter (1996).

31 See, e.g., *Ca. Mi.* 2–3, *Caes.* 1.4, *Alc.* 1–16, *Cic.* 2, *Per.* 1–2 (building programme), *Cim.* 1–2 (theme of euergetism).

32 The return to a more chronologically oriented narrative is signaled by '*alla tauta men hysteron*'.

Cato's household I decided to anticipate it' (*Ca. Mi.* 25.11–13).[33] Sometimes he simply flashes forward to complete a topic, as in the narration of Agesilaus' future aid to the exiled Megabates (*Ages.* 13.3–4). Both of these examples as well as the example of Barsine in the *Alexander* cited above display Plutarch's tendency to bring related information together thematically, even if this disrupts the chronological framework of the *Life*.

External *prolepses*

Plutarch also incorporates external *prolepses* into the narrative, as when he narrates the fate of the protagonist's descendants (see *Ca. Mi.* 73, *Ca. Ma.* 27.7, *Ant.* 87). The *Aristides* contains an external proleptic reference that recounts how Alexander the Great, many years later when he was King of Asia, rewarded the valour and munificence of the Plataeans (who had voluntarily ceded a portion of their territory to the Athenians so that they might defend Greece on their own soil) by building the walls of Plataea (*Arist.* 11.9). In the *Cato Maior* Plutarch rebukes the Censor's anti-Hellenic remark that 'Rome would lose her empire when she became infected with Greek letters' with the terse observation that 'time has certainly shown the emptiness of this ill-boding speech of his, for while the city was at the zenith of its empire, she made every form of Greek learning and culture her own' (*Ca. Ma.* 23.2–3). In the *Lycurgus* he recounts the stability of Lycurgus' laws for 500 years after the statesman's death until their eventual abrogation in the reign of Agis (*Lyc.* 29.6–30.2).

Achronic narration of habitual or characterising behaviours

When describing the general characteristics of an individual's behaviour Plutarch abandons a strictly chronological presentation in favour of iteratively recounting habitual behaviours occurring in various stages of life (childhood, adolescence, adulthood, senescence). The *Themistocles*, for example, contains long sections of iterative narrative usually composed with the imperfect tense. The section recounting Themistocles' childhood contains several proleptic references presented iteratively in the context of his early education that serve to adumbrate the statesman's future rhetorical ability and pragmatic intelligence (*Them.* 2.1–7). The personal qualities of Tiberius and Gaius Gracchus are compared at length in the beginning of their double *Life* (*Ti. and C. Gracch.* 2–3). Agesilaus' habitual behaviour as king, especially the way he dealt with the Ephors, is discussed at length by Plutarch (*Ages.* 4.2–5.4). Plutarch, in his account of Pericles' early political successes, frequently discusses in a timeless

---

33 Cf. also *Ca. Mi.* 30.9–31.1.

iterative way his powerful oratorical ability, unflappable nature, and other tactics he employs to consolidate his power (*Per.* 5.1; 7.5–8.6; 9.2–5; 11.2–12.3).[34]

Plutarch employs another technique in which he strings together various anecdotes without any distinct chronological markers. The *Alcibiades* is one such case in which presumably the relative opaqueness of the chronology in the first half of the *Life* directly follows from the biographer's desire to accentuate his portrayal of Alcibiades' complex character.[35] After describing the influence of *eros* in Alcibiades' relationship with Socrates and his other lovers, a time period extending through Alcibiades' late adolescence and early manhood presumably (*Alc.* 6), Plutarch returns to Alcibiades' early boyhood (*Alc.* 7). The *Life of the Elder Cato* also posed some challenges for Plutarch which he resolves by selectively grouping his discussion of the statesman's career and private life. In particular, it is the Censor's famous sayings which attract his attention due to their characterisation potential. These he presents randomly without, in the majority of cases, any temporal reference points and some he presents as habitual statements, introducing them with the imperfect ἔλεγε (*Ca. Ma.* 8–9).[36]

The analeptic *Synkrisis*

The comparative structure of the *Parallel Lives* is underscored in the analytical postscript or *Synkrisis* appended to all but four[37] of the 22 pairs of *Lives*. In these brief analeptic essays Plutarch weighs the positive and negative sides to the two protagonists' conduct and reiterates the deeds (repeating internal *analepses*) that are significant in this evaluative process, occasionally in a more critical way.[38] The emphasis is placed sharply on moral instruction in the *Synkriseis*.

Occasionally Plutarch inserts in the *Synkrisis* important information that he has omitted in the *Life* proper (completing internal *analepses*). He acknowledges, for example, his neglect to mention that Crassus struck Lucius Annalius in the face with his fist and 'drove him bleeding from the forum' (*Comp. Nic. et Crass.* 2.3). His reference to Crassus' maltreatment of women in the *Synkrisis* (*Comp. Nic. et Crass.* 1.2) is also not supported in the *Life* even though the story of his father's corruption of the vestal virgin Licinia in the proem (*Crass.* 1.4–5) appears to foreshadow this theme (along with avarice).

---

34 Habitual activities ushered in by powerful statesmen such as Romulus, Lycurgus (Spartan *agogē*) and Numa also fall under this timeless iterative category.

35 See above.

36 The exceptions being the reference to King Eumenes' visit to Rome and Scipio's solicitation of aid for exiles from Achaia at the instance of Polybius.

37 *Phocion–Cato Minor, Alexander–Caesar, Themistocles–Camillus, Pyrrhus–Marius.* On the *Synkrisis* in general see Duff (1999) 243–286.

38 See Duff (1999) 59.

Motif of reference to the narrator's own time

Not infrequently Plutarch makes reference to his own time in the *Lives*. This occurs in those situations when he refers to what he has seen with his own eyes (autopsy). Plutarch has witnessed the whipping to death of the Spartan youths at the altar of Artemis Orthia (*Lyc.* 18.2). He has seen Agesilaus' spear still on display in his own day in Sparta (*Ages.* 19.11) and the statue of Lysander on display at Delphi (*Lys.* 1). Plutarch apparently visited the sanctuary of the Nymphs at Mieza, 'where even now people point out the stone seats and shady walks Aristotle used to frequent'[39] (*Alex.* 7.4) and, in Chaeronea, he has seen 'Alexander's Oak', located near the communal grave of the Macedonian dead, against which Alexander pitched his tent (*Alex.* 9.2). When tracing the descendants of Antony, Plutarch has harsh things to say about Nero who came to the throne in his time (*Ant.* 87.9).

Sometimes the reference to the narrator's own time reflects the impact of a past action on present circumstances. This includes for instance aetiological explanations of customs, laws, names, etc., that are still in use in Plutarch's day, such as the origin of the military term 'maniple' (*Rom.* 8.7–8) and the wedding salutation 'Talasio' (*Rom.* 15.1–4). These references all mark the interpenetration of the historical past in the narrator's present.

Space

As an individual's life unfolds not just in a temporal but also in a spatial dimension, it seems reasonable to consider the question as to under what circumstances and to what end Plutarch incorporates the discussion of space, buildings, and other physical objects into his narrative, such as the examples of autopsy cited above, Agesilaus' spear and Lysander's statue. They fall within the narrator's time, but they memorialise the past. The ancient historian John Buckler, in a study entitled 'Plutarch and Autopsy', has clearly demonstrated that Plutarch very often was on site and strove to collect data for his *Lives* whenever his travels took him to those locations.[40] By leaving Chaeronea he also was able to gain access to the holdings of libraries and archives in large cities on these journeys and during his years as a student. We know that he lived in Athens for a time, and visited Sparta, Rome, Africa, and Asia Minor.[41] His association with Delphi as priest at that sanctuary availed him of various important documents and inscriptions.[42] In the proem to the *Life of Demosthenes* (1–2), Plutarch himself acknowledges that his trips to Italy enabled him to experience first-hand monuments and battle sites and thus facilitated comprehension of his Roman

---

39 Translation by R. Waterfield. See also Frazier in this volume.

40 Buckler (1992) 4788–4830. See also Frazier and Alcalde Martín in this volume.

41 See Buckler (1992) 4791–4792.

42 See Buckler (1992) 4792 and Stadter (2015) 70–97.

sources.[43] It is apparent that Plutarch's travels influenced his composition of the *Lives* in a significant and positive way.

The chronotope

Mikhail Bakhtin formulated an important concept that seems to be applicable to Plutarch's most striking narratives of space. He coined the term 'chronotope' in reference to the periodic intersection or fusion of time and space that appears to take place in literary texts. This term, in Bakhtin's words, expresses 'the intrinsic connectedness of temporal and spatial relationships' in literature and he defines it as follows:

> In the literary artistic chronotope, spatial and temporal indicators are fused into one carefully thought-out, concrete whole. Time, as it were, thickens, takes on flesh, becomes artistically visible; likewise, space becomes charged and responsive to the movements of time, plot and history. This intersection of axes and fusion of indicators characterises the artistic chronotope.[44]

Pericles and the adornment of Athens

The narration of Pericles' adornment of the acropolis is one of the most important examples of a 'chronotope' in the *Lives*. The prologue to the *Lives of Pericles and Fabius Maximus* (*Per.* 1–2) alludes to the building program instituted by Pericles and echoes parts of Pericles' funeral oration transmitted to us by Thucydides (esp. 2.41–43). The deeds (*erga*) referred to in the prologue include not just statements and actions worthy of imitation, but also physical monuments attesting to past greatness. This connection is reprised in the body of the *Life of Pericles* describing the building project (13). In this section of the *Pericles* we have one of Plutarch's rare descriptions of buildings (Parthenon, Odeum, Propylaea) interwoven with a discussion of the architects and artists who were directly involved in various aspects of their planning, construction, and adornment. His description betrays an intimate familiarity with the architectural wonders of Athens. Plutarch's own assessment of the

---

43 See also his comments on the inconvenience of living in a small city in *De E* 384E. See the *Life of Demosthenes* (1–2): 'But as for me, I live in a small city, and I prefer to dwell there that it may not become smaller still; and during the time when I was in Rome and various parts of Italy I had no leisure to practice myself in the Roman language, owing to my public duties and the number of my pupils in philosophy. It was therefore late and when I was well on in years that I began to study Roman literature. And here my experience was an astonishing thing, but true. For it was not so much that by means of words I came to a complete understanding of things, as that from things I somehow had an experience which enabled me to follow the meaning of words' (Translation by B. Perrin, Loeb).

44 Bakhtin (1981) 84.

magnitude of Pericles' achievement in the *Synkrisis* of the *Pericles/Fabius Maximus* is glowing to say the least (*Comp. Per. et Fab.* 3).[45]

This example of a chronotope serves a twofold purpose. A physical structure such as the Parthenon represents on one level the enduring legacy of Pericles' virtue (and not that of Phidias or any other artisan or craftsman working on that building, on this Plutarch is clear).[46] On another level the monument (or deed) has didactic significance for the reader who derives inspiration and moral guidance from the representation (*mimesis*) of superlative achievements and is then moved to thoughtful and considered acts of emulation and imitation (*mimesis*). The bivalent nature of both of the Greek words *erga* and *mimesis*, adumbrated in the prologue, are thus realised in the text.[47] The *Theseus*, *Romulus*, and *Numa* also contain significant chronotopes that are too extensive to undergo analysis here.[48]

Solon and Lycurgus: Culture heroes and foundational figures

Solon and Lycurgus are two statesmen whose actions left an indelible impression on their respective cities. Both men left monuments to commemorate their significant activities. We are informed, for example, that Solon founded a temple to commemorate a victory over Megara, a victory that continued to be reenacted by the Athenians:

> There is also a dramatic reenactment of events which seems to corroborate this version. An Athenian ship used to sail up to the island, with the crew initially keeping quiet, but then charging into the attack yelling and screaming, while one man in full armor used to run to cape Sciradium and fetch the men on land. Also nearby is a temple to Enyalius founded by Solon to commemorate his defeat of the Megarians (*Sol.* 9.4).[49]

The continued reenactment of this victory in the same spatial location in which it took place is important. It functions as a dynamic memorial. The site of the victory became the site of the commemorative act for some indefinite time. As so often, a recurring ritual, especially when tied to a physical structure such as a temple, anchors collective or cultural memory. Plutarch himself does not seem to know how long this ceremony had been conducted. The nearby temple served as the anchor point. As we have seen, chronotopes such as these commemorate the virtue of the

---

45 Translation by B. Perrin, Loeb.

46 This is made abundantly clear in the prologue (*Per.* 1.4–2.1).

47 The bivalent nature of these two words employed in the prologue is well discussed by Duff (2001) 351–363.

48 See Banta (2006), (2007a), (2007b), and Beck (2012).

49 Translation by R. Waterfield.

individual concrete incarnation. Space is tied via achievements to character and for this reason is incorporated into the narrative.[50]

Similarly Plutarch reports that Lycurgus founded a temple (*Lyc.* 11). The story of how this came about is related in an anecdote that records how Lycurgus, owing to the unpopular nature of one of his reforms, the *syssitia*, among the wealthy, is forced by an angry mob of fellow citizens in the agora to flee for his life (*Lyc.* 11). One of his pursuers, a certain Alcander, succeeds in overtaking him and knocks out his eye with his staff. Alcander is punished for this by being placed in Lycurgus' personal custody where domestic servitude and his close association with the Spartan lawgiver are meant to encourage Alcander to reform his future behaviour. Plutarch concludes his narrative of these events by recording that Lycurgus founded a sanctuary in honour of Athena with the epithet of Optilletis to commemorate his loss. We know that Plutarch visited Sparta and he most likely encountered this story when he toured the temple.[51] The actions of Lycurgus, his character, and the physical memorial erected in commemoration of these events intersect in a chronotope.

Themistocles and the Persian War

In Plutarch's other biographies the availability of abundant historical sources that provide the biographer with ample material for his narrative may be the cause of infrequent references to space, as the challenge becomes one of selection from and condensation of a multitude of events. In addition, his heavy reliance on written sources rather than autopsy undoubtedly set limits on what aspects of the physical location he could confidently describe. On occasion, however, he does display in the narrative his awareness of the tangible past presence of an individual whose most notable actions are tied to a location he himself has visited. One example of this occurs in the *Life of Themistocles*. Plutarch felt an especial affinity for Themistocles. He was personally acquainted with his distant relative as he reports at the end of the *Life* (*Them.* 32).[52] The great Athenian's involvement in the Second Persian War is subjected to a lengthy treatment in the *Life* (*Them.* 7–18). He presents detailed information about important locations in the war that may not be familiar to all of his readers, locations visited by Plutarch himself. The most vivid example of this, containing both visual and olfactory sensory information, is his description of Artemisium and a temple of Artemis located there:

---

50 See also the report of Solon's encounter with Croesus (*Sol.* 27) that differs quite radically from Herodotus' version. Stadter (2015) 173–176 gives an intriguing interpretation of Plutarch's account of this meeting vis-à-vis Herodotus.

51 On Plutarch's visit(s) to Sparta in general, see Buckler (1992) 4814–4815. Buckler does not mention this sanctuary, however.

52 Translation by R. Waterfield.

> Artemisium is a north-facing beach in Euboea, past Hestiaea; it lies more or less opposite Olizon, which is in the land once ruled by Philoctetes. There is at Artemisium a small temple of 'East-looking' Artemis, as she was known there, which stands in a grove of trees, surrounded by blocks of white marble fixed on the ground. Rubbing this marble on one's hands yields a saffron-like color and smell. One of the blocks of stone has been inscribed with the following elegiac poem:
>
> There was a time when on this stretch of sea the sons of Athens
> In battle overcame a varied host of men of Asian stock;
> To mark their destruction of the army of the Medes,
> They erected these tokens in honor of the maiden Artemis.

One is shown a part of the beach where in the middle of all the surrounding sand the depths throw up a dark, ash-like dust which looks as though it is the result of fire; it is believed that the wrecked ships and bodies of the dead were burnt on this spot (*Them.* 8).[53]

Because of the sensory information he presents that would appear to lend an almost supernatural aura to the place, I think that this description of this temple is one of the most striking chronotopes in all of Plutarch. There can be no doubt that Plutarch was there, rubbing the marble with his hands and smelling the saffron smell. The naval battle of Artemisium and the defense of the pass at Thermopylae was part of a two-pronged strategy to stop or delay the Persian invasion and Themistocles was the author of this strategy. Plutarch knew this and his description of this place honours the man and commemorates his achievement.[54]

Conclusion: Time and space in the *Lives*

From our examination of Plutarch's literary technique in the *Parallel Lives* it is evident that they are symmetrical compositions with unifying themes that serve to modulate the narrative representation and referencing of time. Plutarch generally foreshadows his major concerns quite early in the proem or early chapters of the first *Life* in the form of anecdotes, for example, which foreshadow future greatness, abilities, traits, persistent behavioural patterns, or problems. These themes are most vividly emergent in the 'grand scenes' of the *Lives*, sections in which the narrative is greatly slowed. Longer episodes of narrative deceleration often showcase the main protagonist's specific involvement in major historical events. Plutarch tends to accelerate the narrative through, or omit from it altogether, long presentations of battles. He rarely presents speeches of any length. Plutarch often clusters chronologically disparate events for thematic purposes or he presents such information achronically, by employing the iterative mode of narration, to facilitate characterisation. Another technique involves the retrojection or displacement of information derived from adult-

53 Translation by R. Waterfield.

54 See also *Them.* 19. For other narratives involving space in the *Lives*, see Beck (2012).

hood into the narrative of childhood and youth to supplement the lack of information on these phases of his biographical subject's life. External and internal *prolepses* and *analepses* are often used to highlight behaviour by re-contextualizing it or clustering similar instances thematically. Plutarch concludes most of the pairs analeptically with a brief retrospective essay (*Synkrisis*) in which the major accomplishments and characteristics of the two heroes are recalled and evaluated instructively in an impartial way.

Plutarch's inclusion of descriptions of places, buildings, monuments, and other objects in the narrative serves multiple purposes. The narrative intersection of time and space in the form of chronotopes, for example, functions as a focalising device in his *Lives* reflective of the character and achievement of the biographical subject. From Plutarch's perspective one appropriate task of politicians in leadership positions is to act as patron of the arts. The beautification of Athens is therefore an achievement that reflects on Pericles more so than on Phidias. The *Lives* are meant to serve as behavioural paradigms and their students are to become active patrons themselves in their own communities. Cimon's philanthropic acts and beautification of Athens also deserves mention here, even if done on a more modest scale than Pericles', they were directly funded by his own resources. The significance of foundational figures is discernable from the monuments they left behind or the ones erected to commemorate their service to the city. While my discussion of Plutarch's use of space in the *Lives* is by no means exhaustive, it does show that descriptions and discussions of space, especially man-made space, is a significant narratological component of Plutarch's biographical technique.

2 Time manipulation and narrative signification

†Françoise Frazier

Espace mémoriel et paysage monumental

Plutarque et l'Athènes de son temps

Abstract: Les références de Plutarque à l'Athènes de son temps, toutes faites dans les *Vies* à une exception près (*De E* 384E), ont été lues jusqu'à présent dans une perspective documentaire, pour préciser sa méthode de travail et le rôle qu'y joue l'autopsie, ou pour y scruter le travail de rédéfinition de l'identité grecque et y chercher les prodromes de 'l'athénocentrisme' qui allait prévaloir à partir d'Hadrien, créant une sorte d'Athènes abstraite et intemporelle. Cette étude propose à la fois un réexamen des textes et une focalisation sur l'Athènes concrète avec ses monuments; Plutarque ne travaille certes pas en antiquaire, comme Pausanias, et le lien entre passé et présent procède chez lui d'un mouvement inverse: là où le Périégète s'appuie sur le présent de la vision pour évoquer événements passés et traditions, le biographe est parfois amené dans son récit au passé à évoquer le présent, mais ces mentions ne servent pas qu'à corroborer ses sources livresques. Elles permettent surtout de situer les faits passés dans l'espace athénien, de dessiner un espace mémoriel vivant, lourd de tout un passé historique. Dans ce passé, la *Vie de Thésée* se distingue par un nombre de mentions du présent plus élevé (7): indice peut-être d'un regain d'intérêt pour ces premiers temps, que cristallisera un peu plus tard l'Arc d'Hadrien, mais tout cela n'est encore qu'en germe, et il importe, lorsqu'on aborde la question en spécialiste de Plutarque et non pas en historien de l'Empire de peser ce que lui-même privilégiait et ce qui n'est passé au premier plan que pour les générations suivantes, voire pour les spécialistes d'aujourd'hui.

Dans la 'géographie personnelle' de Plutarque, citoyen à la fois de Chéronée, de Delphes, d'Athènes et de Rome, ou, pour le dire autrement, dans l'espace où il a vécu l'hellénisme au présent, brillent particulièrement la Béotie, Delphes, dont j'ai traité ailleurs,[1] et Athènes, qui sera le sujet du présent travail. Pour l'étudier, je suis partie de la notion de 'paysage monumental', qui fusionne en quelque sorte le 'lieu de mémoire' emprunté par l'historien Pierre Nora à la rhétorique antique[2] et le 'paysage', cher aux géographes et aux urbanistes, mais qu'utilisent aussi les archéologues.[3] La fécondité de l'utilisation des 'landscapes of memory' dans l'étude de Pausanias[4] semble inviter à les essayer sur Plutarque, en étant toutefois attentif à la différence

---

1 'Delphes dans les *Dialogues Pythiques:* Un lieu inspiré', texte à paraître dans les Actes de la Rencontre 'L'image de Delphes dans la littérature, d'Homère à nos jours' (Toulouse, mai 2014).
2 Reisigl (2009).
3 E.g., Palyvou (2007).
4 Alcock (1996) et Alcock et al. (2001); voir aussi Hutton (2005a).

https://doi.org/10.1515/9783110539479-004

des moments historiques qu'ils ont vécus. Or la recherche actuelle, qui, depuis une vingtaine d'années, scrute la redéfinition de l'identité hellénique à laquelle se sont attachés les Grecs eux-mêmes sous le Haut Empire, a parfois tendance à écraser un peu les perspectives temporelles: peut-être les *Vies Parallèles* ont-elles contribué à l'émergence d'un 'athénocentrisme'[5] que parachèveraient, politiquement, l'action d'Hadrien et l'installation du Panhellénion à Athènes, et, idéologiquement, le *Panathénaïque* d'Aristide,[6] mais l'Athènes de Plutarque n'en est pas pour autant l'Athènes d'Hadrien, que célèbre Pausanias en distinguant régulièrement ἀρχαῖα et καινά.[7] La dimension historique doit ainsi s'ajouter aux considérations très 'classiques' d'un des rares articles à avoir envisagé l'Athènes de Plutarque dans sa réalité matérielle et architecturale,[8] en se centrant sur l'autopsie.[9] Selon John Buckler, l'auteur de cet article, Plutarque n'utiliserait sa connaissance personnelle de la ville que pour corroborer ses sources littéraires, traitant les lieux et les toponymes comme les cultes et les rites dont il fait l'étiologie, bien loin donc du travail d'antiquaire de Pausanias. Sans doute, l'éloge de l'Acropole dans la *Vie de Périclès* constitue-t-il une exception remarquable, mais il ne montrerait rien de plus que la conscience qu'avait Plutarque de l'intérêt de l'autopsie pour enrichir et appuyer son propos.[10] Ces conclusions, un peu décevantes, peuvent, je crois, être enrichies par une étude exhaustive des passages où Plutarque évoque au présent l'Athènes d'aujourd'hui, et la simple vision documentaire être ainsi dépassée.

Les mentions du présent: Plutarque et Pausanias

Dans l'œuvre si étendue de Plutarque, il est remarquable que, en dehors de la dédicace du *De E* (384E) à son ami athénien Sarapion, habitant d' une 'grande ville' où abondent livres et conférences, on ne trouve de référence à la réalité matérielle d'Athènes, avec ses monuments divers et ses places, que dans les *Vies*,[11] c'est-à-

5 C'est l'hypothèse de Lamberton (1997), plus discutable que la manifestation d'une communauté de valeurs gréco-romaines.
6 Oudot (2006).
7 Graindor (1934) situe le commencement des travaux dans les dernières années de la vie de Plutarque, vers 124–125.
8 Je discute ailleurs Frazier [2016] Oudot (2012), qui se focalise sur l'éloge de l'Acropole de *Per.* 12–13, pour y voir un de ces 'textes du Ier et du IInd siècle qui tendent à faire d'Athènes une abstraction '*échappant au temps, échappant également à l'espace*', dématérialisation qui irait de pair avec une 'déshistoricisation' opérée par la 'fonction symbolique forte assignée à l'Acropole comme ensemble architectural et comme lieu emblématique de l'hellénisme'.
9 Buckler (1992) 4816–4821 pour Athènes. Alcalde-Martín reprend le thème dans ce volume.
10 Buckler (1992) 4818–4819.
11 Certains *Propos de table* sont certes explicitement situés à Athènes (1.1 et 1.10; 3.1; 5.1; 8.3 et tout le livre 9), mais, hormis la mention au livre 9 du Diogénéion où Ammonios avait présidé à l'examen des

dire dans des récits *du passé*, où vient parfois affleurer le présent. Cette rencontre s'inscrit dans un mouvement en quelque sorte inverse de celui de Pausanias, qui s'appuie sur le présent de la vision pour évoquer événements passés et traditions, mais chez l'un comme chez l'autre, le tissage de ce 'lien qui manifeste la continuité entre le passé et le présent'[12] se marque dans le texte par l'emploi de νῦν, μέχρι νῦν, ἔτι νῦν ou encore καθ' ἡμᾶς. À l'aide de ces marqueurs,[13] sous réserve qu'un passage au présent sans complément de temps m'ait échappé, j'ai relevé trente passages, si l'on compte pour un *Per.* 12–13. La plupart des *Vies* athéniennes comportent deux ou trois passages:[14] seule se distingue la *Vie de Thésée*, qui en comporte sept.[15] S'y ajoutent un passage de *Caton l'Ancien* (5.4), qui mentionne deux tombes d'animaux, le tombeau du chien de Xanthippe et la tombe des mules de Cimon l'Ancien, et un d'*Alexandre* (69.8), qui évoque à nouveau une tombe exotique, celle de l'Indien, qui date de l'époque d'Auguste. Quant aux monuments cités, outre le Parthénon et les Propylées, dont ils ne disent d'ailleurs pas la même chose, Pausanias et Plutarque ne se rejoignent que pour cinq d'entre eux:[16] le Cynosarges (*Them.* 1.3 = Paus. 1.19.3); les restes des *axones* sur lesquels étaient inscrites les lois de Solon au Prytanée (*Sol.* 25.1 = Paus. 1.18.3); la tombe de Thucydide que seul Pausanias localise à la Porte Mélitidè (*Cim.* 4.3 = Paus. 1.23.9); la tombe d'Anthémocritos au Dipylon (*Per.* 30.4 = Paus. 1.36.3) et enfin la tombe de Pythonikè, courtisane aimée d'Harpale, sur la route d'Éleusis (*Phoc.* 22.1–2 = Paus. 1.37.5), sur laquelle de nouveau les deux auteurs divergent. Alors qu'elle donne à Plutarque l'occasion de stigmatiser un monument déshonorant–comme le font Diogénianos et Sarapion pour les broches offertes par Rhodopis et la statue de Phryné à Delphes dans le *De Pythiae* (400F-401B)–, le Périégète la signale comme étant, 'parmi les monuments funéraires qui ont atteint les plus grandes dimensions et qui sont le plus richement décorés, le plus magnifique de tous les tombeaux antiques que l'on peut voir en Grèce'. Ce peu de convergences et les différences mêmes des détails alors donnés s'expliquent aisément par la différence des genres littéraires et des intérêts des deux auteurs: sans surprise, on trouve chez Plutarque des éléments directement en rapport avec le héros dont il narre la *Vie*, tombeaux, ex-voto, statues, lieux dits, qui n'intéressent pas Pausanias.

éphèbes (736D), les indications se limitent à un simple complément de lieu (Ἀθήνησι, 612E, 645D, 720C, 736C–D, ou ἐν Ἀθήναις, 673C), et n'apportent rien à cette étude.

12 Bruit Zaidman (2003) 23. Voir aussi Alcalde-Martín, Oikonomopoulou et Roskam dans ce volume.

13 Bruit Zaidman (2003) 22, en dénombre 280 dans l'ensemble de l'œuvre de Pausanias.

14 3: *Sol.* 1.7; 9.6–7; 25.1; *Them.* 1.3; 10.10; 22.2–3 et *Cim.* 4.3 et 6; 19.5; 2: *Arist.* 1.3; 27.1; *Per.* 12.1 [13.5, ἐπανθεῖ est particulier]; 30.4; *Nic.* 3.3; 13.3; *Dem.* 22.2–3; 31.1–3; *Phoc.* 18.1; 22.1–2; 1: *Alc.* 21.3.

15 12.6; 17.6–7; 18.1; 27.3–5 et 7; 35.5 et 36.3–4.

16 Ils évoquent aussi tous les deux le départ des Lampadédromies de l'Académie, mais Plutarque (*Sol.* 1.7) la réfère à la statue d'Éros consacrée par Pisistrate, tandis que Pausanias (1.30) parle de l'autel d'Éros, consacré par Charmos (§§1–2), et fait partir la lampadédromie de l'autel de Prométhée (§5). Sur les diverses courses partant de l'Académie, voir Étienne (2004) 132–133.

Les formules introductives, dans leur vocabulaire comme dans leur tour, peuvent éclairer le point de vue de Plutarque. On y trouve des verbes qui posent l'existence d'un lieu ou d'un monument (εἶναι, κεῖσθαι), constatent sa conservation (διαμένειν ou διασῴζεσθαι) ou mettent en avant le nom qui lui est donné (καλεῖν / καλεῖσθαι, λέγουσι); il peut encore 'être montré' (δείκνυσθαι) ou 'témoigner' (μαρτυρεῖν): dans le premier cas, c'est la mémoire des Athéniens eux-mêmes qui joue, comme c'est aussi le cas dans l'attribution des noms, tandis que, dans le second, c'est l'auteur qui, comme le dit Buckler, utilise les monuments pour 'appuyer les sources littéraires'. Ainsi le seul sémantisme fait déjà apparaître diverses perspectives, que modulent deux autres éléments: d'abord la structure grammaticale, qui inscrit la référence soit dans un syntagme déterminatif ('c'est ce qu'on appelle aujourd'hui') soit dans un syntagme locatif, proposition relative ('là où aujourd'hui ...') ou complément prépositionnel ('au lieu appelé aujourd'hui'), ensuite l'emploi des temps verbaux, puisque, alors que, *a priori*, on n'attendrait que des présents ou des parfaits, se détachent cinq textes *à l'imparfait*, se référant à 'notre époque' (καθ' ἡμᾶς), qui réclament un examen attentif.

Le cas délicat des passages à l'imparfait: la mémoire de Plutarque?

Ces cinq passages[17] m'avaient semblé *a priori* pouvoir refléter des souvenirs de Plutarque remontant au temps de ses études.[18] À l'examen, cette déduction apparaît des plus fragiles, d'abord parce qu'on trouve ce tour dans des passages de Polybe, Denys d'Halicarnasse et Pausanias, où il n'est pas question de souvenirs de jeunesse,[19] ensuite parce que Plutarque lui-même l'emploie ailleurs. On le trouve ainsi pour le chêne dit d'Alexandre qu'on 'montrait encore de notre temps' (ἐδείκνυτο, *Alex.* 9.3); sans doute se trouve-t-il au bord du Céphise, en Béotie, où Plutarque a dû le voir, mais évoque-t-il pour autant un souvenir personnel? Que dire, bien plus, pour Chalcis, où il est question du prêtre de Titus qui 'était encore élu de notre temps' (ἀπεδείκνυτο, *Flam.* 16.5–6) ou pour Sicyone, où des descendants d'Aratos 'demeuraient de notre temps' (διέμεινε, *Arat.* 54.8), mention à l'imparfait qui conclut une œuvre dédiée à un de ces descendants, Polycratès de Sicyone, dont Plutarque évoque même les fils, associant au présent l'avenir de la lignée? Et la mention des poèmes de Sparte 'qui étaient encore conservés de notre temps' (διεσῴζετο,

17 *Sol.* 25.1; *Arist.* 1.3; *Them.* 22.3; *Nic.* 3.3; *Dem.* 7.6.

18 La suggestion en est déjà faite par Flacelière, in Flacelière et al. (1972) 147, n. 2 à propos de *Nic.* 3.3.

19 Plb. 3.87.6 (à propos de Fabius): ἔτι γοῦν ἐπεκαλοῦντο καὶ καθ' ἡμᾶς οἱ ταύτης τῆς οἰκίας Μάξιμοι; D.H. 4 14.4: ἣν ἔτι καὶ καθ' ἡμᾶς ἑορτὴν ἄγοντες Ῥωμαῖοι διετέλουν; Paus. 1.24.1: τέρατα γὰρ πολλῷ καὶ τοῦδε θαυμασιώτερα καὶ καθ' ἡμᾶς ἔτικτον γυναῖκες. Je dois ces références à l'obligeance de Julien du Bouchet, qui m'a également aidée à préciser les questions linguistiques et que je remercie vivement.

Lyc. 21.4) n'est pas pour clarifier les choses, puisque les références à des *logoi* ou des écrits conservés sont généralement au présent, sans complément temporel.[20] Tout juste peut-on suggérer une tendance à associer καθ' ἡμᾶς et le présent à un usage ou un état d'esprit actuel,[21] et καθ' ἡμᾶς et l'imparfait à des monuments,[22] tour qui, dans une langue sensible à l'aspect, insisterait sur la permanence, la continuité temporelle avec le passé raconté.[23] Peut-être aussi ces variations de temps tiennent-elles à un sentiment particulier qui anime l'auteur au moment où il écrit—ce que montre par exemple l'emploi de l'imparfait épistolaire dans des passages où l'auteur évoque les circonstances de rédaction de la lettre,[24] mais il apparaît délicat de déterminer ce sentiment et plus hasardeux encore de transposer à Plutarque ce que pense un locuteur français lorsqu'il dit 'de mon temps il en était ainsi'—phrase qui suppose un locuteur d'un certain âge faisant référence à ses jeunes années—d'autant que dans le seul passage où il se réfère très précisément à son séjour d'étudiant à Athènes, pour raconter une anecdote pittoresque rattachée à la statue de Démosthène (*Dem.* 31.1–3 qui commence par Μικρῷ δὲ πρόσθεν ἢ παραβαλεῖν ἡμᾶς Ἀθήναζε λέγεται τὸ τοιόνδε συμβῆναι), il n'emploie pas l'imparfait. Le monument est décrit au *parfait*, tel qu'il est encore aujourd'hui (ἕστηκε δὲ τοὺς δακτύλους συνέχων δι' ἀλλήλων, καὶ παραπέφυκεν οὐ μεγάλη πλάτανος, 31.2).[25] Il faut enfin noter que l'imparfait n'apparaît dans aucune des nombreuses références qui émaillent la *Vie de Thésée*: marque d'un intérêt plus livresque, plus distancié en tout cas avec ces temps mythiques?

La plus ancienne référence en tout cas concerne les restes des *axones* de Solon au Prytanée (ἔτι καθ' ἡμᾶς ἐν Πρυτανείῳ λείψανα μικρὰ διεσῴζετο, *Sol.* 25.1) et Plutarque ne donne aucun détail marquant qu'il les a vus, même si c'est probable;[26] la même chose vaut pour la salle d'études souterraine que fit construire Démosthène, dont la conservation est mentionnée sans plus (ἐκ δὲ τούτου κατάγειον μὲν οἰκοδομῆσαι μελετητήριον, ὃ δὴ διεσῴζετο καὶ καθ' ἡμᾶς, *Dem.* 7.6), mais qu'il a dû voir. De même il donne l'inscription que portent les trépieds votifs d'Aristide dans le temple

---

20 *Fab.* 1.9; *Cic.* 2.3; *Luc.* 1.8; *Crass.* 33.3; *Brut.* 13.4; à rapprocher de *Lyc.* 1.2.

21 *Publ.* 19.10: καὶ καθ' ἡμᾶς ... κηρύττουσι; *Ages.* 35.2: ἀτέλειαν, ἣν ἔτι καὶ καθ' ἡμᾶς ἔχει Καλλικράτης; *Comp. Sol.-Publ.* 1.3: ἔτι καὶ καθ' ἡμᾶς Ποπλικόλαι καὶ Μεσσάλαι καὶ Οὐαλέριοι δι' ἐτῶν ἑξακοσίων τῆς εὐγενείας τὴν δόξαν ἀναφέρουσι; *Arist.* 27.7: ἔτι πολλὰ καὶ καθ' ἡμᾶς ἡ πόλις ... θαυμάζεται καὶ ζηλοῦται; voir aussi *De frat. am.* 478C et *De sera num.* 560D.

22 Il existe aussi quelques cas d'aoriste, qui se réfèrent, comme il se doit, à un événement particulier survenu 'de notre temps': *Flam.* 12.13; *Cons. ad Apoll.* 118D; *De Is. et Os.* 380B.

23 Humbert (1982) 138, §235.

24 Koskenniemi (1956) 190: 'Das Imperfekt wird im Briefstil für die Gegenwart des Schreibers gebraucht, wenn beabsichtigt ist, vor allem die Schreibsituation oder ihren Hintergrund zu schildern'.

25 À opposer au plus-que-parfait de *Nic.* 3.3, *voir n. 28.*

26 Paus. 1.18.3 emploie le parfait: πλησίον δὲ πρυτανεῖόν ἐστιν, ἐν ᾧ νόμοι τε οἱ Σόλωνός εἰσι γεγραμμένοι ... .

de Dionysos et l'on peut supposer qu'il l'a lue.[27] Il signale encore, au plus-que-parfait, la présence de deux ex-voto offerts par Nicias.[28] On trouve, dans le même temple, des trépieds chorégiques—offerts en réalité par un autre Nicias, comme l'a établi Wycherley à partir de l'inscription, qui permet de dater l'offrande de 320 – 319 av. J.-C.:[29] l'erreur vient-elle de gens qui les lui auraient montrés en même temps que ceux d'Aristide ou est-ce une interprétation inexacte de sa part provoquée par la mention de trépieds de Nicias dans le *Gorgias* (472b)? C'est en tout cas bien un 'détail vu' qu'il donne pour l'autre offrande de Nicias, sur l'Acropole, le Palladion, qui, précise-t-il, *a perdu sa dorure*. Enfin, dans cette liste, une place particulière doit être réservée au sanctuaire d'Artémis Aristoboulé érigé par Thémistocle à Mélité—ignoré de Pausanias—où il est tentant de penser que Plutarque a eu pour guide le descendant du grand homme qui fut son condisciple à l'école d'Ammonios.[30] L'ensemble du passage mérite une lecture attentive:

> Ἠνίασε δὲ τοὺς πολλοὺς καὶ τὸ τῆς Ἀρτέμιδος ἱερὸν εἱσάμενος, ἣν Ἀριστοβούλην μὲν προσηγόρευσεν ὡς ἄριστα τῇ πόλει καὶ τοῖς Ἕλλησι βουλευσάμενος, πλησίον δὲ τῆς οἰκίας κατεσκεύασεν ἐν Μελίτῃ τὸ ἱερόν, οὗ νῦν τὰ σώματα τῶν θανατουμένων οἱ δήμιοι προβάλλουσι καὶ τὰ ἱμάτια καὶ τοὺς βρόχους τῶν ἀπαγχομένων καὶ καθαιρεθέντων ἐκφέρουσιν. Ἔκειτο δὲ καὶ τοῦ Θεμιστοκλέους εἰκόνιον ἐν τῷ ναῷ τῆς Ἀριστοβούλης ἔτι καθ' ἡμᾶς, καὶ φαίνεταί τις οὐ τὴν ψυχὴν μόνον, ἀλλὰ καὶ τὴν ὄψιν ἡρωικὸς γενόμενος.[31]

La variété des temps est remarquable: la présence de la statue est située dans le passé (ἔκειτο)—au temps où Plutarque l'aurait vue?— mais il en dégage au présent (φαίνεται) ce qu'elle révèle de la physionomie du héros. Des strates temporelles différentes apparaissent aussi à propos du sanctuaire où elle se trouve: aoriste pour sa réalisation à l'époque de Thémistocle, présent pour les usages d'aujourd'hui, lesquels, fait remarquable, ne découlent en aucun cas de la nature de l'édifice. On n'a donc pas ici la mise en avant d'un usage qui témoignerait d'un fait passé, mais la localisation permet d'inscrire ce passé à l'intérieur de la cité d'aujourd'hui.

---

27 *Arist.* 1.3: νίκης ἀναθήματα χορηγικῆς τρίποδας ἐν Διονύσου καταλέλοιπεν, οἳ καὶ καθ' ἡμᾶς ἐδείκνυντο, τοιαύτην ἐπιγραφὴν διασῴζοντες Ἀντιοχὶς ἐνίκα, Ἀριστείδης ἐχορήγει, Ἀρχέστρατος ἐδίδασκε.

28 *Nic.* 3.3: εἱστήκει δὲ καὶ τῶν ἀναθημάτων αὐτοῦ καθ' ἡμᾶς τό τε Παλλάδιον ἐν ἀκροπόλει, τὴν χρύσωσιν ἀποβεβληκός καὶ ὁ τοῖς χορηγικοῖς τρίποσιν ὑποκείμενος ἐν Διονύσου νεώς.

29 Wycherley (1978) 184, repris par Buckler (1992) 4820.

30 *Them.* 32.6.

31 *Them.* 22.2 – 3: 'Il choqua aussi le peuple en fondant le sanctuaire d'Artémis, auquel il donna l'épiclèse d'Aristoboulè ['Excellente Conseillère'], par allusion aux excellents conseils qu'il avait donnés à Athènes et aux Grecs. Il fit construire ce sanctuaire près de sa maison à Mélitè, là où les bourreaux jettent à présent les corps de ceux qui ont été mis à mort et portent les vêtements et les cordes de ceux qui se sont pendus. Il y avait encore de mon temps, dans le temple d'Aristoboulè, une petite statue de Thémistocle qui montre qu'il n'avait pas seulement l'âme, mais aussi la physionomie d'un héros'.

Καλοῦσι / Δείκνυται: La mémoire d'Athènes et ses 'curiosités'

À côté des usages, ce que disent ou montrent les Athéniens est mis en avant dans huit passages,[32] qui constituent une sorte de bref inventaire de curiosités locales, où se côtoient, pour les 'choses nommées', l'Hermès d'Andocide, toujours ainsi appelé contre le témoignage même de l'inscription, qu'il porte,[33] et l'Hermès de la Porte d'Égée,[34] et pour les 'choses montrées', la maison de Phocion, et des tombes, celles, prestigieuses, d'Aristide à Phalère et de Thucydide, ou, plus pittoresques, du Chien, à Salamine, et de l'Indien, non localisée: a-t-on là un aperçu de l'Athènes inattendue que découvrait le 'touriste antique'? En tout cas, au-delà de l'appui que ces mentions peuvent apporter au propos—l'aspect de la maison de Phocion confirme *de facto*, ou plutôt *de visu*, sa simplicité—, mais que Plutarque ne souligne pas explicitement, elles font affleurer une mémoire athénienne, que méconnaît la perspective purement documentaire de Buckler.

Μαρτυρεῖ: Monument, lieux et rites comme preuves

Très remarquable à cet égard me semble, contrairement à ce que ferait attendre l'analyse de Buckler, la rareté des passages comportant le verbe μαρτυρεῖν ou le substantif μαρτύριον, la troisième catégorie que je distinguerai. On ne relève en effet que cinq passages qui mettent en avant cette fonction, quatre si l'on met à part le témoignage très général 'de la puissance et de l'ancienne prospérité de la Grèce' que donne l'Acropole (*Per.* 12.1). Le nom de *Cimoneia* donné à des monuments funéraires est ainsi censé prouver que les restes de Cimon ont bien été ramenés à Athènes (*Cim.* 19.5), tandis que le lieu-dit *Horcomosion* témoigne de l'accord passé avec les Amazones, comme le fait aussi le sacrifice qui précède les *Theseia* (*Thes.* 27.7).[35] Selon la remarque de Buckler, Plutarque associe volontiers toponymie et rituels: la fête des *kybernèsia* et leurs *hérôa* témoignent *ensemble* pour la version de Philochore,

---

32 *Arist.* 27.1; *Cim.* 4.3; *Phoc.* 18.1; *Nic.* 13.3; *Alc.* 21.3; *Ca. Ma.* 5.4; les deux sont associés dans *Them.* 10.10 et *Alex.* 69.8.

33 *Alc.* 21.3: Διὸ καὶ νῦν Ἀνδοκίδου καλεῖται, καὶ πάντες οὕτως ὀνομάζουσι, τῆς ἐπιγραφῆς ἀντιμαρτυρούσης. Nous connaissons tous dans nos villes modernes de vieux habitants qui, lors des changements de noms de rues, continuent à utiliser l'ancien; voir aussi *Nic.* 13.3.

34 *Thes.* 12.6: ἐνταῦθα γὰρ ὁ Αἰγεὺς ᾤκει, καὶ τὸν Ἑρμῆν τὸν πρὸς ἕω τοῦ ἱεροῦ καλοῦσιν ἐπ' Αἰγέως πύλαις.

35 Ἀλλὰ τοῦ γε τὸν πόλεμον εἰς σπονδὰς τελευτῆσαι μαρτύριόν ἐστιν ἥ τε τοῦ τόπου κλῆσις τοῦ παρὰ τὸ Θησεῖον, ὅνπερ Ὁρκωμόσιον καλοῦσιν, ἥ τε γινομένη πάλαι θυσία ταῖς Ἀμαζόσι πρὸ τῶν Θησείων; la liaison est étroitement établie par la particule τε.

qui fait de Phaiax et Nausithoos le timonier et le pilote de Thésée (*Thes.* 17.6–7),[36] comme le sanctuaire d'Enyalios fondé par Solon et le rituel commémoratif de l'attaque de Salamine appuient une des versions de la conquête de Salamine (*Sol.* 9.6–7). Dans ces deux cas, on remarque que les preuves sont mentionnées dans le cadre d'une confrontation explicite entre versions et, pour le dernier texte, à vrai dire, la valeur de preuve n'est même attribuée qu'au rituel. Le texte à nouveau mérite d'être lu dans son entier:

> Ἔοικε δὲ τῷ λόγῳ τούτῳ καὶ τὰ δρώμενα μαρτυρεῖν. Ναῦς γάρ τις Ἀττικὴ προσέπλει σιωπῇ τὸ πρῶτον, εἶτα κραυγῇ καὶ ἀλαλαγμῷ προσφερομένων, εἷς ἀνὴρ ἔνοπλος ἐξαλλόμενος μετὰ βοῆς ἔθει πρὸς ἄκρον τὸ Σκιράδιον ἐκ γῆς <ἐπὶ τοὺς> προσφερομένους. Πλησίον δὲ τοῦ Ἐνυαλίου τὸ ἱερόν ἐστιν ἱδρυσαμένου Σόλωνος. Ἐνίκησε γὰρ τοὺς Μεγαρέας, καὶ ὅσοι μὴ διεφθάρησαν ἐν τῇ μάχῃ, πάντας ὑποσπόνδους ἀφῆκεν.[37]

La description du rituel invoqué comme preuve débouche, au terme du mouvement des célébrants, sur le promontoire de Skiradion, et la phrase suivante, qui s'ouvre sur un adverbe de lieu, 'tout près' (πλησίον), y ajoute aussitôt la mention du sanctuaire d'Enyalios, fondé par Solon, avant que l'explication de la fondation (sa victoire) nous ramène au récit à l'aoriste (articulé par γάρ): c'est ainsi l'événement qui explique le monument, et non l'inverse. Plus qu'une preuve, on a là de nouveau une *localisation*, qui permet de se situer dans l'espace pour qui connaît les lieux, et, pour qui ne les connaît pas, d'apprécier la dimension symbolique du trajet, puisque le mouvement offensif que reproduit le rituel s'achève ainsi sur un monument qui commémore la victoire.

La localisation dans Athènes: une 'ville de mémoire'?

C'est autour de cette localisation dans l'Athènes d'aujourd'hui, déjà rencontrée à propos du sanctuaire d'Artémis Aristoboulè, et présente cinq fois dans le groupe des choses dites et montrées par les Athéniens,[38] que se constitue la dernière catégorie, la plus nom-

---

36 Si on lit μαρτυρεῖν, selon la correction de Ziegler, le témoignage est invoqué par Philochore; si l'on garde l'indicatif des manuscrits μαρτυρεῖ, c'est Plutarque qui l'introduit.

37 *Sol.* 9.6–7: 'Le rituel aussi semble témoigner pour cette version: un vaisseau athénien s'approchait d'abord sans bruit, puis, tandis qu'on se portait à sa rencontre avec de grands cris de guerre, un homme en armes, sautant du vaisseau, courait en hurlant au promontoire de Skiradion à la rencontre de ceux qui venaient de la terre. Près de là se trouve le sanctuaire d'Ényalios, fondé par Solon: il fut en effet vainqueur des Mégariens et libéra par une convention tous ceux qui n'étaient pas tombés au combat'.

38 *Ca. Ma.* 5.4 (sur le promontoire qu'on appelle aujourd'hui 'Tombeau du chien'); *Them.* 10.10 (là où l'on montre aujourd'hui le 'Tombeau du chien'); *Arist.* 27.1 (à Phalère); *Cim.* 4.3 (près de la tombe d'Elpinice, la sœur de Cimon); *Phoc.* 18.1 (à Melitè).

breuse, forte de onze passages,[39] où le sémantisme des verbes permet d'opérer à nouveau quelques distinctions. Proches encore de la simple dénomination, la mention d'un changement de nom, comme 'à la Porte Thriasienne, aujourd'hui nommée Dipylon' (*Per.* 30.4) ou 'au Peisianactéion aujourd'hui appelé Pœcile' (*Cim.* 4.6) ne manifeste pas seulement un intérêt pour la toponymie, mais permet de se situer dans la ville actuelle. L'emploi des verbes εἶναι, κεῖσθαι ou διαμένειν signale l'existence d'un monument à tel endroit: les tombes des juments de Cimon l'Ancien sont près de son propre monument (πλησίον εἰσί, *Ca. Ma.* 5.4), celle de Thésée au cœur de la ville, près de l'actuel gymnase (κεῖται μὲν ἐν μέσῃ τῇ πόλει παρὰ τὸ νῦν γυμνάσιον, *Thes.* 36.3),[40] le Cynosarges se trouve en dehors des portes (ἐστὶν ἔξω πυλῶν, *Them.* 1.3), extériorité spatiale qui correspond à cette forme d'extériorité au corps civique qu'est la bâtardise. La précision est ici donnée en incise,[41] comme l'est aussi la mention que le tombeau de Pythonikè '*subsiste* aujourd'hui encore à Herméion, sur le chemin que *nous prenons* pour aller d'Athènes à Éleusis' (*Phoc.* 22.2). Enfin, cette rencontre entre passé et présent ressort encore mieux dans les passages comportant une relative de lieu, et, fait notable, en dehors de la statue d'Éros, érigée par Pisistrate 'là où aujourd'hui ... les coureurs des lampadédromies allument leurs flambeaux' (*Sol.* 1.7), tous les autres passages, accompagnés ou non de remarques toponymiques, sont tirés de la *Vie de Thésée* et s'ajoutent à la localisation, déjà relevée, de sa tombe: ce sont la tentative d'empoisonnement par Égée et le poison renversé 'là où se trouve aujourd'hui l'enceinte du Delphinion' (*Thes.* 12.6),[42] les prières qu'il va faire au Delphinion avant d'embarquer 'le six du mois de Mounychion, jour où aujourd'hui encore l'on envoie les jeunes filles au Delphinion pour y faire des supplications' (*Thes.* 18.1),[43] les malédictions lancées 'là où se trouve aujourd'hui le lieu-dit *Araterion*' (*Thes.* 35.5)—l'action est à l'origine du nom, mais le nom *situe* l'action plus qu'il ne la *prouve* et l'étiologie importe moins que l'inscription de l'itinéraire de Thésée dans l'espace athénien. Celle-ci culmine dans la description de l'Amazonomachie:

> Ἱστορεῖ δὲ Κλείδημος, ἐξακριβοῦν τὰ καθ' ἕκαστα βουλόμενος, τὸ μὲν εὐώνυμον τῶν Ἀμαζόνων κέρας ἐπιστρέφειν πρὸς τὸ νῦν καλούμενον Ἀμαζόνειον, τῷ δὲ δεξιῷ πρὸς τὴν Πνύκα κατὰ τὴν Χρύσαν ἥκειν. Μάχεσθαι δὲ πρὸς τοῦτο τοὺς Ἀθηναίους ἀπὸ τοῦ Μουσείου ταῖς Ἀμαζόσι συμπεσόντας, καὶ τάφους τῶν πεσόντων περὶ τὴν πλατεῖαν εἶναι τὴν φέρουσαν ἐπὶ τὰς πύλας παρὰ τὸ

---

39 *Thes.* 12.6; 18.1; 27.3–5; 35.5 et 36.3–4; *Sol.* 1.7; *Them.* 1.3; *Cim.* 4.6; *Per.* 30.4; *Phoc.* 22.2; *Ca. Ma.* 5.4 (tombes des mules).

40 Il s'agit du gymnase offert par Ptolémée III (après 229 av. J.-C.).

41 *Them.* 1.3: διότι καὶ τῶν νόθων εἰς Κυνόσαργες συντελούντων—τοῦτο δ' ἐστὶν ... .

42 Le Delphinion inspire à Pausanias une anecdote toute différente (1.19.1): 'À ce qu'on dit, le temple était achevé à l'exception du toit quand Thésée, encore inconnu de tout le monde, arriva dans la cité. Il était vêtu d'un vêtement long, ses cheveux étaient tressés avec élégance; comme il arrivait près du Delphinion, les ouvriers qui construisaient le toit lui demandèrent pour se moquer pourquoi une jeune fille en âge de se marier se promenait toute seule. Pour toute réponse Thésée détela, dit-on, les bœufs du chariot—ce chariot dont ils se servaient—et le lança plus haut que le toit du temple qu'ils construisaient' (traduction Jean Pouilloux).

43 On a ici une variante, avec une proposition temporelle et non locative, mais le lieu y demeure important.

Χαλκώδοντος ἡρῷον, ἃς νῦν Πειραϊκὰς ὀνομάζουσι. Καὶ ταύτῃ μὲν ἐκβιασθῆναι μέχρι τῶν Εὐμενίδων … . (*Thes.* 27.3–5)[44]

Il faut sans doute faire la part du sujet de la narration, qui entraîne de soi la multiplication des références topographiques puisqu'il s'agit d'expliquer les positions des armées; il ne faut pas négliger non plus le fait que Plutarque, de son aveu même, suit le récit détaillé d'un Atthidographe. Cependant rien ne l'obligeait à reprendre les détails et surtout la distinction entre sources livresques et lieux réels n'a pas ici l'importance que lui accorde Buckler. Plutarque, à la suite de Clidémos ou par une inflexion qui lui est propre, s'efforce visiblement de faire réapparaître dans l'Athènes de ses lecteurs les traces du passé, de surimposer le combat du passé sur les lieux actuels, le lieu-dit Amazonion, la rue jalonnée de tombes qui mène à la 'Porte qu'on appelle aujourd'hui Porte du Pirée'.

Quelques conclusions

Dans tous ces passages, Plutarque se focalise le plus souvent sur un seul monument, et l'espace ainsi évoqué ne se distingue pas par son extension, ne dessine pas un ample paysage, mais il apparaît comme *familier*, suggérant des lecteurs qui connaissent Athènes et situent les lieux—mais même pour qui ne la connaîtrait pas, les indications créent une impression de familiarité. Parallèlement, à travers les différentes strates de mémoire, s'établit une certaine profondeur temporelle: Athènes porte tout un passé inscrit en elle. À l'intérieur de ce passé, le temps de Thésée tient une place particulière: on trouve deux à trois fois plus de références au présent dans la *Vie* du roi que dans toute autre, ce qui n'est pas sans évoquer la distinction établie par la Porte d'Hadrien[45] entre l'Athènes de Thésée et la Nouvelle Athènes et dont Pausanias se sert amplement,[46] mais ces références, utilisées comme preuves ou aidant à la localisation, n'apparaissent ni dans ce que 'montrent les Athéniens' ni dans ce qui 'demeurait de notre temps': indice d'un retour progressif de la 'mémoire de Thésée' qu'Hadrien allait exploiter et dont Plutarque nous montrerait les prodromes en écrivant, sans doute dans les dernières, la *Vie* du roi? Peut-être. En tout cas l'Athènes qu'il distingue dans les *Vies* est l'Athènes de Périclès avec l'Acro-

---

44 'Clidémos, qui s'est attaché à en préciser tous les détails, rapporte que l'aile gauche des Amazones s'étendait vers *le lieu qu'on appelle aujourd'hui Amazonion* [*sc.* près de l'Aréopage] et que leur aile droite arrivait jusqu'à Chrysa près de la Pnyx; dans le combat contre celle-ci, les Athéniens lancèrent *du Mouseion* l'attaque contre les Amazones, et les tombeaux de ceux qui tombèrent *sont dans la grande rue qui mène à la Porte appelée aujourd'hui Porte du Pirée*, près de l'*hérôon* de Chalcodon. De ce côté ils furent repoussés jusqu'au *sanctuaire des Euménides*'.

45 On la date ordinairement de 132.

46 Hadrien lui-même est nommé 7 fois; pour les personnages historiques, on a, dans l'ordre, Thésée (27), Cimon et Périclès (4), Solon et Thémistocle (3), Alcibiade (2) et Nicias (1).

pole, et lorsqu'il parle de l'Athènes d'aujourd'hui hors du récit biographique, il célèbre sa richesse culturelle, ses bibliothèques et ses conférences. Dans tout cela, il y a peut-être matière à une relecture ultérieure 'athénocentrée', mais Plutarque ne va pas si loin: il dessine, à travers le récit du passé, l'espace qu'il connaît bien, un espace mémoriel, mais non muséal, où, d'une certaine manière, passé et présent se prêtent mutuellement vie, où le passé réapparaît dans les lieux d'aujourd'hui tandis que le présent s'y enrichit de toute la profondeur de l'Histoire.

Timothy E. Duff

Plutarch and tense: The present and the imperfect

Abstract: This chapter examines two characteristic uses of the imperfective aspect in Plutarchan narrative, and demonstrates how aspectual choices affect narrative perspective and speed. The first is the 'backgrounding' use of the imperfective. It is common in the *Lives* for the broader historical context to be related in the imperfective, that is, through verbs in the present or imperfect tenses, primarily the imperfect indicative and present participle. As a result, events in which the subject of the *Lives* was not involved are presented not as a narrated sequence but as a backdrop against which the subject's actions, narrated in the aorist, stand out. Such a use corresponds with a quickening of narrative time. The second characteristic use of the imperfective is to create vivid, slow-motion tableaux, in which events are narrated as though from the perspective of a participant; that is, in such cases, narrative perspective changes from an 'external' to an 'internal' one. Analysis of a passage from the *Were the Athenians more glorious in war or wisdom?* (346F – 347C) shows that Plutarch himself connected the second technique with vivid, pictorial writing and was aware that it was associated with both an internal perspective and a slowing of narrative time.

This chapter will assess the importance of verbal aspect in Plutarchan narrative: that is, ways in which aspectual choices create meaning. It will do this by concentrating on two, particularly significant, uses of the 'imperfective' aspect in Plutarch, by which I mean verbs or verb forms in the present or imperfect. I begin with what I will call the 'backgrounding' function of the imperfective: that is, the way in which imperfective verb forms in Plutarch may serve to present certain events, usually those in which the subject of the *Life* is not involved, as mere background, against which the activity of the subject stands out. The second, and quite different, use of the imperfective to which I shall draw attention is its function in creating a 'participant perspective': that is, in slowing narrative time and creating the impression that readers 'watch' events as though actually taking place before their eyes. Finally, at the end of the chapter I shall consider Plutarch's own words on the second of these two usages, and will demonstrate that the association of the imperfective with a participant's perspective was one recognised by him and by other ancient critics and seen as an important component of vividness of writing.[1]

1 Fuller discussion of these topics can be found in Duff (2015).

https://doi.org/10.1515/9783110539479-005

The 'backgrounding' function of the imperfective

I begin with what I will call the 'backgrounding' function of the imperfective, a function it shares also with syntactical subordination, with which it is sometimes combined. Plutarch's tendency to concentrate on those events in which the subject of the *Life* plays a part and to skip more quickly over events—often major political events—in which he is not involved, has often been noted.[2] Indeed, Plutarch himself implies that he will do just this in his famous statement in the prologue to the *Alexander—Caesar*; as he declares there, Plutarchan narrative often ignores or passes quickly over important historical events ('battles where thousands die, huge troop deployments or the sieges of cities') and focuses on the biographical subject, and especially on those actions or words of his which might reveal character (*Alex.* 1.1–3).

But this concentration on the subject is not just a matter of how much space is devoted to it: close attention to the syntax of Plutarch's prose, and especially to his use of verbal aspect and subordination, shows that the focus on the subject is even more dramatic than Plutarch's statement about genre in *Alexander* 1 might make us think. What we find, in fact, is a *grammatical* hierarchy, in which the wider political or historical context is often dealt with in subordinate clauses and/or with imperfective verb forms, grammatical features which have the effect of 'backgrounding' the content, while the grammatical main clause, usually in the aorist, is reserved for the action of the subject of the *Life*.[3] As a result, the broader historical context (battles, political developments, etc.) is often relegated to the syntactical background, where it sets the scene against which the action of the biographical subject will burst forth, rather than being presented as part of a sequence of events into which the hero's action is integrated.[4]

A good example of such grammatical 'backgrounding' is provided by *Pelopidas* 20. In chs. 16–17 Plutarch describes the battle of Tegyra (375 BCE), in which the Thebans decisively defeated a more numerous Spartan army. In chs. 18–19 Plutarch gives an extended account of the Theban Sacred Band. He then continues in a long, single sentence:

---

2 E.g., Pelling (1985) 322–324 = (2002) 53–54. Also Pelling (1979) 78–79 = (2002) 5–7; (1999a) 436 = (2002) 180–181; (2011b) 22, 193, 349, 358, 459, 484; Moles (1988) 32–33; Stadter (1989) xlix-li.

3 Cf. Burridge (1992) 114–116, 134–135, 162–163 and 261–274 (and esp. 163 and 269 on Plutarch's *Cato Minor*), who demonstrates that it is a characteristic of many ancient biographies that a high proportion of the verbs have the protagonist of the *Life* as their grammatical subject. (Burridge is not, however, interested in the distinction between foregrounded and backgrounded material, or in subordination).

4 Frazier (1996) 18–19 notes that large-scale historical events are sometimes presented in subordinate clauses, citing as examples *Nic.* 7 and *Fab.* 2. She also notes how the thoughts or perceptions of the subject, expressed with verbs such as πυνθάνομαι, can provide another means to give background information in subordinate clauses.

> Ἐπεὶ δὲ Λακεδαιμόνιοι πᾶσι τοῖς Ἕλλησιν εἰρήνην συνθέμενοι πρὸς μόνους Θηβαίους ἐξήνεγκαν τὸν πόλεμον, ἐνεβεβλήκει δὲ Κλεόμβροτος ὁ βασιλεύς, ἄγων ὁπλίτας μυρίους, ἱππεῖς δὲ χιλίους, ὁ δὲ κίνδυνος οὐ περὶ ὧν πρότερον ἦν Θηβαίοις, ἀλλ' ἄντικρυς ἀπειλὴ καὶ καταγγελία διοικισμοῦ, καὶ φόβος οἷος οὔπω τὴν Βοιωτίαν κατεῖχεν, ἐξιὼν μὲν ἐκ τῆς οἰκίας ὁ Πελοπίδας, καὶ τῆς γυναικὸς ἐν τῷ προπέμπειν δακρυούσης καὶ παρακαλούσης σῴζειν ἑαυτόν, 'ταῦτ'' **εἶπεν** 'ὦ γύναι τοῖς ἰδιώταις [ἀεὶ] χρὴ παραινεῖν, τοῖς δ' ἄρχουσιν ὅπως τοὺς ἄλλους σῴζωσιν'.

I give a fairly literal translation, to try to bring out the grammatical contours of the sentence:

> When the Spartans, having made peace with all the Greeks, began war on the Thebans alone, and when Cleombrotus the king had invaded with ten thousand hoplites and a thousand cavalry, and when the danger confronting the Thebans was not as before but was a direct threat and proclamation of dissolution of their state, and when fear as never before was gripping Boeotia, as Pelopidas was going out of his house—his wife crying as she escorted him and begging him to take steps to save himself—**he said:** 'That advice, lady, is for private citizens. Holders of office should be told to save others'. (20.1–2)

Most of the sentence consists of a series of temporal clauses dependent on an initial ἐπεί and linked with δέ. In these temporal clauses the common peace of 371 BCE, and the Spartan invasion of Boeotia in that year, are mentioned briefly. The reduction of the complex events between Tegyra (375) and the invasion of 371 to a couple of lines is striking.[5] But more important for our purposes is that these events are presented as mere background—note in particular the pluperfect, ἐνεβεβλήκει: Cleombrotus 'had invaded'. More space is given to the danger which 'was' (sc. ἦν) threatening the Thebans, and the fear which 'was gripping' Boeotia (κατεῖχεν, imperfect), though this is still dependent on ἐπεί. Scene-setting imperfectives continue in the present participles which describe the action of Pelopidas and his wife, as he was leaving the house (ἐξιὼν ... δακρυούσης καὶ παρακαλούσης). Finally, we reach the focus of the sentence, and its grammatical culmination: Pelopidas' saying (εἶπεν, aorist and the main verb) which illustrates his courage on behalf of the whole community.

It is clear that Plutarch makes no attempt here at complete coverage. The emphasis is on the subject of the *Life* and his actions or words, and not on the wider military and political events in which he was not involved; they form the backdrop against which he acts. But what is also remarkable is that the 'backgrounding' of these events is inscribed in the grammar of the sentence: they are presented in subordinate clauses and with imperfectives, while the subject's action—in this case a saying—is in a main clause in the aorist.[6]

In many cases such 'backgrounding' subordinate clauses can stretch over a number of lines; the preference is for either genitive participles or temporal clauses intro-

---

5 See Georgiadou (1997) ad loc. on how Plutarch 'omits all the key events that led to the peace of 371'.
6 Cf. the similar remarks of Frazier (1992) 4497 and 4516–4518 on *Alex*. 6.1–2. For the way in which Pelopidas' words pick up the theme of ch. 19, see Duff (2015) 133–134.

duced by ἐπεί or, less frequently, ὡς. There is also a tendency to use imperfect indicatives or present participles within these backgrounded sections, thus giving the impression of a process or state against which the subject acts.[7] Take *Phocion* 12.1:

> Παραδυομένου δ' εἰς τὴν Εὔβοιαν τοῦ Φιλίππου, καὶ δύναμιν ἐκ Μακεδονίας διαβιβάζοντος, καὶ τὰς πόλεις οἰκειουμένου διὰ τυράννων, Πλουτάρχου δὲ τοῦ Ἐρετριέως καλοῦντος τοὺς Ἀθηναίους, καὶ δεομένου τὴν νῆσον ἐξελέσθαι καταλαμβανομένην ὑπὸ τοῦ Μακεδόνος, **ἀπεστάλη στρατηγὸς ὁ Φωκίων** ...
>
> When Philip was infiltrating Euboea, and shipping troops across from Macedonia, and controlling the cities through tyrants, and when Plutarch of Eretria was calling on the Athenians and begging them to rescue the island, which was being seized by the Macedonian, **Phocion was sent out as general** ...

A series of genitive absolutes, with first Philip then Plutarch of Eretria as their subjects, give the background. The participles within these genitive absolutes are all in the present tense: we are given an overview of the situation, of what 'was' happening, not a narrative. Philip 'was' gradually taking over Euboea; Plutarch 'was' begging the Athenians to intervene, the island 'was' being seized. None of these events, crucial as they are for the historian, is narrated in its own right; they are presented merely as a backdrop, what was happening, the situation in which Phocion acts. Into this context, the sending of Phocion, in the aorist (ἀπεστάλη), is presented as the first, and by comparison, decisive action.[8]

A slightly more complex example of this subordination of the broader historical context can be seen in *Phocion* 14.3–5. There we have a particularly long ἐπεί clause, with, from half way through it, a preponderance of imperfect indicatives and present participles:[9] the failure of Chares' mission to the Hellespont in 340 BCE and the Athenian annoyance are not presented as a sequence of events, but as background. The main action of the sentence—in this case, as often, a bon mot—and the grammatical main clause, in the aorist, is reserved once again for Phocion, the subject of the *Life*.[10]

---

7 On the backgrounding function of imperfectives in Thucydides, see Bakker (1997) 13–15 and 26–32; on the backgrounding function in Greek of participles and subordinate clauses, see Fox (1983). For recent work on the aspectual difference between the Greek imperfect and aorist see, e.g., Rijksbaron (1988); (2002) 11–21; Porter (1989); Beetham (2002); Buijs (2007); Mathewson (2008).

8 Compare the similarly structured, but much longer, description of the despatch of Lysander to the Aegean in *Lys.* 3.1–3; cf. Duff (1999) 184–185. For the way in which modern translations 'level out' the grammatical and semantic subordination in such passages, see Duff (2015) 135 and 137–138.

9 ὡρμημένων δὲ τῶν Ἀθηναίων, ἔπραττεν ... ὧν ... ἐπλανᾶτο, χρηματιζόμενος ... καὶ καταφρονούμενος, ἐδέχοντο, παροξυνόμενος ἠγανάκτει καὶ μετενόει.

10 On *Phoc.* 14.3–5, see Duff (2015) 136–138.

Imperfectives as markers of a 'participant' perspective

So far we have considered the 'backgrounding' function both of imperfectives and of grammatical subordination in Plutarchan narrative, and have noted how the effect of backgrounding large-scale political events is to throw attention on to the subject of the *Life*. Such backgrounding also affects the 'rhythm' of the narrative, that is, the relationship between the time the events narrated really took to happen and the time (or amount of text) it takes to narrate them. 'Backgrounding' causes a speeding-up of the narrative as we rush over large-scale political events, which might have taken days or even years to happen, to home in on particular, selected episodes, which always concern the subject.

The second characteristically Plutarchan use of imperfectives to which I shall draw attention is also related to narrative rhythm, though this time it brings about a slowing down. This is where imperfective verb forms, primarily imperfect indicatives and present participles, coincide with a change in narrative perspective, from what we might call an 'external' perspective, where events are narrated from the point of view of a narrator exterior to the action, to an 'internal' one, where events are narrated as though from the perspective of a participant or observer. Or, to put it another way, with a change in narrative mode, from a 'diegetic' one, where narrative is mediated by the narrator, to a 'mimetic' one, where the narrator's presence is less clearly felt. Imperfective verb forms in such cases lend a sense of immediacy, of the readers' 'seeing' events as though happening before their eyes.[11] This usage is particularly common within what Françoise Frazier, in her 1992 discussion of Plutarchan narrative, called 'grandes scènes', that is self-contained passages 'endowed with a certain unity of time', which receive, as she puts it, 'a detailed and mimetic description'.[12]

A good example of Plutarch's tendency both to slow down the pace of the narrative to form large-scale, static tableaux, and to use imperfectives to manipulate the narrative perspective within such scenes, can be found in Dion's march on Syracuse in 356 BCE, narrated in *Dion* 44–46. The situation is that Syracuse had been overrun by the forces of Dionysius II, who were raiding the city from their base on Ortygia. Chapters 42–43 describe how the inhabitants of Syracuse appealed to Dion. The demagogues, however, once night came, began urging (παρεκάλουν) the people not to admit him and his army (44.2). New messages 'were being sent' (ἐπέμποντο)

---

11 I draw heavily here on Bakker's (1997) study of aspect in Thucydides; Bakker associates the 'diegetic' mode with what he calls 'the discourse of the knower', and the 'mimetic' with 'the discourse of the observer'. Buijs (2007) on Xenophon is also very useful. Cf. also Bakker (2007) 118–119; Allan (2007); (2009) 171–186.

12 Frazier (1992). The quotation is from p. 4496: 'Par grandes scènes, j'entends des passages assez longs, dotés d'une certaine unité de temps et qui font l'objet d'un récit détaillé et mimétique'.

to Dion from different groups, some telling him to come, others not to (44.3). Therefore, Plutarch continues, 'he was approaching' (προσῄει) slowly. When night was further advanced, those opposed to Dion 'were holding' (κατεῖχον, not 'took') the gates, while Nypsius, the commander of Dionysius' mercenaries, 'was razing' (κατέσκαπτε) the siege wall [which cut off Ortygia], and 'overrunning and sacking' the city (κατέτρεχε καὶ διήρπαζεν) (44.5).

The violence inside the city is now described, and here we meet the first aorist: the mercenaries under Nypsius' command 'resorted to' (ἐχώρησαν) fire to prevent Dion's arrival, by setting fire to what was around them and scattering burning arrows (ὑποπιμπράντες, διασπείροντες: present participles) (44.8). Imperfectives now continue (44.9): as the Syracusans were fleeing (φευγόντων δὲ τῶν Συρακοσίων), some were being slaughtered in the streets (ἐφονεύοντο); those, or 'that part', of the citizens who were seeking refuge in houses were being forced out (ἐξέπιπτε) by fire. 'This disaster', Plutarch continues, 'most of all opened (ἀνέῳξε) the city to Dion'—the second instance of an aorist verb (45.1). The change of tense in this case seems to mark this as a brief authorial comment, as the perspective switches from that of a participant or immediate witness to that of the narrator.[13] That we have here the narrator's perspective is confirmed by the presence of γάρ, a key marker of external perspective, in the following sentence, 'For he happened to be no longer marching in haste' (ἔτυχε μὲν γὰρ οὐκέτι σπουδῇ πορευόμενος).[14]

We now have a section with a mixture of aorists and imperfectives (imperfect indicatives and present participles) describing further appeals to Dion to make haste. Dion's entry to the city is narrated with an aorist verb (εἰσέβαλε, 'he burst in') (45.5). He 'launched' (ἀφῆκε) his light troops against the enemy and 'began to draw up' (συνέταττε) his hoplites; when he had made these preparations (παρασκευασάμενος), Dion 'was seen' (ὤφθη, aorist) riding through the city (46.1). Further imperfects describe the reactions of the Syracusans to seeing Dion, and the position of Dion's mercenaries, as they tried to advance across the burning city. Aorists recur for the decisive battle: 'When they joined battle' (ὡς δὲ προσέμειξαν τοῖς πολεμίοις), only a few men on each side could fight, but Nypsius' troops 'were broken' (ἐβιάσθησαν) (46.6). But we immediately return to the imperfect: the major part of them were fleeing to the acropolis (i.e. Ortygia) and escaping; those who were left outside, Dion's own mercenaries 'were killing' (ἀνῄρουν).

It is important to note that the distinction in such passages between aorist and imperfective is not related to the event in itself: to what actually happened or might be thought to have happened. It does not, in other words, concern the length of time an event took to unfold (or might be believed to have taken to unfold), or whether it was a repeated or single action. Indeed, the same event could be narrated in either

13 For this phenomenon, i.e. the occurrence of an aorist in the middle or at the end of a sequence of imperfectives to mark the perspective of the narrator, see Bakker (1997) 26; Basset (2009) 212–215.
14 For γάρ as sometimes marking switches backward and forward between the internal perspective of the observer and the external perspective of the narrator, see Bakker (1997) 45–46.

form: in the example just discussed, it would have been equally possible to describe the escape of Nypsius' men to the acropolis and the slaughter of those caught by Dion's men in the aorist. What is at stake—and what is always at stake with aspectual distinctions—is how the event is *conceived* in this particular narrative, the perspective from which it is viewed; by choosing to narrate the flight and slaughter of Nypsius' men in the imperfect, Plutarch draws the reader in, as though he or she were witnessing the event actually happening before their eyes. Choice of aspect, therefore, is one of the key elements in Plutarch's repertoire by which he manipulates narrative perspective.

One more example will suffice to illustrate the tendency for Plutarch to use imperfectives within large-scale tableaux to create a sense that the reader is seeing events unfolding vividly from a participant's perspective. In *Pyrrhus* 28–29 Plutarch describes the attempt of Pyrrhus to take Sparta in 272 BCE. The account is in the imperfect from the start, as it describes how Pyrrhus 'was making' (or 'began to make') a frontal assault on the Spartans (ἐβιάζετο). His son Ptolemy, on the other hand, with a force of Gauls and Chaonians, 'was trying' (or 'began to try', ἐπειρᾶτο) to get across a defensive line of half-buried wagons, but the wagons, which were firmly entrenched, were making (ἐποίουν) this difficult (28.1–3). These imperfectives create a vivid, 'slow-motion' picture of the battle, as though seen from the point of view of the participants.

We then have the first aorists, which describe the heroic sally by the Spartan champion Acrotatus: 'As the Gauls were pulling up (ἀνασπώντων) the wheels and dragging the wagons down (ὑποσυρόντων) into the river, ... the young Acrotatus ... got behind Ptolemy (περιῆλθε τὸν Πτολεμαῖον) ... until he launched an assault (προσέβαλε) on the rear ranks and forced them (ἠνάγκασε) to turn around and start fighting him'. After some desperate combat, described with imperfectives, Ptolemy's men were with difficulty pushed back (ἀνεκόπησαν, aorist) (28.4).[15] The aorists here do not seem to imply a narratorial perspective; rather they mark a decisive action, to be set against the surrounding slow-motion narrative. This sudden centre-staging of Acrotatus is striking, but accords with the valuation of him given in the narrative, where he is described as, in the eyes of Spartan observers, ἀριστεύοντα ('performing a brilliant exploit') (28.5).

Two aorists follow, narrating from an external perspective the fact that night put an end to the battle and that Pyrrhus had a dream. But the dream itself, Pyrrhus' telling of it to his officers, and their ensuing conversation, are all described with imperfects and presents, which mark the return of the internal perspective and a slowing of narrative time once more. Then, Plutarch continues, Pyrrhus 'arose' (ἐξανέστη, aorist) and 'began leading forward' (προσῆγεν, imperfect) his army at dawn (29.4).

15 For ἀνακόπτω as 'drive back', 'check' (LSJ I), cf., e. g., *Pel.* 32.11 (τῶν δὲ μισθοφόρων οἱ μὲν πρῶτοι συμβαλόντες εἰς χεῖρας ἀνεκόπησαν ὑπὸ τοῦ Πελοπίδου, τινὲς δὲ καὶ πληγέντες ἐτελεύτησαν); *Aem.* 32.1 (οὕτω φασὶν ὑπὸ τῶν λόγων τούτων ἀνακοπῆναι καὶ μεταβαλεῖν τὸ στρατιωτικόν); *Caes.* 38.4 (κτύπῳ μεγάλῳ καὶ σκληραῖς ἀνακοπτόμενος δίναις).

The battle now continues (29.5) again with vivid, slow-motion imperfectives: the Spartans 'were defending themselves' (ἠμύνοντο); Spartan women 'were present' (παρῆσαν), 'holding out' (ὀρέγουσαι) missiles. The Macedonians 'were trying' (ἐπειρῶντο) to fill up the trench. Finally, 'when the Spartans began rushing to prevent this' (τῶν Λακεδαιμονίων βοηθούντων), Pyrrhus 'was seen' (ὤφθη),[16] forcing his way into the city (29.7). The aorist marks Pyrrhus' appearance as the next important 'event', which stands out all the more clearly against the background of the slow-motion imperfects.[17]

As this example makes clear, the preponderance of imperfect indicatives and present participles creates here a vivid tableau—that is, they have the effect of presenting not a series of events happening in sequence, but a static picture in which the reader watches events unfolding before his or her eyes. To put it another way, the imperfectives in such scenes are indicators that the perspective is 'internal' rather than 'external', and the mode of narration 'mimetic' rather than 'diegetic'. The aorists that intrude into this context may occasionally mark a return to a more neutral, external, narratorial perspective or even a narratorial comment. But they may also mark the decisive actions, words or appearances that stand out against the background of the imperfectives in the rest of the passage.

The 'participant' perspective explicitly addressed

I have pointed out two characteristic features of Plutarchan narrative: the tendency to pass quickly over events, sometimes major historical events, in which the subject was not directly involved; and the tendency to adopt an internal, participant perspective at key moments. I have also tried to show how these two features are enacted by, and inscribed in, the syntactical structure of Plutarch's prose. Each is frequently marked by the use of imperfective verb forms, though the effect of the imperfectives in each case is entirely different. Similarly, both features involve variation in narrative rhythm, though in opposite directions, one involving a speeding up and one a slowing down.

I want to finish with a question related to the second of these features: was Plutarch, or were ancient critics, aware of the capacity of the imperfective to create the illusion that the reader was 'seeing' events unfold slowly, from a participant's perspective? There is every reason to believe that they were.

---

16 The same verb as used of Dion (above, p. 60). The sudden and decisive epiphany of the hero is a common occurrence in the *Lives:* e.g., *Dion* 46.1; *Cim.* 5.2; *Pyrrh.* 29.7; *Pomp.* 22.6; *Gracch.* 29.2 (all ὤφθη); *Crass.* 1.5; *Alc.* 27.4; *Ages.* 21.3; *Pomp.* 20.5; *Demetr.* 44.10 (all ἐπιφανείς); *Alc.* 5.4 (ἑστώς ... ἄπωθεν). Cf. Luke 1.11 (the sudden appearance of the angel to Zacharias): ὤφθη δὲ αὐτῷ ἄγγελος κυρίου ἑστὼς ἐκ δεξιῶν.

17 A longer discussion of *Pyrrhus* 28–29 can be found at Duff (2015) 141–143.

First, ancient definitions of vividness of writing (ἐνάργεια) stress the 'pictorial' quality of vivid writing, which they link both with its capacity to make the audience feel that they 'see' events unfold before their eyes and with its capacity to arouse their emotions. For example, Dionysius of Halicarnassus (first century BCE) defined Lysias' vividness as consisting in 'an ability to convey the things he is describing to the senses of his audience' (δύναμίς τις ὑπὸ τὰς αἰσθήσεις ἄγουσα τὰ λεγόμενα); Lysias' reader, he continues, thinks 'that he can *see happening* the actions being portrayed (γινόμενα τὰ δηλούμενα ὁρᾶν) and that he is meeting face-to-face whatever characters the orator brings in' (*Lys.* 7). The handbooks of rhetoric make a similar point: Theon, for example, writing probably in the first century CE, defined *ekphrasis* as 'descriptive language, bringing what is portrayed vividly before the sight' (λόγος περιηγηματικὸς ἐναργῶς ὑπ' ὄψιν ἄγων τὸ δηλούμενον). Its virtues are 'most of all, clarity and a vivid impression of all-but-seeing what is being described' (σαφήνεια μὲν μάλιστα καὶ ἐνάργεια τοῦ σχεδὸν ὁρᾶσθαι τὰ ἀπαγγελλόμενα).[18]

Secondly, Plutarch himself makes a similar point in *Were the Athenians more glorious in war or wisdom*, where he compares the most vivid of historians to painters. But in making that comparison, Plutarch also makes explicit the aspectual distinction implied in the passages from Dionysius and the rhetorical handbooks. He begins by stating the difference between painters and writers: 'For the actions which painters portray as *taking place* (γινομένας, present), literature narrates and records as *having taken place* (γεγενημένας, perfect)' (346F). He then claims that, despite this general difference, the most effective historian is he who, by a vivid representation of emotions and characters, makes his narration like a painting. He continues:

> ὁ γοῦν Θουκυδίδης ἀεὶ τῷ λόγῳ πρὸς ταύτην ἁμιλλᾶται τὴν ἐνάργειαν, οἷον θεατὴν ποιῆσαι τὸν ἀκροατὴν καὶ τὰ γινόμενα περὶ τοὺς ὁρῶντας ἐκπληκτικὰ καὶ ταρακτικὰ πάθη τοῖς ἀναγινώσκουσιν ἐνεργάσασθαι λιχνευόμενος.
>
> At any rate Thucydides is always striving for this vividness in his writing, eagerly desiring to make the listener a viewer, as it were, and to produce vividly in the minds of those who read his narrative the emotions of amazement and consternation which were being experienced (γινόμενα) by those who were seeing (ὁρῶντας) them. (*Bellone an pace* 347A)

For Plutarch, then, Thucydides' readers become viewers; they 'see' actions as though 'taking place' before their eyes, and they share the same feelings as the original viewers. That is, to put it in our terms, they see events from an internal, or participant, perspective. And central to Plutarch's point is the aspectual distinction between events portrayed as 'taking place' at the moment (γινομένας), and those represented as 'having taken place' (γεγενημένας).[19]

---

18 Theon, *Progymnasmata* 11, 118–119 Spengel. For ancient definitions of *enargeia*, see, e.g., Zanker (1981); Walker A.D. (1993).

19 In fact, Plutarch makes very similar points in the *Artaxerxes*, when he comments on Xenophon's vivid narrative of the battle of Cunaxa (*Art.* 8.1): 'Xenophon all but shows it before our eyes; he always

Plutarch goes on to quote selectively two passages of Thucydides as illustrations of these qualities: Thucydides' description of the Spartan attack on the Athenian fortification at Pylos in Thuc. 4.10 – 12, and of the battle of the Great Harbour of Syracuse in Thuc. 7.69 – 72. Significantly, these passages show exactly the tendency towards imperfective verb forms which we have noted in Plutarch's own narratives. Indeed, his summary of Thucydides' description of the attack at Pylos, which sticks close to Thucydides' wording, is even more remarkable for its use of imperfectives than is the Thucydidean original:

> ὁ γὰρ παρὰ τὴν ῥαχίαν αὐτὴν τῆς Πύλου παρατάττων τοὺς Ἀθηναίους Δημοσθένης, καὶ ὁ τὸν κυβερνήτην ἐπισπέρχων Βρασίδας ἐξοκέλλειν καὶ χωρῶν ἐπὶ τὴν ἀποβάθραν[20] καὶ τραυματιζόμενος καὶ λιποψυχῶν καὶ ἀποκλίνων εἰς τὴν παρεξειρεσίαν, καὶ οἱ πεζομαχοῦντες μὲν ἐκ θαλάττης Λακεδαιμόνιοι, ναυμαχοῦντες δ' ἀπὸ γῆς Ἀθηναῖοι ...
>
> For Demosthenes drawing up the Athenians at the very edge of the shore at Pylos, and Brasidas urging on his pilot to begin beaching the ship, and hurrying to the landing-plank, and being wounded and fainting and slipping onto the outrigger, and the Spartans fighting an infantry engagement from the sea, and the Athenians a naval one from land ... (*Bellone an pace* 347A–B).

The summary of the fighting at Pylos, which ends abruptly there, without a main verb, is followed immediately by two brief quotations from Thucydides' narrative of the Battle of the Great Harbour (introduced by καὶ πάλιν, 'and also', i.e. 'another example'), which describe the emotions of the infantry on both sides as they watched the battle (347B–C):

> καὶ πάλιν 'ὁ' ἐν τοῖς Σικελικοῖς 'ἐκ τῆς γῆς πεζὸς ἀμφοτέρων, ἰσορρόπου τῆς ναυμαχίας καθεστηκυίας, ἄλαστον ἀγῶνα καὶ ξύντασιν τῆς γνώμης ἔχων' διὰ τὰς συντάξεις <καὶ 'διὰ τὸ ἀκρίτ>ως συνεχὲς τῆς ἁμίλλης καὶ τοῖς σώμασιν αὐτοῖς ἴσα τῇ δόξῃ περιδεῶς συναπονεύων' ...[21]
>
> And also, in his account of the Sicilian expedition, 'the infantry of both sides, while the sea-battle was evenly balanced, enduring an unforgettable struggle and tension of mind' because of the

through his vividness makes the audience share the emotion and the danger of the events, not as though they *had taken place* (γεγενημένοις), but as though they *were taking place* (γινομένοις)'. We have there the same aspectual distinction, the same emphasis on the reader's seeing the action in their mind's eye, and the same stress on their sharing the participants' viewpoint and feeling–that is, that the reader should see and feel much as the original participants did.

20 Babbitt (1936a), like most editors, emends mss. αὐτῆς to αὐτὴν and βάθραν to ἀποβάθραν to bring Plutarch's text closer to Thuc. 4.10.5 and 4.12.1, respectively. Gallo, on the other hand, in Gallo and Mocci (1992), keeps the manuscript readings; see his justification ibid. 89 – 92; also Gallo (1992) 25 – 26 = (1999) 142 – 143. Bernardakis prints αὐτὴν and βάθραν.

21 The text is very controversial. I print that of Babbitt (1936a), who emends in several places to bring the text closer to Thucydides, except that I have not emended ms. ἄλαστον ('unforgettable', 'unbearable') to his ἄληκτον ('unceasing') (Thucydides has πολὺν τὸν); there may be a lacuna. Gallo sticks much more closely to the manuscript readings (see previous note), though in fact translates ἄλαστον as 'incessante'. The exact text adopted does not much affect my argument here, though Babbitt's ἄληκτον would strengthen it.

> fighting and 'because of the continual indecisiveness of the struggle, in their fear swaying with their very bodies according to their opinion of the outcome' ...

Like the summary of events at Pylos, these quotations from Thuc. 7.71 consist of noun phrases with present participles. Unlike the Pylos scene, the subject matter itself of these quotations is the observation of action by (not-quite) participants, and the agonisingly slow-motion nature of the action which they observe, which is 'continually' in doubt. As well as seeing it, these observers share the feelings of those actually fighting, physically swaying their bodies. The quotations from the Battle of the Great Harbour which Plutarch has selected here, in other words, not only give an example of writing made vivid and immediate through imperfective verb forms, but provide a paradigm for the way readers might react to such vivid writing: from their perspective, as they 'see' events 'actually happening' before them, the action may move slowly, in this case agonisingly slowly, the outcome long in doubt; and readers of such vivid narrative may, like Thucydides' internal observers, share the emotions of the actors themselves.[22]

Finally, both the summary of Plutarch's narrative at Pylos and the quotations from his account of the Battle of the Great Harbour lack main verbs. In fact, the main verb for all these nominatives is only finally reached when Plutarch concludes, after the words just quoted: 'in the arrangement and depiction of events happening [they] <are> characterised by pictorial vividness' (347C: τῇ διαθέσει καὶ τῇ διατυπώσει τῶν γινομένων γραφικῆς ἐναργείας <ἔστιν>). Assuming Babbitt's insertion of ἔστιν is correct,[23] the grammatical subject of this final statement appears to be Demosthenes, Brasidas and the forces of the two sides at Pylos, and the infantry watching the Battle of the Great Harbour at Syracuse, i.e. the nominatives in the Thucydides quotations.

The fact that there are no main verbs embedded within the summary and quotation of Thucydides, and that the action is expressed by present participles, is significant: this feature, reinforced by the sheer length of the sentence before we meet a main verb, exemplifies Plutarch's point about what vivid writing consists of. Put simply, what Plutarch gives us in these very selective summaries and quotations of Thucydides is not in fact a narrative at all, in the sense that nothing actually takes place, but a static tableau: a picture, in which the actors, described with present participles, are frozen in the act of always 'doing'—an extreme example of the

---

22 We may note that when Plutarch summarises Thucydides' account of the Battle of the Great Harbour in *Nic.* 25.2–4, although his narrative is much shorter than Thucydides', and although he does not echo Thucydides' vocabulary or phraseology particularly closely (as Pelling [1992] 17 = [2002] 122–123 notes), he does reproduce there the high proportion of imperfectives in it, as well as the presence of internal 'spectators' who experience emotions as varied as the fighters', as they watch the battle.

23 Or, e.g., Bernardakis' <οὐδὲν ἀπολείπουσι>; <ἔστιν> is accepted by Gallo (see n. 20) and by Frazier and Froidefond (1990).

use of imperfectives to slow narrative time and create the impression that the reader is watching events *in progress*, taking place before their very eyes.

Lucy E. Fletcher

Narrative time and space in Plutarch's *Life of Nicias*[1]

Abstract: This chapter takes the themes of time and space and explores them as dimensions of narrative in Plutarch's *Life of Nicias*. It argues that processes of narrative signification, both within an individual *Life* and across paired *Lives*, are not entirely linear. The chapter reveals several techniques of narrative anticipation within the *Life of Nicias* and it argues that these anticipations function in part to generate a structure in the narrative which transcends the merely chronological or successive. Moreover, the chapter also touches on some of the ways in which the paired *Life of Crassus* retrospectively resignifies elements of the *Nicias*, endowing the first *Life* of the book with a broader resolution than is achieved by that *Life* itself. In this way, the chapter takes account of the *Nicias–Crassus* book and demonstrates something of the importance of parallelism for the meaning of the *Nicias*.

This chapter is about Plutarchan narrative; more specifically, it is about processes of narrative signification and the relationship between these processes and narrative temporality. I shall argue two principal points. In the first place, I shall suggest that processes of narrative signification in Plutarch's *Nicias* are not linear; that is to say, these processes do not operate entirely in accordance with the temporal unfolding of the text. Rather, at times the significance of events and circumstances is anticipated by the narrative and at other times significance emerges retrospectively. I shall suggest that anticipating significance affects the way in which a reader approaches the narrative and serves to establish a pattern—beyond relationships of chronological succession—in the events and circumstances narrated. The second main point concerns the effects of the parallel structure of the Plutarchan book on processes of signification and of the generation of meaning in the *Nicias*.[2]

1 I wish to thank Timothy Duff for reading various drafts of the oral and written versions of this chapter, and for all his comments and suggestions. I alone am responsible for any shortcomings. I am also grateful to the audience of the oral version for their comments and questions, and to the editors of the volume for their assistance.
2 On the structure of the Plutarchan book see Duff (2011a), whose approach I follow here.

https://doi.org/10.1515/9783110539479-006

The theoretical model

Fundamental to all approaches to narrative is the idea that it is comprised of two levels: what we shall call 'story' and 'text'.[3] The story level consists of the events, situations, and persons which form the subject of a narrative. In the case of Plutarch's *Nicias*, the story level consists of those events which once occurred in the real world and involved the historical figure of Nicias, for example the famous Athenian expedition against Sicily (415–413 BCE). The textual level is simply the text itself, the text of the *Nicias*.

The two-level structure of narrative engenders in it a peculiar dual temporality owing to the fact that both constitutive levels have their own time. Story events happen in time, having therefore their own duration, order, and number. So, for example, the Sicilian expedition has a duration of two years, it has a position within the broader chronology of historical events, and it has a number, i.e. it occurred only once. The text also has its own temporality on account of the fact that it unfolds in time; that is, it takes time to read a text and thus to move from the start to the conclusion. This time is often measured in terms of textual space, for instance the narrative of the Sicilian campaign in Plutarch's *Nicias* takes up two-thirds of the text. Notably, if the relationship between the duration/extent of text-time and story-time was one of direct proportion, the Sicilian campaign, which lasted for only two years of Nicias' life, would occupy considerably less textual space, or, phrased another way, take less time to narrate. This gives great emphasis to the Sicilian campaign and thus we can see that one key means of signifcation in narrative is to manipulate the relationship between story-time and text-time.[4]

The temporality of the text, the fact, that is, that it takes time for a text to be read, is very important for the way in which narrative produces meaning. This is true not only in terms of manipulating the relationship between story- and text-time, but also for the ways in which meaning and significance therefore emerge in time. It may sometimes be the case that meaning and significance unfold gradually with the temporal unfolding of the text. It is also possible, however, for narrative to anticipate significance and/or to reveal retrospectively the significance of what has come before. This chapter will demonstrate some of the ways in which Plutarchan narrative both anticipates and retrospectively reveals significance in the *Life of Nicias*. I shall suggest in addition that the non-linear processes of signification serve to connect events and circumstances which are dislocated in terms of time (and space), bringing

3 Typical terms for these levels are 'fabula' (our 'story') and 'sjužet' (our 'text'), following the Russian formalists, or, following the French structuralist tradition, 'story' and 'discourse' (our 'text').
4 Genette (1980) 33–160 discusses in detail the relationships which can exist between story-time and text-time in terms of his categories of: 'order', 'frequency', 'duration'. Cf. Genette (1988) 21–40.

them into relationships of coherence which transcend those of mere chronological succession.[5]

The structure of Plutarch's *Nicias*

The *Nicias—Crassus* book opens with a distinct prologue introducing the book as a whole (*Nic.* 1), which is followed by the *Nicias*, then the *Crassus*, and finally the book as a whole is concluded with a formal *Synkrisis*.[6] The *Life of Nicias* begins with what Duff terms a proemial section, which is characterised by its atemporal character (*Nic.* 2–6).[7] This is not an account of Nicias' youth; it functions to give a sense of the character and behaviour of Nicias as a politician. Within this section, anecdotes and descriptions are not given a definite temporal location,[8] and thus relationships of chronology are not made explicit. Two notable themes of the proemial opening are Nicias' fear of the people[9] and his successful, but always cautious, approach as a general.[10]

Following the proemial section, the *Life* is comprised of the following episodes: 7–8, the capture of the Spartan prisoners on Sphacteria (425 BCE); 9–10, the Peace of Nicias and its undoing (422–419 BCE); 11, the ostracism contest between Nicias, Alcibiades and Hyperbolus (c. 417 BCE); 12–30, Sicily (415–413 BCE). These episodes are arranged in accordance with chronology, as we can recognise from our extratextual knowledge of these events. The narrative itself, however, does not establish clear temporal connections between the episodes. Only once (9.1) is any temporal relationship established, which is a vague reference to 'by about this time' (ἤδη δέ που). The narrative of the Sicilian expedition is disproportionately long, in terms of textual time/space, in relation to the time given to narrating the preceding episodes. As we have said above, this serves to place great weight of significance upon these events within the narrative and to make the Sicilian campaign the defining event of Nicias' life. I shall suggest, moreover, that the text makes this campaign the defining event of its subject's life in other ways than simply the time/space devoted to narrating it.

---

5 The issues of the production of meaning, especially in connection with narrative time and processes of reading, is a concern of some theoretical approaches to plot, in particular Brooks (1984). For in-depth treatment of the importance of time in narrative see, for instance, Ricoeur (1984); Kermode (1966).

6 On the structure of the Plutarchan book see Duff (2011a), and on the *Nicias-Crassus* specifically: 218; 221–222; 226; 249–250; 252; 253–254; 258; 271.

7 The terms 'prologue' and 'proemial opening/section' are those of Duff (2008a), (2011a) and (2014).

8 Some anecdotes are datable from our knowledge of the events they mentioned, but Plutarch's narrative does not make the time of the anecdotes explicit.

9 E.g., *Nic.* 4.3–8, 5.1–2.

10 E.g., *Nic.* 2.5, 6.2–7.

Anticipating significance

It takes time for a text to unfold, and thus in the *Nicias*, it takes time to move from the initial portrait of Nicias in the proemial opening, through the individual episodes to the end. I shall demonstrate, however, that meaning is at times *anticipated* in the narrative in advance of its temporal unfolding. I want to suggest that techniques of anticipation function to draw together elements of the *Life* which are otherwise dislocated in terms of textual-time/space and thereby establish connections between these elements which transcend the merely chronological or successive.

One instance of anticipating the significance of events or details of the story in advance of their temporal unfolding in the narrative occurs in the prologue of the *Nicias—Crassus*. The prologue begins,[11] 'Since we seem not inappropriately to place Crassus in parallel with Nicias, and the Parthian with the Sicilian *pathemata* ...' (Ἐπεὶ δοκοῦμεν οὐκ ἀτόπως τῷ Νικίᾳ τὸν Κράσσον παραβάλλειν καὶ τὰ Παρθικὰ παθήματα τοῖς Σικελικοῖς, *Nic.* 1.1). This statement anticipates the significance of the Parthian and Sicilian campaigns for the individual biographies and for the book as a whole. It is commonly a function of prologues to establish the similarities between the biographical subjects which justify their pairing.[12] The fact, however, that prologues often introduce similarities of theme, situation, or events between the protagonists should not obscure the importance of such statements for the ways in which narrative meaning unfolds.

The mention of comparable *pathemata* in the prologue of *Nicias—Crassus* serves to indicate in advance of the narrative the importance of these campaigns, an importance which will be reflected in the time devoted to narrating them: more than half of each *Life* (two-thirds of the *Nicias*) and thus over half of the entire book. Dedicating two-thirds of the *Nicias* to narrating Sicily, a fact of which readers are not yet aware, makes this the defining event of Nicias' life, and the prologue foreshadows that significance in advance. Notably, it is a foreshadowing not so much of content, as of *significance*. The narratee of the *Nicias* is constructed in the prologue as already familiar with events in the life of Nicias, in particular as they are related by Thucy-

---

11 There is a possibility that the opening is corrupt, on which see Duff (1999) 22 n. 28, who comments that the opening seems strange and may be corrupt. Certainly, without any preceding discussion of the reasons for comparing Nicias and Crassus, the assertion 'we seem not inappropriately to place in parallel ...' seems odd and not in keeping with Plutarch's usual modest and well-reasoned narratorial style in the prologues. Even if, however, the opening is corrupt and the campaigns were not mentioned in the very first lines of the prologue, they are mentioned in the prologue—the opening of the book—and made the reason for the pairing. Holden (1887) ad loc. discusses the meaning of the Greek in the opening phrase.

12 As observed by Duff (1999) 256. It is not typical, however, for a prologue to address such similarities in the very first words, as discussed by Duff (2011a) 218 – 222 and (2014) 334 and 339 – 340. On the nature and structure of prologues, see Duff (2008a), (2011a) 216 – 224, (2014) and the earlier work of Stadter (1988).

dides: the opening sentence of the *Nicias*[13] continues with an injunction to readers not to imagine that the narrator is attempting to rival Thucydides by writing about Nicias (*Nic.* 1.1).[14] With the narratee familiar with the events of Nicias' life, mentioning story events in advance is not so much functioning to give advance notice of events themselves, as to accord these story elements special significance by looking ahead to them in advance of their narration.

A more subtle technique of anticipating significance involves the signifying potential of intertextuality. The passage we shall discuss occurs in what Duff calls the proemial opening of the *Nicias* (*Nic.* 2–6). In chapter 5, the Plutarch's narrator discusses how Nicias cultivated a particular image of himself as someone always working on public business, without leisure and thus difficult to access:

> Οὕτω δὴ διακείμενος εὐλαβῶς πρὸς τοὺς συκοφάντας, οὔτε συνεδείπνει τινὶ τῶν πολιτῶν, οὔτε κοινολογίαις οὔτε συνδιημερεύσεσιν ἐνέβαλλεν ἑαυτόν ... εἰ δὲ μηδὲν ἐν κοινῷ πράττειν ἔχοι, δυσπρόσοδος ἦν καὶ δυσέντευκτος, οἰκουρῶν καὶ κατακεκλειμένος, οἱ δὲ φίλοι τοῖς ἐπὶ τὰς θύρας φοιτῶσιν ἐνετύγχανον καὶ παρῃτοῦντο συγγνώμην ἔχειν, ὡς καὶ τότε Νικίου πρὸς δημοσίας χρείας τινὰς καὶ ἀσχολίας ὄντος, καὶ ὁ μάλιστα ταῦτα συντραγῳδῶν καὶ συμπεριτιθεὶς ὄγκον αὐτῷ καὶ δόξαν Ἱέρων ἦν, ...
>
> Being thus cautious with regard to informers, he neither used to dine with any of the citizens, nor did he throw himself into general conversation or social pastimes, ... And if he should have no public business to undertake, he was difficult to access and difficult to speak with, remaining at home and keeping himself shut up, and his friends used to meet those frequenting his door and entreat them to excuse [him], as even then Nicias was at work on public business and without leisure. And the one most especially acting a tragic part with him in these things and who took part in placing around him pride and reputation was Hieron ... (*Nic.* 5.1–3)

In this passage, Nicias is described as playing a tragic part, aided by his attendant Hieron who, most especially, also helped to place around him ὄγκος[15] and δόξα. Part of the idea encapsulated in the mention of playing a *tragic* part and of Hieron's helping to place ὄγκος around Nicias seems to be that Nicias' behaviour was really only so much ostentatious display, giving himself airs and graces, and adding a sort of tragic pretension to his actions.[16] Nevertheless, the passage concludes:

> τῷ δ' ὄντι τοιοῦτος ἦν ὁ Νικίου βίος, ὥστ' <ἂν> αὐτὸν εἰπεῖν τὰ τοῦ Ἀγαμέμνονος εἰς αὐτόν 'προστάτην δὲ τοῦ βίου τὸν ὄγκον ἔχομεν, τῷ δ' ὄχλῳ δουλεύομεν'.

---

13 Quoted above p. 70.

14 See Pelling (1992) = (2002) 117–118 on this sentence.

15 On the range of meanings of ὄγκος, see Gutzwiller (1969). Here, in combination with the idea of tragic role-playing, on which see the next note, the meaning seems to be 'pride'.

16 See de Lacy (1952) and Kokolakis (1960) 85–87 on this sense of 'tragic' in Plutarch's period. For discussions of the 'tragic' in Plutarch's *Lives*, see Pelling (1980) = (2002) 111 n. 27, with further bibliography; Van der Stockt (1992) 162; Duff (1999) s.v. 'Tragic'.

> But Nicias' life really was of such a kind as this, so that he might say the words of Agamemnon about himself', 'as ruler/protector (προστάτην) of life we have pride (ὄγκος) but we are enslaved to the people'. (*Nic.* 5.7, quoting Eur. *Iph. A.* 449–450)[17]

This alignment of Nicias and Euripides' Agamemnon is apt in the context of the proemial section. Plutarch has just stated that Hieron placed around Nicias ὄγκος. The same word then appears also in the quotation, in connection with Nicias playing a tragic role. Moreover, as we have seen, a recurrent theme of the preceding chapters has been Nicias' fear of the people and his resulting servitude to them,[18] which is a theme again picked up by the quotation. The alignment of Nicias and Euripides' Agamemnon draws together these two threads of the proemial section, and—importantly—establishes these characters as comparable.

Establishing the comparability of Nicias and Euripides' Agamemnon in this play is important because, when the wider details of the *Iphigeneia at Aulis* are recalled, Nicias and Agamemnon are alike in other ways than those stated here. Firstly, they are alike in their political role-playing to negotiate public relations and achieve a certain reputation. In the Euripidean speech from which Plutarch quotes, Agamemnon is bewailing his misfortune in being forced to sacrifice his daughter (Eur. *Iph. A.* 440–468). In a speech by Menelaus, however, located a little before Agamemnon's speech (Eur. *Iph. A.* 337–345), Menelaus claims that Agamemnon is himself responsible for his being in his current situation, in which he faces the necessity of sacrificing his daughter. According to Menelaus, Agamemnon had coveted the position of leader of the expedition against Troy, and in order to win the leadership he had made himself always available and open to the people, until he was voted leader and then reverted to his natural policy of being inaccessible and hard to speak with.

In the context of the description of how Nicias cultivated a particular public persona while undertaking public duties, which he does for the sake of gaining a reputation with the people, the two are alike, albeit that the persona adopted by Agamemnon is the opposite of that adopted by Nicias.[19] The idea that Agamemnon's public display is directly responsible for his having been elected leader of the expedition to Troy, and therefore indirectly for his current misfortunes, functions to raise the possibility that Nicias' own misfortunes, the Sicilian *pathemata* foreshadowed in the prologue, also owe something to his ostentatious political display at this time; that is, that Nicias might have been elected leader of the Sicilian expedition in part because of the image he cultivated for himself through the tragic role he is presented as playing in the proemial section.

---

17 The MSS of Euripides has προστάτην δὲ τοῦ βίου τὸν δῆμον ἔχομεν, τῷ τ' ὄχλῳ δουλεύομεν. Editors and commentators generally prefer to emend the Euripides in line with the quotation in the MSS of Plutarch, cf. Diggle (1994), who maintains the MSS reading of δέ and τ', but emends δῆμον to ὄγκον.

18 See above p. 69.

19 Cf. *Nic.* 5.1–3, quoted above p. 71.

Secondly, a further similarity between Nicias and Euripides' Agamemnon also serves a function of narrative anticipation, similar to the example we discussed from the opening of the prologue, whereby specific details of the story are foreshadowed and thereby given additional significance within the narrative. In a passage from the *Iphigeneia* a little after the speech from which Plutarch quotes, Menelaus asks Agamemnon why he feels obliged to go ahead with the sacrifice of his daughter. Agamemnon replies that he will be compelled by the mob of ordinary soldiers (*Iph. A.* 513–514). A few lines later, Agamemnon adds that Odysseus will also stir up the people, by using an oracle, to serve his own ends (*Iph. A.* 524–529), leaving it not even safe for Agamemnon to return to Argos for fear of the people (*Iph. A.* 533–537). In this way, the alignment of Nicias and Agamemnon serves to foreshadow how Nicias will be forced into the Sicilian expedition, the Sicilian παθήματα as Plutarch presents them, by his inability to influence the people and vie successfully with other political figures. Of these political rivals, it is Alcibiades who most especially poses a problem for Nicias. Alcibiades will be presented by Plutarch as inciting the people on to the Sicilian expedition in order to gratify his own ambition and, notably, as making use of oracles to do so (*Nic.* 12–13). In addition, Plutarch, like Thucydides,[20] will present Nicias as unwilling to withdraw from Sicily when Demosthenes urges it (*Nic.* 22) on account of his greater fear of the Athenian people. Aligning Nicias and the Agamemnon of Euripides' *Iphigeneia at Aulis*, for those who recall the wider context of the play, serves to foreshadow details of the story of Nicias' life and therefore has the function of placing additional weight of meaning upon those details within Plutarch's narrative.

Anticipation and structure

As we have said, then, similarities between Nicias and Euripides' Agamemnon foreshadow aspects of the narrative connected with the causes of the Sicilian campaign. The three episodes which are located between the proemial opening and the narrative of Sicily are also presignified by the alignment of Nicias and Agamemnon. These three episodes are the Spartan prisoners on Sphacteria (*Nic.* 7–8), the peace of Nicias and its undoing (*Nic.* 9–10), and the ostracism contest between Nicias and Alcibiades (*Nic.* 11). These episodes share a common theme of Nicias' vying with other politicians, in the first instance Cleon, and in the other two Alcibiades.

The ostracism episode concludes with a comment by the narrator addressing the hypothetical outcomes had Nicias or Alcibiades been ostracised and not Hyperbolus (*Nic.* 11.9). Plutarch's final thought here is that whether Nicias or Alcibiades were o-

---

20 Thuc. 7.47–48. Plutarch reduces Nicias' motives to fear of the Athenians if he should return to Athens; in Thuc. 7.48 Nicias has various reasons, including fear of the Athenians. Cf. Pelling (1992) = (2002) 125, for a discussion of Plutarch's use of Thucydides here.

stracised, things would have been better for Nicias: he might either have been rid of Alcibiades and brought Athens to safety, or been himself ostracised and avoided his final misfortunes (τῶν ἐσχάτων ἀτυχιῶν). Thus the last of these episodes is explicitly connected with Nicias' misfortunes: the Sicilian *pathemata* foreshadowed in the prologue. All three episodes, however, are more subtly connected both with what has gone before in the proemial section, and what is yet to come in the narrative of Nicias' *pathemata* in Sicily by their shared theme of Nicias' inability to vie successfully with his rivals, something which had been anticipated as significant for his misfortunes by the alignment of himself and Euripides' Agamemnon in the proemial section. This creates a pattern of events and circumstances in which themes of the proemial opening and of events prior to Sicily are implicated in bringing about Nicias' *pathemata* which were foreshadowed in the prologue.

Finally the narrative comes to Sicily, the events presignified as the defining events of Nicias' life. The narrative of this campaign begins with Nicias' opposition to the expedition (ch. 12). The role of Alcibiades in forcing it through, which has already been foreshadowed in the text and thereby given additional meaning, is now narrated in its correct temporal location (chs. 12–13). That Nicias was right to oppose the expedition is emphasised with narrative of the many bad omens against the campaign and the opposition of other notable figures, such as Socrates (ch. 13). When the sailing of the expedition is narrated, however, the narrator comments that although Nicias was wise to oppose the undertaking, once he had been forced into going, his excessive caution and hesitancy were inappropriate.[21] It is thus made clear from the start of the narrative of this campaign that, though a victim of others' foolishness, Nicias was also himself responsible for the way things turned out. The account concludes with the news reaching Athens of Nicias' defeat. The very final words of the *Life* are 'thus was it scarcely believed that Nicias suffered that which he had so often forewarned them' (οὕτω μόλις ὁ Νικίας ἐπιστεύθη παθὼν ἃ πολλάκις αὐτοῖς προεῖπεν, *Nic.* 30.3). In this way, the episode ends by looking back to its beginning and tying together the threads of the Athenians' lack of foresight and Nicias' early recognition of probable disaster, a disaster which he nevertheless failed to avert.[22]

The *Nicias* is, then, a very tightly structured narrative in which the character of Nicias and his early actions are incorporated into a pattern of significance which comes to its resolution with Nicias' death in the Sicilian campaign. The sense of resolution arises from the way in which the end seems implicated in earlier events. Human lives do not, however, present themselves in the form of structured wholes. They have a natural structure in that they have a definite beginning and end, but that end need not be explicitly linked with earlier events or the character of an individual. That Plutarch structures the *Life of Nicias* so carefully, then, is not the natural result

21 *Nic.* 14.1–2.

22 See Duff (2011a) 250 on the way in which this sentence recalls the prologue and its functions in concluding the *Nicias* and making the transition to the *Crassus*.

of writing biography. His subtle techniques of anticipating significance serve to create a unified and coherent life narrative for Nicias. The final events of that life are loaded with meaning from the very beginning of the book, and the personality of Nicias, outlined in the proemial section, and other events within his life are all brought into a broad structure which makes his final misfortunes and death seem not simply the end but an appropriate conclusion to his life.

The effects of parallelism

The *Life of Nicias* is, then, a carefully structured narrative. I shall conclude this chapter by demonstrating that processes of structuring and signification do not end, however, with the end of the *Nicias*, but continue to operate across the entire book. Thus the full significance of the story emerges not within the time and space of the *Nicias*, but requires the entire space and time of the book to emerge. Thus we end by observing the importance of the parallel structure of the *Parallel Lives*.

When the narrative of the Sicilian campaign concludes, the next sentence of the book begins the *Life of Crassus* and thus the focus shifts in time and space from fifth century Greece (and Sicily) to first century Rome (and Parthia). Like the *Nicias*, the *Crassus* is predominantly made up of narrative of the great, and final, disaster for the biographical subject: the Parthian Campaign (*Crass.* 16–33). This campaign, like Nicias' in Sicily, was presignified in the prologue of the book. It is these campaigns which justify the parallelism of Nicias and Crassus, and they are given the same broad structure. Narrative of the Parthian campaign begins with the inauspicious circumstances of Crassus' departure (*Crass.* 16.6–8). This directly parallels the beginning of the account of Nicias' Sicilian campaign. As for Nicias in Sicily, things go from bad to worse and end with utter defeat and death for the subject. The structure of the story is, in broad outline, identical: inauspicious beginnings indicate the campaign is destined to fail, poor leadership fails to capitalise on potential opportunities, the campaign ends with great loss of life and total defeat.

And yet, although the campaigns share the same trajectory, that trajectory is differently, indeed inversely, motivated. In *Nicias*, excessive caution and cowardice are the motivating forces behind the course of events. In *Crassus*, by contrast, the motivating forces are excessive boldness and lack of caution. Where Nicias looks back hesitantly, Crassus pushes ahead unreflectingly;[23] where Nicias is cautious and concerned with safety, Crassus shows little concern for safety or the welfare of his troops;[24] where Nicias is fearful and eager to avoid fighting, Crassus is desperate

23 E.g., *Nic.* 14.1; e.g., *Crass.* 17.1.

24 E.g., *Nic.* 15.3–4; 16.1, 16.8–9; e.g., *Crass.* 19.1–3; 22.1–3.

to engage the enemy;[25] where Nicias is excessively concerned with omens, Crassus ignores them entirely.[26]

I suggest, therefore, that the *Life of Crassus* re-signifies the earlier narrative of *Nicias*. After reading the *Crassus*, the focus of the reader's attention has shifted. Now the life of Nicias seems part of a wider pattern—a pattern which reveals the dangers of excess, and the need for reason to control passion. A view Plutarch derived ultimately from Plato.[27] The resolution of the narrative, through the effects of parallelism on signification in the *Lives*, therefore lies in revealing a universal paradigm of the danger of such excess, be it of caution or boldness, of fear or of recklessness. Plutarch thus exploits the narrative time and space of the entire book to draw the maximum significance from his stories.

Conclusion

This chapter considered how processes of narrative signification function in a non-linear way in Plutarch's *Life of Nicias* and with what effects. The focus has been the subtle techniques of anticipating the significance of events and circumstances and how these techniques serve to connect temporally and spatially dislocated elements of the story, and thus generate a broad pattern into which much of the material is integrated. In addition, the chapter has suggested that the parallel structure of the Plutarchan book impacts upon processes of signification, which do not end with the conclusion of the *Nicias* but continue to operate across the book. It therefore demonstrates the importance of parallelism in the *Lives*.[28]

25 E.g., *Nic.* 20.5–8; 21.3–6; e.g., *Crass.* 18.1–2; 23.6–7.

26 E.g., *Nic.* 23–24.1; e.g., *Crass.* 17.10; 18.5; 19.4–8; 23.1–2.

27 Plutarch elaborates his theory of virtue in the *De virt. mor.*; see especially 441D–443D. On Plutarch's views concerning moral virtue, see Babut (1969a); Duff (1999) 72–78 discusses Plutarch's theoretical views on the nature of virtue, and at 78–98 he discusses how this theory is reflected in the *Parallel Lives.*

28 On the importance of parallelism, including of *Synkrisis* and the *Synkriseis*, see Erbse (1956); Stadter (1975); Pelling (1986a) = (2002) 349–363; Swain (1992a) 349–363; Duff (1999), esp. 243–286, and Duff (2011b) 72–75; Humble (2010).

3 Religious locales as places of reflection on language, discourse and time

Frederick E. Brenk

Space, time, and language in *On the Oracles of the Pythia*: '3,000 years of history, never proved wrong'

Abstract: Plutarch's *On the Oracles of the Pythia* awesomely plays with space, time, and language. Besides treating twenty-five Greek city-states, including some in central Italy and Magna Graecia, it reaches out to Macedonia, Rome, Carthage, Egypt, and Persia. The principal speaker twice boasts that the Oracle has existed for three-thousand years (408D), and prophecy itself is a link between past, present, and future. As is typical of the Second Sophistic, both in the monuments mentioned and in the discussions, the essay privileges the distant past at the expense of the present reality. The discussions between an Epicurean, Stoic, and the principal speaker touch on contemporary topics, but cite, from different periods and places in the past, nine famous philosophers from various schools, four astronomers, and four historians. Moreover, citations from nine famous Greek poets and several Delphic verse oracles contribute to the chronology and language. Yet, the essay defeats expectations in surprising ways. Somewhat like Pausanias and Lucian, the speakers tend to undercut the legends and 'miracles' at the shrine. Up to the end, the principal speaker, Theon, has not belittled the importance and greatness of the Oracle in the past (407D). He argues, however, that in former times the Pythia spoke more in prose than in verse, miserable verse at that. So the change is for the better, and the Pythia even today occasionally uses verse (396C, 407A). The past in his view was one of grandiloquence and artificiality, obfuscating the meaning of the Oracle, whereas today simplicity and directness reign. Finally, the present age of peace and prosperity at the shrine and elsewhere is much superior to the miseries of the past. We now learn that most of the glorious monuments on display are, in fact, bloody trophies to the internecine strife of the Greeks.

On the Oracles of the Pythia No Longer Given in Verse, also known as *On the Oracles at Delphi*, or *De Pythiae Oraculis*, is about a very specific topic, 'Why does the Pythia No Longer Speak in Verse?'.[1] Even this turns out to be a false question in this rather deconstructionist dialogue, because we learn that the Pythia has always given most of the responses in prose, and that the present one sometimes speaks in verse (405E, 406F–407A). Plutarch even includes an Epicurean, Boëthus, one who cannot ima-

1 For the relation of this dialogue to the other Pythian ones, see Frazier (2007/2008); Brouillette (2014); Hirsch-Luipold (2014) 164. For Plutarch's role at Delphi, see Stadter (2004).

https://doi.org/10.1515/9783110539479-007

gine any divine involvement in the world, as one of the principal speakers.[2] Not surprisingly, the final and most prestigious speaker, Theon, upends many common opinions about the Delphic Oracle.

As for space, the essay, besides treating twenty-five Greek city-states, including some in the Bay of Naples and Magna Graecia, reaches out to Macedonia, Rome, North Africa, Egypt, and Persia. As for time, like many other works of the Second Sophistic, the essay is structured upon the apparent contrast between the distant past and the present reality.[3] Old is supposedly good. Theon, the principal speaker, boasts that the Oracle has existed for three-thousand years (408D)! Prophecy itself, too, is a link between past, present, and future, since the oracles often refer to the past, are directed to someone in the present, and are supposed to predict the future.[4] Finally, the subject of the dialogue (why the Pythia no longer speaks in verse) pertains also to language. The work includes, moreover, an enormous number of citations from classical authors of different ages and parts of the Greek world. Both space and time are very much in evidence, especially 'lived space' and 'lived time', that is, space and time informed by the activities of human beings.[5] As the speakers stroll through the sanctuary, they hear the monuments explained to a young 'guest' from Pergamum, Diogenianus.[6] He is described as 'intelligent', but is obsessed with one of the first things he sees, the patina on the statues of Lysander and the naval commanders at the opening of their tour. This sounds like what guides might call a 'stupid tourist' and like a bit of irony on Plutarch's part. Ordinary tourists when visiting sites and museums tend to spend an inordinate amount of time (and talk) on the first objects they see. In a similar vein, on another occasion, a speaker, unsure of the author of a citation, attributes it to Pindar, rather than Euripides (405B).[7]

2 On the opening of the dialogue, see Galli (2005) 254–259, who regards memory as highly influenced by *paideia* in this period. Boëthus is very disrespectful of the oracle, and is not a priest. Epicureans and others, however, did not let their philosophy stand in the way of fulfilling their duties as part of the elite, including serving as priests. See Haacke (2008) 145–166. For Plutarch's use of religion and philosophy in the dialogue, see Hirsch-Luipold (2013), esp. 211–212; and Brouillette (2014), esp. on Diogenianus, 43, for progression into the shrine resulting in a deepening of the philosophical argument, 54–55, and for time and memory, 65.

3 E.g., Bowie (1970) and (1996); Whitmarsh (2001a); Giangiulio (2010); Richter (2011); Schmidt and Fleury (2011); de Jong (2012a), and Kim in this volume. For a new approach which embraces less-known works of the period, see Whitmarsh (2013).

4 See also Lucchesi in this volume.

5 For this element in Pindar's description of Delphi, see Eckerman (2014), esp., 21–23, 35–43, and 55–58. For the philosophical and religious implications of space (memory, belonging, commitment, time, narrative), and vs. place, see Sheldrake (2001) 1–32.

6 His father was a friend of Plutarch and is mentioned several times in Plutarch's works. See Flacelière (1974) 42, n. 1, and Kim in this volume. For the organisational aspects of a sanctuary such as Delphi, see Dillon M. (1997) 204–227.

7 However, the discussion, involving natural science, is much like those we find in Plutarch's *QC*.

As the speakers ascend the slope of the mountain, they periodically halt at a dedication.[8] Each has its own space and time (and if we consider the inscriptions, also its language) within the sanctuary and in the history of the city represented. But in their conversations, the speakers reach out beyond the immediate associations, bringing within their scope different parts of the Greek and foreign world, and even the sun and the moon. Each object has for the viewers, as in a modern museum, a date and provenance.[9] For example, as they move in space to the Treasury of the Corinthians, they think of Cypselus, the tyrant of Corinth in the seventh century and the historical events associated with him (400D–E). Random association can also occur. Gazing at the site of the huge iron 'spits' or 'skewers' for roasting cattle, the 'tithes' of the sixth century *hetaira*, Rhodopis, to the god, leads them to cast their gaze farther up the mountain. Here they comment on the statue of Phryne of Thespiae, Praxiteles' controversial model and mistress of the fourth century (400F–401D).[10] The "spits" of Rhodopis also inspire a guide to mention the golden statue of a woman, a sixth-century dedication by Croesus. The woman, a baker, had saved Croesus from his stepmother's attempt to poison him (401E).[11] Such interjections avoid a rather mechanical procedure for introducing details. Alex Purves has noted how the Homeric poets organised their recital of the Catalogue of Ships in the *Iliad* through a kind of mental map in their heads, something also exploited

---

8 They stop first at the 37 statues of Lysander, his officers, and divinities (no. 4 on the map in Amandry [1984]; henceforth referred to as: 'map'); see Bommelaer (2011). For Delphi in general, see Scott (2010) and (2014) and Kindt (2016). For dedications, see Scott (2010) 29–40, and for Delphi in different periods, 41–145. For Delphi in the early imperial period, see Scott (2014) 201–241, including Plutarch at Delphi, 215–222. For a discussion of Theon's theory of the inspiration of the oracles, see Brouillette (2014) 196–199. For differences between Plutarch's and Pausanias' approach to Delphi, see McInerney (2004). Surprisingly, in *De Pyth. or.* Plutarch does not mention the miraculous repulse of the Gauls by Apollo. On this, see Lipka in this volume. For Pausanias on the repulse, see Hutton (2005a) 79, 192, 263. Hutton says little about Delphi itself. For the conflict in Plutarch between science and religion, see Muñoz Gallarte (2010).

9 Many dedications would have carried an inscription from different times and in different forms of the Greek language.

10 Treasury of the Acanthians and Brasidas (map, no. 44), which held the 'iron skewers' of Rhodopis (400F). According to Herodotus, she dedicated a large number of them and they represented a tenth of her earnings. Sarapion says 'Look up there', probably meaning up the hill, to the statue of Phryne. Apparently they were looking up to map nos. 55, 62, etc., where there were numerous statues, but not stopping at the statue (which does not appear on the map). It is rare that the random association has no anchor in anything actually visible at Delphi. For Rhodopis and Phryne, see Glazebrook and Henry (2011) 6–8.

11 In this case the statue is from the past and evidently no longer in the sanctuary. The story of the baker woman leads to other stories of dedications of grain and fruit, and of a statue of Apollo. The speakers do not actually see these, though at least at one time they were in the sanctuary (401F–402C). For new evidence on the unreliability of Herodotus on Croesus' dedications, see Thonemann (2016), esp. 161–165.

by Jenny Strauss Clay.[12] The process here is, similarly, varied by random associations. At more or less exactly mid-point in the dialogue, the speakers stop moving and sit down on the 'south steps' of the Temple of Apollo, gazing at the shrine of Gē and the stream of water below, 'where previously there was a shrine of the Muses' (402C). They can now more easily discuss the principal topic, 'Why the Pythia no longer speaks in verse'. The movement in space, time, and language continues in this section, but is now detached from its link to the dedications. All the references to time, space, and language are now related to the principal theme, though framed by the contrast between past and present.

In the course of ascending, the speakers make 6 stops, at each one engaging in rather extended dialogue.[13] By the time they have reached the Temple, they have skipped an average of 11 objects before each stop.[14] Thus, they halt (1) at statues of the Lacedaemonian ship captains and divinities, taking us to the battle of Aegospotami in 405 BCE (395A ff.).[15] They move (2) to statues of the Argive Kings, apparently spanning the whole history of Argos (396C).[16] They then proceed to (3) the statue of Hiero, tyrant of Syracuse, and his bronze column, representing Sicily of 478–467 BCE (397E–F).[17] After that, they approach (4) (400F) the Treasury of the Acanthians and of Brasidas, a Spartan officer in the Peloponnesian War, who died in 422.[18] The Rock of the Sibyl (5) (398C) belongs to the ancient prophetess who has now become the face on the moon.[19] They then arrive at (6) the Corinthian Treasury of the sixth century, and the bronze palm tree within (399E).[20]

Finally the movement stops as our speakers reach the Temple of Apollo, an edifice originally of the fifth century but destroyed and rebuilt in the fourth. Here, they look down upon the shrine of Gē, presumably from primordial time, and where the shrine of the Muses once stood. Apparently it existed at the time of Simonides (sixth

---

12 See Purves (2010) 36–38 on 'cognitive memory', citing Minchin (2001) 84–87. See also Strauss Clay (2011), esp. 117–118, on the Ship Catalogue in the *Iliad.*

13 The list of 'stations' can be found in Flacelière (1974) 45.

14 The Temple is at map, no. 72.

15 Map, no. 4.

16 Two hemicycles of statues of the Argive Kings (map, nos. 9–12).

17 Map, no. 68; not on map, the statue of Hieron, the tyrant of Syracuse, and the bronze column of Hieron (397E), 478–467 BCE. At 397E–F, the speakers mention the bronze Palm Tree of the Athenians (map, no. 68), an ex voto, surmounted by a gilded statue of Athena, for victory at the Eurymedon River in 468 BCE. Golden dates fell off the palm, and ravens pecked at the shield of Athena (colossal statue of Athena, map, no. 73, but listed as statue bases). Presumably the Crown of the Cnidians, which is mentioned at 397A–B, would have been in their treasury.

18 Map, no. 44. So apparently they were looking at map nos. 55, 62, etc., where there were numerous statues, and did not stop at Phryne's statue (401A) (not on map).

19 Map, no. 37.

20 Map, no. 42.

to fifth centuries) (402C).[21] Thus, rather than a straightforward progression in time, they have moved both forward and backward, partly, but not necessarily, as conditioned by the monuments. Delphi was the center of the earth, marked by the *omphalos*. Now, within this area, they have escaped the long-winded guides and moved into the even more sacred space of Apollo, the Pythia, and the oracles. This movement contrasts with that in the *Dialogue on Love*. In it, the speakers escape the contentious noise of Thespiae, ascending to the sanctuary of the Muses on Mount Helicon, quite a way off. In doing so, they move away from the shrine of the god Eros, who is the subject of their discussion.

In the course of this ascent, the speakers have made reference to virtually every period of Greek history. Though the references are primarily to the archaic and classical periods, they also include the eruption of Vesuvius in 79 CE, the only clear reference to a contemporary historical event (398E).[22] Spatially included (in order of appearance) are Corinth, Athens, Syracuse, Leuctra, Aegospotami, Cnidos, Metapontum in South Italy, the moon (the supposed location of what remains of the Sibyl), Cumae, and Dicaearcheia (Puteoli), Thera, Rome, Carthage, Selinus, Tenedos, Egypt, Olympia, Rhodes, Erythrae, Acanthus, Phocis, Thessaly, Sicyon, Myrina, Apollonia, Eretria, Magnesia, Megara, and Haliartus. After the speakers sit down at the Temple, the references become sparser, but they still include Sparta, Athens, Erythrae, Epidaurus, Aegina, Phocis, and North Africa.

Another aspect of space, time, and language is the literary heritage of classical authors. It has been suggested that in Plutarch's dialogues, through the citations, the great authors of the past reach out and engage with the speakers in the present. The immense spatial, temporal, and linguistic reach is impressive.[23] Homer (Ionia, eighth century) is cited 4 times (396B, 396D, 405A) and is alluded to 3 times (397B, 398A, 405B), while one Homeric verse is adapted to fit a context (409B). Sappho (Lesbos, seventh-sixth century) is alluded to at 397A. Two verses are cited from Simonides of Ceos (sixth to fifth centuries BCE) (402C), Empedocles (Sicily, fifth century) is cited once (400B–C), as is Scythinus, a philosophical poet (402A–B).[24] There is no quote, however, from Aeschylus (Athens, sixth to fifth centuries) nor any reference to him. Sophocles (Athens, fifth century) is cited once from an unknown play (406E). Euripides (Athens, fifth century) is cited 3 times: from the *Thyestes*, *Stheno-*

---

21 Map, no. 72, looking toward the shrine of Gē and the stream of water, that is, of Gē and the Muses (map, no. 34). They really would be looking southeast. This occasions a discussion about a shrine of the Muses.

22 Flacelière (1974) 40 dates the essay to around 125 CE, believing that the καθηγεμών (leader, guide, founder) mentioned as responsible for restoring the shrine (409B–C), was Hadrian. This is rejected, among others by Schröder (1990), who believes (21) that the καθηγεμών was a person whose name was in the lacuna which follows in the ms. Jones (1971) 136, simply puts the date as 'after 95'.

23 During the course of the dialogue, the speakers cite 9 famous philosophers, 4 astronomers, 4 historians, and 9 famous poets from different periods and places. In the final part, Theon cites 3 Delphic oracles, relates the contents of 4 others, and alludes to several more.

24 Fr. 1, West. His verses were based on the teaching of Heraclitus. See Schröder (1990) 303.

boea, and *Phoenissae* (405F, 415B).[25] Pindar (from Plutarch's Boeotia, fifth century) is cited 5 times: from the *Pythian* and *Isthmian Odes*, from a fragment (397A–B, 405F, 406C), and for verses on Boeotia (409B), and is alluded to at 403A. Chaeremon, also a tragic poet (Athens, early fourth century), an unknown Attic comic poet, and an unknown lyric poet are cited once (406B). The verse tribute to Boeotia is now attributed by most modern scholars to Pindar.[26] The poetic space, time, and language, then, has three chief focal points: Homeric epic of eighth century Ionia, tragedy of fifth century Athens, and lyric of fifth century Thebes.

Once at the Temple of Apollo, space, time, and language acquire new characteristics. The references, which no longer are linked to the dedications, relate to Delphic oracles, with only one exception, the verse tribute to Boeotia. Virtually the only speaker now, Theon, cites the verses just mentioned above from Simonides, Heraclitus, Euripides' *Thyestes*, the two places in Pindar's *Isthmian Odes*, Chaeremon, the Sophocles fragment, and the line adapted from Homer.[27] Otherwise the citations are of the three Delphic oracles: to Battus (Thera, seventh century) on the founding of Cyrene, probably taken from Herodotus (408A), to Lysander (fifth century), which possibly refers to his death at Haliartus in central Boeotia (408A), and to the city of Phocis (403F–404A). The contents of 4 other oracles are related to the Spartans before the Peloponnesian War (403B), to Athens before the Sicilian campaign (403F), to Deinomenes of Gela in Sicily (sixth and fifth centuries) (403B–C), and to Procles, tyrant of Epidaurus (seventh century) (403C–D).[28] Our immediate space now is the sacred space of the Temple itself and the oracles, but a new space will emerge at the end of the dialogue. As for time, the Oracle, which has a past, can predict the future, even 500 years in the

25 The speaker, Theon, is not sure at 405B whether the verses are from Pindar or not. Plutarch probably knew that the line was from Euripides' *Thyestes*.

26 Snell-Maehler, fr. 104b. Page rejected the attribution to Pindar, considering it unworthy, and listed it as *adespoton*, *PMG* 997. See Schröder (1990) 443.

27 But Theon also alludes to the following: Orpheus; Hesiod; Parmenides; Xenophanes; Empedocles; Aristarchus of Samos, an astronomer of the fifth century; Timocharis, an astronomer; Aristyllus, an astronomer; Thales, a Greek scientist, seventh-sixth centuries; and Eudoxus of Cnidos, a mathematician and astronomer, fourth century (402F–403E); Pindar (403A, E–F); Homer, *Iliad* and *Odyssey* (405A–B); to Xenophon's statement about the qualities of a bride (405C); Sappho (406A); Herodotus (403E); and Plato and Socrates (406A–B). But he also alludes to lesser known persons: Philochorus, an Athenian historian, third century; Istrus of Cyrene, a Greek historian, who was a slave of the poet Callimachus, third century; Onomacritus, Athenian collector of oracles, sixth century; Prodicus of Ceos, sophist, fifth century; and Cinaethon of Sparta, a genealogical poet of the eighth or seventh century (407B). For these names, see Schröder (1990), ad loc.

28 Deinomenes of Teos founded Gela in the eighth century. Theon also alludes to oracles given to other founders of colonies: Teucer, son of Telamon, founder of Salamis in Cyprus, at the time of the Trojan War; Cretines of Miletus, founder of Sinope, seventh century; Gnesiochus of Megara, one of the founders of Heraclea on the Black Sea, possibly eighth century; and Phalanthus, founder of Taras (Tarentum) in Sicily, end of eighth century (407F–408A). For these names, see Schröder (1990), ad loc.

future, as a speaker remarks earlier (399D).[29] The time is that of the future as predicted in the past. As for language, the thoughts belong to the sacred and divine speech of the god, Apollo, but as transmitted by the very human Pythia.

The new space spoken of at the end is that of the Roman Empire, the new time, the present, and the new language, the prose of the Pythia. We might also note, however, that though Greek was the official language of the East, Latin was the language of the capital, of the West, of the administration in a large part of the Empire, and of many Roman friends of Plutarch who happened to find themselves in Greece. Latin was also the language of the East wherever Roman soldiers were stationed. The Empire was bilingual. To write his works, Plutarch had to learn to read and speak at least some Latin. Surprisingly, and contrary to the usual approach in the Second Sophistic, in Theon's view the present time is better than the past. He views most of the dedications as trophies to the bloody internecine strife over the centuries which brought Greece to its knees. While they were fighting among themselves, Rome was busy carving up the world. The present has brought 'profound peace and tranquility' (408B) and many other blessings, while the present prose of the Oracle is a purer and more appropriate form both for the speech of everyday life and for the sacred utterances of the Pythia.[30] In his view, amazingly, the oracles in verse were appropriate for an age of excess and ostentation (406B–F).[31]

In all this, however, is there not something missing? Is not prophecy supposed to be about the future? Not a single oracle cited or referred to has to do with the future. Those alluded to refer to the humble consultations of farmers and the common people. If they are consultations by cities, they are about crop yields and the like (408B–C). But Theon cites none of these explicitly. Like the dedications and oracles mentioned, the dialogue lives in the distant past. In the end, however, the main speaker seems to cut himself off not only from future prophecy, but also from the dedications made to the Oracle in the past. While speaking about the content and veracity of the oracles from the past ('as yet it has afforded no proof of it being wrong', 408F), he praises, rather, the present, which he certainly hoped would continue into the future. Yet most of Theon's speech, like Plutarch's works, has to do with the past.

In conclusion, Plutarch had a difficult task in composing his *On the Oracles of the Pythia* or, more to the point, *Why the Pythia No Longer Speaks in Verse*. He wanted

---

29 Earlier (399C) he had made the claim that the Oracle made a prediction concerning Agesilaus, the eruption of the volcano at Thera, the defeat of Philip of Macedon by the Romans, and the defeat of the Carthaginians. The five-hundred years refers to the suppression by the Romans of the slave revolt under Spartacus, described as 'the whole world rising up against the Romans'. For divination in general and at Delphi, see Bonnechère (2007); Eidinow (2013), and Ustinova (2013).

30 For the reason for the decline of the Oracle, see Arnush (2005) 100, 104, 108–110, who attributes it to historical changes which affected the neutrality of the Oracle and shifted interest to other oracular sites, such as Didyma.

31 On this, see Kim in this volume. For the very different and somewhat complicated attitude of Pausanias toward the present in Roman Greece, see the very penetrating study of Hutton (2005b), esp. 308–309, 314–317.

to cover a huge temporal and geographic span and to tie the monuments to their history and the persons responsible for them. At the same time he sought to avoid either a monotonous and perhaps trivial guide to Delphi or a dialogue essentially detached from its setting at the shrine. He managed brilliantly to handle the task, sometimes with wit and often defeating expectations, introducing surprising twists, and undercutting the customary reasons for the Oracle's presumed greatness and Greece's best hour. He also managed, while manipulating the past, to present space and time and language through the eyes of his age, the early imperial period, or in other terms, through the eyes of the Second Sophistic. Though exaggerated, Theon's praise of the imperial period, as opposed to the Greek past, is an affirmation of the value of his own age, as he saw it, with a common space, time, culture, two common languages, and relative peace. The successes, which he sometimes rejects as successes, and the culture, which once belonged to Greece alone, had already been absorbed into the Graeco-Roman world and continued to extend their reach outward. Though highly engaged with the past, this is a dialogue which also points to the future, both of the Roman Empire and civilisation to come. In his desire for one world and universal peace, the real prophetic voice is no longer that of the Pythia, but of Theon and ultimately Plutarch.

Lawrence Kim

Poetry, extravagance, and the invention of the 'archaic' in Plutarch's *On the Oracles of the Pythia*

Abstract: Plutarch's dialogue *On the Oracles of the Pythia* centers around the question of why the Delphic oracles are no longer delivered in verse, as they had been in 'ancient' times (*ta palaia*). The answer provided by the character Theon, in a long speech that concludes the text, argues that the ascendancy of prose over poetic oracles is linked to a broader cultural change, and should be celebrated rather than lamented. Moreover, to explain why the Pythias previously prophesied in verse, Theon paints an unflattering picture of the Greek past as an age marked by a penchant for poetry, luxury, and obscurity. In this article I analyse a crucial section of Theon's argument (24.406C–E) to show how he exploits the metaphorical connotations of literary critical terminology to portray the past 'poetic' age as luxurious and extravagant, and to associate prose, and by extension the present, with moderation and restraint. Furthermore, Theon's positive evaluation of the present does not imply a wholesale rejection of the past; rather Theon splits the past into two distinct eras—a 'poetic' followed by a 'prosaic'—and collapses the second with his own imperial present. In doing so, Theon demarcates an era, characterised as poetic, excessive, and immoderate that bears a striking resemblance, I suggest, to that which we call 'archaic'.

The glories of the past cast a long shadow over Plutarch's *On the Oracles of the Pythia.*[1] The dialogue is framed as a reminiscence by Plutarch's close friend Philinus to a certain Basilocles, of a just completed tour of the Delphic sanctuary, probably in the late first or early second century CE;[2] the participants include, beside Philinus, Theon, Sarapion the Stoic, Boëthus the Epicurean (all associates of Plutarch), and a young man, Diogenianus of Pergamon, on his first visit to Delphi.[3] Their conversation is the topic of Philinus' narrative and takes its impetus from the dedications and monuments witnessed by the group as it ascends the Sacred Way to the temple of

1 Much of the research and writing of this article was conducted in Heidelberg during an Alexander von Humboldt-Stiftung Fellowship in 2013–14, but it has also benefited from audience responses in Delphi, Ravello, Charlottesville, and Los Angeles. Thanks to the anonymous referee for his/her comments and especially to Sira Schulz, whose timely editorial interventions have saved me from numerous errors of logic and judgment.

2 On the date, see Valgiglio (1992) 41–42.

3 See Puech (1992) for biographical information on the characters.

https://doi.org/10.1515/9783110539479-008

Apollo.[4] Despite the fact that in Plutarch's era Delphi was filled with statues and memorials from the Hellenistic and Roman periods, all of the sights viewed in the dialogue date from the seventh to the fourth centuries BCE.[5] Such a focus on the cultural and historical highlights of the distant past, and the concomitant occlusion of those of more recent periods, are, of course, typical features of much imperial Greek literature.[6]

The dialogue, however, is also suffused with an acute nostalgia for the past, one that becomes particularly noticeable during the lengthy discussion of the issue that gives the text its name: Περὶ τοῦ μὴ χρᾶν ἔμμετρα νῦν τὴν Πυθίαν (*On the Pythia's not now prophesying in verse*).[7] The title's phrasing μὴ ... νῦν ('not ... now') already underscores the difference between the present and what came before; in the text proper, the interlocutors make it clear that they view the discrepancy as symptomatic of Greece's more general decline from its earlier heights. The abandonment of verse oracles thus functions as a particularly resonant metaphor for Greek culture's passage from a grand, epic past to a prosaic present, and this presumption of decline, familiar from other contemporary works like Ps.-Longinus' *On the Sublime* or Plutarch's own *On the Obsolescence of Oracles*, informs the responses offered in the dialogue by Philinus, Sarapion, and Boëthus. Theon, however, in the long speech that concludes the dialogue (19.403 A–30.409D), completely upends this presumption; rather than indulge the nostalgia of his colleagues for a poetic past, he insists that he prefers the prosaic present.

Although scholars have often noted Theon's unusual defence of the imperial *status quo*, they have rarely explored in any depth the remarkable cultural history he sketches in Greece's development from a 'poetic' to a 'prosaic' culture.[8] While switching from poetry to prose was customarily perceived as a cause for regret, Theon manages, through a series of rhetorical sleights of hand, to overturn expectations and cast the change as positive. In this article I analyse one section of Theon's argument (24.406B–E) to show how he exploits the metaphorical connotations of literary critical terminology to transform an argument about a change in discourse to one about

4 For general literary analyses of the *De Pyth. or.*, see Schröder (1990) 1–124; Babut (1992); Valgiglio (1992) 21–40; Tusa Massaro (2000); Lamberton (2001) 159–165.

5 E.g., the bronze palm tree in the Corinthian treasury (c. 620–600 BCE), the statue of the Syracusan tyrant Hiero (ruled c. 478–467 BCE), the monument commemorating the Spartan victory at Aegospotami (405 BCE). On this tendency, see McInerney (2004) 49–50.

6 Vasunia (2002) on Plutarch's attitude toward the distant past (which Vasunia calls 'archaic'), with a brief mention of *De Pyth. or.* at 378.

7 A more accurate and evocative title than the more conventional Latin *De Pythiae oraculis* (*On the oracles of the Pythia*).

8 The notable exception being the excellent treatment of Van der Stockt (1992) 74–83; 85–88; cf. 78–79 for previous references. Toye (2000) is occasionally insightful but his focus is on archaic and classical history rather than Plutarch. Whitmarsh (2005) 363–365, whose interests in this text are closest to my own, reads Theon's speech as 'a manifesto for a prosaic culture', intimately linked to the downgrading of poetry in the imperial period.

a change in moral outlook, by associating the former verse-dominated age with luxury and extravagance, and the subsequent era of prose with moderation and restraint. To complicate matters, Theon's positive evaluation of the latter, which comprises his own time, does not imply a wholesale rejection of the past; we will see that Theon splits the past into two distinct eras—a 'poetic' followed by a 'prosaic'—and collapses the second with his own imperial present. Although Theon does not explicitly provide temporal parameters for his developmental model, I argue that the 'poetic past' that he characterises as excessive and immoderate bears a striking resemblance to what we call the 'archaic' age.

The poetic past

Theon's account of cultural development depends upon the theory of prophetic inspiration he had expounded earlier in the dialogue (21.404B–22.405D).[9] There, he had insisted that the words of the Pythia are not those of Apollo; rather, the god sends 'images' (*phantasiai*) into the Pythia, who harmonises with and transforms them through a proper movement of her soul. The content of the oracles are the *phantasiai* sent by Apollo, but the form in which they are uttered depends on the Pythia herself. To answer Diogenianus' question, Theon explains that oracles are currently in prose because the Pythia is an uneducated and simple woman, incapable of composing verse; this in no way makes her responses any less divine. Only the form of the utterance, then, has changed at Delphi, not the power of the god.

This argument, however, gives rise to a new question: if the god has no control over the oracular form, why were the Pythias of old, who were presumably also simple and uneducated, able to deliver oracles in verse? To address this complication, Theon presents a bolder theory encompassing far more than just Delphic oracular practice. The ancient (τῶν παλαιῶν) priestesses' propensity for verse prophecy, he declares, was due not to any particularity of their own, but to a more general difference between humans of their age and those of the imperial present: the ancients' predilection for poetic expression.[10]

> ... σωμάτων ἤνεγκε κράσεις καὶ φύσεις ὁ χρόνος ἐκεῖνος εὔρουν τι καὶ φορὸν ἐχούσας πρὸς ποίησιν, αἷς εὐθὺς ἐπεγίγνοντο προθυμίαι καὶ ὁρμαὶ καὶ παρασκευαὶ ψυχῆς ἑτοιμότητα ποιοῦσαι μικρᾶς ἔξωθεν ἀρχῆς καὶ παρατροπῆς τοῦ φανταστικοῦ δεομένην, ὡς εὐθὺς ἕλκεσθαι πρὸς τὸ οἰκεῖον ... ἐν οἴνῳ τε πολλῷ καὶ πάθει γιγνομένους, οἴκτου τινος ὑπορρυέντος ἢ χαρᾶς προσπεσούσης, ὀλισθάνειν εἰς ἐνῳδὸν <καὶ ἔμμετρον> γῆρυν, ἐρωτικῶν τε κατεπίμπλαντο μέτρων καὶ ᾀσμάτων τὰ συμπόσια καὶ τὰ βιβλία γραμμάτων.

9 This theory has been often discussed in modern scholarship: e. g., Schröder (1990) 25–59 (with earlier bibliography); Vernière (1990); Holzhausen (1993); Chirassi Colombo (1996).
10 Translations of *De Pyth. or.* are my own, following the text of Schröder (1990).

> ... *that age* produced *bodily dispositions and natures* possessing something productive and fluent toward poetry, to which straightaway were added desires, impulses, and predispositions of the soul producing a readiness that required just a little external impetus, that is, a disturbance of the imagination; ... the result was that those under the influence of wine or much emotion were immediately drawn toward their own propensities, and, when some feeling of pity or joy arose, they slipped into metered and melodious song and the symposia were filled with love songs, verses, and hymns, and books with similar material. (23.405E–F)

This natural propensity toward poetry meant that ancient discourse was conducted in verse; everything was turned into song.[11]

> ἀμοιβῇ γὰρ ἔοικε νομίσματος ἡ τοῦ λόγου χρεία ... ἦν οὖν ὅτε λόγου νομίσμασιν ἐχρῶντο μέτροις καὶ μέλεσι καὶ ᾠδαῖς, πᾶσαν μὲν ἱστορίαν καὶ φιλοσοφίαν πᾶν δὲ πάθος ὡς ἁπλῶς εἰπεῖν καὶ πρᾶγμα σεμνοτέρας φωνῆς δεόμενον εἰς ποιητικὴν καὶ μουσικὴν ἄγοντες. οὐ γὰρ μόνον <ὧν> νῦν ὀλίγοι μόλις ἐπαΐουσι, τότε [δὲ] πάντες ἠκροῶντο καὶ ἔχαιρον ᾀδομένοις <'μηλοβόται τ'> ἀρόται τ' ὀρνιχολόχοι τε' κατὰ Πίνδαρον, ἀλλ' ὑπὸ τῆς πρὸς ποιητικὴν ἐπιτηδειότητος οἱ πλεῖστοι διὰ λύρας καὶ ᾠδῆς ἐνουθέτουν ἐπαρρησιάζοντο παρεκελεύοντο, μύθους καὶ παροιμίας ἐπέραινον, ἔτι δ' ὕμνους θεῶν εὐχὰς παιᾶνας ἐν μέτροις ἐποιοῦντο καὶ μέλεσιν οἱ μὲν δι' εὐφυΐαν οἱ δὲ διὰ συνήθειαν.

> For *the use of discourse* is like the currency of coinage in trade ... Thus there was a time when people used meter, songs, and lyrics as the coinage of discourse, bringing all history, philosophy, and, in brief, every experience and matter requiring a more noble voice into poetry and music. *Only a few today* listen to poetry without difficulty; *back then everyone* heard and enjoyed songs: 'shepherds, ploughmen, and bird catchers' according to Pindar (*Isth.* 1.48). Because of this aptitude for poetry, the majority of people advised, spoke openly, exhorted, and related myths and proverbs by means of the lyre and song; moreover, they produced hymns to the gods, prayers, and paeans in meter and music, some because of their natural ability, others because of habituation. (24.406B–C)

In a world where poetry was the primary means by which thoughts and ideas were expressed, it follows that the Pythia too would deliver oracles in verse rather than prose.

Theon thus transfers what had been a localised question about the Pythia to a wider arena; the oracle's abandonment of poetry in favour of prose is not simply a matter of Delphic idiosyncrasy, but part of a general change in human modes of communication. Moreover, this transformation is ascribed to the fact that the human beings of the past were essentially different from those of the present in their bodily make-up (κράσεις καὶ φύσεις).[12] It would be hard to imagine a more forceful assertion of the fundamental disparity between present and past, a disparity reiterated in the observation that '*only a few today* (νῦν ὀλίγοι) listen to poetry without difficulty; *back then everyone* (τότε πάντες) heard and enjoyed songs'.

---

11 On the poetic inspiration theory proposed by Theon, see Van der Stockt (1992) 100–121.
12 On the possible medical origins of these terms, see Schröder (1990) 371–372 on 23.405E.

Poetry, luxury, and extravagance

The existence of such a stark rupture between past and present begs the question of change: when, why, and how did human poetic natures become prosaic? Theon, however, neglects to provide any explicit timeframe or reason for the break he has identified between 'then' and 'now', content with merely describing and evaluating the transformation. Rather than lament the loss of the ancient, poetic *kraseis* and *phuseis* as his companions might have done, Theon declares that 'the change has been for the better' (πρὸς τὸ βέλτιον γεγενημένην τὴν μεταβολήν, 24.406B) and celebrates the new circumstances of the present. The intricate and highly compressed network of associations by which Theon manages to transform what had seemed to be a golden age of song into something *worse* than its prosaic successor is worth examining in detail.

The key section of the argument occurs in the middle of chapter 24 (406C–E), immediately after the passage quoted above. Theon concludes his depiction of the age of song with sentence (1), and begins his discussion of its transformation with (2a):[13]

> οὐκοῦν οὐδὲ μαντικῇ κόσμου καὶ χάριτος ἐφθόνει ὁ θεὸς οὐδ' ἀπήλαυνεν ἐνθένδε ⟨τὴν⟩ τιμωμένην μοῦσαν τοῦ τρίποδος, ἀλλ' ἐπήγετο μᾶλλον ἐγείρων <u>τὰς ποιητικὰς</u> ⟨καὶ⟩ ἀσπαζόμενος <u>φύσεις</u>, αὐτός τε φαντασίας ἐνεδίδου καὶ συνεξώρμα <u>τὸ σοβαρὸν</u> καὶ λόγιον ὡς ἁρμόττον καὶ θαυμαζόμενον.
>
> (1) Therefore, the god neither refused to grant ornament and charm to the prophetic art nor expelled the revered muse from the tripod, but rather brought her in, arousing and embracing the *poetic natures*; he himself both inspired the *phantasiai* and encouraged <u>τὸ σοβαρὸν</u> and eloquence, on the grounds that they were appropriate and marvellous. (24.406D)
>
> ἐπεὶ δὲ τοῦ βίου μεταβολὴν ἅμα ταῖς τύχαις καὶ <u>ταῖς φύσεσι</u> λαμβάνοντος <u>ἐξωθοῦσα τὸ περιττὸν ἡ χρεία</u> κρωβύλους τε χρυσοῦς ἀφῄρει καὶ ξυστίδας μαλακὰς ἀπημφίαζε καὶ που καὶ κόμην <u>σοβαρωτέραν</u> ἀπέκειρε καὶ ὑπέλυσε κόθορνον οὐ φαύλως ἐθιζομένων ἀντικαλλωπίζεσθαι πρὸς τὴν πολυτέλειαν εὐτελείᾳ καὶ τὸ ἀφελὲς καὶ λιτὸν ἐν κόσμῳ τίθεσθαι μᾶλλον ἢ <u>τὸ σοβαρὸν</u> καὶ περίεργον ... .
>
> (2a) But when, as life changed along with fortune and *bodily natures*, <u>ἐξωθοῦσα τὸ περιττόν, ἡ χρεία</u> loosened the golden topknots, took off the soft robes, cut the <u>σοβαρωτέραν</u> long hair, and untied the high boot, people rightly accustomed themselves to oppose extravagance with economy and to adopt simplicity and plainness in adornment rather than <u>τὸ σοβαρὸν</u> and over-elaboration ... . (24.406D–E)

As we recall from the passages quoted earlier in section 2 of this article, the explicit subject of Theon's discussion is 'the usage of discourse' (ἡ τοῦ λόγου χρεία, 24.406C), which is distinct from that of the present because of the 'mixtures and natures of [people's] bodies' in 'that past era' (σωμάτων ... κράσεις καὶ φύσεις ὁ χρόνος

13 Untranslated terms will be discussed in the analysis that follows.

ἐκεῖνος, 23.405E). In sentence (1), Theon's reference to 'poetic natures' (τὰς ποιητικὰς ... φύσεις) shows that he is still following this model, as does the mention of changing 'natures' (ταῖς φύσεσι) at the beginning of sentence (2a), which abruptly begins Theon's characterisation of the shift from the 'poetic' era to the present. It is thus only natural that, when one gets to the next phrase in sentence (2a), ἐξωθοῦσα τὸ περιττὸν ἡ χρεία ... (untranslated above), one would presume that ἡ χρεία was shorthand for the aforementioned ἡ τοῦ λόγου χρεία ('usage *of discourse*') and that ἐξωθοῦσα τὸ περιττόν referred to poetry and meant something like 'casting aside the refinements [of verse]'.[14] The discussion would thus remain centered on a change in modes of communication: in the past everyone used poetry, but, as *bios* and *phuseis* changed, 'the usage of discourse, casting aside the refinements of verse' (sadly) turned instead to prose.

These assumptions, however, are soon shown to be mistaken. The rest of sentence (2a) reveals that Theon is no longer speaking of poetic discourse, but of clothing and hairstyle. Ἡ χρεία did not renounce *verse*, but rather 'loosened the κρωβύλους ... χρυσοῦς' (golden topknots), 'took off the ξυστίδας μαλακάς' (soft robes), 'untied the κόθορνος' (high boot), and 'cut off the κόμην σοβαρωτέραν' (*sobarôtera* long hair). It turns out, then, that ἡ χρεία is not 'usage *of discourse*' after all, but just a generic 'usage' associated with styles of dress and appearance.[15] Furthermore, as the following section will show, 'usage' is 'casting off' (ἐξωθοῦσα), not the positively judged 'refinement' (τὸ περιττόν) of verse, but a more abstract and negatively coded 'superfluity' (τὸ περιττόν) associated with extravagant costume.

Three of the examples of clothing and hairstyle that Theon affiliates with those 'poetic natures' (the ξυστίς, the κρωβύλος, and long hair) are cited by fifth and fourth-century BCE authors as symbols of luxury and wealth: wearing a ξυστίς, for example, is seen as a mark of opulence by Aristophanes and Plato.[16] In fact, Theon's reference to κρωβύλους χρυσοῦς ('golden topknots') alludes to a famous passage of Thucydides on the luxuriousness of Athenian elite dress and hairstyle (1.6.3–4):[17]

> The Athenians were the first ... to adopt an easier and *more luxurious* mode of life (ἀνειμένῃ τῇ διαίτῃ ἐς τὸ τρυφερώτερον μετέστησαν); indeed, it is *only lately* (οὐ πολὺς χρόνος ἐπειδὴ) that their rich old men left off the *luxury* (διὰ τὸ ἁβροδίαιτον) of ... fastening a *knot of their hair* with a tie of *golden* crickets (χρυσῶν τεττίγων ἐνέρσει κρωβύλον ἀναδούμενοι τῶν ἐν τῇ κεφαλῇ τριχῶν); this same fashion (αὕτη ἡ σκευή) has prevailed for a long time among the old men of the Ionians. [4] On the contrary the Spartans were the first to *use* (ἐχρήσαντο) a *modest* style

14 *Perittos* denotes either a positive literary quality, e.g., 'refined', or a negative one, '*over*-refined'; given the previous laudatory description of the poetic age's 'ornament and charm' (κόσμου καὶ χάριτος), the former sense would more likely spring to mind.
15 Schröder (1990) 385.
16 Pl. *Resp.* 4.420e; Ar. *Nu.* 70.
17 So Schröder (1990) 385–386; Van der Stockt (1992) 83; Toye (2000) 175. The passage was well-known in the imperial period: e.g., Dion. Hal. *Thuc.* 19 and Luc. *Nav.* 3.

of dressing more in conformity with *modern* ideas (μετρίᾳ δ' αὖ ἐσθῆτι καὶ ἐς τὸν νῦν τρόπον), the rich doing their best to assimilate their way of life to that of the common people.

Thucydides is vague about the timeframe ('only lately'), but Aristophanes attributes this older, more luxurious style to the Athenians of the Marathon generation: in the *Clouds*, Unjust *Logos* characterises them as 'ancient ... and full of crickets' (ἀρχαῖα ... καὶ τεττίγων ἀνάμεστα).[18] A little later, in the mid-fourth century BCE, Heraclides Ponticus similarly identifies those 'who had fought the battle of Marathon' as the ones who 'bound up their locks into *topknots*, put *golden* crickets on their brows and wore their *hair long*' (κορύμβους δ' ἀναδούμενοι τῶν τριχῶν χρυσοῦς τέττιγας περὶ τὸ μέτωπον καὶ τὰς κόμας ἐφόρουν).[19]

Theon thus superimposes what was originally a classical Athenian narrative about fifth-century BCE changes in *fashion* (from luxurious to moderate) onto a different story, that of a change in *discourse* (from poetic to prosaic). While Thucydides refrains from passing judgment on the Athenians' change in fashion, Aristophanes and Heraclides view the luxuries of the older generation sympathetically. By Plutarch's day, however, these habits had acquired an unmistakably negative valence, associated with ostentation, arrogance, and profligacy.[20] By describing the change from poetry to prose with metaphors from fashion, Theon has thus managed to imbue poetry with the negative connotations of luxury, and associate prose with the more positive qualities of moderation. The switch from poetry to prose is no longer imagined as a decline, in which something literarily valuable (the grandeur of verse) is lost, but in positive terms, as progress, in which something morally suspect (luxurious, superfluous costume) is renounced. Theon's approval of this development is made explicit at the end of sentence (2a), when he asserts not only that the transformation in fashion was part of a more fundamental change in *taste*—from 'extravagance' to 'thrift' (πρὸς τὴν πολυτέλειαν εὐτελείᾳ)–but also that this, in his opinion, had occurred 'rightly' (οὐ φαύλως). His use of the charged economic terms πολυτέλεια and εὐτέλεια further suggests that the development has been in a positive direction, from an ostentatious and lavish lifestyle to one more modest and unassuming; now 'simplicity and plainness' (τὸ ἀφελὲς καὶ λιτόν) are preferred to '*to sobaron* and over-elaboration' (τὸ σοβαρὸν καὶ περίεργον). Theon has thus established a parallel between the change in fashion (from luxurious to modest) and

18 *Nu.* 984–986. Cf. *Eq.* 1324 where a personified Athens is imagined as 'wearing the crickets, splendid in his archaic dress' (τεττιγοφόρας, ἀρχαίῳ σχήματι λαμπρός).

19 Heraclid. Pont. F 39, lines 15–17 Schütrumpf (F 55 Wehrli). I retain the manuscript κόμας for the emendation κόρρας; κορύμβος is an Ionic variant of κρωβύλος. Gorman and Gorman (2014) 313–315, argue, however, that this sentence is not part of Heraclides' original text, but a comment by Athenaeus (12.512c), the source of the fragment.

20 On the negative associations of the ξυστίς, cf. Plu. *Alc.* 32.2 and *Bellone an pace* 348F; of the κρωβύλος, Clem. Al. *Paed.* 2.10.3–4, Ael. *VH* 4.22, Ath. 525f; of the κόθορνος, cf. Lau (1967) 128–130; of long hair: see n. 24 below.

that of discourse (from poetic to prosaic); both reflect the new 'usage, which casts aside the superfluous' (ἐξωθοῦσα τὸ περιττὸν ἡ χρεία, 24.406D), prefers moderation to extravagance, and meets with his approval.

From loftiness to grandiosity: σοβαρός

This re-valuation of the change from poetry to prose relies on exploiting the multivalence of certain key expressions. For example, the terms τὸ ἀφελές ('artless', 'simple') and τὸ λιτόν ('frugal', 'unadorned'), which can be used both of literary style *and* of personal character, help Theon assimilate the shift in modes of discourse to that of a more general cultural worldview.[21] Or a word understood as positive in a poetic context, like περιττός ('refined'), becomes negatively charged when seen in the light of fashion ('superfluous').[22] The precise metaphorical move that Theon is making, however, can be best illustrated by the shift in the meaning of the adjective σοβαρός from sentence (1) to (2a). When Theon mentions Apollo's embrace of poetic *phuseis* and his encouragement of τὸ σοβαρὸν καὶ λόγιον in sentence (1), the term σοβαρός, paired with the positive term 'eloquence' or 'erudition', refers to the 'loftiness' or 'majesty' of poetry.[23] A few lines later in sentence (2a), however, he employs it with a clearly derisive connotation: paired with κόμη ('long hair'), σοβαρωτέραν (the comparative form of σοβαρός) means 'very pompous' or 'very arrogant'.[24] When Theon uses the word a third time (at the end of sentence 2a) as a general characteristic of the poetic past, this negative valence is still foremost: the current desire for simplicity and lack of ornament is favorably (οὐ φαύλως) opposed to the 'pomposity' or 'grandiosity' (τὸ σοβαρόν) and 'over-elaboration' (περίεργον) of the past.[25]

This shift in meaning of the word σοβαρός demonstrates how Theon has managed to subvert what had initially appeared to be a traditional lament for a decline from a poetic past marked by 'loftiness' to a prosaic and mundane present.[26] Poetic

---

21 On these terms, see Vischer (1965) 22–27 (λιτός); 30 (ἀφελής).

22 Cf. above, n. 14; Jeuckens (1908) 178 and 180 on Plutarch's use of the term in general, and Schröder (1990) 385 for its negative valence here.

23 Cf. *Per.* 5.1.4, on Pericles' τὸ φρόνημα σοβαρὸν καὶ τὸν λόγον ὑψηλόν: '*lofty* thoughts and sublime speech'.

24 Schröder (1990) 386 cites Luc. *Zeux.* 5 and Ps.-Luc. *Am.* 40. Cf. similarly negative valences in Joseph. *AJ* 14.45, where *sobaros* describes a variety of luxurious styles (including long hair) and Ael. *HA.* 5.21, which links *sobaros* and fancy Persian robes.

25 On *sobaros* as a predominantly negative literary critical term in Plutarch, see Jeuckens (1908) 177–178; a preliminary glance through his corpus suggests that the same holds for the use of *sobaros* in non-literary contexts as well.

26 Babbitt's Loeb translation (1936b), misleadingly in my opinion, maintains the positive connotations of *sobaros* throughout: 'impressiveness' –'stately' – 'ornate'; contrast, e.g., Valgiglio (1992): *la magnificenza – troppo pomposa – la magnificenza eccessiva*, or Flacelière (1974): *sublime – trop fières – faste.*

'loftiness' is now figured as 'pomposity' and paired with 'over-elaboration' rather than 'eloquence', while the adoption of prose is seen as a sign of modesty and proper lack of pretention. In the process, Theon has bestowed upon his original portrait of a poetic past a strong moralising tinge; the ancients possessed a predilection not only for verse, but also for extravagance and ornament more generally, while the contemporary preference for prose matches prevailing moderate values. In Theon's argument, poetry and prose become more than just modes of communication; they function as the ruling metaphors for the cultural values of their respective eras.[27]

Theon ties these three developments—loftiness to simplicity, luxury to moderation, poetry to prose—together in a simile that continues the passage (2a) quoted earlier at the beginning of the previous section.

> ... οὕτως τοῦ λόγου συμμεταβάλλοντος ἅμα καὶ συναποδυομένου κατέβη μὲν ἀπὸ τῶν μέτρων ὥσπερ ὀχημάτων ἡ ἱστορία καὶ τῷ πεζῷ μάλιστα τοῦ μυθώδους ἀπεκρίθη τὸ ἀληθές ... .
>
> (2b) ... so too, since *discourse* underwent a similar change and *stripped itself*, history *descended from meter* just as from *chariots* and once *on foot/in prose* the truth was especially separated from the mythic ... . (24.406E)

While Theon's initial description of 'discourse stripping itself [of its clothes]' (τοῦ λόγου ... συναποδυομένου) links this sentence to the content and motifs of (1) and (2a), he quickly moves on to another familiar image: that of prose (ὁ πεζὸς λόγος) as pedestrian, in contrast to the chariot (ὄχημα) of poetry.[28] The multivalence of the image is fully exploited here. On the one hand the movement from chariot to ground could be characterised as a descent from the loftiness of poetry to a more plain style, signifying a *loss* of something valuable. So Strabo in a famous passage on the development of prose (1.2.6):

> καὶ αὐτὸ δὲ τὸ πεζὸν λεχθῆναι τὸν ἄνευ τοῦ μέτρου λόγον ἐμφαίνει τὸν ἀπὸ ὕψους τινὸς καταβάντα καὶ ὀχήματος εἰς τοὔδαφος.
>
> And the very fact that discourse *without meter* was referred to as *'pedestrian'* indicates its *descent from a height*, or from a *chariot* to the ground.

Aelius Aristides, however, can see the opposition in a different way: to equate poetry with riding a chariot is to suggest its 'cultural' artificiality, while prose, like walking,

---

27 The association between poetry and luxurious clothing was no doubt suggested also by the fact that the ξυστίς and the κόθορνος were closely associated with tragedy; Plutarch refers to the ξυστίς as part of tragic costume at *Alc.* 32.2 and *Bellone an pace* 348F (where he describes it specifically as πολυτελής, 'extravagant').

28 Plutarch also uses the analogy at *De aud. poet.* 16D, where he accuses scientific poets like Empedocles of 'having borrowed from poetry the meter (τὸ μέτρον) and the majesty, just as a chariot (ὄχημα), in order to escape from the prosaic (τὸ πεζόν)'. Cf. Schröder (1990) 387–388 and Norden (1909) 32–35 for a list of similar imagery (imperial examples include Dio Chr. 20.2; Luc. *Bis Acc.* 33; *Pr. Im.* 18; *Menipp.* 1).

is the 'natural' (κατὰ φύσιν) mode of discourse.[29] The tension between these two readings—one privileging the poetic height as 'lofty', the other denigrating it as artificial and unnatural—matches precisely that encapsulated in Theon's own dual use of the term σοβαρός.[30] In the course of just three sentences, Theon has moved from Strabo's position to that of Aristides: prose, which had originally seemed a lamentable step down, or decline, from the 'loftiness' of poetry, is reinterpreted as a positive abandonment of the 'pomposity' of verse and as a step toward a more 'down-to-earth' sensibility.[31]

Out of a series of conventional oppositions—poetry vs. prose, extravagance vs. simplicity, luxurious vs. moderate fashions, chariots vs. pedestrians—Theon has thus conceived a strikingly *un*conventional model of cultural change and progress.[32] What starts out looking like a traditional story of decline from the past grandeur of a poetic age to present prosaic mediocrity is suddenly transformed into one of moral progress, from ancient extravagance to modern moderation. The new past that Theon constructs in his speech is poetic insofar as it is pompous, luxurious, excessive, and thus alien in temperament to what comes afterwards—a present marked by its 'prosaic' appreciation for simplicity, restraint, and modesty.

Conclusion: An 'Archaic Age'?

In the rest of his speech, Theon continues to flesh out his vision of the 'poetic' past, characterizing it not only as extravagant and opulent, but also as captivated by obscurity, ambiguity, and circumlocution (24.406E–26.407E). The speech concludes with a reference to Delphi's renewed prosperity and Theon's insistence that 'I am pleased with the way things are now' (τὰ δὲ νῦν πράγματα καθεστῶτα ... ἀγαπῶ, 28.408B); obscure and lofty verse oracles might have been appropriate for the tumultuous times of the past, but are hardly necessary in the peace and tranquillity of the present.

---

29 Aristid. *Or.* 45.8: 'It is *more natural* for men to use *prose*, just as it is certainly, in my opinion, more natural to walk than to be borne *by a chariot*' (κατὰ φύσιν μᾶλλόν ἐστιν ἀνθρώπῳ πεζῷ λόγῳ χρῆσθαι, ὥσπερ γε καὶ βαδίζειν οἶμαι μᾶλλον ὀχούμενον φέρεσθαι).

30 The chariot image could also be read as reinforcing Theon's negative association of poetry with extravagant habits. Demosthenes describes the youth Phaenippus as 'full of *arrogance*' because he bought himself a chariot 'so that he not travel *by foot*' (ἵνα μὴ πεζῇ πορεύηται· τοσαύτης οὗτος τρυφῆς ἐστι μεστός, *Or.* 42.24); at Ar. *Nu.* 69–70, driving a chariot in a ξυστίς is treated as characteristic of a rich man's lifestyle.

31 For another reading of the metaphor, see Whitmarsh (2005) 370–371.

32 The last change that Theon mentions –'once in prose, the truth was especially separated from the mythic'—recalls the association of 'early' historiography with τὸ μυθῶδες in Str. 1.2.8, Dion. Hal. *Thuc.* 5–7, and Thucydides' famous criticisms of earlier 'logographers' (1.20.1). This idea connects closely with Theon's subsequent remarks on the obscurity of poetry versus the clarity of prose, a topic I will deal with in a future article.

As I mentioned in the introduction, Theon's 'cultural history' is quite distinct from that of his interlocutors, who lament the loss of an idealised Greek past and see the present as deficient in comparison. But his position is also at odds with another prevalent mode of imperial thinking about the past, classicism, which celebrates the present as a revival of the values and style of an ancient 'classical' past, after an intervening period of decline and corruption (cf. Dion. Hal. *Orat. Vett.* 1). Theon also prizes the present, but unlike the classicizers, he neither reveres nor seeks to revive the values of the past, but repudiates them.

In order to interrogate the singularity of his model a little bit more closely, however, one might ask which past Theon has in mind when he speaks about 'that time' (ὁ χρόνος ἐκεῖνος, 23.405E), or when precisely his alleged 'change' (μεταβολήν, 24.406D) in values occurred. Theon never explicitly addresses these matters, but, as scholars have noted, his allusions to changing fashions and the development of historical writing offer some hints as to a possible periodisation.[33] As we noted above, classical Athenian writers associated their ancestors' luxurious fashion choices, to which Theon alludes, with the generation who had fought at Marathon in 490 BCE. Moreover, Theon's remark that 'history descended from meter just as from a chariot' (quoted above as passage 2b) not only recalls Strabo's use of the simile in *Geogr.* 1.2.6, but also invokes the model of historiography's evolution from poetry to prose mentioned by the geographer in the very same paragraph:

> Prose speech of an elaborate kind is very much an imitation of poetical; for poetical elaboration came into the world first and won fame. Then came Cadmus, Pherecydes, Hecataeus, and their followers, writing prose in which they imitated poetry, abandoning meter but in other respects preserving the poetic qualities. Later, others took away, each in his turn, something of these qualities, and brought prose down to its present form, as from a lofty height. (1.2.6)

The historians named by Strabo all date from the late sixth and early fifth century BCE. Theon is thus exploiting the fact that ancient theories about the development of prose from poetry (cf. also Varro F 319 Funke) operated on a roughly similar periodisation as did those concerning the change from luxurious to moderate fashions at Athens; both see the early to mid-fifth century BCE as crucial.[34] By assimilating these two models and extending their purview, Theon imagines a conceptual 'break' between an extravagant and a simple epoch that takes place, not at the end of the period (e.g., 323 BCE) that the imperial Greeks considered their glorious, golden age, but *within* it, in the fifth century BCE. The revered distant past is thereby split into two—an early 'excessive' period viewed with disapproval, followed by a 'modest' one worthy of praise. Moreover, this latter era is assimilated to the imperial present: the entire half millennium from the late fifth century BCE to the early second century

---

33 See Van der Stockt (1992) 79–80.

34 On the ancient theory of prose development, see Schröder (1990) 53–56 for discussion, with earlier bibliography; cf. Toye (2000) 178.

CE is envisioned as a single ‘moderate’ era—in fashion *and* discourse—that stands opposed to what went before, an opulent, poetic, age.

Theon’s periodisation is virtually unique: unlike we moderns, the Greeks of the imperial era did not generally divide the post-Trojan War, pre-Alexander past into separate ‘archaic’ and ‘classical’ ages. I suspect that it is no accident that Theon’s ‘change’ occurs roughly where we identify a break between ‘archaic’ and ‘classical’. Even if Theon’s ‘cultural history’ is vague and operates more through imagery and suggestion than hard chronologies, he is latching on to a feeling among imperial Greeks that the ancient past itself was not a unified, ideologically and culturally consistent entity, and that the ‘early’ part of it was somehow different in its sentiments and practices than what came afterwards. The modern, nineteenth-century invention of the ‘archaic’ period arose from a similar sense among classicists that its literature and culture was fundamentally different from that of the later fifth and fourth century BCE. Can we thus locate the origins of the ‘archaic age’ in the nostalgia of imperial Greece?[35]

35 On the invention of the archaic period, see e.g., Most (1989); Hummel (2003); Payen (2006). Along with Kim (2014), this article represents the initial stages of a forthcoming large-scale project on imperial Greek ideas about the ‘archaic’.

Michele A. Lucchesi

Delphi, place and time in Plutarch's *Lycurgus* and *Lysander*

Abstract: In the *Life of Lycurgus* and in the *Life of Lysander*, Delphi represents a symbolic place associated with extremely important events in Spartan history: the birth of the constitution thanks to Lycurgus and the political and institutional crisis of the fourth century BCE caused by Lysander. In both cases, Plutarch narrates these episodes emphasising the centrality of the oracle of Delphi and the close relationship between the Pythian god and the Spartan leadership. Lycurgus established the strong bond with Delphi by consulting the oracle before beginning his reforms. His journey is described by Plutarch as a separation from Sparta. Lycurgus' second journey to Delphi 'consecrated' the Spartan constitution and won the support of the god to Sparta as long as the Spartans would maintain the Lycurgan laws. This projects the readers towards future phases of Spartan history, when the relationship with the god would in fact be broken. Such a critical period of Spartan history coincides with Lysander's leadership as a navarch and his victories against the Athenians. First, at the beginning of the *Life of Lysander* Plutarch portrays Lysander's ambiguous nature by describing a marble statue placed inside the treasury of the Acanthians at Delphi. Subsequently, later in the *Life* he describes again a monumental complex at Delphi, with which Lysander celebrated his military successes. In both passages, different temporal layers are conflated: the past of Lycurgus, the present time of Lysander, and the future of his later actions; the past of Lysander and the present of both Plutarch/external narrator and the readers. In the narrative, therefore, Delphi constitutes a location where place and time cannot be separated.

The city of Delphi was certainly very important to Plutarch. Not only was he priest at the sanctuary of Apollo for many years, but he also devoted some of his writings to exploring philosophical and theological themes related to the oracle and the cult of the Pythian god.[1] Yet, while the *The E at Delphi*, the *On the Oracles of the Pythia*, and

1 I wish to express my sincere gratitude to Aristoula Georgiadou and Katerina Oikonomopoulou not only for organising the beautiful congress of the I.P.S. at Delphi but also for their hard work in editing this volume. I also would like to thank the anonymous reviewer of my chapter for his/her useful suggestions: any remaining errors or inaccuracies are my own responsibility. For the Greek text of Plutarch's *Parallel Lives* I have followed the most recent Teubner editions, while the translations are my adaptations of Talbert, Scott Kilvert, and Pelling (2005) and Romm and Mensch (2012).

On Plutarch's duties as priest and diplomat of Delphi, see Flacelière (1943); Swain (1991); Stadter (2004); Stadter (2005) 197–198; Talamo (2007); Casanova (2012). Further bibliographical references on Delphi can be found in Stadter (2005) 198 n. 6. In general, on Plutarch's religious spirit, see

https://doi.org/10.1515/9783110539479-009

the *On the Obsolescence of Oracles* probably constitute the most significant results of Plutarch's theological reflection, his religious sensibility permeates all of his works. Not surprisingly, then, Plutarch also attached a great value to Delphi in the *Parallel Lives*. In particular, Delphi often represented a symbolic place, where numerous solemn acts (e.g., the dedication of statues, the making of offerings, and so forth) were performed to celebrate crucial events that carried major implications for Greek and Roman history. Suffice it to recall how in the *Life of Flamininus* Plutarch depicted Titus Flamininus' dedication of silver bucklers, his own long shield, and a golden crown to Apollo after setting Greece free. By mentioning the celebratory inscriptions where the Romans were identified as the descendants of Aeneas and Flamininus defined himself as divine (θεῖος), Plutarch exposed Flamininus' arrogance as much as the ambivalent nature of the liberation of Greece (*Flam.* 12.11–12). Thus, as Stadter thoughtfully notices, 'Delphi for Flamininus is a theatre to display and augment his own fame and to claim special closeness to the divine'.[2]

Indeed, in the *Parallel Lives* what happened in Delphi can illuminate the broader historical circumstances that occurred in the same period. From a narrative perspective, moreover, the episodes concerning the sacred city—primarily those where the protagonists were directly involved—offer a key to the interpretation of the biographies in which they are variously inserted. In this chapter, by adopting this approach, I aim to explore the significance of Delphi for Spartan history. In particular, I concentrate on the *Life of Lycurgus* and the *Life of Lysander*, where this topic becomes highly relevant inasmuch as it connects the two Spartan *Lives* with one another. I shall argue that to some extent Delphi represents a place that allows the readers to go across time within the narrative, moving from archaic to classical Sparta.

The *Life of Lycurgus*

The importance of religion and superstition at Sparta is well-known. As Flower puts it, 'the Spartans arguably paid a more scrupulous attention to religious rituals and acted more often from religious motives than did any of the other Greeks'.[3] This e-

---

Brenk (1977); Brenk (1987); Burkert (1996); Graf (1996); Sfameni Gasparro (1996); Hirsch-Luipold (2014).

2 Stadter (2005) 206. On Plutarch's *Life of Flamininus* and the liberation of Greece, see Swain (1988); Walsh (1992); Scuderi (1996); Pelling (1986a) = (2002) 350–353, and (1995) = (2002) 243–247; Bremmer (2005); Stadter (2005) 206–207. In this regard, I suspect that the contrast between Flamininus' excessive pride, as is indirectly presented through the inscriptions at Delphi, and the general excitement among the Greeks at the announcement that they were being freed by the Romans is sharper than is usually thought by modern scholars (but cf. Swain 1988, 341–343; Pelling (1986a) = (2002) 350–352, and (1995) = (2002) 243–246.

3 Flower (2009) 193. On Spartan religion, see Parker (1989); Spawforth (1992); Malkin (1998); Hodkinson (1983) 273–276; Hodkinson (2000) 271–302; Richer (2007). Flower (2009) argues in favour of con-

lement of Spartan culture was extensively treated by Herodotus and became a central feature of his characterisation of the Spartans (e.g., Leonidas decided to remain at Thermopylae in obedience to the Delphic oracle; Hdt. 7.220). In Plutarch's *Life of Lycurgus* too, the Spartans' religiosity and close connection with Delphi assume great relevance and are viewed as inextricably linked to the reforms conducted by Lycurgus.

As Plutarch narrates, before introducing his new legislation, Lycurgus went to Delphi, offered sacrifices to the god, and consulted the oracle (*Lyc.* 5.4):

> διανοηθεὶς δὲ ταῦτα πρῶτον μὲν ἀπεδήμησεν εἰς Δελφούς, καὶ τῷ θεῷ θύσας καὶ χρησάμενος, ἐπανῆλθε τὸν διαβόητον ἐκεῖνον χρησμὸν κομίζων, ᾧ θεοφιλῆ μὲν αὐτὸν ἡ Πυθία προσεῖπε καὶ θεὸν μᾶλλον ἢ ἄνθρωπον, εὐνομίας δὲ χρῄζοντι διδόναι καὶ καταινεῖν ἔφη τὸν θεὸν ἣ πολὺ κρατίστη τῶν ἄλλων ἔσται πολιτειῶν.
>
> Once Lycurgus had formed this intention, first he travelled to Delphi. After making sacrifices to the god and consulting him, he returned bringing that famous oracle, according to which the Pythia on the one hand addressed him as 'dear to the gods' and 'a god rather than a man', on the other hand, since he had asked for good order, she declared that the god granted this and promised that his constitution would be by far the finest of all.

In the 54 passages of the *Parallel Lives* (counted by the *TLG*) where Delphi is mentioned, as one would expect, verbs of movement or verbs that imply an idea of movement such as ἥκω ('to have come'), ἔρχομαι ('to go') and its cognates, ἀποστέλλω ('to send off'), πέμπω ('to send'), or πορεύομαι ('to march'), and so forth are often employed. Doubtless, for the Greeks as much as for the Romans Delphi was a centre of attraction where they would go in person or send various types of offerings. Only in *Lyc.* 5.4, however, did Plutarch choose the verb ἀποδημέω ('to be or go abroad'), which added to the sense that, in order to change Sparta, first Lycurgus had to visit different places outside the Peloponnese (cf. at *Lyc.* 4 his journeys to Crete, Ionia, Egypt, and possibly Libya, Iberia, and India too, where he studied the local customs and compared the different constitutions). Indeed, the journey to Delphi is viewed from Lycurgus' perspective.[4] As one can also infer, moreover, the passage hints that at this stage of Spartan history Delphi and Sparta were imaginatively more distant than in later periods.

In fact, Lycurgus is portrayed as the initiator of the special relationship between the oracle and the Spartans. The scene described in the *Life of Lycurgus* is an adaptation of the account offered by Herodotus on the same subject: the origin of the Spartan legal and political system (Hdt. 1.65). Yet, while in Herodotus the second part of the response does not have absolute certainty ('only some people say', οἱ μὲν δή τινες λέγουσι) and is recorded as an addition to the first part ('in addition

sidering Sparta's religious practices distinctly different from those of the other *poleis* (a question that, however, does not concern foreign sanctuaries such as the oracle of Delphi).

4 On space and focalisation, see de Jong (2012b) 8–9. Cf. also Beck (2012).

to this the Pythia also declared to him the constitution that now exists at Sparta', πρὸς τούτοισι καὶ φράσαι αὐτῷ τὴν Πυθίην τὸν νῦν κατεστεῶτα κόσμον), Plutarch simply condensed the two sets of information as if the god, recognising Lycurgus' exceptional nature, made a serious commitment to the future of Sparta. This can also explain the presence of an already positive evaluation of the future Spartan constitution in the *Life of Lycurgus*, something that, conversely, Herodotus does not present in these terms.[5] As a result of such a close connection between the god and godlike Lycurgus, the temporal dimensions of the present and the future are conflated (the god's promise *now* anticipates the success of Lycurgus' *future* acts) in what can be defined as an internal *prolepsis*, which the Great Rhetra, prescribing what temples the Spartans had to build, in what groups they should distribute the population, and how and where they should hold their general assembly, fully actualises (*Lyc*. 6).[6]

Indeed, there is historical evidence that in the archaic and classical period Sparta exerted a strong influence on Delphi. Particularly important, in this regard, was the permanent presence at Delphi of the Spartan sacred ambassadors called Πύθιοι, who consulted the oracle on affairs of state and reported directly to the kings (see Hdt. 6.57.2; Xen. *Lac*. 15.5). Yet one should also consider that from a purely historical perspective the role of the Delphic oracle as initiator of political and social reforms is attested not only in the case of Sparta but also in other cities (e.g., Solon's reforms at Athens or the foundation of and the reforms at Cyrene). Furthermore, despite the fact that the literary sources insist on the part played by divine agency in establishing the Spartan polity, they give ampler space to human decisions. In Herodotus, for instance, the Spartan account relates the origin of the Lycurgan legislation to Crete (Hdt. 1.65). Diodorus Siculus (= Ephorus) also offers details about Lycurgus' own initiative (Diod. Sic. 7.12.2–4). In Xenophon's *Polity of the Lacedaemonians*, Plato's *Republic*, and Aristotle's *Politics* too, human agency has a preponderant function.[7]

To some extent, the *Life of Lycurgus* is not very different, considering that, after mentioning the Great Rhetra, Plutarch goes on to discuss the numerous aspects of Spartan society and politics that Lycurgus transformed pursuing *his agenda* for the city: the distribution of land and wealth (*Lyc*. 8–9); the institution of common meals (*Lyc*. 10 and 12); the education of the young (*Lyc*. 14–21); the military disci-

---

5 On the oracle in Herodotus and Plutarch, see Manfredini and Piccirilli (1980) 228–229; Asheri et al. (2007) 127, Nafissi (2010). Another (more complete) version of the oracle can be read in Diod. Sic. 7.12.1 (probably depending on Ephorus), where the author included the god's declaration that the Spartan system of laws would be better than those of all the other Greek cities. Thus, despite closely following Herodotus' text, Plutarch seems to have used Diodorus (= Ephorus) to supplement it. Other echoes of the connection between divine agency and the origin of the Spartan constitution can be found in Xenophon (*Lac*. 8.5) and Plato (*Resp*. 691d-692a).

6 On internal and external *prolepsis*, see de Jong (2007) 5. On Plutarch's use of this technique, see Beck (2007) and in this volume. See also Brenk in this volume.

7 On these topics, see Hodkinson (1983) 273–275; Malkin (1989) 136–142; Hodkinson (2005) 265–266; Nafissi (2010) 93–111.

pline (*Lyc.* 22); the education of the adult population (*Lyc.* 24–25); the election of the Gerousia (*Lyc.* 26); burial norms (*Lyc.* 27). Plutarch, however, employing his typical biographical technique, shapes the narrative so that he can once more give prominence to Delphi and the Pythian god towards the end of the *Life.* Apart from the need to create a conclusion for a biography that contains very few data about the protagonist, Plutarch's choice can be reasonably explained by the fact that the information about the constitution alone was not enough to show the impact of Lycurgus on the history of Sparta and her future generations.

At the end of the long section briefly summarised above, Plutarch portrays Lycurgus returning to Delphi after completing his program of radical changes in order to 'leave the constitution immortal and immovable for the future' (ἀθάνατον αὐτὴν ἀπολιπεῖν καὶ ἀκίνητον εἰς τὸ μέλλον, *Lyc.* 29.1). Before leaving Sparta, however, he made the kings, the elders, and all of the citizens swear that they would continue to apply the existing constitution until his return. In this case too, Plutarch uses a verb (ἀπαίρω) that indicates departure from Sparta rather than a simple movement towards Delphi: 'he set out for Delphi' (ἀπῆρεν εἰς τοὺς Δελφούς, *Lyc.* 29.4). Once again, then, Plutarch presents Lycurgus' journey from the protagonist's perspective as a complete separation from his hometown. This time, moreover, to some extent Lycurgus seems to have tried to 'seal' Sparta off from outside contacts and potentially negative external influences, which could contaminate the purity of the Spartan system, just as anticipated by Plutarch earlier in the *Life* with regard to the prohibition against travelling abroad and admitting foreigners (*Lyc.* 27.6–9).

The scene at Delphi repeats that in chapter 5 of the *Life.* Lycurgus made an offering and subsequently interrogated the oracle (*Lyc.* 29.5–6):

> παραγενόμενος δὲ πρὸς τὸ μαντεῖον καὶ τῷ θεῷ θύσας, ἠρώτησεν εἰ καλῶς οἱ νόμοι καὶ ἱκανῶς πρὸς εὐδαιμονίαν καὶ ἀρετὴν πόλεως κείμενοι τυγχάνουσιν. ἀποκριναμένου δὲ τοῦ θεοῦ καὶ τοὺς νόμους καλῶς κεῖσθαι καὶ τὴν πόλιν ἐνδοξοτάτην διαμενεῖν τῇ Λυκούργου χρωμένην πολιτείᾳ, τὸ μάντευμα γραψάμενος εἰς Σπάρτην ἀπέστειλεν.
>
> Once he reached the oracle and sacrificed to the god, he asked whether the laws that he had laid down were of sufficient quality for the happiness and the virtue of the city. As the god replied that the laws were established well and the city would continue being held in high esteem by using the constitution of Lycurgus, having written the response, he sent it to Sparta.

Despite the characteristic oracular style, one can notice that, in addition to answering Lycurgus' question, the god also gave him a message for Sparta. According to Plutarch's narration, as we saw earlier in this chapter, Lycurgus must have already known that his constitution, being supported by the god, was good. His query concerned the future of Sparta and the preservation of happiness and virtue: the god's reply, then, spoke to the Spartans directly. Just as in the first oracle, here as well two different temporal layers—present and future—are intertwined: as long as the Spartans would maintain their *present* condition, determined by the Lycurgan constitution, Sparta's *future* would be glorious.

Lycurgus' subsequent decision not to return to Sparta and to put an end to his life forced the Spartans to remain faithful to their existing laws, keeping their oath. Thus, thanks to the self-sacrifice of Lycurgus, whose intentions appear to have been clear even before leaving Sparta, Delphi and the oracle became counterparts in a covenant with Sparta guaranteed by their respect of its constitution, a bond that linked them for centuries.[8]

The *Life of Lysander*

The *Life of Lycurgus* already anticipates (with an external *prolepsis*) that the Spartan general Lysander, despatching to Sparta a huge sum of money after his victories in Asia and Greece against the Athenians, later filled the city with love of riches (φιλοπλουτία) and luxury (τρυφή), and irremediably overturned (literally 'subdued', καταπολιτεύομαι) the laws of Lycurgus (*Lyc.* 30.1). The *Life of Lysander* begins with an opening section that confirms the ambiguous character of the protagonist and alerts the readers to his problematic relationship with Sparta, a theme that is emblematically related to Delphi.

Plutarch narrates that still in his time inside the treasury of the Acanthians at Delphi there was a statue of a man with long beard and hair, whom many identified as Brasidas because of the inscription reading 'Brasidas and the Acanthians from the spoils of the Athenians' (Βρασίδας καὶ Ἀκάνθιοι ἀπ' Ἀθηναίων). Plutarch, however, explains that the statue represented Lysander, recalling the traditional custom established by Lycurgus according to which the Spartan soldiers had to remain unshaven so as to make the handsome (καλοί) look more attractive (εὐπρεπέστεροι) and the ugly (αἰσχροί) more fearful (φοβερώτεροι) (*Lys.* 1.1–3).[9]

Modern scholars have convincingly associated these adjectives with moral categories as much as with aesthetic ones, emphasising how Lysander was portrayed as a figure very difficult to interpret: was he good or bad?[10] Indeed, the style of the statue symbolically expresses the idea that Lysander—as atypical a Spartan as Brasidas—

8 Manfredini and Piccirilli (1980) 284 argue that leaving the city after changing its laws, so that the citizens can become accustomed to them, was a common motif for ancient legislators. They also suggest that in this respect the tradition about Solon probably influenced Plutarch's account of Lycurgus. Despite the similarity concerning the separation from their countries, the differences between Lycurgus and Solon are remarkable too. For after completing his reforms in Athens, Solon started travelling, but later returned to Athens (cf. Plu. *Sol.* 29.1–2), while Lycurgus' separation from Spartan was definitive.

9 Cf. Plu. *Lyc.* 22.2 and *Apophth. Lac.* 228 E; Xen. *Lac.* 11.3.

10 Stadter (1992) 42. Duff (1999) 165 suggests that Lysander conformed to Brasidas' pattern, while he rejects the idea that Brasidas represented the Spartan tradition, since he was an 'atypical' Spartan, *contra* Mossman (1991) 111. On Plutarch's use of statues to describe not only the physical traits of the protagonists but also their moral qualities, see Wardman (1967), (1974) 140–152; Mossman (1991); Tatum (1996) especially 135–139; Duff (1999) 163–165.

challenged the Lycurgan tradition, a topic that Plutarch examines throughout the *Life*. The fact that Delphi was the stage where Lysander's ambivalence, crystallised into a statue, was put on display and still generated uncertainty centuries after his death is very significant. Just as in the *Life of Lycurgus*, here too Delphi and Sparta are closely interconnected and their ties prove again to have been very strong.

Similarly, different temporal dimensions—the time of Lycurgus, the time of Lysander, the time of Plutarch/the external narrator, and, indirectly, the undetermined time of the readers of the *Life*—merge into one another. For, on the one hand, the statue links together various historical periods because of Lysander's emulation of (and, as Plutarch seems to suggest, difference from) the Lycurgan tradition. On the other hand, the *Life* encourages the readers to assess and evaluate different phases of Spartan history, that of Lycurgus as much as those of Brasidas and Lysander. Through the pair *Lysander–Sulla*, moreover, Plutarch invites the readers to compare and contrast these periods of Spartan history with the time of Sulla and their present time, in some respects still showing the scars of Sulla's campaigns in Greece. Finally, while in the case of the *Parallel Lives*' immediate audience the time of the readers partly coincides with Plutarch's time, it can also differ if we consider the future readership that Plutarch surely envisaged.

Delphi, therefore, constitutes the ideal centre where different past and present times are made to converge. It also represents the place where the future of Sparta is anticipated. In particular, the first chapter of the *Life of Lysander* makes the readers foresee the development of the narration and Lysander's involvement with Delphi when he was navarch of Sparta (*Lys.* 18.1–3):

> ὁ δὲ Λύσανδρος ἔστησεν ἀπὸ τῶν λαφύρων ἐν Δελφοῖς αὑτοῦ χαλκῆν εἰκόνα καὶ τῶν ναυάρχων ἑκάστου, καὶ χρυσοῦς ἀστέρας τῶν Διοσκούρων, οἳ πρὸ τῶν Λευκτρικῶν ἠφανίσθησαν. ἐν δὲ τῷ Βρασίδου καὶ Ἀκανθίων θησαυρῷ τριήρης ἔκειτο διὰ χρυσοῦ πεποιημένη καὶ ἐλέφαντος δυεῖν πηχῶν, ἣν Κῦρος αὐτῷ νικητήριον ἔπεμψεν. Ἀναξανδρίδης δ' ὁ Δελφὸς ἱστορεῖ καὶ παρακαταθήκην ἐνταῦθα Λυσάνδρου κεῖσθαι τάλαντον ἀργυρίου καὶ μνᾶς πεντήκοντα δύο καὶ πρὸς τούτοις ἕνδεκα στατῆρας, οὐχ ὁμολογούμενα γράφων τοῖς περὶ τῆς πενίας τοῦ ἀνδρὸς ὁμολογουμένοις.
>
> At Delphi, from the spoils of war Lysander set up a bronze statue of himself and each one of the navarchs, and golden stars of the Dioscuri, which disappeared before the events at Leuctra. A trireme two cubits long, made of gold and ivory, which Cyrus sent him as a prize of victory, was placed in the treasury of Brasidas and the Acanthians. The Delphian Anaxandrides also records that a deposit of Lysander lies there, consisting of a talent of silver, fifty-two minas, and, in addition to this, eleven staters too, but he writes something that does not correspond with the information commonly accepted about Lysander's poverty.

In this passage too, then, the presence of a Spartan leader at Delphi is connected with a great change for Sparta (the establishment of Sparta's hegemony over the other Greek states), which is vividly put on display just as at the beginning of the *Life*. Indeed, the complex monumental sculptures described by Plutarch celebrated the Spartan victory at Aegospotami, where the Dioscuri were thought to have assisted the Spartans, as suggested by the vision of two stars shining above the helm of

Lysander's ship (*Lys.* 12.1). At the same time, however, Plutarch projects the readers towards the end of Sparta's imperialism by mentioning the disappearance of the golden stars before the battle of Leuctra (another case of external/extradiegetic *prolepsis*). Lastly, the connection between different historical episodes and time layers also involves the anticipation of Lysander's conspiracy to subvert the Spartan constitution and abolish the dyarchy, traditionally correlated with the Dioscuri (an internal *prolepsis*).[11] For the expensive and luxurious materials employed to create the sculptures and Lysander's very large deposit contrasted markedly with Lycurgus' rejection of any form of luxury. This may be considered an act of arrogance and a serious alteration of the relationship between the Pythian god and the Spartan leadership. Such a distortion of a key aspect of the Lycurgan tradition became evident both when Lysander favoured Agesilaus' accession to the throne of Sparta by interpreting an obscure oracle in his favour (*Lys.* 22.11–12; cf. *Ages.* 3.6–8; *De Pyth. or.* 399 B-C; Xen. *Hell.* 3.3.4) and when, later in his life, he started to fabricate false oracular responses and tried to bribe the Pythia in order to convince the Spartans that the kingship of Sparta should not remain a prerogative of the two Spartan royal families only, but should return to all of the Heracleides or the Spartiates (*Lys.* 24.3–26.6). Indeed, one would hardly imagine a more sacrilegious machination for a Spartan.[12]

Conclusion

In both the *Life of Lycurgus* and the *Life of Lysander*, Delphi and the relationship between the oracle and Sparta play an important role, which is not limited to the time when the encounters between the protagonists and the Pythian god occurred. Rather, as we have shown, Delphi constitutes a symbolic place with which the most important phases of Spartan history (at least, those that Plutarch discusses in his biographies) are associated: the birth of the Spartan constitution and the crisis of Sparta in the fourth century BCE. The centrality of Delphi in Spartan history and the conflation of various temporal dimensions are closely connected with a specific idea of movement (a separation from Sparta) in the *Life of Lycurgus* and with material objects in the *Life of Lysander*. Indeed, in the interrelation between Delphi and Sparta place and time cannot be considered separately.

---

11 On the Spartan dyarchy, see Thomas (1974), Miller (1998), Link (2004), Dimauro (2008), Millender (2009).

12 On the manipulation and abuse of the oracles, see Stadter (2005) 203–205.

Katerina Oikonomopoulou

Space, Delphi and the construction of the Greek past in Plutarch's *Greek Questions*

Abstract: This chapter explores the concept of relative/relational space in Plutarch's collection of *Greek Questions*. It argues that space, rather than geography, provides a key vantage point from which we can interpret the collection's main themes, and understand the ways in which it attempts to link the past with the present. The chapter identifies key spatial experiences that are described within the *Greek Questions*' aetiologies (such as land journey, sea voyage, colonisation, migration), and discusses their role in problematising specific aspects of social or political life in the early Greek past (which, within the work, spans mythical times, pre-*polis* and early *polis* history). It concludes that the *Greek Questions* aligns itself with attitudes to the Greek past found elsewhere in Plutarch's writings (especially in the Pythian dialogues), and which are concerned especially with the negative role civil strife, interstate conflict and political fragmentation played in Greek affairs. The chapter concludes with a discussion of how the special role the enquiries assign to the Oracle of Delphi confirms this picture, given that Delphi seems to function as a sort of hub, but not as a powerful panhellenic centre.

The *Greek Questions:* space versus geography

Plutarch's collection of *Greek Questions* comprises a total of 59 aetiological enquiries on the traditions of various Greek communities. The format of these enquiries is standard: almost all are introduced with the interrogative pronouns τίς/τί..., and as a rule they focus on the meaning of specific terms or titles.[1] The formulation is such (the main verb ἐστί is usually implied) that it allows us to infer that the terms or titles in question were in use in various Greek cities during Plutarch's time. For example, the question 'Who is the πωλήτης ('Seller') among the Epidamnians?' (*Greek Questions* 29) implies that the title 'Seller' was used of a person in the city of Epidamnus during the High Roman Empire.[2] The answer then explains the meaning of such terms or titles by tracing their origins back to local versions of mythical stories, or to events (such as migration and colonisation) that took place in the early history of the Greek city-states. It is presumed that Plutarch, in compiling the collection, widely drew on the Peripatetic tradition of cities' *Constitutions*, although it is also clear that, at least in part, he relied on personal research into local customs and tra-

1 Payen (1998) 40–49; Preston (2001) 95–97.
2 Halliday (1928) 138–139; Payen (1998) 44; Boulogne (2002) 411.

https://doi.org/10.1515/9783110539479-010

ditions.[3] At any rate, the lexical focus, and the role of aetiology as a means of connecting the past with the present ('present' understood as the imperial Graeco-Roman world in which Plutarch lived) are the most distinctive markers of the collection as a whole.[4]

Another key feature, whose importance for the interpretation of the *Greek Questions* was first highlighted by Pascal Payen, is geography: indeed, Plutarch's sample of Greek cities seems to have been deliberately selected so as to draw attention to remote or provincial places, and at the same time underline their geographical diffusion, across different regions of the Greek peninsula, the Greek islands, the Asia Minor coast, Southern Italy, and Cyprus.[5] For Payen, this diffusion is to be contrasted with the focus on the imperial city of Rome in the *Roman Questions* (a collection which, as he rightly argues, is closely related to the *Greek Questions*),[6] and, among other things, puts the stress on the durability of centuries-old local Greek traditions (surviving mostly outside the centres of Hellenism), despite the eventual predominance of imperial power in the wider Mediterranean region.

In this chapter I propose that we pay close attention to the *Greek Questions*' portrayals of space, rather than geography. This is because geographical landmarks (rivers, roads, the coast), despite the fact that they frequently feature in the collection's explanations, are not part of objective descriptions of space. Further, Plutarch does not exploit the geographical theme in order to achieve some sort of cohesive presentation for the material that he has collected: he could have grouped the aetiologies together by region, for example, or employed the theme of the land journey or sea voyage, along the course of which the different Greek cities or communities could have featured as stops.[7] Instead, the haphazard order in which the different Greek cities or communities are discussed encourages a perception of them as independent microcosms, each one with its own linguistic idioms, local culture, history, and religious life.[8] In this context, geographical landmarks are mentioned because they are important to peoples' or communities' experiences of the places in which they live and act—experiences which encompass past as well as present events. The explanations' spatial references thus serve to chart a rich diversity of economic, socio-political and religious activities that shaped individual Greek communities' cultural identity across time.

Contemporary cultural geographers utilise the notions of 'relative/relational space' in order to conceptualise this fluid relationship of people to spaces. The concept of relative space refers to one's subjective experience of space: depending on the type of activities he or she engages in, a human agent might perceive space (for example, distance) in different terms. The concept of relational space, in turn, refers to

3 Giesen (1901); Halliday (1928) 13–15.
4 On the *Greek Questions*' format, see Preston (2001) 95–97.
5 Payen (1998) 49–54; Boulogne (2002) 183–185.
6 Payen (1998) 39–40; see also Harrison (2000) and Boulogne (2002) 183–185.
7 Cf. Hutton (2005a) 54–126, on Pausanias.
8 Payen (1998) 49–54.

the ways in which relations (what is meant by that is the full range of socio-economic and cultural activities of human agents) help shape the meaning and significance spaces carry on the collective level. Relative/relational space is thus contrasted to absolute space, which is the space that the geographer can measure in objective terms, calculating distances, or identifying and describing important landmarks.[9] According to this model, an understanding of how, within the *Greek Questions*, space is perceived and lived can shed helpful light on the terms in which the collection conceptualises key aspects of the early Greek past (in terms of politics, society, economy, or religion), as well as clarify what kind of historical or cultural assessment of the early Greek world it might have invited its imperial readers to make, considering its differences from their own imperial realities.

Spatial experience within the *Greek Questions*

In the Table below I have attempted to make a distinction between different ways in which space appears to be experienced by the Greek communities which feature in the different aetiologies. It should be noted that the different types of spatial experience that I distinguish are by no means watertight, but in fact overlap: for example, as we will see, expulsions or migrations usually involve some kind of land journey or sea voyage. Precisely because of this, however, they help us conceptualise the multiple, intersecting ways in which space operates as a key field of human activity within the collection.

Types of spatial experience	Chapters within the *Greek Questions*
Overland journey	12, 15, 19, 22, 37, 39, 59
Sea voyage	11, 30, 31, 32, 35, 41, 44, 54, 56, 57, 58
Migration	13, 26, 28, 51, 55
Exile or expulsion	5, 12, 14, 21, 26, 28, 33, 37, 38, 39, 46, 56
Colonisation	11, 14, 15, 21, 30, 35
Interstate war and military expeditions	5, 11, 17, 20, 30, 34, 39, 41, 44, 46, 49, 57, 58
Private and public space	2, 8
Pre-*polis* formations (settlements κατὰ κώμας)	17, 37
Religious spaces (shrines, burial grounds)	12, 22, 23, 27, 37, 39, 40, 44, 47, 48
Trade, piracy, plunder, farming	6, 21, 24, 29, 34, 54, 55

Let us consider some characteristic case-studies which will help illuminate some of the main issues involved. My first example is *Greek Question* 2 (291E–F, transl. F.C. Babbitt, Loeb):

9 See Warf (2010); see also Thalmann (2011) 14–41, on relational space in Apollonius of Rhodes. On other methodological approaches to space in ancient Greek literature, see also Purves (2010), esp. 1–23, Gilhuly and Worman (2014). On narratological approaches to space in Plutarch's *Lives*, see Beck (2012) and in this volume. See also the Introduction to this volume.

> Who is the 'woman who rides on a donkey at Cumae'? (Τίς ἡ παρὰ Κυμαίοις ὀνοβάτις;) Any woman taken in adultery they used to bring into the *agora* and set her on a certain stone in plain sight of everyone. In like manner they then proceeded to mount her upon a donkey, and when she had been led about the circuit of the entire city, she was required again to take her stand upon the same stone, and for the rest of her life to continue in disgrace, bearing the name 'donkey-rider'. After this ceremony they believed that the stone was unclean and they used ritually to purify it.

Punishments for adultery committed by women (usually death) are something of a *topos* in imperial Graeco-Roman miscellanistic compilations.[10] The particular punishment that is described here reveals that, in the close-knit communities that comprised the world of the Greek city-state, private conduct was an object of intense public scrutiny. This was especially true of places like Aeolian Cumae, which was regarded in antiquity as a backward society.[11] (The use of the imperfect throughout the answer allows us to infer that the custom that is described was instituted and held in that city at some point in the past, without any clearer indication of the precise historical time). The public nature of the offence of female adultery is underscored by the fact that its perpetrators are made to parade in shame through key civic spaces: the wider geographical precinct of the city, and, more particularly, its agora, and a chosen stone in the agora. Within the narrative context of the enquiry, these locations no longer function as venues for conducting the standard public business (such as commercial transactions, legal and political proceedings), but are re-signified as instruments of public humiliation, in the context of a cruel ritual of social exclusion. The transgressive nature of female adultery is further underlined by the fact that the places in question lay outside a woman's normal sphere of activity (limited within the confines of the *oikos*). All in all, the enquiry's focus on the use of civic space is an effective means of problematising gender relations and the interaction between individual and society in the Greek *polis.*[12]

Greek Questions 13 and 26, next, illustrate a major theme that runs through the enquiries' depictions of space, namely, the way the early Greek communities experienced it through conflict and its outcomes (migration and colonisation). Both enquiries document the continuous migrations of the Aenianians, a Greek tribe (*ethnos*) that was expelled from its area of original habitation by another *ethnos*, the Lapiths.[13] As we learn, after many adventures the Aenianians only acquired a permanent home in the region around the river Inachus after themselves expelling that area's previous inhabitants, the Inachians and Achaeans. Both explanations allow readers to conceptualise the route the nation followed in its wanderings, by making note of key geographical regions and landmarks (the Dotian plain, Aethicia, Molossia, Cassiopaea, the river Auas, Cirrha). But even for readers who are not familiar

10 See, e.g., Plu. *Quaest. Rom.* 6; Gell. *Noct. Att.* 10.23; Ael. *VH* 2.37–38.
11 Halliday (1928) 42.
12 Cf. *Quaest. Graec.* 38–40, 49.
13 Cf. *Quaest. Graec.* 15 (Locrians).

with the regions to which all these names correspond, the cumulative effect of their being mentioned is to draw attention to the Aenianians' constant shifts of location. Moreover, both enquiries' fusion of myth and history, their focus on land conquest, and their mention of kings strongly suggest the stories they narrate are located at a time prior to the establishment of the *polis* as a key unit of political organisation. In this way, they convey a strong impression of the instability and mobility that characterised the life of communities in the early Greek world.

Other explanations shed light on the broader political and socio-economic conditions that underpinned the life and affairs of Greek communities during such unstable or transitional phases: we thus learn that migrations and expulsions were in fact the agents that propelled the Greeks' expansion northwards, to Thrace, to Italy and the West, and to the islands and Asia Minor.[14] In this connection, the enquiries also document different types of economic activity, ranging from piracy and pillaging to trade and farming—the latter developing in stable and organised environments, when communities sought to establish lasting ties with their neighbours.[15] *Greek Question* 29 suggests the challenges involved in the latter instance were considerable (297F–298A, transl. F.C. Babbitt, Loeb):

> Who is the 'Seller' (πωλήτης) among the Epidamnians? As the Epidamnians were neighbours to the Illyrians, they realised that the citizens who had dealings with them (ἐπιμιγνυμένους) were becoming corrupted (γιγνομένους πονηρούς). Fearing a revolution (νεωτερισμὸν), every year they selected one of their fellow-citizens of proven worth, so that he could be in charge of such transactions and exchanges. This person visited the barbarians and provided a market and an opportunity for sale for all citizens, and was for this reason called the 'Seller'.

The enquiry discusses a remote Greek colony, Epidamnus in Illyria, whose closest neighbours and trading partners were barbarians. Contact with them, the answer informs us, increasingly corrupted the citizen body: the term πονηρός probably refers to democratically-inclined citizens: Thucydides (1.24) confirms that in Epidamnus there were political disputes between the oligarchs and the *demos*, which eventually erupted into a full-blown civil war: just prior to the Peloponnesian war, the *demos* expelled the aristocrats, who then formed an alliance with the barbarians, and attacked the city.[16] The story is thus a reminder of the kind of factionalism and political feuding that was chronically plaguing Greek city-states during the archaic and classical periods, often resulting in civil strife. The Epidamnians' solution, according to the explanation, was to appoint one select citizen every year as head of all economic transactions with the Illyrians (who was for this reason named 'Seller'), so that contact between locals and their barbarian neighbours could be minimised or at any rate controlled. A key point of the aetiological story, then, is to place the readers before the difficulties facing Greek colonists

14 See Table above, and Payen (1998) 49–54, 61–64.
15 See Table above.
16 See Payen (1998) 44; Halliday (1928) 138–139.

whose cities, geographically distant from the Greek motherland, lay close to a barbarian hinterland: on the one hand they had to maintain economic ties with barbarians, as they depended on them for their survival; but, on the other, they faced the pressure to preserve and protect local traditions or institutions against external influences that posed a threat. Other enquiries yield additional insights into the question of what territorial proximity to other communities, Greek and non-Greek, meant for Greek city-states: in all cases it is depicted as a factor that often generated interstate conflict, leading to disputes or invasions.[17]

To sum up so far, the experiences of space encoded within the *Greek Questions* point to an underlying perception of the early Greek world as a system of loosely tied *polis*-type structures or tribal regions. It is characteristic of the collection as a whole that it does not ascribe a prominent role to Athens, Sparta or the coastal Greek cities of Asia Minor, which, as we know, were centres of economic and political power in the classical and post-classical period.[18] Instead, what the explanations foreground is the role of regional and inter-city networks,[19] usually formed as a result of geographical proximity, maintained on grounds of old mythical ties, or cultivated through political alliances, marriage partnerships, and joint colonising missions. Thus, the explanations make mention of treaties between Peloponnesian communities (Sparta and Tegea, *Quaest. Graec.* 5), trade or travel between the Ionian islands and the west coast of the Balkan peninsula or Italy (*Quaest. Graec.* 14, 29),[20] and close contacts between the east Aegean islands and the Asia Minor coast (*Quaest. Graec.* 20, 55, 56, 58), or between places such as Boeotia, Megara and Corinth (*Quaest. Graec.* 16, 17). Despite the fact that the existence of networks somewhat mitigates the impression that the early Greek world comprised geographically isolated communities (since it brings the aspect of their interconnectivity to the fore), it does not override the sense that these communities were inherently unstable formations, subject to recurrent outbreaks of civil or interstate warfare.[21]

Delphi as a religious centre

Even though the communities that feature within the *Greek Questions'* aetiologies appear to lack a political centre, they have a clear religious centre, Delphi. It is notable

17 See, e.g., *Quaest. Graec.* 11, 16, 17, 49.

18 Note the relative scarcity of enquiries devoted to Athens, Sparta and Asia Minor cities. See *Quaest. Graec.* 5, 20, 21, 35, 39, 45, 48, 56.

19 On networks in the ancient Mediterranean, see Malkin (2011). On island networks in specific, see Constantakopoulou (2007).

20 But note Hartman's correction at *Quaest. Graec.* 14, 294D: Αἰτωλίαν for Ἰταλίαν, which would link the Ionian islands with the western coast of the Greek mainland, as in *Quaest. Graec.* 29. See Halliday (1928) 79–80.

21 This perception of the early Greek world goes back to Thucydides' *Archaeology* (*Hist.* 1.2–19).

that the *Greek Questions* shows a special interest in Delphic religious customs, as well as in the cultural tradition of the wider region around Delphi, Locris—an interest that probably reflects Plutarch's personal ties to the Oracle, which he had served as a priest.[22] Delphi's religious significance is highlighted by two especially lengthy enquiries, 9 and 12, where aspects of the Delphic oracular and festival calendar are elucidated.[23] In both explanations, the alternation of present and past tenses is characteristic: Plutarch uses the former in a way that suggests the names of Delphian religious officials (the ὅσιος), festivals and calendar months (the Χάριλλα, the Βύσιος, respectively) that are discussed were still in use in the imperial period. He even makes explicit references to his own time twice, through the use of the temporal expressions ὀψέ ('late', 'recently') and ἔτι καὶ νῦν ('now still', 'even now');[24] the past tense on the other hand is used in order to trace the origin of the Oracle's religious activities and customs back to key mythological or legendary stories (such as the myth of Apollo and the serpent Python, itself a story of migration, located in the mythical past), as well as in the context of describing alterations in religious practice, in contrast with how things used to stand. Through these discursive features, the Oracle's diachronic significance in Greek affairs is underscored, subject to an underlying pattern of continuity and change.

Further, various aetiological stories within the collection depict the Oracle as a key player in the affairs of diverse Greek communities, however remote or provincial. *Greek Question* 35, for example, seeks to investigate the origins of an old song-line, chanted by the maidens of Bottiaea (a region in the north of Greece) at festivals, namely, 'let us go to Athens' (ἴωμεν εἰς Ἀθήνας, 298F).[25] According to the explanation, the song-line recalled Bottiaea's colonisation history. Some Cretan men who had been sent to Delphi as a sacred offering (ἀπαρχήν, 298F) set out from the Oracle in order to found a colony (εἰς ἀποικίαν ὁρμῆσαι, 299A). First they settled in Iapygia, in southern Italy, and then they ended up establishing a colony in the northern region of Bottiaea.[26] As it turns out, the colony also comprised some Athenians, survivors from the city's yearly offering of young men to the mythical Cretan king Minos, who had also travelled to Delphi together with the Cretan youths. Sea voyage, colonisation and migration form central themes in the enquiry, all set against an expansive geographical backdrop. What is most notable about the explanation however is that it puts the panhellenic significance of the Oracle of Delphi into relief: Delphi functions in the story as a key

---

22 Note that the Oracle features in other aetiological stories within the *Quaest. Graec.*, such as 15, 35, 48; cf. Boeotia, Plutarch's home region, which also surfaces quite regularly in the aetiologies: see *Quaest. Graec.* 16, 19, 38, 39, 41, 43. On Plutarch's ties to Delphi, see Jones (1971) 10, 26, 28; Lamberton (2001) 52–59; Swain (1991); Stadter (2015) 70–97.

23 See Halliday (1928), ad loc.

24 On the use of such phrases in question-and-answer texts, see Oikonomopoulou (2013) 46–59. See also Alcalde-Martín, Frazier and Roskam in this volume.

25 Cf. Plu. *Thes.* 16.2–3, Arist. fr. 490.1–2 Gigon.

26 See Halliday (1928), ad loc.

point of reference for Greek communities as geographically remote from the mainland as Crete. Its role as a sort of hub that guarantees the cohesion of an otherwise fragmented Greek world is apparent in several other enquiries as well, where it is consulted about local affairs, as well as prior to the foundation of colonies.[27]

Additional testimony to the Oracle's religious authority is furnished by enquiries such as *Greek Question* 59, in which it appears to undertake an actively interventionist role in interstate disputes. The enquiry sets out to explain the meaning of the term 'Wagon-rollers' (ἁμαξοκυλισταί, 304E) in Megara. The explanation typically blends in geographical references and the theme of the overland journey, in order to problematise the violent results of political instability in the Greek *polis*. Most probably relying on anti-democratic sources, the explanation links the theme of violence to democracy, as it locates its account in 'the time of the unbridled democracy (ἐπὶ τῆς ἀκολάστου δημοκρατίας) which brought about both the return-interest (παλιντοκία) and the temple sacrilege' (304E). The statement is an internal cross-reference to *Greek Question* 18, whose explanation narrates the complete collapse of social stability, after the Megarians deposed their tyrant Theagenes and instituted 'freedom without measure' (ἄκρατον ... ἐλευθερίαν, 295D). This loss of control on the part of the Megarian populace, as *Greek Question* 59 explains, culminated in the crime that drunken Megarians committed against some Peloponnesian pilgrims (θεωρία, 304E, ἱερᾶς τῆς θεωρίας οὔσης, 304F) who were travelling through their land on their way to Delphi: the Megarians pushed the religious delegates' wagons into a lake, causing some of them to drown. As the Megarians themselves took no action over this crime due to the politically unstable situation of their city, the Amphictyonic assembly intervened in order to guarantee the attribution of justice for the sacred Peloponnesian mission, determining that the perpetrators be punished with penalties as extreme as banishment and death.[28]

Despite the fact that Delphi is the sole institution that furnishes some sort of centre to an otherwise polycentric Greek world, it is far from fulfilling the role of a power structure capable of uniting the Greek communities, or quelling conflicts between them. Such power structures only surfaced after the conquests of Alexander, and the later rise of Rome.[29] This perception is in line with the manner in which the Oracle's role in Greek affairs is portrayed in texts such as Plutarch's *Dinner of the Seven Wise Men*, also set in the archaic past. There too, the Oracle is mentioned in terms of its function as a religious reference-point (as evinced by its buildings and dedications), as well as in terms of its involvement in colonising expeditions.[30] Yet the world in which it exercises its influence is one that runs the risk of being destabilised

---

27 See Table above.

28 On the Amphictyonic League, see Davies (1988); Richardson (1992) 224, 231.

29 See Payen (1998) 49–64.

30 See *Sept. sap. conv.* 150A, 163B, 164A.

by strife.[31] Further, both the *Greek Questions*' and the *Dinner of the Seven Wise Men*'s portrayal of the Oracle's role in Greek affairs of the pre-classical past can be correlated with Plutarch's dialogue *On the Oracles of the Pythia*. There, the character Theon rebuffs his interlocutor Serapion, who had earlier suggested that votive dedications to the sanctuary by courtesans were morally reprehensible, in the following terms:

> … καὶ σύ μοι δοκεῖς ὁμοίως γύναιον εἴργειν τοῦ ἱεροῦ χρησάμενον ὥρᾳ σώματος οὐκ ἐλευθερίως, φόνων δὲ καὶ πολέμων καὶ λεηλασιῶν ἀπαρχαῖς καὶ δεκάταις κύκλῳ περιεχόμενον τὸν θεὸν ὁρῶν καὶ τὸν νεὼν σκύλων Ἑλληνικῶν ἀνάπλεων καὶ λαφύρων οὐ δυσχεραίνεις οὐδ' οἰκτίρεις τοὺς Ἕλληνας ἐπὶ τῶν καλῶν ἀναθημάτων αἰσχίστας ἀναγιγνώσκων ἐπιγραφάς 'Βρασίδας καὶ Ἀκάνθιοι ἀπ' Ἀθηναίων' καὶ 'Ἀθηναῖοι ἀπὸ Κορινθίων' καὶ 'Φωκεῖς ἀπὸ Θεσσαλῶν', 'Ὀρνεᾶται δ' ἀπὸ Σικυωνίων', 'Ἀμφικτύονες δ' ἀπὸ Φωκέων'.
>
> … so you also seem to me, in a similar way, to be excluding from this shrine a poor weak woman who put the beauty of her person to a base use, but when you see the god completely surrounded by choice offerings and tithes from murders, wars, and plunderings, and his temple crowded with spoils and booty from the Greeks, you show no indignation, nor do you feel pity for the Greeks when upon the beautiful votive offerings you read the most disgraceful inscriptions: 'Brasidas and the Acanthians from the Athenians', and 'The Athenians from the Corinthians', and 'The Phocians from the Thessalians', and 'The Orneatans from the Sicyonians', and 'The Amphictyons from the Phocians'. (401C–D, transl. F.C. Babbitt, Loeb)

Coming from the standpoint of an imperial Graeco-Roman speaker, Theon's remark not only confirms that certain (violent) aspects of the Greek historical past were reflected on in negative terms during the imperial period, it also captures a fundamental unease with Delphi's role as a panhellenic reference-point, given that, for a considerable segment of its history, it amassed wealth and prestige in dedications which were tainted with Greek blood.[32] Such strong explicit sentiments are absent from the *Greek Questions*' aetiologies that mention Delphi. Yet there is no doubt, given the enquiries' overall focus on the migratory and politically unstable nature of early Greek communities, that the aetiologies in question would invite imperial readers at the very least to reflect on the Oracle's efficacy and shortcomings as a centre of sorts at times in history different from their own. In this way, Delphi would play a key role in enabling imperial readers to relate the world of the *Greek Questions* to their own political realities, marked by the presence of a strong political centre, imperial Rome.

---

31 See esp. *Sept. sap. conv.* 149C–E, where Diocles interprets the birth of the hybrid animal (half-human and half-horse) as a portent of strife (στάσις) and discord (διαφορά).

32 On Plutarch's political views, especially on *stasis*, see *Praec. ger. reip.* 823F–825D, and Aalders (1982) 26–27, 51–53; Swain (1996) 173–183; Duff (1999) 89–91, 296–297; Zadorojnyi (2005a) 113–114; on *stasis* in Rome, see Ash (1997); Buszard (2005). See also Brenk and Pelling in this volume.

Conclusions

I conclude that space, in the sense of space that is lived and experienced by subjects (what cultural geographers call relative and relational space), is a key connecting thread between the disparate *aitia* that comprise the *Greek Questions*. As we saw, the experiences of space that are encoded in the different aetiologies vary, and include migrations, colonisations, trade routes, diplomatic missions, or military expeditions across a vast geographical backdrop, namely, the greater Mediterranean world of Greek myth or early Greek history. They also offer accounts of the social processes, singular or collective actions, and political or economic decisions that shaped relations between different communities located in the Greek mainland, or beyond. Last but not least, they convey a strong sense of the inherent instability and lack of cohesion that marked the early Greek world. The technique of the *aition*, applied with a view to explaining the origins of customs, usages or practices, firmly locates these experiences in the mythical past, at a time either prior to the formation of Greek city-states, or during the early phases of *polis*-history. As a result, space functions within the *Greek Questions* as a powerful tool for re-constructing the early Greek past, and understanding its distinctive dynamic.

There is no doubt that this reconstruction betrays the influence of key cultural trends in Plutarch's time: we know that Greek intellectuals of the Second Sophistic sought to locate cultural authenticity in obscure, or local (understood in the sense: 'non-cosmopolitan', or 'unadulterated') versions of Greek myths or customs.[33] However, the *Greek Questions*' exclusive focus on the mythical and pre-classical past diverges from the overall fascination with classical (5th-century) antiquity that runs through Second Sophistic literature. To an extent, this divergence may come down to limitations in Plutarch's source-material, but it may also bespeak a self-conscious attempt to look beyond rose-tinted accounts of the Greek past prior to Rome (for example, the Persian wars; cf. *Praec. ger. reip.* 814B-C), and into much earlier phases of Greek history that often projected a darker, so to speak, image. The mythical and archaic past of Greek communities in particular, characterised by fragmentation and instability, forms a choice that stands in sharp contrast both to the glorious 5th century, marked by the emergence of a coherent 'Greek' cultural identity, and to the Roman imperial present. Plutarch's rather idiosyncratic historical focus[34] in the *Greek Questions* does not make him a historical revisionist, let alone an apologist for the Roman Empire. But it certainly suggests that his view of the 'Greek past' was not singular or single-minded, and that he ascribed value to exploring alternative models of Greek political and cultural life before Rome.

---

33 See Whitmarsh (2001a) 104–108; and the articles in Whitmarsh (ed.) (2010).

34 See also Kim in this volume, discussing Plutarch's notion of the 'archaic' in more detail.

4 Models of the past I: configurations of memory and history for Plutarch's imperial readers

Joseph Geiger

Greeks and the Roman past in the Second Sophistic: The case of Plutarch[1]

Abstract: Since Ewen Bowie's masterly study, Greeks and their past have been explored again and again, yet one feels that Greek attitudes to the Romans have more often than not received serious attention only as far as their relationship to their contemporary Romans was concerned, and their stance towards Roman history has been neglected. Plutarch has not entirely escaped this approach. Yet even after composing the *Lives of the Caesars* Plutarch all but ignored imperial history: compared to the multitude of examples from the history of the Republic he quotes hardly a handful from the Empire. Similarly, in his discussion of the various monuments of the city of Rome he ignores the transformation of the city by Augustus and the later monuments and discusses almost exclusively republican ones, and most notably he chooses for his *Parallel Lives* solely republican personages. Especially this last point is of cardinal importance for understanding Plutarch's view of Roman history, but it has been all but neglected in scholarship. It appears that Plutarch studiously avoided contemporary references or historical allusions that may have been politically relevant. Consequently it is suggested that Plutarch's cautious approach to contemporary politics may have influenced his avoidance of potentially dangerous subjects.

Ever since Ewen Bowie's trailblazing study, the topic of Greeks and their past in the Second Sophistic has been explored again and again. Yet one feels that Greek attitudes to the Romans under the Empire have more often than not received serious attention only on the contemporary plane, exploring the Greek position towards contemporary Roman rule. Plutarch has not entirely escaped this approach. Of course his situation as an upper-class Greek and a Roman citizen of the Empire is of great interest, and indeed much effort has been devoted to determining his exact standpoint vis-à-vis Roman rule. Yet considering his long and serious engagement with Roman history and antiquities a more general appraisal of his attitude to the Roman past is in order, especially since the studies devoted to it dealt mostly with specific problems. In particular, his periodisation of Roman history and his choices, paralleling, as we shall see, to some extent those of the Second Sophistic towards Greek history, are ripe for reconsideration. It has been briefly argued that he was privileging, in the *Parallel Lives*, republican heroes over

1 Thanks are due to Aristoula Georgiadou and Katerina Oikonomopoulou for the exemplary organisation and running of the conference at the wonderful site of Delphi, and to the anonymous reviewer for his many helpful and perceptive suggestions.

https://doi.org/10.1515/9783110539479-011

imperial personages[2] and similarly he was favouring republican monuments in Rome over more recent ones. Let me repeat: while imperial Greek authors', including Plutarch's, dealing only with a specific, 'classical', period of Greek history has been the subject of study and controversy, his choice of only republican heroes for his *Parallel Lives* has been taken as self-evident. In the present study I intend to explore Plutarch's attitude to the Roman past with a view to emphasising the parallels between it and Greek attitudes to the Greek past.

Certainly the assertion that Plutarch 'In the *Parallel Lives* ... reconceived all Greek and Roman history and made it a continuous text on the possibilities and dangers of leadership'[3] ignores the limits Plutarch set himself. Nor can it be maintained that his 'thoughts on political life [were] in as timeless a manner as possible'.[4] Of course the discourse on memory, and on collective memory, is nowadays much in vogue, and I do not deny the usefulness of this approach. However, to maintain that Plutarch's personal and his town's collective memory of Roman history also centred around the events of the late Republic—the Mithridatic wars and especially the battle of Chaeronea, the war with Antony and its culmination at Actium, as argued in a recent book,[5] is of course begging the question—it is the battle of Chaeronea and the times of Antony and Actium that receive particular attention in Plutarch's works, a fact that only instructs us about Plutarch's choices in his writings, not about the collective memory of his townsmen. However, it is the question of how much imperial history Plutarch knew and whether for some reason there was much more of it than can be learned from his works that we must ask. Indeed, the following quotation from a prominent scholar will reveal the conflict between what we assume and what we know for certain: '... il *De fortuna Romanorum* rivela che Plutarco aveva una sorprendente conoscenza sia di Roma repubblicana che della storia imperiale. In questo lavoro Plutarco passa in rasegna le grandi figure del period regio e repubblicano, da Romolo ad Augusto'.[6] It is of course the contrast between the first sentence on republican Rome and imperial history and the specification in the next of the regnal and republican period from Romulus to Augustus to which I wish to draw attention.

To return to my general considerations, it has been observed[7] that the Second Sophistic privileged Greek history from the Persian wars up to the death of Alexander the Great with special emphasis on the history of Athens and Sparta. Plutarch acceded to that attitude, though at a late stage in the composition of the *Parallel Lives*, when expanding the series he added a number of Hellenistic biographies.[8] It is instructive to

2 Geiger (2002) and (2005). Leeck (2010) 64, insists that Plutarch chose his heroes from all important periods of Roman history, but contradicts himself at 69.
3 Stadter (2002a) 7.
4 Ibid. 17, Stadter reviewing his own and Pelling's contribution in the volume.
5 Leeck (2010) 52–56.
6 Stadter (2007) 193.
7 See, e.g., Hamilton (1969) xxii.
8 Geiger (1981).

compare with Plutarch his close contemporary Dio Chrysostom, who shared with him a common background and education. Dio Chrysostom's extant writings, amounting approximately to one fifth or so to Plutarch's, contain not one single reference to Plutarch's republican heroes.[9] By the way, the Plutarchan heroes Numa and Romulus do get mentions and so do the emperors Augustus and Tiberius, as well as of course Nero (with Octavia and Poppaea Sabina), Vespasian, Domitian, Nerva, and Trajan, though, interestingly, not Caligula and Claudius.[10] Just for a wider view of the similarities between the two authors, all of Plutarch's classical Greek heroes, bar Dio of Syracuse, are referred to by Dio Chrysostom, while of the later Hellenistic additions only Demetrius Poliorcetes appears in his writings. To quote an authority on Dio Chrysostom: 'Admittedly, none of the surviving orations commemorate specific events in the ancient history of Greece, but they show a rich sprinkling of references to Pericles, Epaminondas, Plataea, Leuctra, and Thermopylae—in short, the heritage of memories that Greeks cultivated with growing interest under the Empire—and which marked them out distinctly from the Gauls and Britons, subjects of the same monarch'.[11]

The similarity between the attitudes of these two authors to the heroes of Greek history enhances the resemblance of their views on Roman history. A number of other important Greek imperial authors conform to the pattern: '... Aristides fails to mention Roman history and Roman cultural attainments ... Aristides' lack of interest in Rome as a city ... his total lack of interest in Roman history, including even the saintly Numa (who pops up in Dio of Prusa), is noteworthy'.[12] And again: 'Although [Lucian] must have known Latin, he nowhere mentions Roman cultural activities (and has very little interest in Roman history)'.[13]

Yet it is Plutarch's approach to the history of the Roman Empire, not necessarily the contemporary one, that is my chosen theme in this contribution. As we all remember, he authored, prior to the *Parallel Lives*, a series of imperial biographies from Augustus to Vitellius, of which the *Lives of Galba and Otho* are extant. Whatever induced him to undertake that project is not known nor can the two surviving *Lives* be even regarded as a great literary success, compared to the latter series. Nevertheless it is legitimate to compare the two series from certain points of view. The chief one that interests me here are the spin-offs of these biographies in the non-biographical works of Plutarch, commonly labelled the *Moralia*.

In the argument concerning the fact that Plutarch ignored likely subjects from the imperial period for his *Parallel Lives*,[14] Agrippa, Germanicus and Verginius Rufus have been put forward as possible examples for such biographies. Few would argue with

9 Mummius is mentioned for his desecration of Greek statues in *Or.* 37.42, in all probability by Favorinus.
10 Claudius is probably alluded to in *Or.* 41.6.
11 Salmeri (2000) 84.
12 Swain (1996) 275.
13 Swain (1996) 319, n. 75 cont. on 320: 'History: note the confusion of Scipio the Elder, who defeated Hannibal, with Scipio the Younger, who destroyed Carthage, at *Dialogues of the Dead* 25.7'.
14 Geiger (2005) 101.

the proposition that Marcus Agrippa, Augustus' right-hand man, friend, admiral, son-in-law and father of his prospective heirs, or Germanicus, a youthful hero with a tragic end, and indeed Verginius Rufus, a victorious general who insisted on legitimacy in the accession to the imperial throne and later withdrew to private life, would have made most appropriate subjects for Plutarchan *Lives*. Nor would the biographer, with his wide reading in Greek history, despair of finding suitable parallels. Yet it is not only that their biographies were never written, these personages have been studiously avoided: Agrippa and Verginius Rufus are never mentioned in the entire corpus of the *Moralia*—the latter gets of course his due in the *Lives of Galba and Otho*—while Germanicus is mentioned twice with the anecdotal reference that he hated the sight and the sound of the cock (*On Envy and Hate* 537A; frg. 215k). This is all. Now obviously Agrippa could not have been absent from the biography of Augustus, though he is not mentioned in the relatively copious fragments of that *Life*, nor is Germanicus' virtual absence from the *Moralia* proof of his non-appearance in the lost *Life of Tiberius*. From what Plutarch says of Verginius Rufus in the *Galba* and the *Otho* it appears that he should have regarded him as eminently suitable for the subject of a biography. Needless to say, these are typical examples that can easily be multiplied. We must get used to the idea that, despite the series of imperial biographies, Plutarch's interest in Roman history, at least as expressed in his works, was restricted to the era of the kings and the Republic, with the Augustan age as a period of transition. By the way, it seems to me quite instructive that the only examples from Roman history in the two surviving imperial *Lives* are all republican.[15]

A similar picture can be seen when observing Plutarch's concern with Roman monuments. Obviously any sightseer would be astounded by the multitude and variety of the sights in the capital, including our small-town visitor, despite being familiar with Athens and having visited Alexandria. What a scholarly and investigative Plutarch could learn in Rome we may realise from the exceptional case of his studying the temple of Iuppiter Capitolinus and its history in the *Life of Publicola* (13–15): here we get a complete history of the temple from before its consecration, then its destruction by fire, rebuilding by Sulla, destruction again under Vitellius, rebuilding by Vespasian, burning down again and rebuilding by Domitian, with the remark of having seen the columns of this last temple in Athens, and then again in Rome, where they have been made too slender. This is as learned and as inquisitive a tourist as one may wish, even with a dash of architectural criticism—one imagines that learning the history of the shrine, probably making use of his Roman acquaintances or the local guides,[16] followed on having seen the columns first at Athens, and then in Rome. However, in this instance we also learn that temples may have been, besides their obvious functions, also major touristic sites. Almost in the same breath we are told about Domitian's palace, basilica, bath, and concubines' quarters—a re-

15 *Galba* 1: Aemilius Paulus; 29: Scipio, Fabricius and Camillus; *Otho* 13: the Younger Cato and Metellus Scipio; 15: Hannibal, Pyrrhus and the Cimbri.
16 For guides in general see Jones (2001).

ference all the more significant because of the absence of such details elsewhere. In the same *Life* the author also shows acquaintance with the shrines and colonnades of the Tiber isle and the Vica Pota.

This, however, is an exceptional case. For the rest, Plutarch restricts himself to republican and royal monuments,[17] as in the two instances where he enumerates twice the various shrines to Fortuna dedicated by king Servius Tullius, in *Roman Questions* 74.281D–E and again in *The Fortune of the Romans* (322F–323A). Impressive as Plutarch's learning is, these shrines of Servius Tullius certainly did not exhaust Plutarch's acquaintance with temples of Fortuna. Thus, at *Mar.* 26.3 he mentions the Aedes Fortunae Huiusce Diei, dedicated after the battle of Vercelli by Q. Lutatius Catulus. Nowhere do we find descriptions of imperial monuments similar to those in the above mentioned passages: indeed, although such descriptions may have turned up in the lost *Life of Augustus*, where they may have been competing with the description of Pericles' building projects, there is no trace in Plutarch's extant writings of the city's transformation from brick into marble by that emperor. Though it has been conjectured that Plutarch may have realised the possibility of totally separating republican heroes from later imperial ones in the Forum of Augustus, where the marble statues erected by Augustus were clearly distinguished from the later additions in bronze,[18] this conjecture should be regarded with due caution since in our surviving evidence the Forum of Augustus with the Temple of Mars Ultor are ignored together with the rest of the Augustan monuments. Yet I do not believe that anybody will argue that it was possible for a visitor, even one less curious and well-informed than Plutarch, not to notice the Forum of Augustus, the Ara Pacis, the Mausoleum of Augustus, his Sundial or Agrippa's Pantheon, and the prominence of these Augustan building projects in the cityscape of Rome.[19] Of course, the interest in monuments was closely connected to the interest in history. Tourists being guided through Victor Hugo's Paris or Joyce's Dublin will hardly overlook the Eiffel Tower or the Spire of Dublin, though they may disregard them in their description of their tour.

Now it is not advisable to give free rein to historical or literary parallels. Regarding the Greeks' preoccupation with their past in the Second Sophistic, and more precisely with their classical past, the controversy whether this was an outlet for the frustrations of the present or just a frame of literary reference or a channel of communication has by now acquired the status of a classic, including the reply: 'why this literary reference?', not denying the latter solution as a possible additional one.[20] As for the Roman past, obviously there existed for the Greeks no canon of literature to

---

17 See Scheid (2012a), especially 210 on Plutarch's discussing only royal and republican monuments, and disregarding even such monuments as the imperial fora and the buildings in the Campus Martius; Scheid (2012b) is a somewhat simplified version.

18 Geiger (2008), see 'bronze vs. marble' in the Index.

19 See, generally, Favro (1996).

20 Bowie (1970) vs. Millar (1969). The quotation is from Bowie's n. 1.

follow, no 'classics' to adhere to.[21] But one would be ill-advised totally to disregard the political realities and the possible parallels one might draw between political situations, even in an era *ubi sentire quae velis et quae sentias dicere licet* (Tac. *Hist.* 1.1). Conceivably the small-town Greek intellectual may have been less sure about the limits of the permissible than his contemporary, the senatorial historian. Just to stick to the previous examples, the *Lives* of a man, without whom the first *princeps* would have hardly achieved what he did, of a popular prince eliminated by poison, or even of a man happily refusing the imperial throne would have been perhaps too close to the bone. But republican history was ancient history. There is no need to repeat what has been above alluded to, that Plutarch was at liberty to choose republican heroes and to avoid figures from the Empire for subjects of the biographies, whether members of the imperial family or otherwise, but I shall underscore this thesis with the evidence from the *Moralia*. Not only do Plutarch's republican heroes pop up in a variety of places, often in parallel accounts to the *Lives* but also a very wide variety of other republican persons, no doubt spin-offs of Plutarch's wide reading in republican history, make their appearance. On the contrary, the Empire is very poorly served indeed. Even the emperors, whose *Lives* had been written, appear rarely, Augustus being just a partial exception: not only are Galba, Otho and Vitellius never referred to in the *Moralia*, there is also no reference ever to Claudius, and Caligula is mentioned only twice. All the other appearances of politically important personages of imperial times in these voluminous and variegated writings one may count on the fingers of one hand, and even these rare mentions are almost never in a politically significant context. Thus, the *Lives of the Caesars*, composed earlier than the *Parallel Lives*, left no traces to speak of in the assorted writings of the *Moralia*, a remarkable contrast with the numerous repetitions of stories and anecdotes from the *Parallel Lives* in these essays. I do not believe that we should put down this differrence in its entirety to the inferior literary quality of the *Caesars*, and certainly not to a dearth of material in them. Indeed, I find myself in wide-ranging agreement with Rebecca Preston, who maintained that 'Plutarch studiously avoids any contemporary relevance' and that '... all references there to Roman politics concern republican practices and institutions', and again, when she says that 'There is no mention of contemporary political realities'.[22]

In trying to discover the reasons for this peculiarity of Plutarch's works we are somewhat handicapped by the largely insoluble problems of the chronology of these texts.[23] Yet the consensus, according to which probably very few works were written before the *Caesars*, even on a relatively late dating of this work to the reign of Nerva, will eliminate the explanation that Plutarch disregarded imperial persons because he was not yet familiar with imperial history. At any rate, the chrono-

21 It has been suggested to me that Greeks, including Plutarch, may not have been unaware of such canons existing, if virtually, in the minds of some of their Roman friends.
22 Preston (2001) 110 – 111.
23 Jones (1966) is still the best guide for Plutarch's chronology.

logy of Plutarch's works dealing with Roman history may offer some insights. Whichever exact date we assign to the *Lives of the Caesars*, nobody has ever doubted that they are prior to the *Parallel Lives*; another work, the *Roman Questions*, has been dated after 105 CE, that is, it postdates some of the *Parallel Lives* or was composed *pari passu* with some others. One may also mention in this context the apparently late date of composition of the two most political essays, the *Political Precepts* and the *Old Men in Public Affairs*. I discount for the time being the partially extant *On the Fortune of the Romans*, for which in my view no dating is available, since I do not subscribe to the automatic relegation of the so-called rhetorical works to Plutarch's youth. In parentheses I may add that this view dates to the times of the presumed superiority of age and is totally opposed to the present day worship of youth, to which unfortunately I cannot subscribe, even if for only personal reasons. The possibility that Plutarch ignores persons of imperial times in the *Moralia* out of lack of information should be rejected.

To resume, Plutarch not only ignored imperial personages for a feasible inclusion in the *Parallel Lives*, even when he was expanding the series, but he also almost totally, and, it seems, deliberately avoided mentioning such men in his *Moralia*. Now I have alluded to the fact that for the limits of the Greek attitude to the past a political versus a literary solution has been put forward—and I draw attention to the curiosity that the initial proponent of the political solution is a scholar best defined as an authority on the literary and intellectual life of the Empire, while the literary solution was first and most vigorously championed by an historian. As I have maintained, for the Greeks, including Plutarch, no literary canon relevant to Roman history was in existence. It is for the lack of a different solution that I am putting forward Plutarch's political cautiousness—perhaps not unconnected with the possible reactions to his *Lives of the Caesars*, of which we regrettably know nothing—for his avoidance of persons of the time of the Empire both as subjects for his *Parallel Lives* and for mentioning in passing or as *exempla* in the *Moralia*. Admittedly, there is nothing in the surviving *Galba* and *Otho* that will support such a conclusion, yet I would not totally disregard the possibility that it may have been the reactions to the *Lives of the Caesars* that caused Plutarch to be so sparing with later references to that work. Of course he may have been overcautious—once bitten twice shy. Add to this the above-mentioned almost total avoidance of references to imperial persons in general or of their treatment as *exempla*. The question remains open to future investigation.

Joshua Pugh Ginn

Plutarch and the advent of Hellenism in Rome

Abstract: The story of republican Rome's increasing interaction with Greece acquired a particularly strong moralising tone in the historiographical tradition of antiquity, and it came to be closely associated with themes of moral decline and political collapse. The advent of Greek cultural products was commonly described in terms of 'invasion' or 'infiltration' and writers in antiquity liked to focus down on specific moments of 'first contact' or 'first arrival' between the Roman and Greek worlds. This chapter examines Plutarch's treatment of this well-established historiographical tradition in his portrayal of the Middle Republic (late 3rd and 2nd c. BCE). While Plutarch's appreciation of Greek *paideia* as a dynamic agent in moulding the character of individuals, especially Romans, has received considerable attention, and his appreciation of its diachronic appearance in Rome noted, the focus here is on the moment of appearance itself. By considering more fully Plutarch's appreciation of Roman culture, Plutarch's precise interpretation of the dynamics of cultural interaction as Rome first encounters Greece is explored, as well as the manner in which he responds to the prevailing historiographical traditions. Finally, consideration is given to how Plutarch offers his own more original contributions to the story of first contact between Greece and Rome, by changing perspective and looking at Greeks' first impressions of Rome.

Moral decline

In the Roman historiographical tradition, a strong link was formed between Rome's increasing forays into the Greek East and the decline of the republican system. There were different theories about how the process occurred: was it the result of infiltration by corrupting foreign habits, or of the opportunity presented to previously restrained native vices abroad? Or some combination of the two?[1] In either case, the discourse became firmly rooted in spatial terminology, with geographical movement, cultural exchange and moral deterioration being closely equated: Polybius claimed Rome maintained her native moral standards until her wars overseas;[2] Livy won-

1 For more on ancient explanations of decline, see Lintott (1972). Gorman and Gorman (2014) have recently argued that the focus on (foreign) luxury as a cause of decline was a particularly Roman invention, and only subsequently taken up by imperial Greek authors. I would like to thank the anonymous reviewer for drawing my attention to this work, and for their other helpful suggestions.
2 18.35.1.

https://doi.org/10.1515/9783110539479-012

dered in his preface that it had taken so long before greed and luxury immigrated,[3] and noted the Asian booty shown in Cn. Manlius Vulso's triumph as the origin of *luxuriae ... peregrinae.*[4] Both authors also commented on the ways in which M. Claudius Marcellus' import of the artwork from Syracuse into Rome constituted a deleterious new practice, as we shall see later.[5] Sallust blamed the *loca amoena* of Asia for instilling lax habits in the army of L. Cornelius Sulla.[6] Pliny wrote that Asia herself sent luxury to Italy and afflicted Rome's morals, which declined even further in desire of *opulentiam externam* after victory over Achaea.[7] It was not just the arrival in Rome of the alien material wealth of the Hellenistic World, but also the influx of its cultural pursuits that Romans worried about, as the frequent attempts to physically expel their practitioners from the city attest.[8]

Much work has already been done on Plutarch's attitudes to this anti-Hellenic trend in Roman history. In general, Simon Swain has noted how Plutarch 'ignores Roman myths about the effect of Hellenism on Roman society' in talking about republican decline, giving a much more generalised narrative about the corrupting effects of power,[9] and in the *Cato Maior*, he explicitly rejects his protagonist's dire prophesy that Rome would be ruined by 'Greek letters'.[10] Christopher Pelling has observed how Plutarch was careful to distinguish Cato's attitude to luxury from his attitude to Hellenism:[11] they both spring equally from his traditionalism, but are not equally objectionable.[12] Manuel Tröster and Simon Swain have shown how Plutarch does something similar with the luxurious habits and material wealth of Lucullus, dissociating them from his Hellenic cultural achievements and benefactions.[13] Ra-

3 *Praef.* 11: *tam serae avaritia luxuriaque immigraverint.*
4 In 187 BCE: 39.6.7–8.
5 Liv. 25.40.1–2; Polyb. 9.10.1–6.
6 *BC* 11.5–6.
7 *HN* 33.148–149, cf. *Asia devicta* also at 34.14,34.
8 Criticism of Greek culture and philosophy: Liv. 29.19 (Scipio in Sicily); Liv. 40.29.2–14, Val. Max. 1.1.2, Pliny *HN* 13.84–86 (burning 'ancient' Pythagorean books); Suet. *Rhet.* 1.1–2 (expulsion of rhetoricians and philosophers in 161 BCE); Pliny *HN* 7.112–113, Plu. *Ca. Ma.* 22 (hostility to Athenian embassy of 155 BCE); App. *BC* 1.28, Liv. *Per.* 48 (destruction of first permanent theatre); Ath. *Deipn.* 12.547a, Ael. *VH* 9.12 (expulsion of two Epicureans); Polyb. 39.1, Plu. *Ca. Ma.* 12.5–7 (Aulus Postumius' philhellenism mocked); Plu. *Ca. Ma.* 23 (Cato's general hostility). For more on Roman attitudes to Greek cultural activity, see Griffin (1989) and Gruen (1990).
9 Swain (1990b) 126–128 (quotation, 128).
10 23.2–3.
11 Pelling (1989). He cites in evidence *Ca. Ma.* 4.2 (praise of Cato's attitude to luxury, without reference to Greece) and 2.3–6 (omitting, even reversing, anti-Hellenism). One might also compare *Ca. Ma.* 8.2 with Polyb. 31.25.4–5a, as well as *Ca. Ma.* 3.5–7 with Liv. 29.19; in both Plutarch removes the Greek context from the criticisms voiced by Cato. As we shall see, I think this technique extends even further.
12 Plutarch shows how 'strengths as well as weaknesses spring from this exclusiveness', Pelling (1989) 214–215.
13 Swain (1992b); Tröster (2004).

ther than denying that luxury was a problem, or that Rome was in decline,[14] it is worth noting explicitly that Plutarch has accepted the traditional timelines of both moral collapse and increasing Hellenic influence, but does not link them causally. Both Swain and Pelling recognise that Plutarch follows the standard (and relatively accurate) view of the arrival of Greek culture in Rome over the course of the first half of the second century BCE. Furthermore he judges his Roman heroes' cultural achievements accordingly:[15] for example, he criticises Marius more for rejecting the Greek culture available to him, than Coriolanus, who lived in an era before it had reached Rome.[16] Swain in particular stresses that Plutarch was 'conscious that Hellenic culture had been imported to Rome and could never be fully taken for granted among Romans as it could among Greeks'.[17] It is not my intention here to challenge this influential and convincing analysis of Plutarch's general character-portrayal, but rather to take a closer look at Plutarch's presentation of the period of cultural 'importation', that is, the late third and early second centuries BCE. I will consider how Plutarch conceptualised this cultural migration with regard to the prevailing historiographical tradition and arrived at a vision of a Rome that treated Greek παιδεία as its own, at the time when it was at the height of its power.[18]

Military decoration

There can be no doubt that Plutarch does present early Rome as somewhat uncivilised: nowhere is this clearer than *Cor.* 1.6 when Plutarch explains that early Rome lacked a proper word for 'virtue' (*virtus* being properly 'valour'), although some positive elements are admitted among these noble savages, such as the lack of bribery and corruption.[19] War dominates however: it is waged simply as a means of social control, and electoral success depended on displaying one's war-wounds.[20] A similarly under-cultured, though not unredeemed, picture of Rome is painted in the *Pyrrhus*, where we see doughty Romans scoff at the exotic wealth and the unwarlike philosophical reflections of Pyrrhus and his companions:[21] war however is still what brings out the best in the Romans, when refusing a dishonourable victory.[22] A passage of the *Marcellus* seems to continue this presentation, where Plutarch com-

14 See esp. *Ca. Ma.* 4.2–3, *Comp. Cim. et Luc.* 1.1.
15 Pelling (1989) 216; Swain (1990b) 129, 131–133; Swain (1996) 140–143.
16 Swain (1990b) 136–140.
17 Swain (1990b) 126.
18 His rebuttal of Cato at *Ca. Ma.* 23.3. It is unclear exactly when Plutarch considered this to be, but it must fall somewhere between the mid-second century BCE and Plutarch's own day.
19 *Cor.* 14.3–6. Cf. Swain (1990b) 136–137.
20 Campaigns to keep the mob quiet: *Cor.* 12.6, 19.1–2; *Cam.* 9.2–3. War-wounds: *Cor.* 14.2, 15.1.
21 *Pyrrh.* 20. For more on the triangulation of national identity in the *Pyrrhus*, see Mossman (2005). On Pyrrhus, see also Xenophontos in this volume.
22 *Pyrrh.* 21.1–6, cf. *Cam.* 10.

ments on the bloody spoils and unrefined decorations which adorned the city at that time, which was the 'precinct of ever-warring Ares'.[23] Into such a grim setting Marcellus carried the artworks of Syracuse, teaching the Romans, till then ignorant of the amazing products of Greece, to honour and admire them. This striking picture of Rome's traditional bellicose austerity is not, however, amplified throughout the *Life:* Hellenic art may be alien to Rome, but her customs are not so uniformly brutish as this passage might suggest. Indeed, Plutarch defends Rome against such accusations of barbarity earlier in the *Life*, arguing that a gruesome sacrifice on the eve of the Gallic War was an exception forced on them by the Sibylline books. This was not representative of Roman religion, which normally admitted nothing barbaric or unnatural (βαρβαρικὸν μὲν γὰρ οὐδὲν οὐδ' ἔκφυλον), but as far as possible held to beliefs that are perfectly Hellenic and temperate.[24] Here behaving Ἑλληνικῶς (and πρᾴως) is surely not meant as an indication of the improving influence of Greece, but of 'coincidental' conformity of Roman practice with Greek:[25] Plutarch adduces several other examples of fastidious religious observance elsewhere in support.[26] Rome, though she hasn't encountered Greek art yet, still seems pretty civilised. Indeed, another indication that the martial façade is not so comprehensive as we might think is the fate of all those brilliant masterpieces of Hellenic grace: Marcellus intends them for his triumph, but he is instead given an *ovatio*. Rather than a chariot, a laurel-wreath and trumpets, this involves Marcellus going on foot with a myrtle-wreath and pipers, looking entirely ἀπόλεμος καὶ ἡδύς.[27] This is not simply appropriate for Marcellus' character, but also indicates something about Rome: it is, for Plutarch, conclusive proof that the *ovatio* was originally given to generals succeeding through diplomacy and stratagems rather than force.[28] The distinction may have fallen out of use, but it is a sure sign of Rome's less aggressive side.

The *Marcellus*' opening chapter is often used to indicate Plutarch's dim view of Roman culture at the time of the Second Punic War. There, Plutarch notes Marcellus' soldierly ability, but adds:

> τῷ δ' ἄλλῳ τρόπῳ σώφρων, φιλάνθρωπος, Ἑλληνικῆς παιδείας καὶ λόγων ἄχρι τοῦ τιμᾶν καὶ θαυμάζειν τοὺς κατορθοῦντας ἐραστής, αὐτὸς δ' ὑπ' ἀσχολιῶν ἐφ' ὅσον ἦν πρόθυμος ἀσκῆσαι καὶ μαθεῖν οὐκ ἐξικόμενος. εἰ γὰρ ἄλλοις τισὶν ἀνθρώποις ὁ θεός, ὥσπερ Ὅμηρος εἴρηκεν, 'ἐκ

23 A quotation from Pindar, *Marc.* 21.2–3.

24 *Marc.* 3.6.

25 For Plutarch's disagreement with those (such as Juba and Dionysius of Halicarnassus) who tried to trace the sources of Roman habits and the Latin language back to Greece, see Swain (1990b) 126 (esp. nn. 2, 3).

26 *Marc.* 4.2–5.7, 12.1–2. Swain (1990b) 141–142 suggests that this theme of Hellenic religious propriety exists 'to support his proposition that Marcellus is highly interested in Hellenic culture and that the people of Rome are receptive to its introduction'. I disagree only in that Rome's propriety can be admired in itself, without reference to its 'receptivity'.

27 *Marc.* 22.1–2.

28 *Marc.* 22.3–10.

νεότητος ἔδωκε καὶ ἐς γῆρας τολυπεύειν / ἀργαλέους πολέμους', καὶ τοῖς τότε πρωτεύουσι Ῥωμαίων, οἳ νέοι μὲν ὄντες περὶ Σικελίαν Καρχηδονίοις, ἀκμάζοντες δὲ Γαλάταις ὑπὲρ αὐτῆς Ἰταλίας ἐπολέμουν, ἤδη δὲ γηρῶντες Ἀννίβᾳ πάλιν συνείχοντο καὶ Καρχηδονίοις, οὐκ ἔχοντες ὥσπερ οἱ πολλοὶ διὰ γῆρας ἀνάπαυσιν στρατειῶν, ἀλλ' ἐπὶ στρατηγίας πολέμων καὶ ἡγεμονίας κατ' εὐγένειαν καὶ ἀρετὴν ἀγόμενοι.

But he was in other ways moderate, humane and a lover of Greek learning and literature insofar as he honoured and revered its proponents. He himself was kept from immersing himself as much as he wanted by his constant activities. For if God made it any men's fate, in Homer's words 'to endure toilsome wars, from youth to old age', it was certainly the foremost Romans of that day, who in youth warred with the Carthaginians for Sicily, in their prime with the Gauls for Italy itself, and in old age met again with the Carthaginians and Hannibal. They did not have the respite which old age allows many, but were called to lead and command in wars in keeping with their noble and excellent qualities.[29]

While Plutarch does clearly envision a somewhat austere and primitive Rome here, we seem a long way from the archaic city of Coriolanus, where ἀρετή lacked a name and war-wounds were a criterion for leadership. With this generation, it is their very virtue and nobility that mark them out for leadership in the succession of wars which befall them. It is not Roman bellicosity that draws Marcellus from cultural endeavours, but circumstances:[30] events dictated that leaders in war were needed, and Marcellus' qualities were called for. The point is reiterated in the *Flamininus*, where the hero is said to have received a military education because Rome's many wars at that time meant her generals learnt to lead by serving in the ranks from a young age.[31] Again we find it is historical circumstances, not inherent Roman barbarism, calling her leading men to war.

Cradles of civilisation

If Rome herself seems different by the time of the Second Punic War, what then of her inhabitants? Are they still noble, but unreconstructed, savages? It doesn't appear so: Flamininus, despite his military upbringing, forms a reputation for justice as much as campaigning, and is identified as someone able to use persuasion and negotiation rather than warfare and force to achieve his ends in Greece.[32] We have already seen the moderate qualities that Marcellus displayed as well as being a fierce soldier and general. Swain and Pelling both connect Marcellus' incomplete education with his

29 *Marc.* 1.3–5.

30 Note how wars start with abrupt, chance invasions in *Marcellus* (3.1–2, 9.1; cf. *Fab.* 2.2) in comparison, for example, with *Cor.* 12.6, 19.1–2; *Cam.* 9.2–3.

31 *Flam.* 1.4.

32 *Flam.* 1.5, 2.3–5. Scuderi (1996) sees Plutarch as playing with cultural stereotypes here in presenting a cultured Roman and a Greek man of action, which corresponds with a mission of cultural reconciliation often claimed for the *Parallel Lives* as a whole. There is some truth in this, no doubt, but that Flamininus' qualities are *typically* Greek does not also mean they are owed to Greek influence.

eventual inability to restrain himself,[33] while Swain also identifies Flamininus' 'lack of any deep education' as the cause of his later shameful ambition.[34] While Greek *paideia* might well have helped both men in restraining their passions,[35] this does not also mean that it is equally responsible for their finer qualities. Indeed, the *Marcellus* passage has recently been used by Tim Duff as an example of his 'static/illustrative' model in Plutarchan descriptions of education, whereby a person's approach to education reflects underlying character, rather than explains it: in Marcellus, 'we are invited to see in his attitude to his studies confirmation of a good character.'[36] Plutarch's comments on Flamininus' Hellenism are in a similar vein, I would argue, just as Plutarch's reworking of Polybius' and Livy's comments on Marcellus' introduction of the Syracusan art to Rome[37] is not so much designed to mute criticism, but to highlight that (whatever the pros and cons) Marcellus took pride in having brought appreciation of Greek art to Rome, which in turn indicates his character.[38] That Marcellus and Flamininus are characterised by their attitudes to Greek culture is not controversial, nor does it deny some explanatory force in their *paideia*,[39] but putting the emphasis on its illustrative aspect highlights that their humane and admirable characteristics have arisen largely independently. No doubt both men could have benefitted from some Hellenic refinement, but by this time Rome could produce such noble characters on its own.

This is most clear in the *Aemilius Paulus*, where Plutarch tells us his subject grew up in an age 'blossoming with great men most distinguished in fame and quality'.[40] While Paulus is also later identified by Plutarch as a philhellene, his early career is based on firm adherence to Roman tradition, both in his duties as augur and as a commander.[41] When Paulus retires from politics, devoting himself to religious and familial duties, Plutarch notes that he passed on his native and ancestral (ἐπιχώριον

33 Pelling (1989) 199–208, 230–232; Swain (1990b) 132, 142. This, as we shall see, may well be the case, though I strongly disagree that these passions are to be identified as 'fatal ambition' (Swain, ibid.), as I believe a considered reading of *Marc.* 28.6 shows. Pelling, speaking of 'rashness' and 'bellicosity,' is somewhat closer to it, especially in his focus on the 'how the same qualities contribute both to a man's greatness and his flaws'.

34 Swain (1990b) 132, 135. This deficiency is based on *Flam.* 1.4, with an interest in Hellenic studies inferred from 5.7: this latter passage does not, strictly speaking, indicate any shortcomings.

35 As is the implication of *De virt. mor.* 452D, a passage oft cited by Swain.

36 Duff (2008b) 16. This is as compared to a 'developmental' approach (predominant in the *Moralia*) which tries to *explain* adult character through early attitudes.

37 By recasting it into the mouths of senators: see Polyb. 9.10.1–6; Liv. 25.40.1–2; cf. *Marc.* 21.4–6. For a fuller comparison with Livy, see Pelling (1989) 199–205.

38 The characterising force is even clearer when compared with *Fab.* 22.6–8, where Plutarch shows no reticence in praising Marcellus explicitly for his πρᾳότης and φιλανθρωπία, but there it serves to characterise Fabius, since these are qualities in which he would normally excel. Cf. Swain (1990b) 140–141; Xenophontos (2012) 174–176.

39 Duff (2008b) 23.

40 *Aem.* 2.5–6.

41 As augur: *Aem.* 3.2–5 (cf. *Marc.* 4.7). As general: *Aem.* 3.6–7 (cf. *Marc.* 4.7), 13.6–7.

and πάτριον) education to his children.[42] Plutarch explicitly credits Paulus' moral rectitude to this upbringing in the *Synkrisis:*

> Αἰμίλιος μὲν ὑπὸ τῶν νόμων καὶ τῆς πατρίδος οὕτως ἔοικεν εὐθὺς ἀφικέσθαι παρεσκευασμένος … Τούτου τεκμήριον, ὅτι Ῥωμαῖοι μὲν ὁμαλῶς ἐν τῷ τότε χρόνῳ πάντες ἦσαν εὔτακτοι καὶ ὑποχείριοι τοῖς ἐθισμοῖς, καὶ τοὺς νόμους δεδιότες καὶ τοὺς ἄρχοντας … .
>
> For Aemilius seems to have been brought up already so by the strictures of his country … Proof of this is that all Romans alike at that time were disciplined and restrained in habits and fearful of the laws and magistrates … .[43]

However, this is not the narrow-minded conservatism of a Cato: Paulus displays the sound basis of his character by choosing to supplement Roman education of his sons with the Greek cultural pursuits now available. Similarly, after traditional Roman discipline brings Paulus and his army success abroad against Perseus, the general can turn aside to appreciate his civilised surroundings[44] and demonstrate both an appreciation for Greek culture and a Roman aptitude for justice and good governance.[45] This harmonious fusion of Greek and Roman practice is best demonstrated by Paulus' distribution of the Macedonian spoils: demonstrating typically Roman restraint,[46] he did not even look at Perseus' treasure—his only concession derives from his philhellenic sensitivities, giving his sons the king's library.[47] A traditional Roman upbringing can work in harmony and on an equal footing with Greek *paideia*.

With this in mind, I return to Plutarch's presentation of Cato the Elder. I have already mentioned Plutarch's deliberate separation of Cato's attitudes to luxury and to Greek culture. Now we observe another clear separation in the *Life:* Cato's rejection of Greek *paideia* is largely kept distinct from discussion of how he raised his son. Given that Cato's dire prophecy about the influence of Greek learning on Rome was explicitly addressed to his son,[48] Plutarch might have reflected this in the upbringing Cato was giving him; instead, he calls Cato a πατὴρ ἀγαθός for the care he took in overseeing his education.[49] Despite Cato's thorough programme of reading, jurisprudence, and physical exercise, Plutarch could easily have criticised it for the bias against Greek literature and philosophy that it must have displayed.[50] He might also have made more of Cato's preference that his son wasn't indebted to a slave for something as important

---

42 *Aem.* 6.8.
43 *Comp. Aem. et Tim.* 2.1–2.
44 Cf. *Ca. Ma.* 12.5–7.
45 Alternating perfectly in Plutarch's account: restoring government and distributing supplies (*Aem.* 28.2–3), appreciating Greek art (28.4–5), confirming political settlement and tribute payments (28.6), organising games and banquets (28.7–8).
46 Cf. Polyb. 18.34.8–35.12.
47 *Aem.* 28.10–11.
48 *Ca. Ma.* 23.2.
49 *Ca. Ma.* 20.1.
50 *Ca. Ma.* 12.6, 22–23.

as education: Marius' more vicious quip provides us a model for what he could do.[51] The virtues of such a Roman upbringing were evident to Plutarch, the benefits of Greek culture aside: the period which marked the first arrival and uptake of Hellenic *paideia* in Rome also seems to have represented the apogee of her traditional methods, producing the home-grown virtues that brought her an empire.

First among equals

I have thus far dealt with how Plutarch interprets the prevailing historiographical tradition of Rome's first encounters with Greek culture. We have focused on the arrival of Greek artefacts and education in Rome, but I will end with Plutarch's own more original twist on the tradition, namely the Greek perspective on Rome. We see this briefly but pithily expressed in the *Pyrrhus*, in the protagonist's comment on first seeing the Roman army, that there was nothing barbarous about these barbarians.[52] Plutarch's redeployment of this story in the *Flamininus* has already been well expounded,[53] but is worth considering in the current context. Here, Plutarch has the local Greeks echo Pyrrhus' words of Flamininus' army as it advanced after the retreating Macedonians. Despite Macedonian slanders, they find Flamininus 'young in age, humane in appearance, Greek in voice and accent and a lover of true honour'.[54] Plutarch's sleight of hand makes Flamininus (apparently) the first Roman encountered in Greece; he had stressed the unfamiliarity earlier in the *Life*[55] and reinforces it here by applying the words of the first Greek invader of Italy to this first Roman incursion into Greece. Then it had been Rome's military organisation that had been praised; now, on foreign soil, it is her more humane virtues—the very virtues that were required to endear Greeks to this ἀλλόφυλος ἀρχή.[56] This phrase recurs again during the liberation decree at Corinth, when again Plutarch focalises the scene through the eyes of Greeks, even more elaborately than Polybius or Livy, and dwells on the altruism of Rome.[57] Through these first impressions, Plutarch presents a new version of the growth of Roman power—not one of military strength and increasing avarice, but of moral strength and the spread of justice. Martial circumstances had raised great generals for Rome but their native virtues were also required if she was to gain an empire deservedly.[58]

---

51 *Ca. Ma.* 20.6; *Mar.* 2.2.
52 *Pyrrh.* 16.6–7.
53 Schepens (2000) 354–355; Mossman (2005) 512–513.
54 *Flam.* 5.6–7.
55 *Flam.* 2.5.
56 Ibid.
57 *Flam.* 10.4–12.1; cf. Polyb. 18.46; Liv. 33.32–33.
58 Cf. *Flam.* 2.3–5, 12.8–10.

Viewing Roman conduct abroad through the eyes of locals (usually Greeks) is a regular technique of Plutarch. Aemilius Paulus' fastidious disposition of games and banquets astonished the contemporary Greeks,[59] and it was not just sympathy for Greek culture that was celebrated: even Cato's harsh justice on Sardinia made Roman rule more beloved to the inhabitants (albeit also more forbidding).[60] After sacking Syracuse, Marcellus—that great advocate of Greek culture in Rome—also acts as an ambassador of Roman virtues to Greeks. Before Marcellus' settlement of Sicily, Plutarch tells us:

> Τῶν δὲ Ῥωμαίων τοῖς ἐκτὸς ἀνθρώποις δεινῶν μὲν εἶναι πόλεμον μεταχειρίσασθαι καὶ φοβερῶν εἰς χεῖρας ἐλθεῖν νομιζομένων, εὐγνωμοσύνης δὲ καὶ φιλανθρωπίας καὶ ὅλως πολιτικῆς ἀρετῆς ὑποδείγματα μὴ δεδωκότων, πρῶτος δοκεῖ τότε Μάρκελλος ὑποδεῖξαι τοῖς Ἕλλησι δικαιοτάτους Ῥωμαίους.
>
> The Romans were thought by foreigners as a warlike and fierce-fighting people and had not given examples of consideration, humanity and political virtue in general: Marcellus himself was the first to demonstrate to Greeks that the Romans were a most just people.[61]

Coming between the conquest of Syracuse and that first importation of Greek goods, this passage tempts the reader to see Plutarch playing with the historical tradition. Marcellus was not just the first philhellene in Rome, but the first to show the attraction of Rome to the Greeks.

The story Plutarch tells of first contact between Greece and Rome is not a straightforward one; he is interested in both sides of the story and the interaction of cultures which resulted was ultimately a positive process, allowing Rome to flourish and Greece to be restored from the decline which was evident even before the depredations of the likes of Sulla. To Plutarch, looking back from the peace and prosperity of Greece under the Empire, the history of the mid-Republic was not just the story of Greek culture arriving in Rome, but of Roman virtues spreading to Greece and it was little wonder that he considered Cato to have badly erred with his apocalyptic prophesies regarding Roman philhellenism. Greek culture had certainly not destroyed Rome, but nor was this simply a case of *Graecia capta* conquering her rustic and war-like conquerors; rather Rome had indeed only reached her pinnacle when combining Greek παιδεία with her own native virtues.

---

59 *Aem.* 28.7–8.

60 *Ca. Ma.* 6.4.

61 *Marc.* 20.1. Some read δικαιοτέρους for δικαιοτάτους, but the comparative would have to be used in a more or less absolute sense here any way ('rather just'). I do not agree with Swain (1990b) 141, that this is denigrating Marcellus' contemporaries; rather, the unfamiliarity of the two cultures is stressed.

Susan Jacobs

Creating paradigms for the *politikoi*: Bridging the gap in political space and time with pre-imperial heroes

Abstract: Plutarch's *Parallel Lives* are often seen as vehicles for comparing the moral character of statesmen and gaining new insights into issues of moral virtue and vice. However, the political careers of the heroes could also provide lessons in meeting the practical challenges of the political arena. Like the subjects of the *Lives*, Plutarch's readers were often involved in managing a populace, administering cities and overseeing relations between Roman authorities and the provincial cities. Given this overlap, to what extent do Plutarch's *Lives* bridge the gap in political space and time between the statesmen of the past and those operating within the constraints of imperial rule? Does Plutarch create models not only of moral traits, but also of effectiveness in performing the duties of a *politikos* under Rome? In this chapter, I argue that the *Lives* provide paradigms for resolving the practical problems routinely encountered by men in political office. Incidents from *Pericles*, *Cato the Younger*, *Phocion*, *Aemilius Paulus* and *Timoleon* are used to illustrate some of the techniques Plutarch employed to magnify the parallels between the past and his own day in the challenges statesmen faced in administering cities and conducting diplomacy within an imperial structure. Without changing the sequence of historical events, Plutarch linked certain incidents to motives for action that his readers shared, as well as to practical strategies they could imitate in performing their official duties—often echoing the pragmatic political concerns voiced by Dio in his city speeches and by Pliny in his *Letters*. By presenting the historical record in this way, Plutarch transformed the great leaders of the independent Greek states and the Roman Republic into role models for men managing political affairs in the Roman Empire of his own day.

The *Parallel Lives* are often interpreted as vehicles for evaluating the moral character of statesmen and gaining new insights into issues of moral virtue and vice[1]—topics of interest to members of the educated elite in their private and public lives.[2] With regard specifically to political affairs, the *Lives* also addressed the potential conflict between moral principle and expediency for the *politikos* acting on behalf of the common good.[3] This chapter explores another area in which the *Lives* held lessons for the

1 See, for instance, Pelling (1980) 102–103 = (2002) 135–136; Pelling (1995) 215–216 = (2002) 246–247; Duff (1999) 70; Geiger (2014) 294–298.

2 On Plutarch's audience, see Jones (1971) 48–64; Stadter (2002b); Duff (2007) 8–9; Duff (2011b).

3 See discussions of Cato the Younger in Pelling (1997a) 228–229 or Duff (1999) 70, 131–160 and of Phocion in Tritle (1992) 4267.

https://doi.org/10.1515/9783110539479-013

politikos, namely, pragmatic insights into effectively performing the functions of public offices in Rome and the provinces. Deriving such lessons from the political experiences of famous statesmen of the past created a challenge because those statesmen operated not only in a distant historical and cultural setting, but also in a different political 'space' in terms of the scope for independent action in the political arena. To portray these heroes as paradigms of political effectiveness for the imperial era, Plutarch had to 'bridge the gap' between both the cultural practices of historical and current times and the political roles played by leaders in the two periods.

To this end, Plutarch crafted *Lives* that would capture the historical challenges his subjects faced, while also amplifying the parallels between the political arenas in which they operated and those of his own day. He simplified the contours of the political landscape to an opposition between aristocrats and the people[4] and focused on the practical problems shared by the city leaders of the past and present—such as controlling the people, resolving factional disputes, dealing with alliances and rivalries and maintaining a positive rapport between ruler and ruled.[5] By crafting his narratives in this way, Plutarch narrowed the gap in 'political space and time' between the experiences of his heroes and of the *politikoi* in his audience and transformed the historical figures into role models of effective political action under Roman imperial rule.

In Plutarch's day, men active in political life—such as Sosius Senecio, who advanced along the senatorial career path, or Euphanes, who was active in his province[6]—held a variety of administrative posts overseeing matters such as city finances, law courts or the provision of basic services (e.g., roads or water supply).[7] Positions could also carry diplomatic responsibilities if a man served as a provincial governor, procurator or emissary between his city and the Emperor or his representatives.[8] In the *Lives*, Plutarch often provides a detailed description of how the heroes performed similar functions.

The incidents examined below, which are drawn from *Pericles*, *Cato the Younger*, *Phocion*, *Aemilius Paulus* and *Timoleon*, illustrate some of the techniques Plutarch employed to magnify the parallels between the political challenges of the past and his own era. Without changing the sequence of historical events, Plutarch included details about the motives and practical strategies of the hero that mirrored the experience of readers[9]—often highlighting concerns raised by other writers offer-

4 Pelling (1986b) 160–165 = (2002) 211–216 discusses Plutarch's imposition of a *boule-demos* framework on Roman politics.

5 This overlap is illustrated in the use of statesmen of the classical and Hellenistic periods to exemplify the precepts for contemporary politicians in *Praec. ger. reip.*

6 On Sosius and Euphanes, see Jones (1971) 55–56 and 110 and Swain (1996) 144–145.

7 See Jones A.H.M. (1966); Garnsey and Saller (1982); Reynolds (1988); Trapp (2007) 166–168.

8 See Jones A.H.M (1966) and Braund (1988a).

9 Pelling (1980) and Pelling (1986b) discuss Plutarch's adaptation of sources; de Blois (2008) notes that the overlap between the treatises and *Lives* helped readers apply the framework to their own situations.

ing pragmatic advice to statesmen or to cities, such as Cicero, Dio and Pliny.[10] By describing the process of choosing and implementing strategies—and the attributes and practical insights called into play—Plutarch portrayed great leaders of the independent Greek states and the Roman Republic as exempla of how to manage political affairs under the Roman Empire.

Pericles: Rebuilding the Acropolis

In the *Political Precepts*, Pericles is cited as a positive role model in many areas,[11] including public spending that benefitted the community. The principle at issue is stated at 818C–D: while spending that encourages wantonness is to be avoided, expenditures in a religious context on public spectacles, slight distributions of funds or spending on public areas are entirely appropriate. In *Pericles*, Plutarch uses the rebuilding of the Acropolis to provide a more complete treatment of the benefits of public construction and the political challenges with which they were associated. Both aspects would resonate with the elite of Plutarch's day, who were often engaged in euergetism involving construction of public edifices.[12]

Plutarch devotes two lengthy chapters (*Per.* 12–13) to the rebuilding of the Acropolis from the project's inception through its completion, while also touching on controversies surrounding the use of funds collected from the allies (*Per.* 12.1–4) and the accounting for how the funds were spent (*Per.* 14.1–2; 31.3). While recognizing that Pericles' public spending in other areas was criticised by contemporaries for making the people 'luxurious and wanton' (πολυτελῆ καὶ ἀκόλαστον) 'instead of frugal and self-sufficing' (ἀντὶ σώφρονος καὶ αὐτουργοῦ) (*Per.* 9.1),[13] Plutarch uses a speech by Pericles and authorial comments to place the spotlight on the economic benefits of his expenditures. At *Per.* 12.4, Pericles explains the positive impact of the Acropolis project on employment and prosperity in Athens:

> And it is but meet that the city ... should apply her abundance to such works as by their completion will bring her everlasting glory and while in process of completion will bring that abundance into actual service, in that all sorts of activity and diversified demands arise, which rouse every art and stir every hand, and bring, as it were, the whole city under pay, so that she not only adorns, but supports herself as well, from her own resources. (*Per.* 12.4)

---

10 Especially Cicero's *Ad Quintum* 1, Pliny's *Letters*, and Dio's city orations, including *Or.* 34 to Tarsus, *Or.* 38 to the Nicomedians and *Or.* 40, 44, 45 and 48 to Prusa. Each of these works addressed challenges related to administering provincial cities, overseeing public projects and managing relations among groups within a single city and between different cities in a province.

11 Pericles is also a positive exemplum in oratory (803A, 803B, 803F), sharing of power (812D, 812E) and avoiding abusive speech to enemies (810C).

12 On euergetism, see Jones A.H.M. (1966) 172, 247; Lendon (1997) 84–89; Zuiderhoek (2009).

13 In this chapter, all Greek and Latin citations and English translations are from Loeb volumes.

Elsewhere, Plutarch's comments similarly emphasise the positive effects of Pericles' military expeditions and public projects (*Per.* 11.4–5; 12.5–7), which provided all groups with a 'pretext for getting a beneficial share of the public wealth' (πρόφασιν ἀπὸ τῶν δημοσίων ὠφελεῖσθαι καὶ μεταλαμβάνειν, *Per.* 12.5).

Plutarch's description of Pericles' motives for rebuilding the Acropolis is not found in any surviving source. This lack of corroboration raises the question of whether or not such motivation actually belonged to the 5th century: while Plutarch may have drawn on a source now lost, he may also have inserted details that reflected the thinking of his own age in order to add contemporary resonance to Pericles' actions.[14] However, regardless of the origin of Plutarch's account, the recurrence of economic benefits as a theme in Dio and Pliny[15] suggests that Plutarch included these details about the impact of public policies on employment and prosperity because the topic was immediately relevant to his readers. Dio, for instance, in *Oration* 35, emphasises the economic advantages of the courts in Celaenae, which 'bring together an unnumbered throng of people' (ξυνάγεται πλῆθος ἀνθρώπων ἄπειρον), enable those with goods 'to sell at the highest prices' (πλείστης ἀποδίδοσθαι τιμῆς) and leave 'nothing out of work' (μηδὲν ἀργὸν) (*Or.* 35.15).[16] Moreover, in *Or.* 40 and 45, Dio describes his own project and its problems. Pliny also often raises issues related to construction in his letters to Trajan.[17] Against this backdrop, the motivation attributed to Pericles—and the controversies over funding and accounting—are likely to have echoed those faced by men in Plutarch's audience serving as archons or overseeing their own projects for public benefit.

Cato the Younger: Preparation for the quaestorship

While Plutarch may well use authorial comments and a speech by Pericles to tie the reconstruction of the Acropolis and other spending to contemporary concerns, in *Cato the Younger* he characterises Cato as a 'modern day' *politikos* by illustrating in action the principles of political effectiveness expressed as precepts in the *Moralia*.[18] For instance, Plutarch attributes to Cato the same reasons for public service that he recommends to Menemachus in the *Political Precepts* (798A–799A) and claims as his own in

14 Stadter (1989) 154–155.

15 On topics in Dio's city speeches, see Swain (1996) and Swain (2000). For Pliny's *Letters*, see Sherwin-White (1969) and Radice (1969).

16 Dio links his own sponsorship of a construction project to the benefits of making his city more impressive (*Or.* 40.5–7) and attracting as many inhabitants as possible to Prusa (*Or.* 45.13).

17 Topics include engineering (*Ep.* 10.37; 10.39; 10.41–42), locations (10.49; 10.70), efficient spending of funds (10.17.3–4; 10.24; 10.47) and collection of funds subscribed by the citizens (10.40).

18 See especially *Praec. ger. reip.*, *An seni*, *Maxime cum principibus* and *Ad princ. iner.* Swain (1996) 161–185, Beck (2004) and de Blois (2004) provide summaries of the advice in these treatises and links to the *Lives*.

Old Men in Public Affairs (783C): Cato chose public life not for reputation, or riches, nor by chance, but as 'the proper task for a good man' (ὡς ἴδιον ἔργον ἀνδρὸς ἀγαθοῦ), believing he should be 'more attentive to the common interests than the bee to its honey' (μᾶλλον ... προσέχειν τοῖς κοινοῖς ἢ τῷ κηρίῳ τὴν μέλιτταν) (*Ca. Mi.* 19.2).

Plutarch illustrates the meaning of 'being attentive to the common interests' in a description—more detailed than in other sources[19]—of how Cato prepared for his duties as quaestor (*Ca. Mi.* 16.1–18.5),[20] a position still held by statesmen in Plutarch's day. Cato refused to become quaestor until he had acquired knowledge of the office by studying the law and talking to men who had held that post (*Ca. Mi.* 16.1). This preparation enabled him to 'humble the clerks and make them submissive' (τοὺς γραμματεῖς ταπεινώσας καὶ ποιήσας ὑποχειρίους, *Ca. Mi.* 17.1) as soon as he assumed office, unlike many newly-elected, but inexperienced, quaestors who, as Plutarch explains, had to defer to their clerks:

> When [these clerks] received as their superior officers young men whose inexperience and ignorance made it really needful that others should teach and tutor them, they would not surrender any power to such superiors, but were superiors themselves. (*Ca. Mi.* 16.2)

In contrast, Cato immediately imposed strict standards and cleaned-up corrupt practices—actions that initially met great resistance, but ultimately won praise for their positive results (*Ca. Mi.* 16.3–18.3).

Plutarch's treatment of Cato's quaestorship reflects the practical challenges of his own day. Sextus Iulius Frontinus, whose career spanned the period between the reigns of Nero and Trajan, expressed the same concern about his inexperience when he was appointed water commissioner (*curator aquarum*) under Nerva: he did not want to depend on the experience of subordinates:

> I deem it of the first and greatest importance to familiarise myself with the business I have undertaken, a policy which I have always made a principle in other affairs. For I believe that ... there is nothing so disgraceful for a decent man as to conduct an office delegated to him, according to the instructions of assistants. Yet precisely this is inevitable whenever a person inexperienced in the matter in hand has to have recourse to the practical knowledge of subordinates. (*De Aquis* 1.1–2)

To avoid such dependence, Frontinus collected details about maintaining aqueducts and published them in *On the Water Supply of Rome* as a reference for himself and his successors. Moreover, because inexperience often impeded newly-appointed officials, he routinely wrote a practical handbook for his successors after he had gained experience in a particular administrative post (*De Aquis* 1.2). Tacitus similarly praises Agricola's concerted efforts as tribune 'to know the province and to make himself

---

19 Such as Cassius Dio (37.22–43.13) or Appian (*Civil Wars* 2.1–14).

20 Pelling (1980) 136 = (2002) 103; Geiger (1988) 251; and Swain (1990a) 197–198 see Plutarch's description of Cato's conduct in office as reinforcing the depiction of Cato's moral qualities.

known to the army, to learn from the experts and to follow the best men' (*noscere provinciam, nosci exercitui, discere a peritis, sequi optimos*) (Tacitus, *Agricola* 5.1). Such passages suggest that setting high standards of performance in political positions—as demonstrated by Cato in the *Life*—remained a meaningful goal for Plutarch's contemporaries.

Phocion: Mediator between Athens and Macedonian overlords

In addition to supplying models of political effectiveness in city administration, Plutarch's heroes also illustrate strategies for managing diplomatic relations between the provincial cities and the Emperor. In the *Political Precepts*, the principles of interacting with Roman rulers are discussed at 813E–816A, where the city leader is advised: (1) to stay within the limits set by Rome (813E); (2) to show himself and his state blameless towards the rulers (814C); (3) to have friends in high places (814C);[21] and (4) to maintain harmony in his city (815B). In *Phocion*, Plutarch supplements these general precepts with a demonstration of practical strategies that put them into effect. In particular, Plutarch uses Phocion's relations with the Macedonians to address two diplomatic challenges of his own day: establishing a friendly rapport with Emperors and governors and keeping their intervention to a minimum.[22]

With regard to 'having a friend in high places', Plutarch quotes two of Phocion's principles for engaging overlords: first, 'It is better to supplicate and try to persuade the victors for both you and them, and not to fight' (βέλτιόν ἐστιν ὑπὲρ ἀμφοῖν πείθειν καὶ παραιτεῖσθαι τοὺς κρατοῦντας ἢ μάχεσθαι, *Phoc.* 17.3), and, second, 'Either be superior in arms or be friends with those who are superior (ἢ τοῖς ὅπλοις κρατεῖν ἢ τοῖς κρατοῦσι φίλους εἶναι, *Phoc.* 21.1). Plutarch links Phocion's friendly rapport with both Alexander (*Phoc.* 17.4–18.5) and Antipater (*Phoc.* 26.3–30.6) to his ability to adapt his advice to the moral character of each leader. Phocion softened Alexander's attitude toward Athens 'by saying many things that suited with Alexander's nature and desires' (καὶ πολλὰ καὶ πρὸς τὴν Ἀλεξάνδρου φύσιν καὶ βούλησιν εὐστόχως εἰπὼν) and, ultimately, is welcomed by Alexander as a friend (*Phoc.* 17.5–6).

Phocion's approach is different with Antipater, who exhibited 'a certain ruthlessness and hatred of goodness' (ἀγνωμοσύνῃ τινὶ καὶ μισαγαθίᾳ) (*Phoc.* 27.2) and was a 'more burdensome' (ἐπαχθέστερος) master and tyrant, whose wrath was harder to appease (*Phoc.* 29.2). Phocion did not provoke Antipater by resisting his demand for a garrison—unlike Xenocrates, who adamantly opposed it as unworthy of free

---

21 Van der Stockt (2002) provides an extensive discussion of this passage.

22 Duff (1999) 155 contrasts Phocion's willingness to mediate to ease the harshness of Macedonian rule to Cato's inflexible stance. Beck (2004) 111–113 highlights Phocion's ability to adjust to the demands of the times, again in contrast to Cato.

men (*Phoc.* 27.4–5); instead, Phocion preserved a friendly rapport and later obtained more lenient terms for exiles (*Phoc.* 29.3) and a delay in the payments due from Athens (*Phoc.* 30.4). Phocion's influence was reinforced by his success in maintaining harmony in Athens, where he managed affairs 'mildly and according to the laws' (πρᾴως καὶ νομίμως) and kept 'busybodies and innovators' (τοὺς πολυπράγμονας καὶ νεωτεριστάς) out of office (*Phoc.* 29.4).

Phocion is more prominent in the embassies to the Macedonians in Plutarch's account than he is in those of Diodorus or Nepos. Whereas Plutarch presents Phocion as the only influential negotiator with Alexander and Antipater, Diodorus names Demades as the envoy to Alexander (Diod. 17.15.4–5) and attributes the results of negotiations with Antipater to the entire embassy (Diod. 18.18.1–6). Nepos simply reports that Phocion incurred Athens' hatred for scheming with Demades to turn the city over to Antipater (Nepos, *Phocion* 2.1–3). By presenting Phocion as the leading ambassador making decisions on behalf of Athens, Plutarch creates a portrait that more closely parallels the challenges facing city leaders interacting with Rome in his own day.

The importance of constructive relations with Rome is reflected as well in Dio's city speeches. Dio often points to his own role as ambassador[23] and highlights the benefits that can flow to a city from imperial favour, including buildings, festivals, assize courts and autonomy in administering justice.[24] Dio further advises cities to cultivate a reputation for internal harmony and peaceful relations with other cities in order to keep Rome at bay.[25]

Aemilius Paulus and Timoleon: Conducting relations with subject cities

Plutarch also creates role models for provincial governors or other imperial appointees engaging with the provincial cities.[26] The general principles for exercising supreme authority—as they would apply to Roman imperial rule—are found in *To An Uneducated Ruler* and in *Philosophers and Men in Power*, which define the primary duties of a ruler as setting a pattern for virtue, administering justice and overseeing the well-being of subjects.[27] In *Aemilius* and *Timoleon*, these principles are illustrated in action when the heroes, as representatives of powerful states, establish relations with cities only recently freed from tyrants.

---

23 *Or.* 40.5 and 45.3.

24 *Or.* 40.10; 44.11; 46.14; and 48.11.

25 *Or.* 34.9–10; 34.40–41; 38.33–37; and 48.12–13.

26 In the *Lives*, proper (and improper) conduct of the ruler exercising supreme power over subjects is illustrated by generals engaging allies or subject states, or in the actions of kings, like Demetrius, Pyrrhus or Alexander.

27 See, for instance, *Maxime cum principibus* 799B; *Ad princ. iner.* 780B, 781D.

In *Aemilius*, Plutarch's description of Aemilius interacting with the Greeks after Perseus' defeat—which offers more detail than Livy (32.8–11) and Diodorus (31.8.10–13)—illustrates how liberality and courtesy wins the goodwill of subject states, a primary objective of any ruler. When Aemilius tours Greece (*Aem.* 28.1–13), he not only provides benefactions that parallel those still sought by Greek cities in Plutarch's day—including gifts, festivals, games and banquets—but also respects the rankings of local people, 'paying to each one that degree of honor and kindly attention which was properly his due' (τὴν πρὸς ἕκαστον αὐτοῦ τῆς κατ' ἀξίαν τιμῆς καὶ φιλοφροσύνης ... αἴσθησιν ... ἐνδεικνύμενος, *Aem.* 28.7). Courtesy and liberality as the means of creating goodwill towards a provincial governor are also important themes in Cicero's advice to Quintus (*Ad Quint.* 1.16–24), while Pliny instructs Maximus 'not to detract from anyone's dignity, independence, or even pride' (*nihil ex cuiusquam dignitate, nihil ex libertate, nihil etiam ex iactatione decerpseris*, *Ep.* 8.24.3) when he takes up his position in Achaea.

In *Timoleon*, Plutarch moves beyond the issue of establishing initial relations with subject cities to the challenges of setting up institutions enabling a city to manage its affairs with a minimum of oversight.[28] At the close of the *Life*, Plutarch includes a 'snapshot' of Timoleon taking a limited role in Syracusan affairs, illustrating how supreme authority should be exercised in an imperial system:

> While they [the Syracusans] decided other matters by themselves, for the more important deliberations they summoned him ... And he, after returning their greetings and allowing some time for their felicitations and praises, would then listen carefully to the matter under debate and pronounce opinion. And when this opinion had been adopted, ... the citizens, after sending him on his way with shouts of applause, would proceed at once to transact the rest of the public business by themselves. (*Tim.* 37.5–7)

This vignette is more expansive than in other accounts. Diodorus reports that Timoleon remained fully engaged in Syracusan affairs until his health declined (Diod. 16.65–90), while Nepos describes Timoleon living as a private citizen whose advice was sought on public actions because of the goodwill felt towards him (Nepos, *Timoleon* 3.5–4.4). Plutarch's version brings the independent actions of the Syracusans to the fore as well, underscoring the importance both of the ruler's duty to respect the autonomy of cities and of the citizens' responsibility to manage their affairs independently to the extent possible.

The goal of limiting intervention of the overlord is repeatedly stressed in Dio, including in his advice to the men of Prusa to hide their internal differences from the new governor so that he would visit them 'with joy and eagerness' (ἡδέως καὶ προθύμως) like a physician visits the healthy, rather than 'with apprehension and worry

28 Over the *Life*, Timoleon is shown freeing Sicily from tyrants, attracting new settlers (*Tim.* 23.2–5), arranging land distribution (23.3–7) and overseeing the administration of justice (24.3–4)—and ultimately is revered as a new founder of the polity for which he served as a 'master-builder' (δημιουργός) (*Tim.* 35.4–5).

over their treatment' (ὑπόπτως καὶ ἀηδῶς θεραπείας ἕνεκεν, *Or.* 48.2–3). Pliny's letters to Trajan, in turn, reflect the wide scope of potential Roman intervention—both solicited and unsolicited—in a city's administrative, financial and legal affairs, with *correctores* being appointed with increasing frequency to address local problems.[29]

Conclusion

Although this chapter links contemporary political challenges to Plutarch's depiction of specific events in only five of the *Parallel Lives*, throughout the entire series Plutarch employs similar techniques of attaching motives and detailing methods of exercising authority in order to amplify the parallels between the practical problems faced by his heroes and by men in his audience.[30] In addition to direct authorial comment and the correspondence between precepts in the *Moralia* and actions in the *Lives*, Plutarch incorporates the priorities of the imperial era into his depiction of the motives and strategies of his historical heroes and thereby facilitates comparison with the contemporary political landscape.

Importantly, Plutarch provides detail on 'why' and 'how' famous actions were undertaken, without distorting the sequence of major events or detracting from the intended lessons in virtue and vice. Indeed, these details often reinforce the portrait of his hero's moral character. However, by linking certain incidents to motives for action that Plutarch's readers shared, as well as to practical strategies they could imitate in performing their official duties, Plutarch closed the gap in political space and time between the *politikoi* of different eras: he brought the great statesmen of the past into sharper focus as paradigms of how to administer cities and conduct diplomacy in his own day.

29 Pliny addresses new construction (*Ep.* 10.23, 10.37, 10.39, 10.49, 10.70), prisons (10.19, 10.31), legal precedents (10.47, 10.56, 10.58, 10.65, 10.79) and city finances (10.17, 10.54, 10.81). He offers advice to a newly-appointed *corrector*, Maximus, at *Ep.* 8.24.

30 Plutarch integrates similar elements into his treatment of other contemporary concerns, such as exile in *Alcibiades* and the perils of excessive honours and neglect of justice in *Demetrius*. Plutarch goes beyond Nepos (*Alc.* 9.4–5) and Diodorus (14.11.1–4) in ascribing Alcibiades' actions during his second exile to a desire to benefit his country (*Alc.* 36.2–37.4), while *Demetrius* includes extensive authorial comment to tie Demetrius' downfall to his acceptance of extraordinary honors (*Demetr.* 10–13; 30.4–5) and neglect of justice (*Demetr.* 42.1–6)—pitfalls also discussed by Cicero (*Ad Quint.* 1.1.31 and 1.1.13–14) and by Dio (especially in *Orations* 1 and 3).

Eran Almagor

Greatness measured in time and space: The *Agesilaus–Pompey*

Abstract: This chapter explores how time and space are presented in the parallel biographies of Agesilaus and Pompey and the relevance of this presentation for Plutarch's age. It offers two propositions: (1) The respective failures in the careers of the two protagonists (in terms of their failure to achieve spatial expansion and the absence of an enduring memory for their accomplishments) stem from the heroes' flawed perceptions of space and time. In these misconceptions, space is either envisioned in terms of the privileged position of one place (Agesilaus' Sparta) or as having no centre at all (Pompey's world); time is perceived as either a dimension in which different time-levels are confused (as in Agesilaus' world) or as lacking a balance between different time-frames (as in Pompey's world). (2) There are historical as well as moral dimensions to these conceptions of space and time in the two *Lives*, which relate to the position of Greece in imperial Rome. The deeds of the statesmen Agesilaus and Pompey ultimately contributed to Greek political weakness under Rome as well as to the cultural flourishing of Greece alongside the thriving Latin literature.

The parallel structure Plutarch adopts for his *Lives* of Greek and Roman heroes is unique.[1] The two biographies are usually of heroes separated by a gulf of time and space, yet are closely linked and should be read as one.[2] It is to the credit of Plutarch's ingenuity that he succeeds in creating an entirely new composite universe with its own spatiotemporal features out of these two distinct life stories.[3] In each of the *Lives*, the physical traits of the fictional world are closely related to the character of the protagonists in question, their moral qualities and their political careers, as if space somehow reflects their psychological make-up writ large.[4] The compound universe of the paired *Lives* similarly echoes the common features of the two person-

1 I follow the Teubner edition's section numbering throughout this contribution. On Plutarch's parallel pattern of the *Lives* and its significance, see the edited volume of Humble (2010). Cf. Erbse (1956); Larmour (1992); Walsh (1992). Among previous examples of parallelism, but not of biographies, cf. that of Caecilius of Calacte (Plu. *Dem.* 3.2); cf. Roberts (1897); Innes (2002), esp. 277. I would like to thank Aristoula Georgiadou and Katerina Oikonomopoulou for a *time*less conference that occupies a special *place* in our hearts.
2 See Erbse (1956); Stadter (1975); Pelling (1986a).
3 See Beck (2007) and in this volume; Verdegem (2010) 80 – 81 on questions of frequency; Zadorojnyi (2012) on *mimesis* of past models through time; and Beck (2012) and in this volume on spatial locations and physical structures in space across biographies.
4 Following the Platonic analogy of the soul and the *polis:* See *Resp.* 2. 368c–9a; 4.434d–445e; 5.462c–e; 8.544d–545c. See Neu (1971); Williams (1973); Ferrari (2003).

https://doi.org/10.1515/9783110539479-014

alities exemplified in the historical heroes. Since the parallelism is of a Greek hero (and environment) and a Roman figure (and his own surroundings), the complex structure allows Plutarch to consider the way the two spaces intermingle as well as the way Greek history (namely, Greek time) is interwoven into the Roman imperial heritage.

In this brief study, I will explore the way time and space are presented in the *Lives* as echoing the worldview of the heroes and their perception of life. I will address the implications of these approaches for the interrelation of the two parallel biographies. In the second section I will show that the way the two *Lives* represent time and space is relevant to Plutarch's age as an allegory of the place of Greek culture and history within the Roman Empire. I will focus on one pair as an example, namely, the biographies of two failed heroes, the Spartan king Agesilaus II and the Roman statesman and commander Pompey.[5]

I

Something funny happened to Agesilaus and Pompey on their way to greatness. Theirs were stories that should have been tales of success and glory—for example, for winning victories in three continents—but they were not. Their careers were rather failures in terms of spatial expansion during their lifetime and in terms of the lack of an enduring memory of their feats. In these two respects, the duo was to be eclipsed by Plutarch's another pair—respectively, Alexander and Caesar. This failure seems to stem from their intriguing misperceptions of time and space.[6] Artistically, Plutarch accomplishes the integration of the two men's careers, which were visibly divided into two periods (*Ages.* 40.3, *Pomp.* 46.1), the physical asymmetry of the lame Spartan king (cf. *Ages.* 2.3, 27.1, 30.1), Pompey's constant movement (below) and other elements into a compound picture in which time and place have singular features.

The reign of Agesilaus (401–359 BCE) is characterised by Sparta's struggle for hegemony in Greece (*Ages.* 16–19, 22–35, 40.3) and the clash between Greeks and Persians (*Ages.* 6, 8–15, 23). It is basically a story of a series of territorial disputes and

5 On Agesilaus and Pompey as a pair see Hillman (1994); Harrison (1995); Nevin (2014).

6 The two *Lives* belong to a group that may be termed as the 'Alexander cluster' in the *bioi*, that is, *Lives* that reflect on the Macedonian king, particularly with respect to the ambition to conquer, and especially the east. On the Greek side of the pair, the heroes shown are figures who mostly precede Alexander chronologically, or even his successors, but all are 'failed Alexanders'—they fall short of Alexander's achievement. On the Roman side, they can be termed as 'Alexander wannabes'. Thus, besides the *Alexander—Caesar,* this cluster incorporates *Agesilaus—Pompey, Cimon—Lucullus, Demetrius—Antony, Nicias—Crassus, Sertorius—Eumenes, Pyrrhus—Marius* and *Lysander—Sulla*. This comment has nothing to do with the order of composition of the *Lives*, which is a vexed question in itself. On some studies which suggest a combined reading of several of these *Lives* see, e.g., Mossman (1992); Harrison (1995); Beneker (2005); Buszard (2008).

clashes (e.g., *Ages.* 6.1–4, 8.3, 9.2–4, 10.1–5, 11.1–4, 12.3–4, 15.4, 16.1, 17.1–4, 18.1–4, 21.3, 22.1–11, 26.2–9, 27.3, 28.1–5, 30.7, 31.1–6, 32.1–13, 34.1–2, 35.2–6, 39.7) and of an attempt to create a regional empire or achieve greatness by gaining supremacy and terrain. When the leadership in the campaign against Persia in 396 BCE is given to Agesilaus, he is certain that territorial gains will help him obtain reputation in Greece.[7] Even the gradual expansion of his command gives him great honour.[8] Yet the campaign is brought to an end because the so-called Corinthian War in Greece (395–387 BCE) forces Agesilaus to return with his task unfinished (ἀτελευτήτῳ ἐπὶ ἔργῳ; cf. *Il.* 4.175).[9] At this point the narrator exclaims (*Ages.* 15.5):

> Ἀγησιλάῳ μέντοι οὐδὲν κρεῖσσον ἢ μεῖζόν ἐστι τῆς ἀναχωρήσεως ἐκείνης διαπεπραγμένον, οὐδὲ γέγονε παράδειγμα πειθαρχίας καὶ δικαιοσύνης ἕτερον κάλλιον.
>
> Agesilaus never performed a nobler or a greater deed than in returning home as he now did, nor was there ever a fairer example of righteous obedience to authority.[10]

Not even Alexander performed as great an act as this, when he was away and Macedonia was confronted by Agis III the Spartan, claims the narrator with no small measure of irony.[11]

To better understand this outcome, let us explore the way space and time feature in the biography. Echoing the territorial structure within the Peloponnese, where Sparta is central and adjacent areas (like Messenia) are peripheral, the political prominence of Sparta in the Greek world is reflected in Agesilaus' stance towards it, lending it a similar centrality. While this centrality would make sense as a political metaphor (cf. *Ages.* 27.7), it would be quite inappropriate and awkward if this image were to be perceived as having spatial significance as well. Imagining Sparta as a centripetal force around which the entire world revolves is of course, misguided; it attributes to space some peculiar features it does not possess. Instead of being an even dimension, in which every place is in theory equal to the other, space in this *Life* is repeatedly described as uneven.[12] The inconsistency between Agesilaus' demand of autonomy for

---

7 Cf. *Ages.* 6.1–5: Agesilaus believes that Lysander's procuring him the command against Persia was a greater deed than raising him to the throne; *Ages.* 9.2: Agesilaus wishes to do something memorable in the eyes of the Greeks, like the sojourn of the Ten Thousand in Persia. Cf. Xen. *Hell.* 3.4.2. See Shipley (1997) 116–119, 146–148.

8 *Ages.* 10.9: the honour of assuming control of the navy as well as the army was given to no Spartan previously. Cf. Xen. *Hell.* 3.4.27, Shipley (1997) 44, 163–164. Also. Plu. *Ages.* 37.1.

9 On the causes and course of this war see McKay (1953); Bruce (1960); Kagan (1961); Perlman (1964); Hamilton (1979); Lendon (1989).

10 Perrin translation in the Loeb. Cf. Xen. *Ages.* 1.36 (vs. *Hell.* 4.2.3) and Nepos *Ages.* 4.3. See Shipley (1997) 206–208.

11 Cf. Diod. 16.63, 68, 17.62.4–63.4; Arr. *Anab.* 2.13; Curt. Ruf. 6.1; Just. 12.1; See Badian (1967) and Lock (1972).

12 Note the mention of high or inaccessible places: 'most sacred' places (ἐν τοῖς ἁγιωτάτοις ἱεροῖς: 14.2), Mount Helicon (18.9), Heraeum (22.1), Cadmea (23.6–11), the Acropolis in Megara (27.1), the Issorium (32.6–9).

the cities of Boeotia and his unwillingness to consent to a similar request for Laconia is well known (*Ages.* 28.1–3). Note the retort of a Spartan (*Ages.* 31.8) to an Argive who has told him 'Many of you lie buried in the lands of Argos'. The Spartan answers: 'But not a man of you in the lands of Laconia' (Ὑμῶν δέ γε οὐδεὶς ἐν τῇ Λακωνικῇ).[13] Similarly, in *Ages.* 39.5–6, the trench which Agesilaus' enemies build in Egypt offers Agesilaus freedom of movement into their own held territory but not the other way around, allowing him to fight on fair and equal terms against a superior number of troops (cf. Diod. 15.93.3).[14] The spatial imagery of imbalance is the basis for the perception that all political events are ultimately gravitated towards one privileged place, that is, Sparta (*Ages.* 6.2).[15] Agesilaus is thus repeatedly drawn back to his city.[16] This appears to be the cause and source of his strength, but also of his weakness, vice and failure.[17] He never succeeds in accumulating territories abroad for Sparta and in achieving glory for himself while doing that, because he has to return to his home base (unlike Alexander, for instance: *Ages.* 15.6). Similarly, Sparta pulls Alcibiades (*Ages.* 3.1–5) with disastrous effects for the royal line and draws the Thebans (*Ages.* 31.1–6, 34.3–5), who bring irreversible calamity to Spartan might and reputation.

As a corollary to the well-defined features space seems to possess in the *Agesilaus*, time is also awkwardly presented. It would appear that the Spartan king fails to fully appreciate the nature of time: he misperceives the differentiation between temporal units and treats all of them as equal, regardless of the passage of hours, seasons or years (cf. 14.3). This misconception implies that theoretically the transition between completely different periods of time while staying at the same place is possible. Indeed, Agesilaus is portrayed as visiting the same moment in the past again: for instance, in his identification with Agamemnon of the distant past (*Ages.* 6.4). In his sleep he hears a voice comparing his expedition with that of Agamemnon (both commanding the *same* army, waging war against the *same* enemy and setting out from the *same* places).[18] Agesilaus' misapprehension is echoed in the biography se-

---

13 There are many instances of asymmetry in the *Agesilaus*; cf. 33.5, 34.10. See the punishment inflicted on those who had shown cowardice in battle (*Ages.* 30.4): half of their beards were to be shaven, and half were left to grow. Cf. the physical simile in 33.3: ῥοπὴ τὴν πᾶσαν ἔκλινεν, 'turned the entire scale'.

14 Cf. Agesilaus' scheme in *Ages.* 26.7, where Spartan fewer soldiers are demonstrated to paradoxically outnumber the allies.

15 Cf. the general movement towards Sparta in the biography: 20.2 [Xenophon], 24.9, 27.5, 31, 37.7.

16 Cf. *Ages.* 22.8, 27.3: κομισθεὶς εἰς Λακεδαίμονα πολὺν χρόνον ἔσχεν ἀρρώστως, and 40.2.

17 His obedience to laws and authority and his defence of his country in 370 BCE (*Ages.* 4, 15, 34) are virtues, while his treachery (23, 37), transgression of justice (24–6), bias against Thebes (6, 22, 26, 28, 30) and love of victory (33)—all for the sake of Sparta (cf. Xen. *Hell.* 5.2.32)—are vices. Cf. Hamilton (1992) 4205–4207, who finds 'balance between acts to emulate and those to avoid'.

18 Ἀγαμέμνων καὶ σὺ νῦν μετ' ἐκεῖνον, ἐννοεῖς δήπουθεν· ἐπεὶ δὲ τῶν μὲν αὐτῶν ἄρχεις ἐκείνῳ, τοῖς δὲ αὐτοῖς πολεμεῖς, ἀπὸ δὲ τῶν αὐτῶν τόπων ὁρμᾷς. Cf. 5.6, 9.7 and Xen. *Hell.* 3.4.3. See Hamilton (1992) 4213–4214 and Nevin (2014) 49–59. The falsity of this claim is obvious to readers aware of the Dorian invasion taking place after the Trojan war (cf. Pind. *P.* 1.62–66, Hdt. 1.56.2, Thuc. 1.12.3:

veral times; for example, the Isthmian games in Corinth are held twice in the same place—once when Agesilaus holds the city, and once when the Argives reclaim it—and consequently some contestants win victories a second time, while others win the first time and lose the second time (*Ages.* 21.5). A parallel can be seen in the repeated expeditions of the Spartans in the same Theban territories (*Ages.* 26.4).

Ignoring the essential difference between time phases leads to Spartan failure and to Agesilaus' eventual death. He assumes that he can still act as the younger version of himself when he is much older, something that both the narrator (in *Ages.* 36.4: τοῦ γὰρ καλοῦ καιρὸν οἰκεῖον εἶναι καὶ ὥραν ..., 'honourable action has its fitting time and season'; cf. 33.5, 34.5) and the Egyptians (in *Ages.* 36.9) find ludicrous. Disregarding the difference between moments in the day causes the Spartan Sphodrias at one point to fail, when, though initially he intends to attack the Piraeus during the night, in effect he does so only after sunrise, in the light of day (*Ages.* 24.7). At the end, when Agesilaus attempts to leave Egypt in haste, the winter season—as part of the year's cycle—causes him to keep close to the shore, to drift and never reach home (*Ages.* 40.3).

Indeed, it is the passage of time, a fact ignored by Agesilaus, that brings about his death at the age of eighty four (*Ages.* 40.3). Agesilaus' failure is echoed in the unsuccessful attempt of Agis IV to revive previous traditions (*Ages.* 40.5); Agis' Sparta cannot go back to the older Sparta, just as Agesilaus is incapable of really returning to the days of Agamemnon:[19] the irreversible nature of time does not allow the negation of what has been done and the restoration of what is long since gone. In fact, time is not symmetrical, and the 'arrow of time' (so named by physician Arthur Eddington) goes in one direction only. This is also seen in the ridiculous symmetry brought in the first and last chapters: the triad Archidamus—Agis—Agesilaus is repeated in the same sequence.[20] The conclusion is not that history repeats itself, as

---

eighty years after the Trojan War, Eratosthenes, *FGrH* 241 F 1a, Strab. 13.1.3, Hall (1997) 114–128, making Agesilaus' Sparta distinctly different from Agamemnon's Sparta. This theme is ironically repeated in Agesilaus' following in the footsteps of Menelaus (*Ages.* 40.3 ~ Hom. *Od.* 4.351–592). The narrator seems to allude to this Spartan belief in circularity, by citing the paradox that philosophers present (*Ages.* 5.5), namely, that if there was an end to strife and discord, the heavenly bodies would stand still, and all generation and motion would cease in consequence of the general harmony. Following this conviction, the narrator continues, the Spartan lawgiver introduced the spirit of ambition and contention into his civil polity, with the desire that the citizens should always be at variance and conflict with each other. See Shipley (1997) 109–112.

19 The notion of 'two cities' of Sparta (cf. Cartledge and Spawforth [1989]) basically comes from Plutarch.

20 *Ages.* 1.1: Ἀρχίδαμος ὁ Ζευξιδάμου βασιλεύσας ἐπιφανῶς Λακεδαιμονίων, κατέλιπεν υἱὸν ἐκ γυναικὸς εὐδοκίμου, Λαμπιδοῦς, Ἆγιν, καὶ πολὺ νεώτερον ἐξ Εὐπωλίας τῆς Μελησιππίδα θυγατρός, Ἀγησίλαον. *Ages.* 40.5: τὴν δὲ βασιλείαν Ἀρχίδαμος ὁ υἱὸς αὐτοῦ παρέλαβε, καὶ διέμεινε τῷ γένει μέχρις Ἄγιδος, ὃν ἐπιχειροῦντα τὴν πάτριον ἀναλαβεῖν πολιτείαν ἀπέκτεινε Λεωνίδας πέμπτον ἀπ' Ἀγησιλάου γεγονότα. Cf. Shipley (1997) 401, on a sense of closure here.

Agesilaus would have us believe; quite the contrary, this odd repetition rather shows the absurdity and the unfeasibility of the notion.

The two periods of Sparta, of Agesilaus II and of Agis IV, are thus mentioned as two strata of time. Agesilaus' own lifetime can also be divided into two periods of time—as a young *privatus* and as an elder man in office: *Ages.* 40.3; cf. 2.3–6, 3.4–6, 24.3, 36.3, 36.9. The narrator takes into account the lapse of time, allowing the comparison of the period of Agesilaus with the following ones—that of Alexander, for instance (*Ages.* 15.4: the triumph over the Persians was left for Alexander and the Macedonians). In a way, the narrator's position is similar to that of the Roman protagonist of the pair, who is remote from the Greek world in yet another stratum of time. Pompey, almost three centuries later than his Plutarchan Greek counterpart, makes the same comparison between his own age and the Greeks of old, living in the shadow of the Greek past, to which he compares himself.

The *Life*, and indeed the career, of Pompey involves a comparison with past Greek models, especially with Alexander in the first half of the biography, chapters 1–46.[21] Pompey is seen to engage in a competition for glory, attempting to surpass his (real and mythic) predecessors.[22] Constantly turning his attention backward in time, Pompey emerges as the mirror image of Agesilaus, and in many respects his personality ostensibly rectifies the character flaws of his counterpart protagonist. In particular, Pompey appears to have a more realistic view of time, which may explain his initial success. Unlike Agesilaus, Pompey appears to acknowledge that there are different temporal frameworks and to emphasise this fact. At the beginning of the biography, his youthful age is presented as an asset (*Pomp.* 8.3). Pompey repeatedly uses time to his advantage, like when he manages to muster three complete legions in a short period of time (*Pomp.* 6.6). He takes pride in the fact that it has taken him only forty days altogether to defeat his enemies in Africa and settle the situation there (*Pomp.* 12.8), and then reiterates how his young age benefits him (ἔτος ἄγων ἐκεῖνο τέταρτον καὶ εἰκοστόν, 'though he was but twenty-four years of age'; cf. *Pomp.* 54.1). When Pompey eradicates piracy in the east, it takes him less than three months (*Pomp.* 28.3: οὐκ ἐν πλείονι χρόνῳ τριῶν μηνῶν). He also utilises particular times during the day, like the dark hours, when the moon's light is at the Romans' backs (*Pomp.* 32.11; cf. *Pomp.* 12.1), unlike Agesilaus' Sphodrias, who fails to make good use of the times of day. Pompey thus seems to be sure that time is on his side, a fact which is explicitly stated (*Pomp.* 41.3: τὸν χρόνον εἶλκεν) when he faces Mithridates.[23]

Pompey thus appears to display an awareness of time which guides his political (and military) career that Agesilaus seems to lack. This awareness aids him in his ini-

---

21 See Nevin (2014) 60–67; Greenhalgh (1980) passim, esp. 6, 11, 104, 122, 126, 134, 137, 168–173.

22 Alexander: *Pomp.* 2.2–3, 34.7, 36.2, 46.2, *Comp. Ages. et Pomp.* 2.6; Amazons (allusion to Alexander): 35.5–6; cf. 1.1 (Heracles); 63.2 (Themistocles and Pericles) and 67.5 (Agamemnon).

23 Cf. *Pomp.* 13.9: time made the epithet 'Magnus' familiar and not unpleasant; 21.1: he remained in Spain long enough time to quell disorders.

tial successes, as well as in his later life.[24] Pompey does so by acknowledging a focal point in the temporal dimension (the 'arrow of time' again). Yet, while the temporal dimension in Agesilaus' *Life* is flawed in one way, Pompey's is shown to be imperfect in another. Whereas Agesilaus makes no essential difference between moments in time and is under the impression that he can easily manoeuvre his way through them, Pompey differentiates between the past and present but is constantly torn between them. His temporal dimension is thus split between an emphasis on the normal passage of time and an awkward denial of it and a focus on the past. The clearest simile for this temporal division is Pompey's famous saying to Sulla, that more people worship the rising sun than the setting sun (*Pomp*. 14.4: τὸν ἥλιον ἀνατέλλοντα πλείονες ἢ δυόμενον προσκυνοῦσιν): the two temporal foci are presented as two different suns. This element appears in the structure of the *Life* as two time-frames, one of growth and fame, and the other of stagnation and decay. The narrator marks this transition explicitly (*Pomp*. 46.2):

> ὡς ὤνητό γ' ἂν ἐνταῦθα τοῦ βίου παυσάμενος, ἄχρι οὗ τὴν Ἀλεξάνδρου τύχην ἔσχεν· ὁ δὲ ἐπέκεινα χρόνος αὐτῷ τὰς μὲν εὐτυχίας ἤνεγκεν ἐπιφθόνους, ἀνηκέστους δὲ τὰς δυστυχίας.
>
> How happy would it have been for him if he had ended his life at this point, up to which he enjoyed the good fortune of Alexander! For succeeding time brought him only success that made him odious, and failure that was irreparable.[25]

For Pompey, time has no obvious continuity between its different parts (beginning/past and end/future), and he seems to bounce between them.[26] For instance, Pompey is told that only a thrush 'out of season' can save him (*Pomp*. 2.11)—a passage which, ironically, is itself taken out of its chronological sequence (ταῦτα μὲν οὖν ὕστερον, 'this was later') and physically placed in a paragraph in the first half of the *Life*.[27] In another example, Pompey is reported to be thirty–four years old when he is back from the east (*Pomp*. 46.1), but this, the narrator claims, was merely an image, as his real age was forty (actually, forty–four in 62 BCE). Again, we see one temporal frame looking forward (progressing with the flow of time), another looking backward. Similarly, Pompey is said (*Pomp*. 79.5) to have lived 59 years—one year *short*

24 *Pomp*. 58.3: he refuses to give Caesar the same temporal advantage by declining to prolong Caesar's time in office; 66.1: Pompey uses delays in order to win; 75.1: Pompey's fortune stays with him for a longer period of time than customary.
25 Both existing commentaries on this *Life* end at this point: Watkins (1984) and Heftner (1995).
26 Archaic Rome: 4.6–10; 25.9 (Romulus). See also 64.7 and 80.4 (τὰς δὲ πρώτας στρατείας ἔτι νέος Πομπηΐῳ συνεστρατευμένος, '... in his youth had served his first campaigns with Pompey'). Pharsalus affects its spectators and victor to adopt this approach of bouncing between historical moments: see Zadorojnyi (2012) 193–196, on *Pomp*. 72.5–6 and 70.1–7 ('metahistorical spectacle'). Cf. *Comp. Ages. et Pomp*. 3.3.
27 Cf. *Luc*. 40.1 and Watkins (1984) 26.

of 60 (ἑξήκοντα μὲν ἑνὸς δέοντα), but, on the other hand, to have died one day *later than* his birthday (μιᾷ δ' ὕστερον ἡμέρᾳ).[28]

The inability to balance these two temporal frameworks has some interesting impact on the representation of Pompey's spatial world. In short, his universe, unlike Agesilaus', contains no distinct centre; it has no fixed place to which he is pulled over and over again. This explains why Pompey is able to successfully conduct several expeditions at far corners of the empire and to have territorial gains. For instance, in *Pomp*. 38.4–5 the narrator recounts how Pompey had a great desire to recover Syria and march through Arabia to the Red Sea to reach the Ocean from all its sides, just as earlier, in Africa, he arrived at the Outer Sea (τῆς ἐκτὸς θαλάσσης), and in Spain he reached the Atlantic Ocean. This notion forces Pompey to be constantly nomadic and always on the move.[29] Pompey never succeeds in seizing Mithridates, and instead chooses to engage in other fronts (Albanians and Iberians: *Pomp*. 34–35).[30] Later on, however, this mode of life leads him away from political success. For example, Pompey shuts his door before those who come on Cicero's behalf and slips away by another (*Pomp*. 46.9); well-known is Pompey's saying *navigare necesse est*, *vivere non est necesse* (*Pomp*. 50.2); Caesar is amazed (*Pomp*. 63.2) that Pompey gave up Italy when he was in possession of a strong city, mastered the sea and was expecting reinforcements. Here lies a reason for his ensuing failure. Because Pompey has no centre of operations, none of the territories he is able to gain and hold add up to produce a cumulative achievement. Eventually, his life's project does not end in a territorial expansion of his power base.

In the first time period or the first half of the biography, Pompey's itinerant life appears adventurous and courageous, and seems to spell success. In the second half, once it is realised that Pompey has no centre of operation, this nomadism portrays him as misguided and fleeing from reality.[31] Note how the depiction of reaching Africa in *Pomp*. 38.5 as a token of success is markedly different from the ill-advised and miserable end of his life on the African shore. The presentation of the sea is another case in point. With its tides and ebbs, the sea is notoriously unreliable.[32] In the first half of the biography, the sea contributes to Pompey's rise to glory, aiding him with the suppression of the pirates (*Pomp*. 26.5) and enabling him to speedily arrive at different scenes of war. Near the end of the *Life*, however, it is the same sea in which Pompey finds his death. Ever wandering, Pompey is killed on his way to Egypt (*Pomp*. 79.1–5) incapable of fleeing the boat he is on. As opposed to Agesilaus,

---

28 Cf. the contrast with *QC* 8.717C; *Cam*. 19.11 (Pompey dies on the very day of his birthday or a day before). Cf. Velleius 2.53.3. See Pelling (2011a) 227–229, on this contradiction.

29 Cf. ἀναστρεφόμενος (*Pomp*. 3.5). The wandering character of Pompey in his biography is markedly different from his portrayal in other biographies, and corresponds to indecision noted by Pelling (1980) 134: 'He is, indeed, a man to whom things happen—and he lets them'.

30 See Almagor (2013b) 163–165, on *Pomp*. 35 and the elusiveness of the Amazons.

31 Note his awkward idea to arrive in Parthia (*Pomp*. 76.6–8).

32 Cf. Pelling (1988) 22 on this imagery in *Demetr.–Ant*.

who was led *from* Egypt towards his centre in Sparta but was defeated by the seasonal stormy sea in a fatal miscalculation of time, Pompey was travelling *to* Egypt, having no centre to return to, led by the persistent comparisons with the past,[33] and equally unable to perceive reality properly.

II

What is the significance of this analysis and how is it related to Plutarch's own age? The direct conclusion that can be reached from this examination would be that both Agesilaus and Pompey present two variations of a personality whose views of time and space are perverted. In this worldview, time and space are either given a superfluous focal point or are deprived of one. The result is either an image of a centripetal space (in which one place is unduly turned into a centre, attracting and drawing historical agents and thus hindering a successful political action beyond it), or else a perception of split time (with two focal points, in the present and in the past), violating the notion of 'time's arrow', with a corresponding centrifugal space (in which historical agents move away from the centre). Agesilaus and Pompey thus present two extremes, and should both be avoided. A more balanced type of person is needed, one that could combine the merits of both schemes. This ideal person's worldview would have both a space with a clear centre linked with other parts of the geographical cosmos and a realistic notion of time which would hamper any false illusions of living in (or competing with) the past. By getting rid of the grounds for such destructive delusions, such an ideal person would not only be morally better but also politically successful, in a way that Agesilaus and Pompey were not.

Yet, there is more to this analysis than a mere simple moralistic claim. There is a sense in which the actions of Agesilaus and Pompey have paved the way to the political situation in which Greece and Rome are found in the imperial period. The Greek-speaking world is politically subdued by Rome, but—one would imagine, in the eyes of Plutarch as well as other Greek men of letters—is culturally thriving, alongside a rich Latin cultural milieu. How could this come about? The answer is presumably to be observed in the Greek and Roman histories presented by Plutarch.

In a way, Agesilaus' perception of Sparta as central to the point of limiting his activity abroad had enormous impact on Spartan and Greek history. His vision dictated reality, and his understanding of history and time, according to which several periods of time could be located in the same place, was a self-fulfilling notion. Note that Agesilaus himself in fact undergoes the burial ceremony twice: once as covered by wax abroad (in Egypt) and once as inhumed in Sparta (*Ages.* 40.5); the imagery actually appears earlier in the biography, in the advice given by one of the elders

---

33 The city of Alexander in Africa and Pompey's own dealing with Ptolemy XII, cf. *Pomp.* 13.8, 49.9–13.

to Agesilaus 'not to dig Lysander up again' (*Ages.* 20.5: μὴ τὸν Λύσανδρον ἀνορύττειν).[34] Thus, two moments of time are captured in one place. At the end, his Sparta was indeed to become another layer of time, later coated by subsequent strata, like Alexander's era and Roman imperial reality. Thus, in the world-embracing Roman Empire, Greek history is just another layer, or indeed, in the geographical composite, which is the Roman Empire, Greece is spatially the same, but laden with Roman presence. The different layers, which existed in different periods, are present in a simultaneous display, as if they are all concurrent in one place, like in an archaeological site.[35] This is the way space freezes time, as it were.[36] Through his actions and notions, therefore, Agesilaus was one of the key figures who has weakened Greece politically by causing it eventually to be laden with the presence of subsequent superior forces. In general, it would seem that to Plutarch, the cause of Greek loss of political independence and the destruction of Greece appears to be the internal dissent, which made Greeks unable to repel Macedonian and later Roman intervention from the *outside*.[37] Thus, through the presentation in one text of the bygone Greek world and Roman history prior to the contemporary Roman imperial reality, Plutarch is able to explore the historical significance of Greek culture within Roman space in a more subtle and sophisticated manner.[38]

As we have seen, Pompey is not only living his own time, but is equally engaged with Greek (actually Greek and Macedonian) history. His love of things Greek is exemplified in many instances; most of the last words and phrases Pompey uttered in his life are said to have been in Greek: cf. *Pomp.* 79.2, where he is described as taking a little roll containing a speech written by him in Greek, which he prepared for his use in addressing Ptolemy, and beginning to read in it. In fact, it may be seen that this predilection leads him to have two places co-existing in his mind. Since Pompey's existence involves several concurrent places, it is only symptomatic that this would find a spatial equivalent in his burial: his body is cremated in one place, buried in another (his Alban villa), but his head later finds its way to Caesar

---

34 Incidentally, an Egyptian practice: Hdt. 2.41.6: ἀνορύξαντες δὲ τὰ ὀστέα ἀπάγουσι καὶ θάπτουσι ἐς ἕνα χῶρον πάντες.

35 One image that may be advanced to illustrate this vision of layers of historical periods is the image of the archaeological *tell*. As is well known, a *tell* (a Semitic a loanword; see Wright et al. [1974] 123–124) is a mound created by the cumulative building and rebuilding of progressing generations on the same physical place (particularly in the Near East). On the archaeological concept see Davidson (1976); Miller Rosen (1986). The superimposed layers of the *tell* often overlap horizontally.

36 The image of the *tell* is, I argue, exactly what Plutarch is doing in his paired biographies, in most cases where the first layer of this literary *tell* is the Greek *Life* on which the Roman biography is added as a later stratum. The existence of both the Greek and the later Roman layers together in one unified space (the literary text) is tantamount to building on the same place in a *tell*. I think the metaphor of the archeological *tell* is appropriate, as it discloses the way different periods of the past co-exist with each other and with the historical appreciation of later generations.

37 *Ages.* 15.3–4. Cf. *Flam.* 11.3–7; *Cim.* 19.3. Cf. Almagor (2014) 284, 288.

38 See Boulogne (1994) 57–61; Swain (1996) 137–144.

(*Pomp.* 80.7–10). Pompey is thus buried in several different places at the same point of time. This again may be a metaphor for the importance of Greek culture in Rome. Through Pompey's resurrection of Greek images and texts (admittedly, mostly related to Alexander), he was pivotal in reviving Greek culture, and setting it as a contemporary cultural centre alongside Rome.

Plutarch, however, may also be including a warning for the future. It would appear that to the biographer the cause of the fall of the Roman Republic was internal discord that brought devastation from the *inside.*[39] There may be a point in suggesting that this internal place Romans should be aware of is Greece. This is in fact the place where Pompey's political career ends at the Battle of Pharsalus (48 BCE). Moreover, it is the Greek East where he finds his death.[40] The Greek world may not be so beneficent to the Romans after all. Living the life of a different person, like Alexander, has cost Pompey his own life. This is a powerful idea, alluding to the incompatibility of Greek ideals and images with those of Rome and portraying them as potentially harmful, as it were, to the Romans.[41] The ensuing weakness of the Roman Empire would make it vulnerable. This idea entirely corresponds to popular common beliefs circulating at the time, namely, that the destruction of Rome would come from the east.[42]

---

39 Cf. *Pomp.* 70.1–7. This point is seen in Roman moralists of the time. Cf. Sallust, *Cat.* 10.1, 37.7, *Jug.* 41.1, 7–8, *Hist.* fr. 11, 12; cf. Florus, 1.47. See Lintott (1972) 627 n. 10.

40 In the *Comp. Ages. et Pomp.* 4.6, Plutarch likens the Pharsalus plain to the Greek στάδιον and the θέατρον where Romans fight.

41 This corresponds to the bias of some traditionalist Romans. Cf. Cato *ap.* Livy 34.3.3; Plu. *Ca. Ma.* 23; Plin. *HN* 29.7.14 and Gruen (1992) 54–55, 75–80. Also Livy, 39.16.8, Pliny, 24.5.5, 29.9.19, Juvenal, 3.60–72, Swain (1996) 319–322. Note the expulsion of Greek philosophers from Rome: Plin. *HN* 7.30.113, Plu. *Ca. Ma.* 22.5; see Balsdon (1979) 33–38, 99–100. Expulsion of Greek physicians: Plin. *HN* 29.8.15.

42 For the Roman fear of the east see Charlesworth (1926). The east as corrupting Rome: Sallust, *Cat.* 11.5; Plin. *NH* 33.53.148: *Asia primum devicta luxuriam misit in Italiam*; cf. Florus, *Ep.* 1.47.7: *Syria prima nos victa corrupit, mox Asiatica Pergameni regis hereditas.* The idea of successive world empires (four or five) found in authors from the Roman period (e.g., Polyb. 38.22.1–3; Vell. Pat. 1.6.6 [Aemilius Sura]; Dion. Hal. *Ant. Rom.* 1.2.1–4; cf. Tac. *Hist.* 5.8; Appian, *Praef.* 9; See Mendels [1981]) was used by groups opposed to the imperial rule and predicting its fall. See Swain J.W. (1940) 12–22. Cf. Lact. *Div. Inst.* 7.15.11. Some Greeks, whom Livy derisively calls *levissimi* ('light-headed': 9.18.6), have cast their eyes and hopes on Parthia (cf. Justin, 41.1.7); cf. Jos. *BJ* 2.388 and *Ant.* 10.213, together with Daniel 2: 34–35.

5 Models of the past II: Plutarch and the classical era

Geert Roskam

Discussing the past: Moral virtue, truth, and benevolence in Plutarch's *On the Malice of Herodotus*

Abstract: The importance of παιδεία in Plutarch's works has long been recognised. Thorough familiarity with the rich Greek tradition is felt to be crucial for a man's self-understanding and 'Greekness'. In such a context, a work such as *On the Malice of Herodotus* is obviously significant, and we may be inclined to think that Plutarch's reaction against Herodotus is primarily motivated by 'patriotic' reasons and a concern for Greek identity. Several scholars have adopted this position, and their view in fact gains some support from a short programmatic sentence at the beginning of the work, where Plutarch claims to react both for his ancestors and for the truth (854F). Yet on closer inspection, the matter is not so simple. In this chapter, I would like to argue that Plutarch's main purpose throughout On *the Malice of Herodotus* is an ethical one and that the key of the work is not to be found in a concern for the historical truth or for Greek identity, but in Plutarch's moral approach towards literature. That, however, need not imply that the programmatic sentence mentioned above should simply be ignored: I suggest that Plutarch there considers his project from a broader perspective.

The scandal that is called Herodotus

Questions concerning Greekness occur in many of Plutarch's works.[1] In the *Roman Questions*, for instance, he explores different aspects of the Roman culture, trying to make sense of them from a Greek point of view,[2] whereas his lengthy treatise *On Isis and Osiris* provides an attempt at an *interpretatio Graeca* of Egyptian religion.[3] In all such cases, the cross-cultural discussion of course stimulates the (re)consideration and (re)definition of Greek identity. Briefly, in Plutarch's view, as in that of many contemporary authors, Greekness is no longer based on birth—a Roman such as Numa can even be called more Greek than his Spartan counterpart Lycurgus (*Comp. Lyc. et Num.* 1.10)—but on a thorough familiarity with the rich cultural, intellectual and historical tradition. Παιδεία

1 The secondary literature on this topic has increased during the last decades. Ground-breaking general studies on identity in imperial Greek literature are those of Swain (1996), Whitmarsh (2001a), and Goldhill (2001).

2 Cf. Boulogne (1992) and (1994); Preston (2001).

3 Richter (2001) argues that Plutarch in *On Isis and Osiris* fails to show genuine respect for Egyptian culture, and rejects everything that cannot be appropriated into his own Greek outlook.

https://doi.org/10.1515/9783110539479-015

is now crucial for a man's self-understanding and for what it means to be Greek.[4] And this παιδεία expresses itself in an elaborate dialogue with the great, canonised authors and a discussion of the magnificent achievements of the past. The great military victories and the brilliant career of famous heroes can and should always be revisited and be used as a mirror and a source of inspiration.

Now if all this is true, a work such as *On the Malice of Herodotus* cannot but be highly relevant in such a context. For Herodotus had long been canonised in Plutarch's age as one of the most important Greek historiographers. He had become a classical model, and his literary qualities are more than once praised by Plutarch himself.[5] Moreover, his subject matter is directly constitutive of Greek identity. For he did not only discuss the earliest periods of Greek civilisation but also the Persian Wars, that tremendous clash between two different cultures, between Greeks and barbarians, which was also the absolute triumph and culmination point of Greek history. Historical events, in short, that were of the utmost importance for the definition of Greek identity. It is probably no coincidence that Plutarch's *Themistocles* is characterised by a heavy ἔτι καὶ νῦν strand, as has recently been observed by Pelling:[6] the relevance of these great days continues to be felt even in Plutarch's time.

This does not mean, of course, that nothing has changed. Plutarch realises very well that such heydays definitively belong to the past. Greece has long lost its freedom and a repetition of the scenario at Marathon is simply impossible. In a celebrated passage from the *Political Precepts*, Plutarch makes this perfectly clear: 'Marathon, the Eurymedon, Plataea, and all the other examples which make the common folk vainly to swell with pride and kick up their heels, should be left to the schools of the sophists' (814C).[7] This passage has often been regarded as evidence of Plutarch's sound judgement and good political insight, and rightly so, but we should not forget that Plutarch does not say that this brilliant tradition should be entirely forgotten. It should not be relegated to some dusty corner, where it can be safely ignored, but should be transported from the political domain to that of rhetoric, that is: it can be embraced in the world of culture and παιδεία. Even Plutarch's contemporaries can still derive lessons from it—though not subversive ones—and this is precisely what Plutarch is doing himself in his *Themistocles* and *Aristides*.

---

4 On the importance of παιδεία in Plutarch, see the seminal articles of Pelling (1989) and (2000), and Swain (1990b); cf. (1996) 139–145.

5 See *De Her. mal.* 854E, 855A and 874B; *Non posse* 1093B; Pelling (2007a) 155–162 and Schorn (2011) 184–199.

6 Pelling (2007a) 150–151. See also Alcalde-Martín, Frazier and Oikonomopoulou in this volume.

7 All the translations of Plutarch's works are from the Loeb, except for those taken from *On the Malice of Herodotus*, which are borrowed from Bowen (1992).

In this light, it is of course important that historical events are presented correctly and constructively, for that will allow the reader to appropriate history in a meaningful way within his own context and will thus help him in constructing his own identity. And it is exactly at this point that Herodotus becomes problematic. For in Plutarch's view, Herodotus provides his readers with a highly biased, malicious account of Greek history, which interferes with such a meaningful appropriation of the past and thus with the self-definition of their Greek identity. In that sense, we can conclude that Plutarch's decision to write a work such as *On the Malice of Herodotus* is motivated by his concern to obtain a correct, morally appropriable account of this significant period in Greek history. After all, an important part of his Greek identity is at stake. This point seems too obvious to be made, but every Socratic knows that even the most obvious point may turn out to be wrong.

Plutarch's moral concern in *On the Malice of Herodotus*

And in fact, on closer inspection, things prove much more complicated. No one can deny, I think, that Plutarch's principal point in this work is a moral one. The title (whether authentic or not) is fully accurate in this respect: it is all about Herodotus' supposed κακοήθεια, that is, about the author's bad character (ἦθος)—a moral point indeed. The first sentence of the work shows the same moral interest: many, so Plutarch argues, have been deceived by Herodotus' smooth style, but even more by his character (τὸ ἦθος αὐτοῦ). From the very beginning, Plutarch thus introduces several key themes that are part and parcel of his general approach towards literature. The possible dangers of a beautiful style are often mentioned and discussed at length in the treatise on *How the Young Man should study Poetry*, whereas the focus on character is no less typical of Plutarch's thinking about literature. We may here recall a famous passage on Sophocles' evolution (*De prof. in virt.* 79B):[8] just as Sophocles finally turned towards the 'most moral' (ἠθικώτατον) style, so a student of philosophy makes moral progress if he turns to the kind of discourse that deals with character and passion (τὸν ἁπτόμενον ἤθους καὶ πάθους λόγον). Equally relevant is Plutarch's advice to young politicians to use a discourse that is full of unaffected character (ἤθους ἀπλάστου ... ὁ λόγος ἔστω μεστός, *Praec.* 802F–803A).

Read against this background, the first sentence of *On the Malice of Herodotus* thus already suggests that there exists a complex relation between the author's style, his character, and the content of his work, and this approach, as I said, is typical of Plutarch. That is the way he reads literature, the historians, and Herodotus. And thus, it is not surprising that while reading, he also pays attention to indications

8 Discussed in Bowra (1940); Pinnoy (1984); Van der Stockt (1992) 62–72; Roskam (2005) 270–271; Pelling (2007b).

that could reveal an author's character. In Herodotus' case, he discovered traces of a bad character, κακοήθεια, and that caused his reaction. There we, in all likelihood, come across the most important key to a good understanding of the work. In other words, this key is neither to be found in historiographical polemics nor in a concern about Greek identity, but in a moral reading of literature.

Moreover, we should not pass over the charge of κακοήθεια too lightly. In so many of his works, Plutarch time and again emphasises the importance of virtue, moral progress, a good and harmonious disposition, which are in his view the necessary conditions for happiness and the good life. Accusing somebody of a κακὸν ἦθος from such a perspective is no trivial matter, and it is worthwhile to pause here for a moment and examine in more detail the precise meaning of the concept of κακοήθεια in *On the Malice of Herodotus*.

Near the beginning of the work, Plutarch issues a series of clear and useful criteria that allow a reader to judge whether a narrative is written with malice.[9] This list has often been regarded as 'precepts for historians,'[10] although strictly speaking, the scope is more general, since the criteria can in principle be applied to every kind of narrative (διήγησις, 855B). Plutarch also takes care not to illustrate these general indications of malice with concrete examples taken from Herodotus, which is methodologically sound. It is true, however, that most examples are derived from the field of history[11] (which facilitates the application to Herodotus later on) and even that Plutarch primarily had the historian in mind while drawing up his list (cf. 855E: σημεῖον οὐκ εὐμενοῦς ἐν ἱστορίᾳ τρόπου; 855F: ὁ δ' ἱστορίαν γράφων, διήγησις ἱστορική).

Throughout the work, the concept of κακοήθεια is further refined. In the second sentence, it is already opposed to εὐκολία and ἁπλότης (854F). The latter term denotes uncomplicated simplicity (characteristic of women; *Comp. Ar. et Men.* 853D) and absence of all chicaneries, the former a good temper which Plutarch quite often connects with mildness (πρᾳότης).[12] Much has been written about the significance of this virtue in Plutarch's works.[13] In this context, it suffices to note that Herodotus proves blameworthy in this respect as well (868A).

9 There is a good discussion of the list in Marincola (1994) 195–196.
10 Theander (1951) 32–33 and 35; Russell (1966b) 182; Bowen (1992) 3; Teodorsson (1997) 444; Pelling (2002) 152.
11 Even the example from the comic poets (855F–856A) illustrates a historical point. A special case is that of Aristoxenus' evaluation of Socrates in 856C–D. See on this Schorn (2011).
12 See, e.g., *Crass.* 3.6; *De virt. et vit.* 100D; *De coh. ira* 461A, 462A and 462C; *De tranq. an.* 468E; *An vitiositas* 499B; *Cons. ad ux.* 608D.
13 See, e.g., Martin H. (1960); Bucher-Isler (1972) 21–22 and 30; de Romilly (1979) 275–307; Roskam (2004) 250–254 and (2005) 256–258.

Further down, κακοήθεια is often understood as a lack of εὐμένεια,[14] and this is an interesting position. Prima facie, one may be tempted to think of εὐήθεια as the most obvious opposite of κακοήθεια,[15] but εὐήθεια is quite an ambivalent term.[16] It also has a negative connotation of naïveté and as such does not really qualify as a straightforward (moral) ideal opposed to κακοήθεια. Plutarch therefore resorts to the notion of benevolence, thus killing no less than four birds with one stone: he avoids the above mentioned difficulty, keeps loyal to a distinguished philosophical tradition,[17] takes the opportunity to underline the importance of a benevolent attitude,[18] and introduces an important aspect of the method of working that he claims to use himself in his *Lives.*[19]

Plutarch also repeatedly connects κακοήθεια with κακολογία.[20] Hence the numerous references to Herodotus' slander (διαβολή)[21], defamatory language (βλασφημία)[22] and false accusations (συκοφαντεῖν).[23] Plutarch thus again establishes a direct link between Herodotus' λόγος and his ἦθος, which reflects both the traditional conviction that an author's character can indeed be detected in his work[24] and Plutarch's own thinking about literature as discussed above.

A last important characteristic of a malicious person is, in Plutarch's view, his delight in another's misfortune (ἐπιχαιρεκακία).[25] Such a base attitude is juxtaposed to envy, and both, so Plutarch argues, are born of one and the same vice (858B). Once again, all this illustrates that Plutarch's preoccupation in this work is with a moral issue.

In the main part of *On the Malice of Herodotus*, Plutarch lists different traces of the great historian's malice, proceeding in a fairly systematic way. Although the general list

---

14 See, e.g., 855B (μὴ καθαρᾶς μηδ' εὐμενοῦς ... ἀλλὰ κακοήθους); 855B (οὐκ εὐμενής); 855E (οὐκ εὐμενοῦς ... τρόπου); 855F (δυσμενής ... καὶ κακοήθης); 865B (ἰδίαν τινὰ ... ὀργὴν καὶ δυσμένειαν); 866D (οὐκ εὐμενής).

15 Rather than ἀγαθοήθεια (thus Harrison [1992] 4664), which is more current in Neoplatonism (but cf. Plato, *Resp.* 509a3); cf. Pearson, in Pearson and Sandbach (1965) 2.

16 Cf. already Plato, *Resp.* 348c11–12 versus *Epist.* 360c7.

17 See esp. Aristotle's definition of κακοήθεια in his *Rhetoric* (1389b20–21): ἔστι γὰρ κακοήθεια τὸ ἐπὶ τὸ χεῖρον ὑπολαμβάνειν πάντα.

18 On Plutarch's positive appreciation of goodwill, see Bucher-Isler (1972) 20–21.

19 In this context, scholars often point to the famous proem of *Cimon* (2.4–5); cf. Holzapfel (1884), 26 and 52; Russell (1972) 62; Hershbell (1993) 154.

20 See, e.g., 855C (ἡδόμενος τῷ κακολογεῖν); 864B (κακῶς εἴπῃ); 874B (κακολογίαν; against Fletcher (1931), who defends μικρολογίαν). The link between κακοήθεια and κακολογία returns in the *Corpus Platonicum* too; see *Resp.* 401a6–7.

21 See 859E; 862D; 863A; 863B; 863E; 865B; 867A; 868E; 870C; 870D; cf. also 856A and 856C in the list of general criteria.

22 See 858A (κακοήθειαν τῇ βλασφημίᾳ προστίθησι); 864A; 874B; cf. 855D; 870D.

23 See 863A; 866D; 868D.

24 See esp. Russell (1981) 161–164; cf. Homeyer (1967) 184–185.

25 See 855B and 858B.

of criteria soon fades into the background, it is never forgotten.[26] For it is not because Plutarch indeed refrains from applying his criteria one by one, as a pedantic schoolmaster, that the list has immediately lost all of its relevance. In fact, most of Plutarch's arguments throughout the work can easily be regarded as a direct application of one of these general criteria that are listed at the beginning,[27] and this observation is important for a correct interpretation of the work as a whole. It shows that Plutarch never abandoned his specific goal, that is, the disclosure of Herodotus' malice.

And this specific purpose is also clearly circumscribed in the small chapter 10, which is often ignored and which indeed presents itself as little more than a formal conclusion of the list of criteria and a transition towards the main part of the work. Yet here, Plutarch again makes it very clear what he precisely intends to do: he wants to observe the agenda and manner of Herodotus (856D: κατανόησιν τἀνθρώπου τῆς προαιρέσεως καὶ τοῦ τρόπου). This short programmatic sentence rings several bells. The term κατανόησις recalls the famous proem to *Nicias*, where Plutarch states that he has collected the material that serves the κατανόησιν ἤθους καὶ τρόπου (1.5),[28] whereas προαίρεσις introduces one of the most important and basic concepts of Plutarch's ethical thinking.[29] In light of this, the word τρόπου is here best understood as 'manner' (thus Bowen) or 'character', rather than as 'method' (thus Pearson in the Loeb). Plutarch is not so much concerned with Herodotus' historical method or even with historical truth, as with the historian's character. The whole work deals with an ethical topic.

Plutarch the patriot

a) The provisional conclusion, then, is that it is all about ethics. This has little to do with history and even less with identity, but everything to do with a moral reading of literature[30] and, more generally, with Plutarch's view of παιδεία.[31] However, an interesting phrase from the introduction considerably complicates this picture. After the above discussed sentence on κακοήθεια and εὐκολία, there follows a lengthy lacuna. When the text resumes, we read (854F):

> … μάλιστα πρός τε Βοιωτοὺς καὶ Κορινθίους κέχρηται, μηδὲ τῶν ἄλλων τινὸς ἀπεσχημένος, οἶμαι προσήκειν ἡμῖν ἀμύνεσθαι ὑπὲρ τῶν προγόνων ἅμα καὶ τῆς ἀληθείας, κατ' αὐτὸ τοῦτο τῆς γρα-

26 *Contra* Lachenaud (1981) 112: 'la partie principale de l'ouvrage ne répond guère, par la démarche qui est suivie, aux promesses du préambule'; Bowen (1992) 4: 'when the list is done, … Plutarch more or less ignores it. It provides neither the intellectual nor the formal framework of the attack that ensues'.
27 Cf. Homeyer (1967) 181–182.
28 See also *Pomp.* 37.1; *Ca. Mi.* 37.10; *Reg. et imp. apophth.* 172C; *Praec. ger. reip.* 799B.
29 See esp. Wardman (1974) 107–115, Pérez Jiménez (1995) and Roskam (2005) 350–351.
30 Cf. Ragogna (2002) 29–30.
31 Hershbell (1993) 154.

φῆς τὸ μέρος· ἐπεὶ τά γ' ἄλλα ψεύσματα καὶ πλάσματα βουλομένοις ἐπεξιέναι πολλῶν ἂν βιβλίων δεήσειεν.

<Since> he has employed <malice>[32] against the Boeotians and against the Corinthians in particular while sparing no one else either, I think it becomes my duty to come to the rescue, both for my ancestors and for truth, as far as that aspect of his work goes. It would take many volumes to work through all his fictions and fabrications.

Here Plutarch strikes a completely different note. We here find another familiar aspect of Plutarch's authorial persona, that is, his patriotism which we know from his enthusiastic interest in the history of Thebes and its great heroes Epaminondas and Pelopidas, from his detailed account of an episode from the history of his hometown Chaeronea at the beginning of *Cimon* (1.1–2.2), and in the first place from his famous proem to *Demosthenes.*[33] And this motif of a patriotic spirit, strategically placed near the beginning of the work, apparently introduces the issue of the locally embedded identity. Plutarch seems to suggest that it is here that we can find the most fundamental motivation behind the work, and accordingly, the patriotic reflex has often been regarded as the main key to *On the Malice of Herodotus.*[34] Plutarch claims to react for the sake of the Boeotians and Corinthians, and indeed, the sections dealing with Thebes (864B; 864D–865F; 866D–867B) and Corinth (859E–860C and 870B–871C) are among the longest of the work. Moreover, we may add that Plutarch also defends the oracle of Delphi, where he himself served as a priest.[35] All this obviously casts some doubt on the above conclusion that the whole work only deals with a strictly circumscribed moral topic, that is, Herodotus' κακοήθεια. There seems to be more at stake, and this sentence may well invite us to return to the suggestion with which we began, namely that Herodotus' account is a direct menace to Plutarch's (locally embedded) identity and that Plutarch indeed reacts ὑπὲρ τῶν προγόνων.

The question, however, is how seriously we should take Plutarch's assertion here. Unfortunately, the lacuna at the beginning of the sentence does not allow for confident conclusions, but it seems that the motif of patriotism is here primarily introduced as a convenient means in order to confine the focus of the following discus-

32 Bowen's translation is based on Turnebus' conjecture <ἐπειδὴ δὲ κακοηθείᾳ>, which makes perfect sense with regard to content, but is far too short to fill the entire lacuna.

33 Discussed by Mossman (1999) and Zadorojnyi (2005b).

34 Weissenberger (1895) 77: 'zudem darf auch nicht verkannt werden, dass hier der in seinem Nationalstolze durch den herodoteischen Bericht tief beleidigte Boeotier spricht, weshalb ja auch die Veranlassung zur Abfassung einer solchen Schrift für Plutarch sehr nahe lag'; Legrand (1932) 535: 'cette animosité de Plutarque a pour cause principale, lui-même n'en fait point mystère, son particularisme béotien'; Teodorsson (1997) 440: 'It is obvious that Plutarch's strong patriotism is the main reason, if not the only one, for his attack on Herodotus'; Dognini (2007) 482: 'Le motivazioni di queste critiche sono da ricercare senza dubbio nell'orgoglio beotico di Plutarco che esplicitamente dichiara di mal sopportare le critiche mosse da Erodoto a Beoti e Corinzi'.

35 Cf. Holzapfel (1884) 27–28; Lachenaud (1981) 122; Hershbell (1993) 160–161.

sion. Since a comprehensive refutation of all Herodotus' lies is simply impossible in a work of a limited size, Plutarch prefers to deal only with the issues that concern Boeotia and Corinth. The reason of this seemingly random selection Plutarch then finds in the care for his ancestors. This, briefly, is patriotism in the service of feasibility. But there is more. No sooner is this patriotic purpose introduced than it is almost completely forgotten. Plutarch does not at all confine himself to this aspect of Herodotus' work, but deals with a wide variety of topics, gathered from nearly all of the books of the latter's *History* (except the fourth one). In that respect, the chapters about the Thebans and the Corinthians are only part of a much broader discussion. Of course it cannot be denied that Plutarch enters at length into these topics, which may well suggest that he is far from indifferent to the matter, but that does not alter the fact that his spirited defences of Thebes and Corinth are in the end only two sections in a much larger whole. And thus, we end up with the embarrassing conclusion that patriotism has, at best, a very limited role in *On the Malice of Herodotus*, when it is proclaimed at the outset as the ultimate motivation of the whole work. How should this awkward tension be understood?

b) Before trying to answer this question, I would first like to broaden the perspective. Although Plutarch's mention of his ancestors primarily points to the Boeotians, we may, for the time being, venture to interpret it more broadly as a reference to all the Greeks. After all, Herodotus spares nobody (μηδὲ τῶν ἄλλων τινὸς ἀπεσχημένος). He himself claims to write the history of Greece (862A) and some of his readers even think that he glorifies Greece (867C),[36] but in Plutarch's view, this conviction is completely erroneous. It may in fact be regarded as the result of Herodotus' malice that disguises itself as εὐκολία (cf. 854F). For Herodotus is really a 'barbarophile' (φιλοβάρβαρος),[37] and it is against this pro-barbarian stance that Plutarch time and again reacts in *On the Malice of Herodotus*. Near the beginning of the text, for instance, he refutes Herodotus' assertion that the Persians learned pederasty from the Greeks (857C) and that the Greeks took over their venerable religious traditions and gods from Egypt (857C–E). Further on, he repeatedly opposes Herodotus' alleged custom of minimising the great military achievements of the Greeks. Moreover, Plutarch observes that Herodotus more than once transposes his own wickedness to his heroes, for both the Spartans (861E) and Themistocles (869F) appear in the Herodotean account as malicious. The evaluative summary that Plutarch makes in the last chapter is particularly revealing: if one has to believe Herodotus, so Plutarch argues, nothing great or glorious is left from the four great battles which the Greeks fought against

36 Cf. Lucian, *Herodotus* 2, where it is said that Herodotus is known by everyone as the author who celebrated the Greek victories (ὁ τὰς νίκας ἡμῶν ὑμνήσας); cf. also Hermogenes, *On types of Style* 408.9–25 Rabe (on Herodotus as the most panegyrical of the historians other than Xenophon), and Aelius Aristides, *Or.* 28.69.

37 Philipp (1984) 332: 'Aus dem Vorwurf der Barbarenfreundlichkeit schließlich spricht ein gekränkter Nationalstolz'. The term φιλοβάρβαρος may have been coined by Plutarch himself; thus Schmidt (1999) 6 n. 21, 44 n. 90 and 280 n. 33.

the barbarians (874A), whereas, on the other hand, the tradition (φήμη) represents these events as excellent accomplishments (τῶν τότε κατορθωμάτων) (874B). One may be tempted to conclude that Plutarch's defence of this renowned tradition is in that respect also a defence of his ancestors, now taken in a broader sense, and thus we come back to the above quoted sentence from the first chapter: Plutarch is indeed reacting ὑπὲρ τῶν προγόνων.

Yet even here, this conclusion is premature and problematic. For first of all, Plutarch occasionally also defends barbarians against the criticisms of the 'barbarophile' Herodotus, such as Croesus (858D–F) and Deïoces the Mede (858F). At least in these passages, the point is not a patriotic but an ethical one. And in fact, on closer inspection, this may well be true for the whole work. For in most cases, Plutarch's defences of the Greek achievements can perfectly be understood as direct applications of the criteria he formulated at the beginning of his work. Every argument aims to lay bare Herodotus' κακοήθεια, and therein precisely lies its very *raison d'être.* Moreover, the direct reference to the ancestors comes as an isolated issue. Throughout *On the Malice of Herodotus*, Plutarch is, quite systematically, concerned with Herodotus' bad character; the patriotic or Panhellenic agenda is secondary at best. And yet, Plutarch himself underlines this patriotic goal at a programmatic place, near the beginning of the work. We remain confronted with the same awkward tension mentioned above.

The importance of historical truth

Although I shall come back to this tension in due course, for the time being, I prefer to bracket it and stay with my basic contention that the primary goal of the work is an ethical one. But the programmatic sentence still contains a further difficulty: Plutarch also claims to react ὑπὲρ τῆς ἀληθείας. This raises the question of truth, and notably the historical truth. Are we, then, entitled to consider *On the Malice of Herodotus* as a historiographical polemic after all?

Here we should bear in mind Marincola's pertinent remark that ethics and historiography cannot be separated in antiquity.[38] This is absolutely correct, and it is illustrated by Plutarch's own *Parallel Lives* and by so many other extant and non-extant historiographers. At the same time, however, Marincola's remark reflects the point of view of a historian. It remains to be seen whether Plutarch himself writes *On the Malice of Herodotus* as a historian who carries on a historiographical polemic and thus evidently takes ethical considerations into account. Or is he rather a moral philosopher who focuses on a historiographer's work and thus deals with historical matters from an ethical perspective? Of course, one may object that this distinction is to a certain extent artificial and even anachronistic, yet it may still be helpful in as-

38 Marincola (1994) 192–193.

sessing the precise character and goal of the work. And in this context, it is important to note that there already existed a rich historiographical tradition of anti-Herodotean polemic before Plutarch,[39] but that the particular charge of κακοήθεια should apparently be traced back to Plutarch himself. And this reflects a moral point of view. Once again, the key to the work thus proves to be the moral approach towards literature.

Yet it would be unwise to discard Marincola's remark too quickly. Plutarch's programmatic sentence at the beginning of *On the Malice of Herodotus* unmistakably shows an interest in the problem of historical truth, and Plutarch explicitly blames Herodotus for his many ψεύσματα καὶ πλάσματα. Those lies are refuted throughout the corpus, often by means of the argument from historical plausibility (τὸ εἰκός).[40] Nevertheless, we should always keep in mind the specific orientation of Plutarch's general argument. Particularly interesting here is Plutarch's explanation that he does not deal with Herodotus' lies but with his malicious lies (870A: ἀλλ' ὃ μὲν ἔψευσται, λόγος ἡμῖν οὐδείς· ἃ δέ γε κατέψευσται μόνον ἐξετάζομεν). Ordinary lies, then, are not the subject of this work: these are the ἄλλα ψεύσματα the refutation of which would require many books (854F). In *On the Malice of Herodotus*, Plutarch is only interested in a specific kind of lies: ἃ κατέψευσται. The verb καταψεύδεσθαι is translated somewhat misleadingly in LSJ as 'tell lies against' or 'speak falsely of': in this case at least, the prefix κατα- has a derogatory connotation. The lies belittle the value of the achievements, literally bring them down to earth. And this notion fits in very well with one of the principal themes of the work, viz. the reply to Herodotus' unjustified criticism of the Greek victories. As such, the prefix κατα- implies a clear moral component. Such καταψεύσματα are relevant because they provide direct evidence for Herodotus' character. If that is true, Plutarch is not so much interested in the correction of Herodotus' neutral historical mistakes as in his malicious distortions of the truth rooted in his bad ἦθος.

This insight may allow us to nuance an often heard criticism of Plutarch's position in *On the Malice of Herodotus*. Since Hauvette, scholars have often argued that Plutarch in this work shows an over-simplified view of Greek history, in that he sticks to the conviction that everything in the history of the Persian Wars was good and that

---

39 Ctesias already called Herodotus a liar (Photius, *Bibl.* cod. 72, 35b42–36a1). Manetho wrote a book against Herodotus (Eustathius, *In Il.* III, 238.9 Van der Valk; *Etymologicum Magnum*, s.v. Λεοντοκόμος (560.22–23); cf. also Josephus, *Ap.* 1, 73). Later, Favorinus presented Herodotus as untrustworthy in his *Corinthian Oration* (7 and 18). Aelius Harpocratio wrote a work *On Herodotus' lies* (*Suda* I, 367.3 A.), whereas Valerius Pollio wrote *On Herodotus' thefts* (Eusebius *PE* 10.3.23) and Libanius composed a polemic *Against Herodotus* (Libanius, *Epist.* 615.3). Aelius Aristides, finally, refutes several of Herodotus' statements about Egypt in his *Egyptian discourse* (*Or.* 36.41–63). For a survey of the reception of Herodotus, see esp. Priestley (2014); cf. also Dognini (2007).

40 See, e.g., 865C and D; 870D; 871A; 873D. On the importance of τὸ εἰκός as a criterion in Plutarch's works, see, e.g., Nikolaidis (1997) 336–339 and Cook (2001) 336–337. Significant is also 859D, where Plutarch refers to other authors who have written 'more accurately' (ἀκριβέστερον)—a criterion that focuses on historical truth rather than on a moral point (cf. 861D).

he has a blind respect for the great heroes of the past.[41] It is true that some such bias may be felt in *On the Malice of Herodotus*, but it is very important to realise that it is the direct consequence of the specific focus of the work. Plutarch is in this context evidently only interested in those passages where Herodotus seems to downplay the great achievements of the Greeks, *because such passages are, according to Plutarch's own criteria listed at the beginning, indications of malice*. These are the passages which Plutarch selects, because they are grist to his mill. But we should take care not to generalise and we should not conclude from this specific focus in *On the Malice of Herodotus* that Plutarch always showed the same positive bias. As a matter of fact, he was not naively uncritical towards every detail of the history of the Persian Wars. In his *Themistocles* and *Aristides*, for instance, he was not blind to the character flaws of his heroes. But in *On the Malice of Herodotus*, this is simply not the point. Of course, he there confines himself to those passages where Herodotus καταψεύδεται. It is the direct consequence of his specific methods and goal and should not be explained by his uncritical attitude towards Greek history.

A few words should finally be said about a different kind of argument that repeatedly returns in *On the Malice of Herodotus*. Plutarch often notes that Herodotus in fact contradicts himself (see, e.g., 856F: καίτοι ... φησὶν αὐτός; 857C–D; 858D; 863B; 865B). This argument from inconsistency, which can very often be found in the *Corpus Plutarcheum*, especially in the philosophical polemics, adds a rhetorical flavour to the work.[42] This is not the place to enter at length into the significant parallels between the eristic strategies that are used in *On the Malice of Herodotus* and the polemical treatises against the Stoics and Epicureans, a topic that would definitely repay further study. Here, the point is that these arguments from inconsistency suggest a theoretical interest and a concern for the historical truth. Is Plutarch, then, in such sections arguing ὑπὲρ τῆς ἀληθείας? Well, probably yes, although this need not imply that his arguments are irrelevant to the general ethical thesis concerning Herodotus' malice. In an interesting passage (861A), Plutarch in fact directly connects Herodotus' inconsistencies with the general topic of κακοήθεια, and this is not unin-

---

41 Hauvette (1894) 101: 'Le seul principe qu'il suive, sans d'ailleurs l'exprimer formellement, est celui-ci: 'Tout est beau dans l'histoire de la lutte victorieuse des Grecs contre les Perses; les ancêtres n'ont laissé que de grands exemples; ce qui tend à faire tache dans le tableau lumineux de cette brillante époque est contestable, et doit être effacé''; cf. Ziegler (1951) 871: '... den leichtgläubig-optimistischen P., der insbesondere das hellenische Altertum allein durch die rosenrote Brille des schwärmenden Romantikers betrachtet und nicht nur bei den einzelnen Persönlichkeiten, sondern auch bei den griechischen Staaten der klassischen Zeit von nichts anderem als von Ruhmestaten und überschwänglichem Edelmut hören will', and Barrow (1967) 157: 'Plutarch is fanatically biased in favour of the Greek cities; they can do no wrong'.

42 Seavey (1991) has argued that *On the Malice of Herodotus* should be regarded as an epistolary forensic oration. His view has gained some support (see Hershbell [1993] 158–159 and Ragogna [2002] 28; Cf. Bowen [1992] 4: 'There is a semi-forensic air to the whole work'), but Marincola (1994) 198 n. 44 objects, correctly to my mind, that the many judicial characteristics of the work are not without parallel in other historiographical criticism; cf. Schorn (2011) 192.

telligent. Plutarch's point is that Herodotus' malice slips into his account on any excuse and that he thus always καταψεύδεται. It is unavoidable that this entails crooked distortions and thus inconsistencies, even more so because in Plutarch's view, harmony is the work of virtue.[43]

Conclusion

Let us now, by way of conclusion, try to weave the different threads of our argument together. Plutarch did not remain indifferent towards Herodotus' account of Greek history. He discovered in him clear indications of a bad character, and this he wants to lay bare in his work *On the Malice of Herodotus*. In this work, Plutarch's principal aim is an ethical one, and his basic argument is rooted in his moral and pedagogical approach towards literature. Yet at the very beginning of the work, one short programmatic sentence introduces a completely different perspective: Plutarch claims to react out of patriotic motives and because he is concerned for the truth. Now it is quite striking that this perspective, which throughout the rest of the work entirely fades into the background, all of a sudden surfaces at the very end. After having admitted that Herodotus is a good writer, Plutarch warns one more time against his malicious defamations (874B–C):

> ὥσπερ ἐν ῥόδοις δεῖ κανθαρίδα φυλάττεσθαι τὴν βλασφημίαν αὐτοῦ καὶ κακολογίαν, λείοις καὶ ἁπαλοῖς σχήμασιν ὑποδεδυκυῖαν, ἵνα μὴ λάθωμεν ἀτόπους καὶ ψευδεῖς περὶ τῶν ἀρίστων καὶ μεγίστων τῆς Ἑλλάδος πόλεων καὶ ἀνδρῶν δόξας λαβόντες.
>
> As in roses we must watch out for the rose-beetle, so in Herodotus we must watch for the mean and partisan attacks that are disguised by a smooth and soft appearance. Otherwise we shall accept all unawares opinions which are false and out of place about the best and greatest of Greek cities and Greek heroes.

In this concluding sentence, the two poles of ancestors and truth indirectly return, in reverse order. The reference to the absurd and false opinions points to the problem of historical truth, whereas the latent enthusiasm about the best and greatest cities and men of Greece recalls Plutarch's *pietas* towards his ancestors.

And thus, the work concludes with a marvellous circular structure, but what is the point? What is the final relevance of this different perspective for the whole work? Is this a mere rhetorical *flosculum* that helps in constructing Plutarch's own persona as an author? A beautiful exercise in image building? I think there is more to it. I suggest that Plutarch, both at the beginning and at the end of his work, takes some distance and considers the relevance of his work from a broader point of view. *Mutatis mutandis*, he is doing what he usually does at the end of the *Lives*, in the concluding *Synkriseis*. There too, he evaluates the matter from a new, more distant perspective—ἀποθε-

43 *Maxime cum principibus* 777C; parallels can be found in Roskam (2009) 170.

ωρεῖν, as he calls it himself.[44] Here too, we find a similar approach to ἀποθεωρεῖν that considers the whole careful, systematic and lengthy study of Herodotus' κακοήθεια from a broader perspective. A perspective where moral virtue and truth finally meet each other in a benevolent, constructive discussion of the past and where the greatest and best achievements of the Greeks get the praise they deserve. The perspective, in short, that is typical of Plutarch himself.

44 See Roskam (2011) 221–223 and (2014) 186–191.

Paolo Desideri

Solon on the road

Abstract: In the Greek historical tradition from Herodotus on, Solon is presented as a man of travels and as one who established relationships with other peoples: his meeting with the Lydian king Croesus is emblematic of this. Plutarch's *Solon* preserves this traditional element: the *Life* connects Solon's travels with the very essence of his political activity. Furthermore, Plutarch treats travel as an invaluable source of knowledge. The comparison between Solon's final political failure and Publicola's much honoured death underlines how Solon, who shared, in a way, the same destiny as Croesus, also benefited from the lesson he had taught the Lydian king.

The historical tradition regarding Solon lays stress on the trips the great Athenian lawgiver had undertaken in the East, both before and after the period of his political activity in his native city. My aim in this contribution is to discuss the importance Plutarch assigns to this biographical detail, when he resolved to dedicate a *Life* to Solon, within the scope of his 'biographical restructuring'—to use Christopher Pelling's words—of Greek and Roman history at the turn of the first and second centuries CE. I will not address the general problem of the relationship between Near East and Greece in Greek archaic age, which constitutes the natural background of such trips,[1] nor take a position on their historicity.[2]

Since Plutarch's general plan in the *Lives* was to proceed by pairs of heroes, I shall first of all seek to interpret the function Plutarch believed he could assign to the pair of Solon and Publicola; in fact, as we shall see, Plutarch explicitly declares that the relationship between the two heroes must be considered peculiar. Then, turning to Solon's *Life*, I intend to show that, according to Plutarch, Solon's early experiences as an overseas trader must be considered a necessary prerequisite for his political activity as reformer of the Athenian social and constitutional organization—not so much in terms of the concrete measures he passed, but, rather, with regard to the distinctive characteristics of his political personality. Finally, I shall discuss the overseas travels Solon made during the ten-year period when, after his reforms, he voluntarily abandoned Athens, visiting Egypt, Cyprus and Sardis. Taking into account the situation he encountered in Athens after his return, and the retirement from public life to which he was forced in the end, I shall show that, in a sense, Solon himself had to take advantage of the lesson he had taught Croesus. This can

1 On this topic, which is the subject of intense debate among the historians of archaic Greece, I will limit myself to referring to Raaflaub (2009).
2 Bibliography in Raaflaub (2009) n. 6.

https://doi.org/10.1515/9783110539479-016

be seen as the last, but not least, positive effect of his experiences in a foreign world, the validity of which is confirmed by the parallel *Life of Publicola*, as well.

The pair of Solon and Publicola

It is not easy to find the right place for the *Life of Solon* within Plutarch's *Parallel Lives*, as far as both the relative and absolute dating of its composition (and/or publication), and its role within the general project of the *Lives* are concerned. On the question of its absolute dating, we have only two indications: one, very general indeed, from *Solon*, and the second one from the *Life of Publicola*. In *Sol.* 32.2 Plutarch says that the Olympieium remained unfinished—which gives as *terminus ante* the dates of 131–132 CE, when the huge temple was finally completed under Hadrian's reign.[3] In *Comp. Sol. et Publ.* 1.2 he mentions Publicola's death as an event of six hundred years ago, which fixes the date of this *Life* at 97 CE or so: this could be, therefore, a *terminus post* for the composition of the pair perfectly congruent with *Publ.* 15.3–5, in which Plutarch 'refers to Domitian with a hostility which shows him to have died' (as Christopher Jones put it).[4] It is suggested that the pair *Solon-Publicola* could have been written towards the middle, more probably than in the final part,[5] of Plutarch's composition of the *Lives*.[6]

As regards the role of the pair in the *Lives*' general context, we can rely on just one indication—though an important one—provided by Plutarch. This is the passage in the *Comparison*, in which Plutarch mentions that 'there is something peculiar in this comparison, and something that has not been true of any other thus far, namely that the second (i.e. Publicola) imitated the first, and the first (Solon) bore witness to the second'.[7] In order to explain what he means, Plutarch, referring to Solon's conversation with Croesus, points out that Publicola's life, evaluated according to Solon's own criteria, was even happier than that of Tellus;[8] secondly, he says that 'Publicola, in his political activities, enhanced the fame of Solon, by making him the fairest of examples for one who was arranging a democracy'.[9] Here we have perhaps the best example of what I have called elsewhere the 'generazione congiunta' ('con-

3 Piccirilli (1977) 100; Manfredini and Piccirilli (1977) ad loc.

4 Jones (1966) 69.

5 As it is quoted in Coriolanus' *Life* (33.2), the date of which is, in any case, likewise uncertain.

6 Jones (1966) 68.

7 *Comp. Sol. et Publ.* 1.1: Ἆρ' οὖν ἴδιόν τι περὶ ταύτην τὴν σύγκρισιν ὑπάρχει καὶ μὴ πάνυ συμβεβηκὸς ἑτέρᾳ τῶν ἀναγεγραμμένων, τὸν ἕτερον μιμητὴν γεγονέναι τοῦ ἑτέρου, τὸν ἕτερον δὲ μάρτυν (this, and all the other English translations to be found in my text are by B. Perrin, in the Loeb).

8 Ibid. Ὅρα γάρ, ἣν Σόλων ἐξήνεγκε περὶ εὐδαιμονίας ἀπόφασιν πρὸς Κροῖσον, ὡς Ποπλικόλᾳ μᾶλλον ἢ Τέλλῳ προσήκει.

9 Ibid. 2.1: Οὕτω μὲν ὁ Σόλων κεκόσμηκε τὸν Ποπλικόλαν, τὸν Σόλωνα δ' αὖ πάλιν ἐκεῖνος, ἐν τῇ πολιτείᾳ παραδειγμάτων κάλλιστον ἀνδρὶ κοσμοῦντι δημοκρατίαν θέμενος; the term δημοκρατία is also used in *Publ.* 1.2 and 10.7 (in *Publ.* 6.6 and 21.10, P. uses the more general term πολιτεία).

current creation') of a pair—a type of parallelism also represented by other pairs, such as *Dion and Brutus* and *Demosthenes and Cicero*.[10] In the other cases, the Roman *Life* in the pair is functionally connected with the large, homogeneous group of the late-republican Roman *Lives*. In our case, however, even though one can accept the idea that Publicola may have been suggested to Plutarch by Sosius Senecio,[11] or that Solon is the first in the sequence of the pair's composition,[12] both heroes appear to have sprung out of Plutarch's mind at one and the same moment, as if they were naturally connected. It is therefore reasonable to think that such an intimate connection can best be explained by the fact that Plutarch decided to include the two *Lives* in his general plan in order to underline the importance he attributed to the founding of democracy in Athens and Rome respectively, the two 'capitals' of the Graeco-Roman world.[13]

Though Plutarch's idea that Publicola followed the example of Solon in founding a democracy is questionable—as is his definition of the Roman Republic as a 'democracy'—it is clear that it was important for him to underline the birth in both cities of a political system in the true meaning of the term, as opposed to whatever authoritarian form of government there might have been before. The institution of such a system was not only a real point of departure for the historical success of both cities, but at the same time—more importantly from a biographical point of view—the beginning of real political personalities in Greek and Roman history. In fact, in these two heroes Plutarch sees a particularly close interdependence between ethical values and political impulses and objectives, perhaps aiming in this way at underlining that democracy necessarily requires noble moral qualities in its leaders. At any rate, the figure of Solon—which will be, from now on, at the centre of our attention—has been constructed by Plutarch in such a way as to represent the ideal moral qualities a political leader ought to be equipped with in order to be able to create and ensure a democracy. Special attention must be paid to the variety of human experiences Solon had when setting out to achieve his political reforms, as well as to the consequences of his decision to leave Athens for a long journey once he had accomplished this important task.

'He travelled to get experience and learning rather than to make money' (*Sol.* 2.1)

The first aspect of Solon to draw attention to in Plutarch's *Life of Solon* is his intellectual pursuits, which had already been underlined by Herodotus and Plato, as well as by the old tradition which assigned Solon a place in the group of the

10 See Desideri (1992a) = (2012a) 238 ff.
11 Nikolaidis (2005) 304 (cf. n. 48).
12 See *Publ.* 1.1: Τοιούτῳ δὲ γενομένῳ τῷ Σόλωνι τὸν Ποπλικόλαν παραβάλλομεν.
13 See Desideri (2013) 19 ff.

'Seven Sages'.[14] Besides, when composing Solon's *Life*, Plutarch himself had already written—or was in the process of writing—his *Symposium of the Seven Sages*, where Solon appears not only as one of the Seven, but also as a prominent figure in the discussions which are staged therein.[15] It comes as no surprise, therefore, that at the beginning of his *Life*, immediately after giving the basic information about his family and origins, talking about Solon's trading when still very young, Plutarch says that he travelled in order to 'get experience and learning rather than to make money' (πολυπειρίας ἕνεκα μᾶλλον καὶ ἱστορίας ἢ χρηματισμοῦ πλανηθῆναι τὸν Σόλωνα, 2.1). This is a departure from previous authors, who had attributed these travels to the financial difficulties Solon had after his father's death.[16] 'For he was admittedly', Plutarch concludes, 'a lover of wisdom (σοφίας ἐραστής), since even when he was well on in years he would say that he "grew old ever learning many things"' (*Sol.* 2.2).[17] The inclusion of this famous verse by Solon—which is quoted again at the end of the *Life* (*Sol.* 31.7)—confirms the idea that Solon's interest in travel was essentially due to a true love of wisdom: curiously, Plutarch attributes the same motivation to Cleombrotus, one of the prominent speakers of *On the Obsolescence of Oracles*, about whose travels we are told that they aimed not at amassing wealth by trade, but at 'getting together information' (ἱστορία) 'to serve as a basis for a philosophy that had as its end and aim theology' (410B). Plutarch specifies, moreover, that 'in those earlier times, to use the words of Hesiod, 'work was no disgrace' (*Works and Days*, 311) nor did a trade bring with it social inferiority, and the calling of a merchant was actually held in honour, since it gave him familiarity with foreign parts (βαρβαρικά), friendships with kings, and a large experience in affairs' (*Sol.* 2.3).[18]

In writing about Solon's trading, Plutarch had to work within a tradition on Solon's trading activity in his youth, with which he was not at ease. Therefore he tried to 'justify' Solon in one way or another in what appeared to be an unacceptable breach of the unwritten Greek law against 'banausic' works. Even though elsewhere (*Per.* 1.4–1.6) Plutarch appears to agree with this law, he exhibits in this case a certain willingness to make an exception for overseas trade, as long as it is conceived as a tool for increasing not one's riches, but, rather, one's knowledge and intellectual skills. He goes so far as to list in the same category of 'sage traders', so to speak, founders of great cities, such as Protis, the founder of Marseille, as well as Thales,

---

14 On the legend of the 'Seven Sages' see Snell (1971), and now Ramelli (2005).

15 See Desideri (1985).

16 *Sol.* 2.1: Καίτοι φασὶν ἔνιοι πολυπειρίας ἕνεκα μᾶλλον καὶ ἱστορίας ἢ χρηματισμοῦ πλανηθῆναι τὸν Σόλωνα. Plutarch is probably thinking of Hermippus, on which see Manfredini and Piccirilli (1977).

17 *Sol.* 2.2: Σοφίας μὲν γὰρ ἦν ὁμολογουμένως ἐραστής, ὅς γε καὶ πρεσβύτερος ὢν ἔλεγε 'γηράσκειν αἰεὶ πολλὰ διδασκόμενος'.

18 *Sol.* 2.6: ἐν δὲ τοῖς τότε χρόνοις καθ' Ἡσίοδον 'ἔργον οὐδὲν ἦν ὄνειδος', οὐδὲ τέχνη διαβολὴν ἔφερεν, ἐμπορία δὲ καὶ δόξαν εἶχεν, οἰκειουμένη τὰ βαρβαρικὰ καὶ προξενοῦσα φιλίας βασιλέων καὶ πραγμάτων ἐμπείρους ποιοῦσα πολλῶν.

Hippocrates the mathematician, and even Plato, who, Plutarch says 'defrayed the expenses of his sojourn in Egypt by the sale of oil' (*Sol.* 2.4).

The second problem Plutarch faced concerned the extent of Solon's own wealth. Plutarch, following Aristotle's lead, firmly believes that Solon was a μέσος, that is, a member of the so-called middle class, as his father, though of noble origins, 'had impaired his estate in sundry benevolent charities' (*Sol.* 1.2).[19] This financial position was, in his opinion, the best precondition for attempting and carrying out such important reforms as those planned by Solon, which aimed at restructuring the city socio-economically and politically. Here the difficulties arose from Solon's own testimony: in fact, some of his poems reveal a certain contempt for wealth, whereas elsewhere he affirms, with Plutarch's firm endorsement, that 'wealth I desire to have, but wrongfully to get it, I do not wish, as justice, even if slow, is sure' (*Sol.* 2.3). On the other hand, as Plutarch acknowledges in another poem, Solon even 'classes himself among the poor, rather than the rich' (*Sol.* 3.2), which could legitimise the serious allegations of bias or even conflict of interests that his political enemies made against some of his measures, above all the σεισάχθεια, the cancellation of debts (*Sol.* 15).

The third problem, and the most important one from our point of view, was that of the relations with foreign people which Solon established through his travels abroad. As we have already seen, Plutarch positively evaluates at the beginning of the *Life* the broadening of intellectual horizons gained through such a wide-ranging set of experiences. He resumes this topic at the end of the *Life*, when recalling Solon's decision to leave Athens for a period of ten years after having completed his political reforms. Though Solon claimed, as Plutarch says, that his choice was for reasons of business, what he really hoped was 'that in this time the Athenians would be accustomed to his laws' (*Sol.* 25.5).[20] However, we are not informed about any trading activities Solon carried out during his travels. Plutarch simply mentions three places where Solon stayed, Egypt, Cyprus, and Sardis, and summarises his activities there in the following way: in Egypt Solon attended a high-level philosophical seminar, as it were, with two very learned Egyptian priests, Psenophis of Heliopolis and Sonchis of Sais. Making explicit reference to Plato's *Timaeus*, Plutarch says that the two priests told Solon the story of the lost Atlantis, which he planned to introduce to the Greeks in the form of a poem (*Sol.* 26.1). When in Cyprus, he persuaded Philocyprus, one of the kings of the island, to move his city to a better place, and helped to arrange it in the best possible manner both with regard to convenience of living and safety (*Sol.* 26.2–3). And finally in Sardis, which Solon visited at the invitation of Croesus, he had the famous interview with the king, whose reliability, as Plutarch observes, is questioned by some authors for chronological reasons. It is well

19 *Sol.* 2.1: τὴν οὐσίαν τοῦ πατρὸς ἐλαττώσαντος εἰς φιλανθρωπίας τινάς, ὥς φησιν Ἕρμιππος.
20 Cf. Herodotus (1.29): κατὰ θεωρίης πρόφασιν ἐκπλώσας, whereas Aristotle (*Ath.* 11.1) states the following about Solon: ἀποδημίαν ἐποιήσατο κατ' ἐμπορίαν ἅμα καὶ θεωρίαν. See Camassa (1996) 572ff.

known that Plutarch strongly opposed what he deemed to be a hypercritical attitude towards tradition, refusing to reject an episode which 'comports so well with the character of Solon, and is worthy of his magnanimity and wisdom' (*Sol.* 27.1). What we conclude is that Solon's journey, at least in Plutarch's view, was undertaken out of exclusively intellectual interests—θεωρία, in his words, but that the hero took advantage of it for political objectives as well. Business, on the other hand, was completely out of the picture.

From Croesus to Peisistratus: Solon's final political failure

The most important of Solon's three travel experiences abroad was his conversation with Croesus, which from Herodotus on became a symbol of the superiority of Greek values over their barbarian counterparts. In this episode, Solon is represented 'as giving to his host Croesus far more than he gains', as Judith Mossman puts it.[21] But I do not believe that the reason why Plutarch retold this story was just to underline this aspect, that is, to reaffirm pride in Greek identity. In fact, Plutarch simply says that the episode 'comports well with the character of Solon, and is worthy of his magnanimity and wisdom' (*Sol.* 27.1). I suggest that the entire story should be interpreted within the context of Solon's final years, upon his return to Athens after his ten-year absence. In fact, the last chapters of Plutarch's *Life* are devoted to the disappointing political situation Solon found in a city that, far from growing accustomed to his laws, as he expected, had not yet achieved peace, but, rather, was slowly sliding towards Peisistratus' tyranny. Solon realised that he was unable to govern the city any longer, and at the end was forced to abandon it to its tyrannical destiny. The final act of Solon's political activity is emblematic. Plutarch says that, after trying in vain to resist Peisistratus' political pressure, 'as no one had the courage to side with him, he retired to his own house, took his arms, and placed them in the street in front of his door, saying: 'I have done all I can to help my country and its laws' (*Sol.* 30.5). In a sense, this conclusion of Solon's political activity can be considered as a sort of historical nemesis. The tears that Croesus shed on the pyre prepared for him by Cyrus were not the end of the story, as it were: Solon too was forced to recognise, through personal experience, that it was unwise 'to be puffed up by the good things we have, or to admire a man's felicity while there is still time for it to change' (*Sol.* 27.6).

Plutarch cannot conceal that the final part of Solon's life was not happy. And it seems only fitting that we find a sort of acknowledgement of this nemesis, on the part of Plutarch, when he tells the story of the final part of the life of Solon's parallel, Publicola. I have already stated that in the *Comparison* of this pair Plutarch attributes to Publicola an even greater happiness than that of the good Athenian citizen Tellus,

---

21 Mossman (2006) 292.

who had just died defending his country; in fact: '... Publicola, while he lived, was foremost among the Romans in influence and repute for virtue, and since his death the most illustrious family lines of our own day ... have for six hundred years ascribed the glory of their noble birth to him' (*Comp. Sol. et Publ.* 1.2). Moreover, Plutarch says that when he died, after defeating Rome's enemies, '[H]is loss filled not only friends and kindred, but the entire city, numbering many tens of thousands, with weeping and yearning and sorrow' (*Comp. Sol. et Publ.* 1.4). At this point the comparison is no longer between Tellus and Publicola, but between Solon himself and Publicola. Solon remains in fact the wisest of men, but Publicola 'was the most happy, since what Solon prayed for as the greatest and fairest of blessings, these Publicola was privileged to win and continue to enjoy until the end' (*Comp. Sol. et Publ.* 1.5). Truly 'he had brought his life to perfection', as Plutarch says at the end of his *Life* (*Publ.* 23.2). In the new Roman world, Solon's overseas experiences—especially the most painful one, that of the precariousness of life, to which in the end he was himself witness—could be said to have not only continued to be a source of moral teaching, but also to have found their most rewarding affirmation.

Elisabetta Berardi

Modelli del passato in due conferenze di Plutarco: *De gloria Atheniensium* e *De audiendo*

Abstract: Due testi dei *Moralia* nati come conferenza, l'epidittico *De gloria Atheniensium* e il trattato etico-pedagogico *De audiendo*, mostrano una evoluzione nel rapporto di Plutarco con i modelli dei *palaioi*, segnando un plausibile percorso cronologico. Il rapporto con la cultura classica, che passa attraverso una rielaborazione dei suoi testi fondanti, si traduce in una maggiore libertà espressiva sul piano linguistico, nel raggiungimento di una *koinē* alta, aperta sempre agli influssi della retorica e dell'atticismo, ma con un diverso grado di appropriazione dei testi, e indica una linea evolutiva di un autore che ha saputo far proprio e interpretare in modo autonomo il passato letterario dei Greci.

La lingua di Plutarco: una *koinē* alta

Nella seconda metà del II secolo d.C. il lessicografo Frinico, gettando uno sguardo retrospettivo sugli scritti di Plutarco, esprime un giudizio censorio di non conforme uso attico sui termini δυσωπία e σύγκρισις, impiegati dal Cheronese in una accezione mai attestata presso gli antichi e quindi, a suo giudizio, scorretta.[1] L'obiezione lascia intendere come potesse essere percepito il linguaggio plutarcheo in età imperiale: poiché il fenomeno dell'atticismo si esprime nella sua forma più matura attraverso la scelta dei vocaboli, la disapprovazione per i termini adoperati da Plutarco comporta di necessità una valutazione negativa sulla sua lingua. In effetti, il vocabolario dell'autore è molto ampio, riflesso di vasti interessi e vastissime letture di testi non rigorosamente 'classici', aperto anche a ricezioni di ciò che pare neoformazione ellenistica e capace di innovazioni linguistiche che per lo più non sono coniate su radici attiche.[2] Il giudizio di Frinico sul fatto che l'*usus* di Plutarco non sia conforme all'attico è confermato nella storia degli studi soprattutto a partire dalla seconda metà del '900. Molti lavori precedenti sulla lingua di Plutarco sono ancora oggi preziosi per singole osservazioni, ma è necessario rovesciare il loro punto di vista: la lingua di Plutarco non è, come un tempo si riteneva, un attico che inclina talora alla *koinē*; al contrario, secondo l'opinione ora largamente condivisa, è una *koinē* alta

1 Phryn. *Att.* 160, 243, Fischer (1974); Pade (2014) 533.
2 Weissenberger (1895) = (1994) 22–27; Fernández Delgado (1990); Swain (1996) 137; Teodorsson (2005).

https://doi.org/10.1515/9783110539479-017

aperta agli influssi dell'atticismo, che non presenta unitarietà e uniformità nelle diverse opere del *corpus.*[3]

Conferenze epidittiche, conferenze etico-pedagogiche: dal *Del gloria Atheniensium* al *De audiendo*

Nella percezione della *élite* la *koinē* elevata appare comunque prossima all'atticismo[4] e proprio in quella *élite* va individuato il pubblico delle *Vite* e dei variegati *Moralia.*[5] Anche in occasioni di discorso di fronte a un uditorio, Plutarco non si indirizza a grandi folle né con declamazioni di tipo epidittico, quale il *De gloria Atheniensium* (345C–351B), né con conferenze divulgative di carattere etico-pedagogico quale il *De audiendo* (37C–48D), destinata a un pubblico ristretto di uditori abituali (una σχολή[6]) e poi trascritta e dedicata al giovane Nicandro. Egli si rivolge ad ascoltatori e lettori attenti al classicismo di contenuti, meno sensibili a un atticismo sintattico e lessicale, anche se con ogni probabilità competenti in tal campo; Plutarco non ignora affatto il contemporaneo fenomeno atticista, ma ne prende le distanze e in molti testi, tra cui il *De audiendo,* tratteggia a beneficio del pubblico l'immagine negativa del neosofista sterilmente consacrato al culto del linguaggio degli antichi, così da delineare per contrasto la figura dell'autentico filosofo, che trasmette contenuti morali fondati sulla sapienza degli antenati.[7] D'altra parte Plutarco nella sua prima giovinezza è passato attraverso la medesima preparazione degli aspiranti sofisti, che prevede lo studio dei poeti e dei prosatori del passato e la produzione di testi epidittici.[8] Memorie di poeti e prosatori testimoniano quindi negli scritti di Plutarco una eredità della sua formazione, più o meno incisiva a seconda del genere dell'opera che egli compone, e si affiancano ai testi dei filosofi antichi, recepiti anch'essi come classici.[9]

Il *De gloria* e il *De audiendo,* accomunati da una originaria dimensione di conferenza, segnano momenti significativi di un percorso di maturazione culturale di Plutarco proprio nel rapporto con i testi e la lingua dei παλαιοί, rapporto che si

3 Giangrande (1988); Giangrande (1990); Redondo (1990); Giangrande (1992); Swain (1996) 137; Torraca (1998); D'Ippolito (2010) 89.
4 Kim (2010) 471.
5 Gallo (1998) 3517 e n. 26; Santaniello (2000); Muccioli (2012) 44–53.
6 La Matina (2000) 180, 184.
7 Schmitz (2014) 40.
8 Bowie (2014); si veda anche La Matina (2000) 177–178.
9 Schmitz (2012) 83. Inoltre, in un suo specifico progetto paideutico, Plutarco propugna la lettura dei poeti svincolata dal tradizionale legame con la retorica, come primo gradino per la formazione del filosofo: Xenophontos (2010).

esplica sotto forma di citazione, allusione, influssi lessicali. I due testi appaiono evolversi, dal *De gloria* al *De audiendo*, nella direzione di un classicismo in cui il dettato si affranca sempre di più da una ripresa letterale dei παλαιοί acquisendo una maggiore dimensione allusiva ed evocativa. I modelli vengono rielaborati, dal punto di vista della lingua, in modo tale da fare avvertire un senso di familiarità e quindi di autorevolezza nella comunicazione, senza che tale sensazione sia riconducibile a una espressione precisa, e nel contempo di frequente piegati per esprimere contenuti nuovi o sottoposti a una nuova interpretazione.

Il *De gloria Atheniensium*

Il *De gloria* appare composto per celebrare Atene nella *polis* stessa[10] e presuppone un gruppo di ascoltatori simpatizzanti in un contesto di scuola; a giudizio di Plutarco infatti non è opportuno, nel presente dell'impero, esaltare in una dimensione civica le vittorie dei Greci contro il barbaro, Maratona, l'Eurimedonte, Platea; tali menzioni avrebbero forse potuto destabilizzare le folle e di certo indisporre Roma nei confronti delle *poleis*.[11] Nel testo Plutarco afferma la superiorità dell'azione sulla cultura: la domanda su chi abbia maggiormente contribuito alla gloria di Atene ha come risposta gli uomini che ne fecero la storia, statisti, generali, comandanti, piuttosto che storici, pittori, autori di tragedie, oratori. Una dichiarazione in apertura condensa efficacemente l'assunto: ἄνελε τὴν Περικλέους πολιτείαν ... καὶ Θουκυδίδης σοι διαγέγραπται, 'elimina l'azione politica di Pericle ... e hai cancellato Tucidide'.[12] Senza gli uomini di azione non esisterebbero quelli che ne hanno narrato le imprese, in particolare gli storici, paragonati per il loro rapporto con chi ha compiuto le azioni ad attori di vicende create da altri, a messaggeri o a pittori di eventi.

Il *De gloria* è mutilo dell'inizio e della fine e lo stato di conservazione pare dipendere dalla natura di brogliaccio in cui l'autore stesso l'avrebbe lasciato.[13] È arduo stabilire una collocazione temporale per il testo:[14] nel *corpus* troviamo espresso il convincimento che la retorica si addica ai giovani,[15] ma anche nella prima maturità Plutarco si dedica alle declamazioni[16] né si può parlare di un percorso rettilineo dalla retorica alla filosofia; con certezza si può solo concludere che per Plutarco la produzione epidittica resti collaterale e di minore rango.[17] Si è ipotizzato tuttavia che la data di composizione del *De gloria* si collochi intorno agli anni 60 del

10 *Bellone an pace* 345F; Frazier (1990) 166; Mocci in Gallo e Mocci (1992) 7.
11 *Praec. ger. reip.* 803A; 814A–C. Gascó (1990); Oudot (2011).
12 *Bellone an pace* 345C–D.
13 Gallo in Gallo e Mocci (1992) 29 n. 32; Schmitz (2014) 32–33.
14 Frazier in Frazier e Froidefond (1990) 165–167.
15 *De soll. an.* 959B–C.
16 Frazier in Frazier e Froidefond (1990) 15–17.
17 Gallo (1998) 3535–3536; Schmitz (2014) 33.

I secolo, quando Plutarco completava la sua istruzione in Atene; la declamazione avrebbe potuto costituire un ringraziamento solenne di un giovane per la formazione ricevuta.[18] Anche non tenendo conto di tale congettura, il testo appare in effetti fortemente ispirato da una impostazione scolastica che ne condiziona dettato e ordito. Nell'esaltare la gloria portata ad Atene dalle πράξεις rispetto ai λόγοι, Plutarco si rivela influenzato da un classicismo che trionferà nei testi retorici della Seconda Sofistica proprio in tale immagine di Atene, culla della storia della grecità, una immagine ereditata dalla tradizione del λόγος ἐπιτάφιος e ben testimoniata dall'Epitaffio di Pericle tucidideo.[19]

Il *De audiendo*

Ben diversa è la cura autoriale che mostra di aver ricevuto il *De audiendo*, conferenza rielaborata, secondo quanto l'autore stesso dichiara, per esser dedicata in forma di trattato etico-pedagogico al giovane Nicandro; questi, indossata da poco la toga virile, si avvia a completare un processo di formazione filosofica, sperimentando per la prima volta autonomia e indipendenza, con i pericoli che ciò comporta. Molti sono i personaggi con nome Nicandro nei *Moralia*, tutti in qualche modo riconducibili all'ambiente di Delfi;[20] il giovane parrebbe identificabile con il figlio di Eutidamo, collega di sacerdozio di Plutarco al santuario di Apollo, e quindi plausibile destinatario negli anni 80–90 di un testo che espone quali siano le corrette modalità di ascolto di una lezione filosofica, dalle disposizioni esteriori a quelle interiori. Il *De audiendo* verrebbe a collocarsi a quasi vent'anni dopo la presunta data del *De gloria*.[21]

Nel prescrivere le norme per l'ascolto, Plutarco tratteggia i difetti tipici del conferenziere e soprattutto del pubblico, di cui raffigura una gamma di tic e comportamenti dannosi; il discorso, costellato di riprese dal patrimonio antico, risente di una generale ispirazione platonica che informa tutto il testo, a partire, in apertura, da una vigorosa appropriazione e modifica delle parole di Socrate nell'ottavo libro della *Repubblica*.[22]

---

18 Ziegler (1949) = (1965) 115; Thiolier (1985) 10–12; Mocci in Gallo e Mocci (1992); Gallo (1998) 3525.
19 Frazier in Frazier e Froidefond (1990) 175–177; si veda anche *infra*, p. 189.
20 Puech (1992) 4862–4863.
21 Hillyard (1981) XXXIV-XXXVIII (80–100 d.C.).
22 Hillyard (1981); Jazdewska (2013).

Lingua e modelli classici, tra Tucidide e Platone

Gli aspetti linguistici del *De gloria* e del *De audiendo* sono stati oggetto di analisi approfondite dei rispettivi ultimi editori.[23] L'epidittico *De gloria*, che ha subito in passato dagli studiosi interventi di tipo normativo, mostra una commistione di elementi attici, termini di derivazione poetica, e aspetti di *koinē* lessicale e sintattica, che Plutarco propone a un pubblico probabilmente più sensibile alle immagini classiche che non al dettato atticista.[24] La presenza della *koinē* è senza dubbio più forte nel *De audiendo:* Plutarco smaschera qui i pericoli di una dizione artefatta di sofisti conferenzieri e dispiega una notevole ricchezza di vocabolario che spesso attinge all'attico; tuttavia si tratta di termini che non incontrano il successo degli oratori atticisti o che sono impiegati da Plutarco in accezione non attica; parimenti la sintassi mostra costrutti di frequente sviluppati nella prosa di *koinē*.[25]

In entrambi i discorsi si riscontra un comune atteggiamento nei confronti dello iato: sia il *De gloria* sia il *De audiendo* sono infatti trascrizione di un parlato e in questo genere, a causa della dimensione dell'ascolto, si rileva la tendenza all'eliminazione dell'incontro vocalico.[26] Il *De gloria* tuttavia mostra come Plutarco ritenga il rifiuto a priori dello iato un atteggiamento lontano dal senso della misura: compare qui la derisione di Isocrate che, paragonato ai condottieri impavidi nello scontro delle armi, è invece intimorito dal pensiero del cozzo delle vocali.[27]

Come appena ricordato, il *De gloria* non si fonda su una adesione all'atticismo linguistico; rivela comunque classicismo nell'ispirazione filosofica di stampo platonico e nella notevole presenza di Tucidide, fonte di molti eventi menzionati in modo cursorio e oggetto di una lunga citazione adattata da Plutarco ai suoi fini.[28] Il testo è inoltre intessuto di *topoi* ed *exempla* di carattere scolastico, quali l'elenco di storici che presero parte agli eventi e storici che si limitarono a narrarli;[29] la celebrazione della vittoria all'Artemisio; l'esaltazione dei Maratonomachi e del messaggero Tersippo, l'encomio di Cimone, Milziade, Temistocle, Pericle;[30] la derisione da Isocrate, lento nell'elaborazione di opere con cui voleva spingere i comandanti all'azione;[31] un giudizio canonico sullo

---

23 *De gloria Atheniensium:* Thiolier (1985); Frazier in Frazier e Froidefond (1990); Gallo in Gallo e Mocci (1992); *De audiendo:* Hillyard (1981).

24 Gallo in Gallo e Mocci (1992) 28–31, 82–88; Torraca (1998) 3495–3501.

25 Hillyard (1981).

26 D'Ippolito (2010) 207.

27 *Bellone an pace* 350E.

28 Sul platonismo del *De gloria* Frazier in Frazier e Froidefond (1990) 168–172; Van der Stockt (1990); su Tucidide nel *De gloria*, Gallo in Gallo e Mocci (1992) 89–92, n. 40; Payen (2011).

29 *Bellone an pace* 345D–E; lo studio di 'canoni' dava origine a simili elenchi: si veda Nicolai (1992) 298–299.

30 *Bellone an pace* 348C–D: l'apologia dei quattro politici accusati da Socrate nel platonico *Gorgia* (514c–516e; 503c) di aver adulato il popolo ateniese era frequente nei discorsi retorici; si veda per esempio *A Platone, in difesa dei Quattro* (*or.* 3) di Elio Aristide.

31 *Bellone an pace* 350E–351A: sul *topos*, Gallo e Mocci (1992) 110 nn. 118–119.

stile dei tre tragici Eschilo, Sofocle ed Euripide.[32] Le imprecisioni storiche del testo tradiscono quella matrice filoateniese del IV secolo che elogiava retoricamente la grandezza di Atene, le cui radici andavano rintracciate già nell'Epitaffio di Pericle.[33] Il convincimento plutarcheo che nella composizione il contenuto debba essere preminente sulla forma trova espressione in un aneddoto elogiativo su Menandro,[34] che rivela un apprezzamento per il commediografo anch'esso in linea con le tendenze scolastiche della prima età imperiale: tra I e II secolo d.C. Menandro è un autore-modello in tutti i livelli dell'istruzione per molteplici fini.[35]

Del resto Plutarco, che mostra come elemento costante della sua produzione il rifiuto dell'eccessiva cura formale, apprezza la genuina chiarezza attica e, stando a Isidoro di Pelusio, avrebbe indicato in Gorgia il responsabile della sua corruzione: Gorgia, introducendo una lingua densa di figure retoriche e contaminandone la purezza, la σαφήνεια, produsse una malattia che finì per contagiare addirittura Platone.[36] Su questa linea si muove il rimprovero ai sofisti nel *De audiendo:* Plutarco, che intende mostrare come il corretto ascolto della lezione filosofica sia utile al progresso morale, smaschera in più punti l'insidia costituita da quegli oratori che celano sotto una forma atticista ossessivamente rifinita contenuti sterili.[37]

Il duro attacco alla mania dell'atticismo coinvolge anche il pubblico delle conferenze. L'ascoltatore che non si cura del concetto se non è espresso in puro stile attico è assimilato a chi non vuole πιεῖν ἀντίδοτον, 'bere un antidoto', se non da una coppa di argilla attica, o a chi non brama un indumento qualunque nemmeno in inverno, preferendo rimanere seduto immobile e inerte, come avvolto ἐν τρίβωνι Λυσιακοῦ λόγου λεπτῷ καὶ ψιλῷ, 'nel mantello leggero e sottile del linguaggio di Lisia'.[38] Come è stato notato, Plutarco rielabora qui in modo originale, con una fine *detorsio*, quel passo del *Gorgia* in cui il sofista si vanta di poter persuadere meglio dei medici un paziente a 'bere un farmaco', φάρμακον πιεῖν.[39] A mio giudizio è plausibile che anche nel secondo paragone si celi un sapiente richiamo a Platone: l'immagine dell'uomo seduto inerte nel τρίβων di Lisia pare ricordare alla memoria del pubblico con un effetto di *mise en abîme* l'esordio del *Fedro.*[40] Come il *Gorgia*, la cui cornice prende le mosse da una conferenza di successo—Gorgia ha appena compiuto quella

32 Thiolier (1985) 80, n. 3; Gallo e Mocci (1992) 98–99 n. 66.
33 Muccioli (2012) 143, 165–166.
34 Menandro, poiché ha concepito la trama della commedia, non è preoccupato del poco tempo per la sua stesura (347E–F).
35 Su Menandro in Plutarco da ultimo Karavas e Vix (2014) 185–187. Il primato di Menandro è messo in discussione dalla lessicografia atticista: Blanchard (1997); per una posizione più sfumata sul ruolo dell'atticismo nel tramonto della fortuna del poeta si veda Tribulato (2014).
36 Plu. fr. 186, Isidoro di Pelusio, *Epistola* 2, 42.
37 *De aud.* 41D, 46E, 48D; D'Ippolito (2010) 94–95.
38 *De aud.* 42D.
39 Pl. *Grg.* 456b–c e Schmitz (2014) 39.
40 Nel *De audiendo* sono stati rilevati alcuni rimandi al *Fedro* (*De aud.* 40E; 44D–E: Hillyard (1981) 185). Sulla ricorrenza del *Fedro* nelle scuole di retorica Trapp (1990).

ἐπίδειξις che Plutarco considera pericolosa nel *De audiendo*–, anche il *Fedro* muove da un contesto di ascolto: conoscendo la passione di Socrate per i discorsi, Fedro gli racconta di aver seguito con entusiasmo un discorso di Lisia su *eros* e di aver trascorso la mattina seduto a rileggerne il testo. Socrate esorta allora Fedro a mostrargli lo scritto di Lisia, che a ragione sospetta Fedro celi sotto l' ἱμάτιον, in modo che possano a loro volta sedersi a esaminarlo, lungo le rive dell'Ilisso.[41] L'inerzia di chi trascorre tempo seduto avvolto nel τρίβων di Lisia pare quindi riscrivere in modo allusivo l'esordio del *Fedro* con una notazione maggiormente critica sull'attività retorica.

Una rielaborazione allusiva

Come già accennato,[42] in modo analogo al *De audiendo*, che trova nel *macrocorpus* di Platone il suo modello di riferimento, anche il *De gloria* appare ispirato nel suo classicismo soprattutto a un testo: nella declamazione riecheggia in più punti l'Epitaffio di Pericle per i caduti del primo anno di guerra, un discorso frequentatissimo dalle scuole di retorica e quindi ben noto alla *élite* che costituisce il pubblico di Plutarco.[43] Pericle vi proclama che la superiorità di Atene è dimostrata dalla verità dei fatti; non i λόγοι ma gli ἔργα denunciano l'eccellenza della *polis:* Atene non ha bisogno dei canti di Omero. L'adesione piena al discorso di Pericle non è l'unica declinazione possibile su cui orchestrare il motivo della gloria di Atene: due generazioni più tardi rispetto a Plutarco Elio Aristide nel suo *Panatenaico* (*or.* 1) recepisce l'Epitaffio rovesciandone sottilmente l'argomento. Per Aristide la grandezza di Atene non è data dalle πράξεις ma dai λόγοι; la lingua attica, madre del dialetto di Omero, il padre della cultura greca, dilaga per tutto l'impero riportando un trofeo incruento superiore alle vittorie militari degli antenati. Come è stato notato, l'interpretazione di Aristide manifesta l'atticismo pieno del II secolo d.C., orchestrato sulla centralità indiscussa della lingua e degli autori del passato;[44] Plutarco invece, ispirandosi al medesimo passo, ne ripropone sostanzialmente l'assunto della superiorità degli ἔργα, secondo un classicismo di maniera.

In effetti, il rapporto del *De gloria* e del *De audiendo* con lingua e testo dei παλαιοί rivela una linea evolutiva che rafforza l'ipotesi di loro differenti date di composizione: il *De gloria* aderisce in modo scolastico alle pagine tucididee, facendone risuonare concetti ed espressioni. Il *De audiendo* mostra come un più maturo Plutarco sappia appropriarsi dei riferimenti culturali soprattutto platonici in

---

41 Pl. *Phdr.* 228a–e; nel testo platonico sono molto frequenti i termini dell'ascolto ἀκοή e ἀκούειν, cruciali nel περὶ τοῦ ἀκούειν plutarcheo.

42 *Supra*, p. 186.

43 Thuc. 2.35–46; su Tucidide nel *De gloria* si veda ancora Thiolier (1985) 13.

44 Oudot (2008).

modo altamente allusivo, proponendo al suo pubblico una fittissima rete di richiami profondamente rielaborati al patrimonio letterario degli antichi.

Myrto Aloumpi

Shifting boundaries: *Philotimia* in democratic Athens and in Plutarch's *Lives*[1]

Abstract: This chapter explores Plutarch's relationship to the Greek past, by discussing his use of a key concept, that of *philotimia*, in contexts related to Athenian democracy. To this end, I examine *philotimia* in Thucydides and Demosthenes vis-à-vis Plutarch's readings of *philotimia*, especially in his *Lives* of Athenian heroes of the classical period. The comparison between the sources focuses on two topics. Firstly, I consider differences in the conceptual construction of *philotimia*. In Plutarch, I argue, *philotimia* seems to be more of a trait that inhabits the hero and is part of his character as well as of his personality. Its manifestation, good or bad, is presented firstly as a matter of the individual's choices which are undoubtedly, yet only secondarily, informed by external stimuli, while at the same time specific socio-political norms do not seem to delimit the range and quality of *philotimia* manifestations. *Philotimia* within Athenian democracy, on the other hand, should be seen more as a civic virtue and as a social construct, the quality of which is primarily informed from without: its manifestation and, more importantly, its representation in public discourse is delimited within specific contexts, metaphorical and literal. As I argue, this is how democratic ideology managed to tame this dangerous as well as much needed virtue. Secondly, I look closely at two separate ways in which *philotimia* can be manifested in democratic Athens, that is, as private and public *philotimia*, and I examine how the individual's private motivation is represented in each case. As we will see in Thucydides and Demosthenes, the fields of private and public *philotimia* are defined quite clearly, and different connotations are attributed to *philotimia* in each respective case. On the contrary, *philotimia* in Plutarch, as an individual quality informed from within, is not further defined as private or public: the motivation of the agent who is *philotimos* can be negative or positive according to his good or bad quality and depending on the specific occasion of manifestation at hand. As I conclude, the difference between Plutarch and democratic sources is a product of different moral purposes and generic demands.

1 An earlier version of this chapter was delivered at the 10th Congress of the International Plutarch Society, May 2014. I am very grateful to Aristoula Georgiadou and Katerina Oikonomopoulou for giving me the opportunity to contribute to this volume. I should also like to thank Rosalind Thomas, Tim Rood, Chris Pelling, Chrysanthos Chrysanthou, and Rovertos Heller for much helpful advice. This chapter is dedicated to the memory of my godmother, Ioanna Kastani.

https://doi.org/10.1515/9783110539479-018

This chapter explores Plutarch's relationship to the Greek past, by discussing his use of a key concept, that of *philotimia*, in contexts related to Athenian democracy. Specifically, I examine *philotimia* in Thucydides and Demosthenes vis-à-vis Plutarch's readings of *philotimia*, especially in his *Lives* of Athenian heroes of the classical period. The comparison focuses on two topics.

In the first section, I identify conceptual shifts in the construction of *philotimia*. In Plutarch, I argue, *philotimia* seems to be more of a trait that resides in the hero, and is part of his character as well as of his personality. Its manifestation, good or bad, is presented firstly as a matter of the individual's nature and choices which are undoubtedly, yet only secondarily, informed by external stimuli. In this sense, *philotimia* can be understood more as an individualistically expressed ethical/ moral quality rather than as a civic virtue primarily informed by specific socio-political norms and expectations. By contrast, in classical Athens, the manifestation of *philotimia* as well as its representation seems to be quite strictly regulated and controlled. *Philotimia* is presented neither as a flexible trait that can be directed by the individual *philotimos* towards any end, honourable or dishonourable, nor does *philotimia* appear in association with any kind of ambitious activity and objective. *Philotimia* within Athenian democracy should be seen more as a civic virtue, as a social construct, the quality of which is primarily informed from without: its manifestation and, more importantly, its representation in public discourse is delimited within specific spaces, metaphorical and literal, while excluded from others. In this way, the Athenian democratic ideology managed to tame this dangerous as well as much needed virtue.

In the second section, the discussion looks closely at the distinction between private and public *philotimia*, which is evident in sources from democratic Athens, and the relation of each of them to personal and public interest. As we will see, such a distinction does not appear to be present in Plutarch: when private interest and motivation are discussed, *philotimia* does not need to be qualified as ἰδία (private) and to be set against δημοσία (public) *philotimia* in order to express private motivation. *Philotimia* in Plutarch, as an individual quality informed from within, can be negative or positive according to the good or bad qualities of the agent himself and manifested in a wide variety of contexts. In Thucydides and Demosthenes, by contrast, as we shall see, private and public *philotimia* are sometimes clearly defined and carry different connotations. Thus, within democratic settings, the range of acceptable *philotimia* manifestations is closely regulated and public demonstrations of *philotimia* seem to be checked by the city.

Such a differentiation brings out the fact that *philotimia* as a concept is constructed by and employed within distinctive social contexts and according to different generic demands each time. This examination endeavours to bring to the fore the distance that exists between certain models and values of the classical Greek past, and Plutarch's understanding and adaptation of such ideas according to the generic and moral purposes of his work.

Construction 'from within', construction 'from without'

At the beginning of *On the Bravery of Women* and of *Phocion–Cato the Younger*, Plutarch makes a general point on the nature of virtues. He claims that in their expression virtues have many nuances and differentiations, depending on the character and the personal nature of the agent (*Phoc.* 3.3–4, 7–8; *Mul. virt.* 243C); yet this situation does not produce many different 'braveries and wisdoms and justices' (*Mul. virt.* 243D). It has been noted that such statements run against an approach that sees Plutarch's biographical work as having been produced 'predominantly from a character-viewpoint', according to which Plutarch's aim is to see the heroes of his *Lives* almost exclusively as embodiments of ethical qualities.[2] By admitting a degree of variety in his heroes' expression of ethical qualities, Plutarch opens the ground for understanding these great men not merely as examples of fixed virtues or vices, but, more importantly, as individuals whose specific expressions of certain ethical qualities are peculiar to their personality. Thus, Plutarch may indeed have been primarily interested in the *ēthos* of great men, but the various perspectives from which he approaches character and the details of and comments on a hero's life also shed light on the unique personality of each man.

Following this line of thought, Nikolaidis pursued the question whether *philotimia* in a group of *Lives* should be understood as a matter of difference of character or as a corollary of the socio-political conditions peculiar to each hero's case.[3] Nikolaidis concluded that *philotimia* is indeed expressed according to the character of each hero, and that the socio-political environment does not affect the behaviour of the *philotimoi*; the various expressions of *philotimia* are primarily a matter of an individual's choices and *philotimia* should be seen not so much as an ethical virtue, but as 'a means or a motive':[4] depending on 'its dosage', 'on the quality as well as on the quantity of the honours desired and sought after',[5] *philotimia* may lead to virtuous or vicious behaviour.

---

2 For a mainly 'character viewpoint' approach, see Gill (1983), quotation from p. 472. *Contra* Gill: Pelling (2002) 321–329; Nikolaidis (2014) 362–365.

3 Nikolaidis (2012) mainly examines *Philopoemen–Titus Flamininus, Themistocles, Alcibiades–Coriolanus* and the *Gracchi*.

4 Nikolaidis (2012) 52–53, quotation from p. 52. *Philotimia* in Plutarch has been approached and understood in various ways. The most thorough examination of the concept is Frazier (1988), with a brief, schematic representation of its meanings on pp. 126–127; Also, Wardman (1974) 115–124; Duff (1999) 72–89; Pelling (2012); Aristotle includes *philotimia* in his discussion of ethical virtues in the *Nicomachean Ethics* (1125b1–26), but not in a clear-cut sense, as he prefers to leave unnamed the middle ground regarding the virtue of pursuing honour. Depending on the occasion, the virtuous middle ground borrows its name from one of the two extremes.

5 Nikolaidis (2012) 53, with Arist. *EN* 1125b7–26.

Nikolaidis is right when saying that *philotimia* in Plutarch is primarily a matter of dosage and quality of honours sought after, in other words, that the *philotimos* should also be *philokalos* if his behaviour is to be praised. Selecting a specific activity to demonstrate one's *philotimia* in proper or improper dosages seems to be in Plutarch primarily a matter of individual choice.[6] At the same time, though, we cannot disregard the role that the socio-political environment plays in the moral and ethical development of an individual: character and personality and, by extension, personal taste and preference, virtue and/or vice are all informed to a wide extent by social norms and are all developed 'in dialogue with' specific socio-political circumstances.[7] Frazier has shown that in the *Lives* an individual's ambition may be affected by the historical, political and ethical contexts in which he develops and thus ambition and *philotimia* are not one and the same thing in Sparta, in classical Athens, in Hellenistic times and in Rome.[8] Nonetheless, the specific generic demands of biography place the individual in the limelight and lead indeed Plutarch to give priority to individual initiative and to internal, psychological factors in his presentation of *philotimoi* individuals. *Philotimia* in Plutarch is a moral quality that inhabits in the soul of an individual and can go right or wrong depending on the agent's nature, choices and responses to external stimuli.

The situation was not the same in fifth and mostly fourth century Athens where the focus shifted from the individual to the collective: the socio-political norms of democracy demanded from the citizens specific manifestations of *philotimia* beneficial to the collective and consequently the range and quality of the activities in which an individual could demonstrate his *philotimia* were more clearly defined. In democratic Athens, *philotimia* as a civic virtue is a social construct and thus its quality and the range of its manifestations that are considered appropriate is primarily informed and regulated by the socio-political standards of democracy in accordance with collective interest and not by the nature and psychological condition of individual agents. Thus, if we want to understand the conceptual shifts in the meaning of *philotimia* in Plutarch vis-à-vis democratic Athens, we should endeavour to explain the way in which *philotimia* is constructed and functions in each case.[9]

---

6 In many cases, though, *philotimia*/*to philotimon* seem to control the individual rather than the other way round. See, e.g., *Pyrrh.* 30.1; *Alc.* 34.2; *Comp. Lys. et Sull.* 4.2.

7 On the idea of the 'self in dialogue', see Gill (1996), esp. 14–16 (for a brief introduction to the idea); Also, Gill (1996) 85–86 on his model of 'self' as 'a psychological agent' and as 'an ethical agent'.

8 Frazier (2014) 493–498.

9 What interests me is how *philotimia* is *presented* as being constructed and functioning in the sources, not the probably unanswerable and empirical question of how it is constructed and functioning 'in reality'. See Skinner (2002) 85: '... as soon as we see that there *is* no determinate idea to which various writers contributed, but only a variety of statements made by a variety of different agents with a variety of different intentions, what we are seeing is that there is no history of the idea to be written. There is only a history of its various uses, and of the varying intentions with which it was used. ... the persistence of particular expressions tells us nothing reliable about the persistence

In Plutarch *philotimia* is quite often seen as a psychological condition and is recognised as a quality pre-existent within an individual that is then manifested in diverse ways in response to external stimuli.[10] In the *Life of Alcibiades*, for example, we find *philotimia* in this 'stand-by' mode as a characteristic of the hero that can go right and wrong.[11] Alcibiades' flatterers throw him into political activities that do not suit his age by appealing to his *philotimia* and *philodoxia* (*Alc.* 6.4). Socrates, on the other hand, when it was to be decided whether an award of valour would be made to him or to Alcibiades following a battle where both of them excelled, insisted that the prize should go to Alcibiades 'because he wanted to increase his [Alcibiades'] *philotimon* in honourable undertakings' (*Alc.* 7.5).[12] What is assessed is not whether Alcibiades is *philotimos* or not—this is taken for granted—, but whether his *philotimia/to philotimon* is going to be manifested virtuously or viciously. *Philotimia/to philotimon* is presented as a quality that is *a priori* inherent in the hero and which is directed by external influences either towards inappropriate ends (by the corruptors) or to noble undertakings (by Socrates).[13] Socio-political expectations and norms are not presented as delimiting *a priori* the acknowledgement of a particular individual as *philotimos* nor do they put *a priori* restrictions on what actions should count as manifestations of *philotimia*;[14] it is rather the creative virtue and/or vice of the agent in response to external stimuli, as presented in a *Life*'s narrative, which opens up the field of *philotimia* activities and of *philotimia*'s meaning.

In democratic Athens, on the other hand, during the fourth century and especially the time of Demosthenes where *philotimia* is more frequently seen as a civic vir-

---

of the questions that the expressions may have been used to answer, nor of what the different writers who used the expressions may have meant by using them'.

10 E.g., *Them.* 3.3–5, 18.1; *Lys.* 23.2–5; *Ages.* 2.2, 8.4; *Sull.* 3.4; *Comp. Ag. Cleom. et Gracch.* 5.5; *Flam.* 20.1–4.

11 Pelling (2002) 351 says that '[I]t is characteristic of Plutarch, in his best work, to bring out how the same qualities contribute both to a man's greatness and to his flaws'.

12 Translations of Plutarch are taken from Loeb. In all instances, I prefer keeping *philotimia* vocabulary untranslated.

13 On *Alcibiades*, see also Nikolaidis (2012) 43; Duff (1999) 215–218. On the relation of *philotimia* with flattery, Nikolaidis (2012) 44.

14 In a similar manner, the praiseworthy conduct of Titus when he served for a second time as military tribune as well as his contemptible chasing down and killing of an old and helpless Hannibal are both related to Titus' τὸ φύσει φιλότιμον (*Flam.* 20.1–4, 21.1). The conduct of Titus in the latter case is explained in terms of his inability to restrain his passion as he was full of desire for δόξα and youthful enthusiasm even though he was old.

In democratic contexts, it was expected that one's praiseworthy behaviour related to military or administrative offices could be related to his *philotimia*, but it is less likely that undertakings such as chasing down a helpless enemy in order to associate one's name with his death would have been described as motivated by innate *philotimia/to philotimon* and passion for δόξα that cannot be restrained by the agent. Cf. the representation of Alcibiades the elder's and Alcibiades the younger's inappropriate *philotimia* in Lys. 14.2, 35–40, 42–43 and note the interplay among *dunamis, ponēria*, and *philotimia* throughout the speech.

tue,[15] what counts as a manifestation of *philotimia* is more or less pre-defined in accordance with public interest and democratic ideology.[16] In this sense, *philotimia* in public discourse is not approached as a characteristic pre-existent within the individual, which can go right or wrong depending on the agent's creative virtue or vice, but is usually a positively evaluated term attributed to an individual only if his behaviour abides by the democratic *philotimia* standards of benefiting the city.[17] Thus, not any kind of activity and, more importantly, not any kind of activity *where ambition was expected to be found* was at the same time a proper occasion for demonstrations of a citizen's *philotimia*.[18]

In Dem. 21.158 – 159, Demosthenes comments on Meidias' private luxuriousness, which includes a huge house that overshadows the whole neighbourhood, a carriage drawn by a pair of white horses, and a group of three or four slaves that accompany him in the Agora and talk loudly about their master's extravagant symposia. Then, he reflects on what should count as *philotimia* and exhorts the jurors to judge Meidias' management of his finances by this standard:

> οὐ δεῖ δὴ τὰ τοιαῦτα ἑκάστοτε τιμᾶν οὐδὲ θαυμάζειν ὑμᾶς, οὐδὲ τὴν φιλοτιμίαν ἐκ τούτων κρίνειν, εἴ τις οἰκοδομεῖ λαμπρῶς ἢ θεραπαίνας κέκτηται πολλὰς ἢ σκεύη καλά, ἀλλ' ὃς ἂν ἐν τούτοις λαμπρὸς καὶ φιλότιμος ᾖ, ὧν ἅπασι μέτεστι τοῖς πολλοῖς ὑμῶν· ὧν οὐδὲν εὑρήσετε τούτῳ προσόν.
>
> That surely isn't the kind of conduct you should honour and admire when it occurs; nor should you judge *philotimia* by these criteria—whether a man builds a distinguished house or possesses a lot of maidservants or fine furniture: you should look for a man who is magnificent and *philotimos* in things of which the majority of you have a share. You'll find that none of this applies to Meidias.[19] (Dem. 21.159)

---

15 But also in the time of Thucydides, even though in an implicit way, as we will see in the next section.

16 Such a strictly civic-oriented evaluation of *philotimia* is not reflected in the world of the *Lives* from classical Athens and this seems to belong to the same kind of manipulation that Plutarch employs with fifth century sources when exploring the role of the masses in that time. As Saïd (2005) 17 points out, Plutarch 'minimises the role of the masses in the making of history by systematically substituting an individual to a collective agent' (see Saïd's comparison of Thuc. 2.22.1 with *Per.* 33.6 on p. 18). In a similar vein, *philotimia* as a civic virtue in democratic Athens gives way to *philotimia* as an individualised (and individualistic) quality, virtuous and/or vicious, in the *Lives*.

17 The majority of the examples concerns the Attic orators. As we will see in section 2 below, ἰδία *philotimia* acquires negative connotations in Thucydides. See also Aesch. 2.176 – 177 for a very Plutarchan use of *philotimia* vocabulary: the Athenians who are fooled into choosing war over peace are the ones who have 'ambitious and passionate souls' (τὰς ψυχὰς τὰς φιλοτίμους καὶ λίαν ὀξείας).

18 Thus, the rhetoric of praiseworthy *philotimia* of the individual is absent from the surviving deliberative speeches of the fourth century. Advising the city at the Assembly and participating in policy-making are not standardized as *philotimia* activities in the orators, even though one would fairly think that politics and the Assembly were replete with honour and ambition. This is probably explained as a matter of prioritizing in democracy collective effort and achievement in democracy instead of individual contribution when politics and policy-making are the issue. See below.

19 Edition of the text and translation (adapted) are by MacDowell (1990).

Philotimia should not be judged by the magnificence and ostentation of any kind of expenditure and activity but only of those that are beneficial to the community.[20] Such undertakings would be indices of one's power and excellence demonstrated in a legitimate and publicly beneficial way.[21] By being wealthy and ostentatious without benefiting the city, Meidias, again, manifests his power. But this is an illegitimate manifestation of power that does not make one *philotimos* in an egotistical or in a negative sense—a comment that we would expect to find in a *Life*—,[22] but makes one a *hybristēs* and not *philotimos* at all,[23] precisely because the public manifestation of *philotimia* within democratic contexts is usually positively evaluated according to established norms.[24]

Demosthenes avoids chastising Meidias as being viciously *philotimos*, even when Meidias is presented as demonstrating excessive power in unacceptable ways. This rhetorical tactic, followed throughout the speech, would be effective only if *philotimia* within democratic ideology is understood as a positively charged civic virtue.[25] On the other hand, in the *Life of Alcibiades*, we saw *philotimia* as a motivational power that may lead to, evaluatively speaking, diametrically opposed actions. It holds a prominent position amidst ideas and notions that form the wider nexus of ambition, a nexus seen by modern scholars 'first and foremost as a psychological re-

20 Cf. also [Dem.] 42.24–25, where we find what I would characterize as an ironic use of *philotimia* (in 24), with all the uncertainty that such a statement carries. The speaker calls his opponent *philotimos* and then takes it back in a strikingly ironical way, when it is showed that the activities initially associated with *philotimia* are not publicly beneficial. On the relation of irony, meaning of a text and intention of its author, see Skinner (2002) 111–113.

21 In this framework, Demosthenes turns Meidias' trierarchy into a self-interestedly motivated expenditure that is not an index of *philotimia* but of cowardice and unmanliness (Dem. 21.160–167) with Roisman (2003) 131. Again, egotistical manifestations of power and selfishness are not described as negative expressions of *philotimia* and Demosthenes never accuses Meidias of being a bad *philotimos*. If one is bad, then he is not *philotimos*. On the prerequisite of public spiritedness when spending for the city, see Ober (1989) 226–230.

22 Cf. for example, Sulla's *philotimia* that reached such a level that he made a ring, which he wore all the time, representing his capturing of Jugurtha (*Sull.* 3.4).

23 Throughout the speech, Demosthenes builds his profile as *philotimos* and *metrios* vis-à-vis Meidias who is *hybristēs*, non-*philotimos* and non-*metrios* (e.g., Dem. 21.67, 69, 101, 128–129, 134–135, 160–167, 186).

24 *Philotimia* as a civic virtue is very often explicitly expressed towards the city/people/homeland. E.g., πρὸς ὑμᾶς (Lys. 29.14; Dem. 19.173, 223; 20.69; 21.67; [Dem.] 47.54; 50.64; Aesch. 3.19, 220), πρὸς τὴν πόλιν (Isoc. 18.61; Lyc. 1.140), πρὸς τὴν πατρίδα (Lyc. 1.15). On the occurrence of this formula in honorific decrees, see in general Brock (1991) 164 and especially on *philotimia* vocabulary Whitehead (1983) 63. Note also a similar association in *Cim.* 4.7: φιλοτιμούμενος πρὸς τὴν πόλιν, ὡς οἵ τε συγγραφεῖς ἱστοροῦσι (*philotimoumenos* towards the city, as the historians narrate). If Plutarch quotes directly from Athenian sources, this phrasing enforces my point that such a representation of *philotimia* is peculiar to classical Athens.

25 Whether Dem. 21 was delivered or not does not change its significance as a source for fourth-century Athens. On the debate, see Harris (1989); MacDowell (1990) 23–28.

ality'.[26] Such an understanding of the concept which can go either right or wrong depending on the agent points towards an 'individual-centred' approach to *philotimia*.[27] Springing from within, *philotimia* may indeed be activated in response to external stimuli, but at the same time it is not presented as being pre-defined as a concept by strictly defined socio-political norms and expectations. *Philotimia* in Plutarch remains a power whose quality is primarily informed by the quality of the agent as a person.

ἰδία and δημοσία *philotimia* in Athenian politics

If we turn to Plutarch's discussions of fifth-century Athenian politics, we see that at least twice strong dissensions threatening to set the whole city in turmoil are attributed to this dominant passion, *philotimia*, which drives the behaviour of powerful citizens involved in public life. *Philotimia* is not presented as being delimited by any kind of social conventions that would control its manifestations and make it serve the city. Quite the opposite: the only way that *philotimia* can serve the well-being of the community is by its absence from political life. The elimination, or not, of *philotimia* is a matter of individual initiative.[28] In the *Life of Pericles*, what seemed to be a non-threatening distinction between 'the people' and 'the aristocrats' turned into an extreme polarisation of the citizen population because of Pericles' and Thucydides' *hamilla* and *philotimia* (*Per.* 11.1–3). In the *Life of Cimon*, Pericles' initiative in introducing the decree proposing the reinstatement of Cimon from exile[29] shows that in those days public dissensions remained political and personal hatred had no place in them: 'Even *philotimia*, that master passion, paid deference to the country's welfare' (*Cim.* 17.9).[30] If the city is to prosper, powerful individuals should make sure they leave their *philotimia* out of the picture.

If we turn to Thucydides for an account of the role of *philotimia* in fifth-century Athenian politics, it appears that, in a similar manner, *philotimia* is a negative factor

26 Frazier (2014) 491, who at the same time notes that *philotimia* is also presented as a principle that stands outside an individual's soul, as 'an essential factor in politics'.

27 Here, I do not attempt to engage in a theoretical-philosophical approach to the individual *philotimos* seen either from a 'subjective-individualist' or from an 'objective-participant' perspective. See Gill (1996) who convincingly argues for an 'objective-participant' conception of self in Greek epic, tragedy and philosophy.

28 Also, *Comp. Arist. et Ca. Ma.* 5.3–4.

29 For a different account of this episode which also hints at private motives and personal reasons behind Pericles' initiative, see *Per.* 10.1–5.

30 Similarly, Agesilaus saved Sparta by renouncing his inherent passions (*Ages.* 33.1): *philoneikia* and *philotimia* are presented as *emphuta pathē*, inner qualities in a stand-by mode the handling of which, even in public-political contexts, is a matter of individual initiative. Cf. Frazier (1988) 122, who sees *philotimia* in *Cim.* 17.9 and *Sull.* 4.4 not so much as an individual characteristic, but as an autonomous factor disastrous for public life; also Nikolaidis (2014) 360 with nn. 40–42.

in politics (Thuc. 2.65.7; 8.89.3).[31] There exists, however, a significant difference between the two authors: whereas Plutarch dismisses *philotimia* single-handedly as a factor that can be proven devastating for public life, for Thucydides it is not *philotimia* in general, but *idiai philotimiai* that should be eliminated from politics.[32] If *philotimiai* had to be qualified by *idiai* in order to denote personal interest contrary to the well-being of the city, this could indicate that already in the time of Thucydides there was another kind of *philotimia/philotimiai*—δημόσιαι or κοιναί—[33] that was accepted as advantageous for the community. Such a phrasing could mean that channeling *philotimia* towards specific publicly beneficial activities as well as defining the different fields of *philotimia* should not be seen as a fourth-century Athenian phenomenon (manifested most clearly at the time of Demosthenes), but may also be traced back to the end of the fifth century.[34]

In Thucydides, then, the qualification of *philotimia* as ἰδία points towards different types of *philotimia*, private and public, which colour accordingly the meaning of the concept. Here, *philotimia* is not evaluated as positive or negative depending on the way an individual expresses it nor is it a matter of personal initiative to exclude *philotimia* from specific spaces, as it appeared in *Cim.* 17.9 and *Per.* 11.1–3;[35] quite differently, an individual who exploits politics for private considerations is not castigated for acting out of *philotimia*, but out of ἰδία *philotimia.*[36] In the specific field of politics, *philotimia* is described as ἰδία in order to be negatively charged as a personal, egotistical power and set against public interest. This qualification is indicative of a pre-evaluated understanding of the concept and shows that *philotimia* in general, that is to say unqualified *philotimia*, is not *ex definitione* perceived as a quality that can be detrimental to the public interest.

In Demosthenes' time, *philotimia* is a 'cardinal virtue' of Athenian democracy, and its quality is generally positive,[37] even when it is qualified as ἰδία. The field with-

---

31 The only positive mention in Thucydides comes in Pericles' *Funeral Oration*, 2.44.4 (*to philotimon*).

32 Similarly, in *Cam.* 31.2 (φιλοτιμίας ἕνεκα καὶ δόξης ἰδίας) it is the notion coupled with *philotimia* and not *philotimia* itself that is qualified as private.

33 Whitehead (1983) 59.

34 I do not argue here for a chronological development of the phenomenon that started in the fifth century and was 'completed' in the fourth. What I wish to point out is that in fifth-century Athenian sources, just as in fourth-century, there does not seem to exist one broad and uniform sphere where *philotimia* may be manifested in different ways, but the meaning of *philotimia* is coloured differently according to the activity with which it is associated.

35 In *Fab.* 25.3, *philotimia* and *philonikia* denote a shift towards private motivations that explain Fabius' behaviour.

36 Note, though, that similarly to Plutarch, in 3.82.8 Thucydides colours unqualified *philotimia* negatively by association with *pleonexia*.

37 On 'cardinal virtues' of Athenian democracy, see Whitehead (1993). Here, I am referring to the place that *philotimia* holds strictly within democratic contexts and within democratic ideology. Once we move outside democratic settings the situation changes and *philotimia* can be presented as a vice; for example, Demosthenes' presentation of Philip and his *philotimia* sounds quite Plutarch-

in which *philotimia* emerges as a civic virtue is specifically that of benefactions, especially monetary ones, whereas it is not explicitly associated with active political participation. We do not find the vicious version of ἰδία *philotimia* denoting private exploitation of political participation, neither a virtuous version of *philotimia* related to political contributions, such as participation in Assembly debating. In an instance where Demosthenes discusses his benefactions to the state, both kinds of *philotimia*, δημοσία and ἰδία, are positively charged and at the same time dissociated from his active political participation:

> ἐξελθόντι δ' ἐκ παίδων ἀκόλουθα τούτοις πράττειν, χορηγεῖν, τριηραρχεῖν, εἰσφέρειν, μηδεμιᾶς φιλοτιμίας μήτ' ἰδίας μήτε δημοσίας ἀπολείπεσθαι, ἀλλὰ καὶ τῇ πόλει καὶ τοῖς φίλοις χρήσιμον εἶναι, ἐπειδὴ δὲ πρὸς τὰ κοινὰ προσελθεῖν ἔδοξέ μοι, τοιαῦτα πολιτεύμαθ' ἑλέσθαι ὥστε καὶ ὑπὸ τῆς πατρίδος καὶ ὑπ' ἄλλων Ἑλλήνων πολλῶν πολλάκις ἐστεφανῶσθαι, καὶ μηδὲ τοὺς ἐχθροὺς ὑμᾶς, ὡς οὐ καλά γ' ἦν ἃ προειλόμην, ἐπιχειρεῖν λέγειν.
>
> When I reached adulthood, I performed duties suitable to my background—I furnished choruses and triremes, paid taxes, lost no opportunity for private or public *philotimia*, but was of service to the city and to my friends. And when I decided to enter state affairs, I chose policies which earned me many crowns from both my country and the rest of the Greeks, and even you, my enemies, did not try to say that those policies were not honourable.[38] (Dem. 18.257)

δημοσία *philotimia* is the good done towards the city, ἰδία *philotimia* is private *philotimia* in the sense of benefiting one's friends, whereas political activity beneficial to the state is carefully dissociated from both. It seems that not every field of activity where ambition was expected to be found was at the same time a suitable place for demonstrations of a citizen's *philotimia*.[39]

Finally, if we want to explore further the relation between *philotimia* and the expression of negatively coloured private motivation, we may focus on the different usages of *philotimia* in association with *kerdos* in Thucydides, Demosthenes and Plutarch. Thucydides, referring to the successors of Pericles, mentions that, κατὰ τὰς ἰδίας φιλοτιμίας καὶ ἴδια κέρδη κακῶς ἔς τε σφᾶς αὐτοὺς καὶ τοὺς ξυμμάχους

an. In fact, Demosthenes' Philip could be seen as a reversed *alter ego* of Plutarch's Caesar (cf. Dem. 2.15–21 with *Caes.* 17).

38 Edition of the text and translation (adapted) are by Usher (1993).

39 Thus, in *On the Chersonese*, a deliberative speech that offers a clear example of political participation in action, Demosthenes, in order to stress the publicly beneficial character of the advice he has been giving in the Assembly-meetings, has to leave any considerations of *philotimia* out of the picture: 'nor have I been prompted either by *kerdos* or by *philotimia*, but continue to offer advice which does indeed lower me in your esteem, but which, if you will follow it, would contribute to your greatness' (Dem. 8.71, translation from Loeb, adapted). This exclusion of *philotimia* from the Assembly does not make *philotimia* a vice; it rather shows that, at least at the level of public rhetoric, the field within which the individual could legitimately present himself as pursuing *philotimia* was delimited.

ἐπολίτευσαν;[40] Demosthenes, projecting his incorruptibility, declares to his fellow citizens that οὐκ ἀνταλλακτέον εἶναί μοι τὴν πρὸς ὑμᾶς φιλοτιμίαν οὐδενὸς κέρδους;[41] and Plutarch puts in the mouth of Nicias an interpretation of Alcibiades' desire to sail against Sicily: ἰδίων ἕνεκα κερδῶν καὶ φιλοτιμίας.[42] *Philotimia* without any further qualification can be employed in Plutarch for the demonstration of private motivation, but this is not so in the other two cases. In Thucydides (ἰδίας and set against the collective: ἕς τε σφᾶς αὐτοὺς) and in Demosthenes (towards the collective: πρὸς ὑμᾶς and set against κέρδος) the scope of appropriate and inappropriate *philotimia* is defined and its quality is evaluated accordingly.

Conclusions

When examining *philotimia* in Plutarch's *Lives* and in Athenian sources, we notice a wide discrepancy in what counts as a manifestation of *philotimia* in each case. What I tried to show in this chapter is that such a differentiation is a corollary of different understandings of *philotimia* as a concept in general. Thus, in Plutarch *philotimia* is very often seen as primarily springing from within and its manifestation as a matter of the agent's creative virtue and/or vice and, therefore, the range of *philotimia*'s activities is very broad. In Athens, on the other hand, *philotimia* as a civic virtue is restricted within specific places and related activities while excluded from others and its meaning is qualified accordingly (thus we saw the differences between ἰδία and δημοσία *philotimia*).

Such a differentiation seems to be a matter of different priorities: from Thucydides' and Demosthenes' examples emerges that what was important within democratic ideology was to specify as far as possible the range of honourable *philotimia* activities and provide the community with a more specific idea of *philotimia* as a civic virtue in harmony with democratic prerequisites and socio-political norms. Plutarch, on the other side, was interested in presenting to his readers how *philotimia* works as an internal force, how it can be harnessed or, conversely, get out of control

---

40 Thuc. 2.65.7: 'For the sake of private *philotimiai* and private profit they pursued policies which were bad for themselves [the Athenians] and for the allies', Rhodes (1988) adapted.

41 Dem. 19.223: 'no profit was worth giving up my *philotimia* in your service', MacDowell (2000) adapted. *Philotimia* is qualified by 'towards you' (πρὸς ὑμᾶς), which shows the public character of *philotimia*, and is set against private interest, denoted by *kerdos*. In the same passage, Demosthenes says that he is deprived of his private *philotimiai* (τῶν ἰδίων φιλοτιμιῶν), referring to the honours that the city usually awarded to ambassadors upon completion of their duties. Here *idiai philotimiai* is almost identical in meaning with the reward ('personal/private honours') that Demosthenes expects; *idiai philotimiai* does not stand for calculated interest as the motivational power driving one to participate in politics and/or to benefit the city, but is rather the personal benefit that one enjoys as a reward after having done so. See MacDowell's (2000) commentary on paragraph 40.

42 *Nic.* 12.4–5: 'of satisfying his own private greed and *philotimia*', Loeb adapted.

and destroy ambitious statesmen and broader communities, thereby providing his readership with examples of 'a more descriptive style of moralism'.[43]

Literary sources on Athenian democracy, it seems to me, also provide a good example of how *philotimia* is successfully harnessed, but this time not so much by individual initiative but by the democratic city itself, which had managed to direct different kinds of *philotimia* to specific activities. Plutarch, reflecting on his heroes from a temporal distance, has the advantage of examining them retrospectively: judging from the outcome of their actions, he can attribute ethical-psychological traits, such as *philotimia*, to their character and personality in order to explain behaviour and motives. Fifth- and fourth-century sources on Athenian democracy, especially in public discourse, speak the language of the city: the orators, talking about themselves and their opponents publicly, had to fulfil certain socio-political expectations; and even Thucydides, who is reflecting on past events and is not the most ardent supporter of democracy, seems to be following the flow.

Unveiling the workings of democratically informed *philotimia* was a task that fell outside Plutarch's scope. It, after all, revealed more the collective intelligence encoded in democratic ideology than that of the individual. And probably that was something that, even if Plutarch had recognised it, he did not care to stress. The broad field of *philotimia* that is constructed in the *Lives* and includes flexible and often contrasting manifestations of *philotimia* brings to the fore the gap between the stability of socio-political models of behaviour promoted by democratic ideology and 'the dissolution of an apparently stable sense of selfhood' that came with the rise of imperial ideology.[44]

43 Pelling (2002) 237–251 distinguishes between two kinds of moralism in Plutarch, 'protreptic' and 'descriptive'; quotation from p. 242.
44 Larmour (2005) 47.

6 Philosophy and religion between past and present

Bram Demulder

Is dualism a Greek word? Plutarch's dualism as a cultural and historical phenomenon[1]

Abstract: Throughout his oeuvre, Plutarch offers different presentations of dualism (the notion that reality ultimately consists of two non-reducible principles). This contribution looks at the function of space (Greek vs. non-Greek) and time (pre-Platonic past vs. Plutarch's Platonic stance) in some of these accounts. The presentation of dualism through a comparison with music in the works *On Tranquillity of Mind* and *On the Generation of the Soul in the Timaeus* serves as a yardstick. Next, Plutarch's use of the Greek (esp. Euripides, Heraclitus and Plato) and non-Greek (viz. Egyptian and Zoroastrian) past in the works *On Tranquillity of Mind*, *On the Generation of the Soul in the Timaeus* and *On Isis and Osiris* is considered. It is argued that Plutarch reverts to a more rudimentary dualism in order to create room for non-Greek material without shaking the foundations of his Platonism, in this way combining cultural introversion (i.e. hellenocentrism) with extroversion (i.e. openness to other cultures).

In his seminal study of Plutarch's dualism (the notion that reality ultimately consists of two non-reducible principles[2]), Ugo Bianchi quotes and affirms the conclusion of Julien Ries that, in Plutarch's time, 'la pensée grecque, fatiguée, s'ouvrait largement aux religions de salut venues d'Egypte et du Proche-Orient'.[3] Given this perspective, it is not surprising that Bianchi's discussion starts from and focusses on the influence of non-Greek religion found in Plutarch's treatise *On Isis and Osiris*. Although, eventually, Bianchi does not fail to point out that Plutarch platonises the Egyptian and Near Eastern material, this focus risks underplaying the importance of Plutarch's identity as a Greek and a Platonist in his reception of the history of dualism, which will be the focus of this chapter.

In the first section, I will bear with the rather short-sighted sympotic advice of Florus and 'leave the Egyptians'—and the Near Eastern Zoroastrians for that matter—'out of it',[4] while discussing the different presentations of dualism in Plutarch's so-called 'practical ethics' (especially in the work *On Tranquillity of Mind*) and in his interpretation of Plato's *Timaeus* (in the work *On the Generation of the Soul in the Timaeus*). This exploration of different brands of dualism will serve as a yardstick for

1 I would like to thank Daniel S. Richter and Michiel Meeusen for their Delphic comments and Liesbeth Schulpé for correcting my English.

2 This is, of course, an imperfect working definition of a philosophical concept which is notoriously hard to define. See, e.g., Alt (1993) 10–11.

3 Bianchi (1987) 350.

4 *QC* 5.10, 684F (transl. Hoffleit): ἐᾶν ἐκέλευε τοὺς Αἰγυπτίους. On this *quaestio* and the involvement of the Egyptians, see Meeusen's contribution in this volume.

https://doi.org/10.1515/9783110539479-019

the second part, in which I will address the role played by the reception of the history of early Greek philosophy and non-Greek religion in Plutarch's Platonism by including *On Isis and Osiris* in the discussion. This approach will yield a clearer view of how considerations of space (Greek vs. non-Greek) and time (pre-Platonic past vs. Plutarch's Platonic stance) interact to form the multi-layered, culturally and historically informed notion that is Plutarch's dualism.

Dualism(s) in Plutarch's practical ethics and in *On the Generation of the Soul in the Timaeus*

In the Plutarchan texts which can be labelled as 'practical ethics', writings in which Plutarch is less concerned with the technical subtleties of Platonic ontology[5], we encounter several passages which assume a plain dualistic worldview. From the work *How to tell a Flatterer from a Friend* (61D–E) we learn that we should always keep in mind that our soul has two opposed parts—the one rational, the other irrational—and that friends side with the better part, whereas flatterers try to appeal to the bad part. In *Concerning Talkativeness* (506F–507A) Plutarch suddenly introduces the monad and the dyad, not, as one would expect when such terms are mentioned, in a technical discussion of their ontological status,[6] but in an explanation of how gossip works: while a secret stays with its possessor in a monadic fashion, a dyadic rumour tends to be dissipated incessantly and unstoppably.

A passage from *On Tranquillity of Mind* is more complicated and deserves a closer look. Plutarch's advice to Paccius on how to achieve εὐθυμία includes a warning not to get bogged down in worries about past misfortunes. This is illustrated by a comparison: in our souls, we should combine the inevitable bad with the good like the musician combining low notes and high notes or the grammarian combining vowels and consonants.

> [L]ike colours in a painting, so in the soul it is right that we should place in the foreground bright and cheerful experiences and conceal and suppress the gloomy; for to wipe them out and be rid of them altogether is impossible. 'For the harmony of the universe, like that of a lyre or a bow, is by alternatives,' [Heraclitus fr. B51 DK] and in mortal affairs there is nothing pure and unmixed (καθαρὸν οὐδὲν οὐδ' ἀμιγές). But as in music there are low notes and high notes (βαρεῖς φθόγγοι καὶ ὀξεῖς), and in grammar there are vowels and consonants (φωνήεντα καὶ ἄφωνα γράμματα), yet a musician or a grammarian is not the man who dislikes and avoids the one or the other, but rather the man who knows how to use all and to blend (μιγνύναι) them properly, so also in human affairs, which contain the principles of opposition to each other (since, as Euripides [*Aeolus*, fr. 21 *TrGF*] has it, 'The good and bad cannot be kept apart, / But there's some blending, so that all is well'), we should not be disheartened or despondent

5 Van Hoof (2010) 37–40.

6 Cf. the significantly different approach in the more technical works *Quaest. Plat.* 1002A and *De an. procr.* 1012E; 1025C–D.

> in adversity, but like musicians who achieve harmony by consistently deadening bad music with better and encompassing the bad with the good (ὥσπερ ἁρμονικοὺς ἀμβλύνοντας ἀεὶ τοῖς κρείττοσι τὰ φαῦλα καὶ τὰ χείρονα τοῖς χρηστοῖς ἐμπεριλαμβάνοντας [7]), we should make the blending (μῖγμα) of our life harmonious (ἐμμελές) and conformable to our own nature. For it is not true, as Menander [fr. 500 *PCG* VI.2] says, that 'By every man at birth a Spirit stands / A guide of virtue for life's mysteries'; but rather, as Empedocles [fr. B122 DK] affirms, two Fates, as it were, or Spirits, receive in their care each one of us at birth and consecrate us … . (*De tranq. an.* 473F–474B, transl. Helmbold[8])

This comparison might perplex the reader in two ways: (1) Plutarch introduces the musician alongside the grammarian, but in the second part of the comparison the grammarian is left out. The musician is indeed the more fitting analogy when one wants to describe the result as a harmony (ἁρμονικούς, ἐμμελές); (2) The exact nature of the resulting mixture (οὐδ' ἀμιγές, μῖγμα, μιγνύναι) remains unclear. Is it in the objects and events themselves, as is suggested by the quotations from Heraclitus, Euripides and Empedocles and Plutarch's comments on them? Or is it only in the psychological perception of objects and events, as the comparison itself and the general message of the essay suggest?[9] Plutarch does not really seem to be concerned with the distinction in this case. What matters is that we are presented with a clear-cut opposition between good and bad and with some kind of immediate mixture of these opposed elements in human life.

Turning from practical ethics to more technical philosophy, a passage from *On the Generation of the Soul* seems to provide a parallel with the *On Tranquillity of Mind* passage.[10] Plutarch is again talking about matters of the human life in a dualistic fashion. The same fragment from Heraclitus is quoted and Empedocles is mentioned again. Once again, it is the comparison which throws us off balance:

> As some sound (φωνή τις) is not speech and not significant but speech (λόγος) is an utterance in sound (ἐν φωνῇ) that signifies thought (διανοίας), and as concord is what consists of sounds and intervals (ἐκ φθόγγων καὶ διαστημάτων) and a sound is one and the same thing, an interval the diversity and difference of sounds, and the mixture of these results in song and melody (ᾠδὴ … καὶ μέλος), so the affective part (παθητικόν) of the soul was indeterminate and unstable and then was bounded when there came to be limit and form (πέρατος ἐγγενομένου καὶ εἴδους) in

---

7 This is the reading in the Loeb. Pohlenz' Teubner edition follows a group of manuscripts reading ὥσπερ ἁρμονικοὺς ἀμβλύνοντας τοῖς κρείττοσι τὰ χείρονα καὶ τὰ φαῦλα τοῖς χρηστοῖς ἐμπεριλαμβάνοντας. Since the different *ordo verborum* does not yield a different meaning, this should not concern us here, but see the comments by Martinelli Tempesta reported by Volpe Cacciatore (2007) 100, n. 16.

8 All translations are taken from the Loeb, except for *De Is. et Os.*, for which Griffiths (1970) is used.

9 Cf., e.g., *De tranq. an.* 469A: we should, 'mingling (μιγνύντας) good with bad, cause the better to outshine the worse'; 473B: 'But that every man has within himself the storerooms of tranquillity and discontent, and that the jars containing blessings and evils are not stored 'on the threshold of Zeus', but are in the soul, is made plain by the differences in men's passions'.

10 Cf. Broecker (1954) 149–151, who also mentions the parallel with *De Is. et Os.* 369B–D, which will be discussed below.

> the divisible and omnifarious character of the motion. And, once having comprehended sameness (τὸ ταὐτόν) and difference (τὸ θάτερον) with the similarities and dissimilarities of numbers that produced consensus out of dissension, it is for the sum of things rational life and concord and reason guiding necessity that has been tempered with persuasion and which by most people is called destiny, by Empedocles [fr. A45 DK] love together with strife, by Heraclitus [fr. B51 DK] concord of the universe retroverse like that of lyre and bow, by Parmenides light and darkness, by Anaxagoras intelligence and infinitude, and by Zoroaster god and spirit, the former called by him Oromasdes and the latter Areimanius. Euripides [*Troiades* 886] has erred in using the disjunctive instead of the copulative conjunction in the prayer, 'Zeus, whether natural necessity / Or the intelligence of mortal men', for the power that pervades all things is both necessity and intelligence. Now, the Egyptians in a mythical account say enigmatically (Αἰγύπτιοι μὲν οὖν μυθολογοῦντες αἰνίττονται) that, when Horus was convicted, the breath and blood were assigned to his father and the flesh and fat to his mother. Of the soul, however (τῆς δὲ ψυχῆς) nothing remains pure or unmixed or separate from the rest. (*De an. procr.* 1026A–C, transl. Cherniss)

At first sight Plutarch seems to be doing exactly the same thing as in *On Tranquillity of Mind:* explaining dualism by comparing it to the production of harmonic music from opposed elements. However, in this case, the *comparans* is not simply music. The element of language, used in *On Tranquillity of Mind* only to be dismissed, is reintroduced here. Plutarch is talking about the production of a *song*, i.e. of music *with* words. This complicates the previous, straightforwardly dualistic scheme. Suddenly, there are four elements instead of two; their combination is presented in two stages. First, φωνή receives διάνοια and thus becomes λόγος. At this point, there is meaningful speech, but not yet a song: harmony is still lacking. Therefore, in a second move, λόγος (the resulting mixture from the first stage) is mixed with monotone noise (φθόγγος) and interval (διάστημα). The result is a song. This more complex comparison reflects Plutarch's interpretation of the generation of the soul in Plato's *Timaeus*, according to which, as he clarifies, the affective part (παθητικόν, which is called divisible being elsewhere), receives limit and form (πέρας and εἶδος, the features of indivisible being), before being inserted in the mix with sameness (τὸ ταὐτόν) and difference (τὸ θάτερον).[11] This is not simply a more awkwardly worded presentation of the dualism found in *On Tranquillity of Mind* but an ontologically different view. In this case, the mixture of the extremes is not immediate. Although they share an affinity with them, the middle terms inserted between the extremes (the lyrics of the song) are not *reducible* to these extreme terms (monotone noise and interval): they have their own nature and specific functions, as is the case for the four entities involved in the composition of the soul, which the comparison intends to clarify.[12] Moreover, it would be too simple to identify the extremes from *On the Generation of the Soul* with good and bad, as was the case in the comparison

---

11 Cf. the analysis of this passage by Opsomer (1994) 40–41.

12 Plutarch criticises Xenocrates precisely for not distinguishing being (the first mixture which is put in the middle) from sameness and difference (*De an. procr.* 1013D). He makes it clear that this first mixture is needed as 'some third principle', since mutual participation of sameness and difference alone cannot have a 'fruitful' (γόνιμον) result (1025F).

from *On Tranquillity of Mind*. Although difference is associated with evil, it should be pointed out that it *is* not evil and that it has its own merits. Without difference, there would be no motion, perception of sensible reality, or practical activity—in fact, no γένεσις at all (e.g., 1025E–F).[13]

The detail of the presentation of dualism in musical terms has revealed crucial differences between the dualism of *On Tranquillity of Mind* and that of *On the Generation of the Soul*. In the former work, as in the other works of practical ethics mentioned earlier, the dualism is plain: two opposed entities enter into combination. In the latter work, on the other hand, we are confronted with a mitigated dualism, with non-reducible middle terms complementing the extreme entities. Moreover, the dualism of *On the Generation of the Soul* dovetails with the rest of the treatise and with other passages devoted to Platonic exegesis, whereas in *On Tranquillity of Mind* we indicated some degree of carelessness as to the exact nature of the mixture. Then, which dualism is really Plutarch's dualism? Perhaps this is not the right question. It is undeniable that Plutarch found a basic truth in the dualistic scheme. He considered this scheme to be relevant to questions pertaining to ontology and epistemology (as in *On the Generation of the Soul*), but also to moral questions (as in *On Tranquillity of Mind* and in the other works of practical ethics). At the same time it is clear that Plutarch allows himself to do different things starting from this basic scheme: while it is presented in its most rudimentary form in *On Tranquillity of Mind* by reference to Heraclitus and Empedocles, it is developed more subtly in *On the Generation of the Soul*, where the doctrines of the past play a role, but the focus is much rather on Plato's fine tuning (or at least on Plutarch's interpretation of it). Neither version is to be preferred overall: a detailed exegesis of a sentence in Plato and a friend needing moral advice are two different things—and luckily so. The reader is reminded of this by the fact that, at the beginning of *On Tranquillity of Mind*, Plutarch, although he mentions that Paccius requested this from him, does not send him elucidations on certain subjects in the *Timaeus*: *On Tranquillity of Mind* is not *On the Generation of the Soul*.[14]

Contextualising dualism in space and time

After this sketch of Plutarch's dualism, which has shown that perhaps we should rather speak of Plutarch's *dualisms* (plural) or of a dualistic *spectrum*,[15] we can finally discuss where negotiations with space (Greek vs. non-Greek) and time (pre-Platonic history vs. Plutarch's post-Platonic stance) fit in this general picture. The passages from *On Tranquillity of Mind* and *On the Generation of the Soul* share some informa-

13 Cf. Opsomer (2007) 383.

14 See Van Hoof (2010) 113–115 for a more detailed analysis of this issue.

15 Cf. Almagor (2013a), although I think I am more prepared than him to see a coherent *substratum* underneath Plutarch's different dualistic endeavours.

tion on this matter, namely, Heraclitus' fr. B51 and the testimony of Empedocles. Moreover, Euripides occurs in both passages, but—and this is not entirely unexpected for a dramatic poet[16]—he turns out to be a philosophical flip-flop: in *On Tranquillity of Mind* he endorses dualism (in this case, Menander serves as the ignorant playwright who only recognises one principle), whereas in *On the Generation of the Soul* he is on record as speaking out against it. In the latter treatise, Plutarch offers us more authorities: along with the other pre-Platonic philosophers Parmenides and Anaxagoras, the Persian tradition is represented by Zoroaster and the Egyptians are mentioned. Plutarch's way of presenting the latter is remarkable.

First of all, Plutarch stresses the non-philosophical character of the Egyptian tradition by saying that they present us with a mythical account (μυθολογοῦντες).[17] After this reservation, which contrasts with the mention of Zoroaster, who is apparently included among the philosophers, a dualistic doctrine involving the story of the dismemberment of Horus is mentioned.[18] As it turns out—and as is suggested by the μέν-δέ structure—the Egyptian dualism is not adduced in support of Plutarch's Platonic exegesis. On the contrary: the essence of Plutarch's dualism—and the conclusion of his historical overview of dualism—is that all elements get thoroughly mixed and very little dualism remains after the intervention of the demiurge, whereas the Egyptians are presented as reversing the cosmic order by first assuming unity, which later gets disrupted. Compared to the Platonic harmony, in which—so Plutarch writes just after mentioning the Egyptians—'god, making the mixture, sank and concealed the differences and the diversities' (1026C), Egyptian dualism remains too extreme for Plutarch's taste since it does not allow for any mixture.

The conclusion drawn from this particular episode of Egyptian mythology is confirmed when we look for the same story in Plutarch's main account on Egyptian culture, *On Isis and Osiris*. As it turns out, the story of the dismemberment of Horus is explicitly excluded from 'the main points' of the myth; Plutarch omits it by calling it one of the 'most outrageous' of the tales (358E). However, we should not reject the Egyptian religion *tout court* as an important predecessor in the dualistic doctrine just yet. Further on (369A–371A), Plutarch gives a doxographical overview of dualistic doctrines very similar to the one in *On the Generation of the Soul*. Although it has been convincingly argued that both overviews point to the same source, it would be misguided to assume that Plutarch thoughtlessly copied this list on different occasions.[19] Accordingly, we should consider how he adapted the doxographical information for his own purposes.

---

16 Cf. Van der Stockt (1992) 164, on the connection of tragedy with falsehood in Plutarch.

17 Although myth can have philosophical potential, Plutarch tends to distinguish it clearly from philosophy (and the philosophical explanation of myth). Cf. Hardie (1992) 4743–4749.

18 On Plutarch's use of this myth, see Hani (1976) 102–104.

19 See the references to Mansfeld and Donini in Opsomer (2007) 385, for a doxographical analysis of this passage.

Introducing dualistic doxography in the context of his interpretation of the Egyptian myth of Isis and Osiris, Plutarch starts off with a pair of quotations found also in the *On Tranquillity of Mind* passage (Heraclitus fr. B15 DK and Euripides, *Aeolus*, fr. 21 *TrGF*). Interestingly, by being linked to Euripides' verses in both passages, the function of the Heraclitean fragment changes compared to *On the Generation of the Soul*. In the latter treatise it was used to count Heraclitus among those who discerned an opposition in the cosmos, whereas in *On Isis and Osiris* and *On Tranquillity of Mind*, the focus is on the harmony of these opposites. Both in *On Isis and Osiris* and in *On the Generation of the Soul* this focus is balanced out later on by reference to other Heraclitean fragments (fr. B53 DK and fr. B54 respectively). This corresponds to the general development of the passages: in *On the Generation of the Soul*, there is a shift from opposition to harmony, whereas the shift in *On Isis and Osiris*, just like in *On Tranquillity of Mind*, is the other way around.

After the Heraclitus-Euripides sequence, Zoroaster's doctrine is not just summarised in a few words, as it was in *On the Generation of the Soul:* it is treated at relatively great length (369E–370C). Notwithstanding the amount of space awarded to this doctrine and the ranking of Zoroaster among 'the majority and the wisest' (369D), that is, those who rightly maintained a dualistic world view, John Dillon's interpretation that Plutarch 'bestows high praise on the Zoroastrian theology' in this passage might not tell the whole story.[20] The mention of, for instance, the Zoroastrian practice of apotropaic offerings to the negative divine force and their superstitions about plants and animals is not followed by a reference to a philosophical interpretation of these habits, as is the case with the Egyptian material.[21] While Egyptian rituals and beliefs—somewhat paradoxically—contain nothing 'fabulous' (μυθῶδες, 353E), the Zoroastrian material is presented as containing 'many fabulous stories' (πολλὰ μυθώδη, 369E).[22] Therefore, it is likely that the excursus on Zoroastrianism serves as a reminder of what had been said in the first chapters of *On Isis and Osiris:* that rituals and myths are in need of such philosophical explanation (355B–D), an effort at the very heart of *On Isis and Osiris* which reaches its climax shortly after this reminder. Consequently, the presentation of Zoroastrianism here contrasts to some extent with the mention of it in *On the Generation of the Soul*, where Zoroaster was ranked among philosophers without further ado: in *On Isis and Osiris*, Zoroastrianism plays the rather negative role the Egyptians played in the doxography in *On the Generation of the Soul*.

After the Zoroastrian excursus, and after briefly mentioning the Chaldeans and the beliefs of the Greeks concerning the Olympians (370C), Plutarch turns to a group which he labels 'philosophers' (thus, again as opposed to what he suggested in *On the Generation of the Soul*, implying Zoroaster's separation from the group)

20 Dillon (1996) 203.

21 On the similarities of Egyptian and Zoroastrian myths and rituals in *De Is. et Os.*, see Brout (2004) 73–79 (with further references on p. 74, n. 10).

22 Cf. 376E, but see 365D for a more neutral use of the term. See also, e.g., *De Pyth. or.* 406E.

(370D–371A). The group consists of Heraclitus (fr. B53 DK), Empedocles (parts of fr. B122, which we know only from the *On Tranquillity of Mind* passage otherwise), Pythagoras, Anaxagoras, Aristotle and, ultimately, Plato. After the mention of the principles of sameness and difference, which, as we saw in *On the Generation of the Soul*, are familiar candidates for a dualistic opposition in Plato, Plutarch explains his interpretation of Plato's *Laws* (10, 896d–897d), which he takes to provide a more explicit account of what was stated in the *Timaeus*.

> He [sc. Plato] states that the cosmos is moved not by one soul, but probably by more, and at least by no fewer than two. Of these he states that one is beneficent and the other is opposed to it and is the creator of opposed qualities. He leaves room for a certain third nature also to exist between them, one which is neither inanimate, nor without reason nor unable to move of itself, as some think, but which depends on the other two, and constantly desires and longs for and pursues what is better. (*De Is. et Os.* 370F–371A, transl. Griffiths)

If we take the interpretation Plutarch provided in *On the Generation of the Soul* to be his definitive account on the issue, as he himself indicates we should (1012B), we are faced with some problems: (1) In *On Isis and Osiris* Plutarch equates the principles of sameness and difference with the beneficent soul and its opposite, respectively. In other words: principles that were painstakingly distinguished there, are conflated: the maleficent soul is in fact divisible being (*De an. procr.* 1014E), which—together with divisible being—forms the substrate for sameness and difference without being reducible to the latter; (2) In *On Isis and Osiris* Plutarch treats the souls from the *Laws* as two different, simultaneously active cosmic souls, whereas in *On the Generation of the Soul* it seems as if we should take Plato to be talking about two successive states of soul, i.e. pre-cosmic and cosmic (*De an. procr.* 1015E);[23] (3) In *On Isis and Osiris*, Plutarch states that Plato leaves (ἀπολείπει) a middle nature, which acts in the cosmos as a mediator between sameness and difference, whereas in *On the Generation of the Soul*, as we have seen, the resulting world soul is a mixture, *including* sameness and difference.

What these issues have in common, is that they facilitate Plutarch's respectful interpretation of the Egyptian myth without shaking the foundations of his philosophical views.[24] The first two adaptations render a clear-cut scheme of oppositions in which Osiris and Typhon fit more easily than in the scheme of *On the Generation of the Soul*. The description of the middle nature allows for a philosophical interpretation of the figure of Isis, who combines features of matter and soul. And this is indeed what follows, as Plutarch announces: 'we shall relate the theology of the Egyptians especially to this [i.e. Plato's] philosophy' (371A).

---

23 For a more charitable interpretation regarding Plutarch's consistency on this point, see Opsomer (2007) 385.

24 Griffiths (1970) and Hani (1976) generally point out Plutarch's truthful presentation of the Egyptian material. For an interpretation which contrasts the dualisms in *De an. procr.* and *De Is. et Os.* more than I do, see Alt (1993) 16–29.

Actually, the real discussion about dualism in *On Isis and Osiris* and its relation to Plutarch's Platonism begins here, but these preliminary points have yielded some important indications as to how space and time play a significant role in how Plutarch judges earlier formulations of dualism. First, non-Greek is distinguished from Greek: whereas the history of Greek philosophy is invariably adduced in support of Plutarch's thesis, the non-Greek influence is ambivalent and depends on the possibility of a Greek interpretation (Egyptian religion being treated favourably in *On Isis and Osiris* but receiving a more critical treatment in *On the Generation of the Soul*, Zoroastrianism the other way around). Second, while accounts from the past are appreciated as unveiling the rudimentary dualistic scheme, it is only since Plato, who explicitly distinguished non-reducible middle terms between the extremes, that the whole truth has been brought to light. However, this Greek and Platonic superiority should not be seen as a rejection of non-Greek influence and pre-Platonic history. As was shown in the first section of this chapter, Plutarch's notion of dualism allows for a whole spectrum of valid presentations. What is more, even the account of Platonic dualism turned out to be adaptable in order to facilitate the insertion of non-Greek religion into the account. Therefore, the question why Plutarch makes such an effort to adduce non-Greek and pre-Platonic support if Plato is most truthful after all, deserves some further thought by way of concluding remark.

Cultural extroversion *and* introversion

Plutarch's interpretation of the myth of Isis and Osiris has generally been described as either cultural extroversion or introversion, that is to say, either a genuine interest in Egyptian matters or a subordination of Egyptian theology to Greek and Platonic standards.[25] However, in Plutarch's mind, the one did not exclude the other. Indeed, Plutarch's final interpretation of the Egyptian myth is thoroughly Platonic and this causes problems for the Egyptian material. Isis, for instance, gets dragged to the middle perhaps more than she should be and, in any case, more than she was in the demonological part of the treatise.[26] But at the same time Plutarch chooses to omit some of the important points of *On the Generation of the Soul* and reverts to a more basic, less subtle dualistic scheme, with a focus on opposition rather than on mixture. This is not due to the ethical character of the text, as it was in *On Tranquillity of Mind.* Rather, by opting for the less detailed scheme, Plutarch leaves more room for the peculiarities of Egyptian theology.

This combination of extroversion *and* introversion occurs at a time when this is a culturally significant dynamic, as other chapters in this volume point out. This dynamic goes hand in hand with what George Boys-Stones describes as the essence of the

---

25 See, e.g., the excellent analysis by Brenk (1999) and Richter (2001) respectively.
26 Cf. Griffiths (1970) 26–27 and 58.

development of dogmatic Platonism: the exploration of ancient wisdom, which contained basic truths from which non-Platonic philosophers had deviated and of which myths and rituals still preserved traces.[27] The most complete reconstruction of this ancient wisdom, however, is to be found in Plato, at least according to the Platonists. An exploration of traditions like the Egyptians' was a way to prove Plato right *and* to broaden and deepen the understanding of Plato. Accordingly, in his interpretation of the Egyptian tradition Plutarch creates room for Plato, but also leaves room for Egypt. Plutarch's negotiations with space and time in developing his notion of dualism are thus multi-layered and significant: while Greekness and the revelation of Platonic truth are unquestionably superior to non-Greek influences and the pre-Platonic past, Plutarch's use of the latter endows the basic truth of dualism with a spatial and temporal universality, which in turn again raises the importance of Greece and Plato as offering the acme of this universal truth. As it turns out, dualism is at the same time 'a Greek word', as Plutarch says about Isis at the beginning of *On Isis and Osiris* (351F), *and* 'a very ancient opinion', in circulation 'among barbarians and Greeks alike' (369B)[28] and it is precisely this combination which makes it so valuable in Plutarch's Greek and Platonic philosophy.

27 Boys-Stones (2001) 99–122.

28 In this sentence as an exception the Loeb translation is used.

Michiel Meeusen

Egyptian knowledge at Plutarch's table: Out of the question?

Abstract: Among all 'barbaric' cultures, that of Egypt seems to have been of specific appeal to the Chaeronean. Plutarch's knowledge of Egyptian religion and mythology is most obviously attested in the *On Isis and Osiris*, but his interest in Egyptian matters radiates throughout his entire oeuvre, including the *Table Talk*. This study aims to investigate how the *Table Talk* reflects Plutarch's attitude towards Egyptian culture. Such an analysis seems relevant, because recent scholarship has especially underlined the panhellenic point of focalisation in this work, arguing that the sympotic setting of the discussions expresses a strong sense of hellenocentrism and Greek cosmopolitanism. This contribution takes an open stance towards Plutarch's hellenocentrism, arguing that the Chaeronean was primarily concerned with the philosophy, rather than the socio-cultural dynamic, behind non-Greek cultural manifestations.

Now do we know that there is no mid-centre of earth or of ocean;
Yet if there be, it is known to the gods, but is hidden from mortals.
De def. or. 409F, transl. F.C. Babbitt

Plutarch's vision of non-Greek culture

Plutarch is commonly known as a well-educated Greek of Boeotian descent, who lived and worked in the socio-political reality of the Roman Empire.[1] His intellectual interests were not, however, confined to an exploration of the dominant Graeco-Roman culture. In several writings, the Chaeronean demonstrates an intellectual openness to the habits and achievements of other—for that matter 'barbaric'—nations (understood as unities both in geographical and cultural terms). From an entry in the Lamprias Catalogue we know, for instance, that he had actually composed a collection of *Barbarian Questions*.[2] He there probably sought to explain specific non-Graeco-Roman phenomena in a zetetic-aetiological fashion after the manner of the *Greek* and *Roman Questions*. This is not to say that Plutarch's stance comes even near to the standards of the contemporary ideology of multiculturalism. As is well known, the Chaeronean was not at all shy to provide an *interpretatio Graeca*

1 See, e.g., Boulogne (1994).
2 Nr. 139: Αἰτίαι βαρβαρικαί. See Schmidt (2008). For a study of the 'barbaric' as a *locus* throughout the *corpus Plutarcheum*, see Schmidt (1999).

https://doi.org/10.1515/9783110539479-020

of a great many non-Greek phenomena, which is mostly interpreted as a hellenocentric attempt to reduce foreign cultures to Greek models.[3]

This contribution aims to address this very issue of Plutarch's hellenocentrism, taking an open stance towards its actual range and purpose. To this end, I argue that Plutarch was primarily concerned with the philosophy, rather than the sociology, behind cultural manifestations, be they Greek or non-Greek. This implies that his interest was not just motivated by scholarly interests but had a primary philosophical end. As such, non-Greekness (predicated on a broad category of foreign knowledge, which is in turn linked to specific geographical locations outside of the Greek territory) operates as an integral *topos* in Plutarch's wider philosophical programme. The fact that it both attracts and repels—just as the truth does—is very germane to this idea.

The concept of empire may be key here. A link can be drawn with Alexander's imperialism and cosmopolitanism as pictured in *On the Fortune of Alexander* 329B–D. Plutarch there reports on Alexander's criticism of Aristotle's categorial thinking about the human race.[4] Aristotle divided all mankind in two groups—Greeks and barbarians—and advised his pupil to treat the first as friends, after the manner of a leader (ἡγεμονικῶς), and the second as enemies, after the manner of a tyrant (δεσποτικῶς). Alexander opposes this classification by positing that all peoples who strive for virtue (ἀρετή) are basically Greek, whereas those who strive for vice (κακία) are barbaric, thus subordinating their different cultural habits and customs as incidental attributes to the category of morality—Greekness, however, still being the common denominator for the positive side of the ethical spectrum.[5]

This contribution will try to demonstrate that a similar re-categorisation lies at the basis of Plutarch's view on Egyptian culture, which among all 'barbaric' cultures seems to have been of specific appeal to the Chaeronean. Plutarch's knowledge of Egyptian religion and mythology is most obviously attested in the *On Isis and Osiris*, but his interest in Egyptian matters permeates his entire oeuvre, including the *Table Talk*.[6] The institution of the symposium was a central space for convivial interaction, where intellectual discussions of multifarious topics among learned gentlemen served as a means to promote the participants' sense of socio-cultural and intellectual-philosophical community. Accordingly, this study aims to investigate how the *Table Talk* in particular reflects Plu-

---

3 Scholars have argued that the regular allusion to Greek values in a work like the *Roman Questions* or in the biographies of Roman statesmen reveals a clear tendency, on Plutarch's side, to perceive Roman culture in a Greek way. For the *Lives*, see, e.g., Swain (1996) 137–145 (esp. 139); Frazier (1996) 279–281; Duff (1999), esp. 287–309; Schmidt (1999) 327. For the *Roman Questions*, see, e.g., Boulogne (1992) 4701–4703; Preston (2001). However, Boulogne (1992) 4703–4707 also points to the underlying philosophical motivation of Plutarch's study of Roman culture.

4 = Arist., fr. 658a Rose.

5 See Schmidt (1999) 283; Humm (2013) 452–453.

6 Some passages may contain actual remnants from Plutarch's *Quaestiones Aegyptiacae*. See Boulogne (2005) 197–198.

tarch's attitude towards the Egyptian culture. Such an analysis seems relevant, because recent scholarship has especially underlined the panhellenic point of focalisation in this work, arguing that the sympotic setting of the discussions expresses a strong sense of hellenocentrism and Greek cosmopolitanism.[7] Greek culture was, of course, very influential within the confines of the Roman Empire, but in this study the perspective will shift to its cultural periphery.

Analysis of the Egyptian passages

In Plutarch's symposia people speak Greek, the *lingua franca* of the elite living in and around the Mediterranean region, but not everyone can, therefore, call himself a Greek by birth. Ammonius, notably, Plutarch's teacher in Athens, was of Egyptian origin, and Plutarch also had several other Egyptian friends.[8] One may presume that the friend-making (φιλοποιόν) atmosphere at the symposium would certainly require a certain intellectual openness and friendliness towards their domestic culture as well. In any case, a true πεπαιδευμένος would automatically be inclined to demonstrate or acquire some outlandish knowledge to or from his fellow symposiasts, be they Greek or not. As Jean Hani has noted, moreover, it seems unlikely that Ammonius did not draw his students' attention to Egyptian things.[9] Plutarch had actually made a study journey to Egypt himself (following in the footsteps of numerous Greek wise men like Plato and Pythagoras, to mention only his philosophical favourites),[10] and in the city of Alexandria he probably encountered and questioned several men of letters, philosophers and hellenised Egyptian priests.[11] Besides these real-life Alexandrian informants, Plutarch presumably also had a number of books at his disposal, especially Herodotus, from which he could draw information on specific Egyptian topics.

On the basis of this variety of Plutarch's Egyptian sources of information the analysis below will distinguish between all-round intellectual and broadly philosophical-religious topics. Reference to Egyptian matters is often made in passing, as is the case, for instance, with the allusion to Alexandrian scholarship in the very first problem of the collection (*QC* 1.1, 614A–C). The more general intellectual topics deal especially with natural phenomena that are related to the exotic Egyptian locale. But the majority of the Egyptian passages concerns more philosophical-religious matters, which are related to Egyptian mythology and the practices and beliefs

7 This ties in generally with the dynamic revival of Greek self-awareness in the Second Sophistic. See, e.g., Klotz and Oikonomopoulou (2011) 3; Vamvouri Ruffy (2012) 218–220.

8 E.g., Theon ὁ γραμματικός (*De facie* 939C–D; *QC* 1.9, 626E; 8.8, 728F), probably to be distinguished from Theon ὁ ἑταῖρος. See Clement and Hoffleit (1969) 48–49, n. b; Puech (1992) 4886.

9 Hani (1976) 10. For a philosophical profile of Ammonius, see Opsomer (2009).

10 *De Is. et Os.* 354D–E. Cf. Ziegler (1951) 654.

11 *QC* 5.5, 678C. See Hani (1976) 10–11.

of Egyptian priests, namely, their worship of certain animals and their abstention from certain kinds of food—an appropriate topic for discussion at the table, indeed.

Intellectual topics

Apart from a number of more anecdotic *faits divers* connected with Egypt as a region/province,[12] there is also particular interest in the natural phenomena related to its locale. These phenomena often have a specific paradoxographical implication, stressing the exotic and unusual character of the place where they occur. A botanical difference is, for instance, attested in the growth of palm trees, which bear sweet dates in Egypt and Syria, whereas in Greece their fruit is inedible (*QC* 8.4, 723C–D). It is also reported that the earth generates mice in Egypt (*QC* 2.3, 637B). But of all the natural wonders occurring in Egypt it is certainly the Nile that has attracted most attention among ancient Greek authors. Herodotus' aetiological account of the summer flooding of the Nile and the location of its sources is well-known (Hdt. 2.19–34). A similar strand of Nile science is found in *QC* 8.5, 725A–E, where the symposiasts discuss why sailors draw water from the Nile before daybreak. The problem is of a natural philosophical kind and looks for plausible physical explanations.[13]

The discussion of such natural problems was a much appreciated intellectual pastime at convivial events.[14] Other such topics that were considered suitable for sympotic deliberation are discussed by Plutarch in the very first problem of the collection, namely, *QC* 1.1, 614A–C, where allusion is made to the activities of the Alexandrian scholars in the Museum. Plutarch says that these topics for discussion are supplied by history, current events, philosophy, piety and deeds of men. When these topics are interwoven in the sympotic discussion, so he says, they will take away the troubles caused by drunkenness. Similarly, those who mix alkanet in the wine and sprinkle the floors with infusions of vervain and maidenhair (in their belief that it provides some cheerfulness and gaiety to the guests)—imitate Homer's Helen, who secretly added a drug to the undiluted wine. They do not see, however, so Plutarch adds, that that story too, having travelled a long way from Egypt (ἀπ' Αἰγύπτου), should not be taken literally, but finds its end in the uttering of appropriate and suitable words (λόγους ἐπιεικεῖς καὶ πρέποντας). The reference is to Homer, *Od.* 4.220–264, where Helen, before she starts telling a tale (μῦθος) about Odysseus to the drinking Achaeans, mixes a drug (φάρμακον) in the wine, which she received, so Homer writes, from an Egyptian woman called Polydamna (the Egyptians were,

12 See, e.g., the account of the heavydrinking, heavyweight Heraclides, nicknamed Heraclous by the Alexandrians (*QC* 1.6, 624B), or of the date of Pompey the Great's death in Egypt (*QC* 8.1, 717C–D).
13 These are found in the heat of the sun corrupting the water and evaporating the lightest parts from it, and in the mixing of the water with earth by the human and animal activity in it.
14 See, e.g., Teodorsson (1996) 211; Oikonomopoulou (2011); Meeusen (2017).

indeed, noted for their profound knowledge of medicine and pharmacopoeia).[15] Plutarch's allegorical reading of this Homeric passage, interpreting the φάρμακον as the bewitching eloquence of Helen's story-telling itself, originates with the Alexandrian grammarians (cf. ἀπ' Αἰγύπτου), but he gives an interesting twist to it.[16] He explains that a timely conversation (λόγος, rather than μῦθος) suiting the moods and the situation at the dinner table serves as a sedative drug. Therefore, men of taste, even if they talk straightforward philosophy, do their talking with persuasive arguments rather than with cogent proofs (τοῦ πιθανοῦ μᾶλλον ἢ βιαστικοῦ). As Maria Vamvouri Ruffy observes, the transition from Helen's μῦθος in Homer to philosophical λόγος in Plutarch suits the situation in the *Table Talk* very well. After all, Plutarch's symposium is concerned with philosophical λόγοι rather than with the unreliable μῦθοι of the poets.[17] Philosophy is the real φάρμακον here.

Nevertheless, several passages throughout the *Table Talk* demonstrate a specific interest in mythological matters, including Egyptian myths. Arguably, these mythological accounts do not just serve as redundant scholarly diversions, but are mostly incorporated to serve a specific argumentative purpose in the discussions.

Philosophical-religious topics

According to Philip Hardie, '[t]he interpretation of myth is often handled [by Plutarch] as an exercise in solving problems.'[18] This is, indeed, generally true, seeing that it often nicely contributes to the development of Plutarch's λόγοι. However, if we call to mind the Chaeronean's fundamental devotion to the πάτριος πίστις[19] in combination with the fact that it is his intention to practice philosophy in guarded terms at the symposium,[20] it is not unlikely that these mythological accounts implicitly hint at a higher level of explanation, thus in a way 'mystifying' the zetetic discourse at issue (which in itself settles for plausible and persuasive arguments as we saw). Hardie notes that Plutarch in many passages, indeed, 'describes myth as a faint reflection of a transcendental truth. ... Myths act as ladders to the truth, which may then be kicked away.'[21] This means that the myth is not an explicit record of the truth as such, but contains a deeper meaning that

15 For further literature on the nature of Helen's drug, see Teodorsson (1989) 52.

16 The same allegory is found in Callimachus (fr. 178.15–20 Harder), who, as Teodorsson (1989) 52 remarks, 'was probably influenced by contemporary allegorizing commentaries on Homer'.

17 Vamvouri Ruffy (2012) 92.

18 Hardie (1992) 4751.

19 Cf., e.g., *De Pyth. or.* 402E, *Amatorius* 756B. See Flacelière et al. (1987) cli–clii.

20 That is, *without* seeming to do so, cf. *QC* 1.1, 614A.

21 Hardie (1992) 4754 (see also 4746–4749 for the relationship between myth and truth in Plutarch more generally). For Plutarch's ambivalent attitude towards myths, see Van der Stockt (1992) 88–97. For the theory that the incorporation of mythological material (in Plutarch's natural problems) implicitly hints at a higher level of aetiology, see Meeusen (2013).

can be brought to light by an adequate interpretation.[22] Let us see how this works out with the Egyptian myths recorded in the *Table Talk.*

With regard to the problem of the number of letters in the Greek alphabet, the geometer Hermeias in *QC* 9.3, 738E lifts the discussion onto a mythological level by referring to the divine origin of Egyptian writing and the use of hieroglyphs. He argues that Hermes (= Toth, cf. Pl. *Phdr.* 274d) was the πρῶτος εὑρετής who introduced the art of writing in Egypt, and that the Egyptians, therefore, write the first letter with an 'ibis', the bird that belongs to this deity. In what follows, Hermeias notes, however, that, in his opinion, the Egyptians err in giving precedence among the letters to one that is inarticulate and voiceless, but with this specification he does *not* as such aim to reject the divine origin of the hieroglyphs. On the contrary, he actually tries to save their attribution to Hermes in what follows on grounds of a numerological account. A link can be drawn here with the previous problem in *QC* 9.2, 738A, where Plutarch agrees with Protogenes in accepting that Cadmus, the mythological founder of Thebes and author of the Greek alphabet, placed the letter alpha first. The underlying idea thus seems to be that the Egyptian myth errs in the particulars of the matter, but that it is correct on a more fundamental level, by virtue of accepting the divine origin of letters and writing (as is also acknowledged in Greek mythology).

A second reference to Egyptian mythology is found in *QC* 8.1, 718B, in the context of the divine character of the principle of generation. There, with regard to the birth (day) of the 'divine' Plato, Tyndares the Lacedaemonian (a Platonist) alludes to Platonic cosmogonic theories in arguing that God, who is the father and creator of the cosmos and of other created things, begets the principle of generation in matter not simply through semen, but through another divine potency (cf. Pl. *Ti.* 28c). In order to illustrate his argument that a god does not consort in the same (sexual) way as mortals do, but impregnates mortal nature by some other kind of contact or touch (that is, by other interventions, rather than simply through semen), Tyndares refers to the Egyptian myth of Apis (the bull-deity and incarnation of Osiris), who, according to the Egyptians, was brought to birth by a touch of the moon.[23] In general, so Tyndares adds, the Egyptians allow that a male god has intercourse with a mortal woman, but conversely they do not believe that a mortal man can provide the principle of birth and conception to a female divinity, since they believe that the substance of the gods is located in air and breath and in certain heats and moistures.[24]

The introductory phrase, signalling that 'the story is not mine' (οὐκ ἐμὸς ὁ μῦθος), puts the plausibility of Tyndares' account in perspective. Indeed, in a parallel account in *Num.* 4, Plutarch takes a more critical position in the philosophical dis-

22 This is the case also, e.g., with the closing myth in the *De facie*. See, e.g., Donini (1986) 207; Meeusen (2013).

23 Reference to this myth is also made in *De Is. et Os.* 368C.

24 This is not further explained. Teodorsson (1996) 161 notes: 'The concise description that Plut. gives here of the highest Egyptian god, Amon, who was identified with Zeus by the Greeks, corresponds to his interpretation of the myth at *De Is. et Os.* 365D ...'.

cussion of divine companionship.[25] If we confront these two passages, it turns out that Plutarch does not simply take the Egyptian account for granted, but as Teodorsson notes he 'does not develop his doubts in a regular rejection [either]. Instead it seems that he maintains an agnostic position towards the problem'.[26] The Egyptian account certainly serves as a plausible addition to Tyndares' main argument, but its incorporation in his λόγος turns out to be somewhat *ad hoc* (in comparison to *Num.* 4.4). This does *not* imply, however, that there is nothing fundamentally valuable in the Egyptian account, since it can illustrate the divine character of the principle of generation in the material world, which is central to Tyndares' argument at issue. It turns out that the Egyptians are again erroneous in the particulars of the matter, but their myths still contain a certain aspect of reliability, so that their beliefs are not implausible (cf. *Num.* 4.4: οὐκ ἀπιθάνως). And it is precisely the persuasiveness of the arguments, rather than the cogency of proofs, that suits philosophical discussions at symposia (as we saw earlier). Note, moreover, that this emphasis on the aspect of persuasiveness is in line with Plutarch's loyalty to Plato and Academic Scepticism, and more precisely with the principle of εὐλάβεια, which demands an epistemic caution towards divine matters.[27] Therefore, Tyndares' argument is not just a noncommittal rhetorical ingenuity that cannot at least hint at the truth.

It can be added that in the social context of the symposium and the convivial protocols that it involves a person's caution in avoiding taking up a dogmatic position in matters that lack cognitive certainty also serves as a useful means to maintain a zetetic attitude in the philosophical discussion and as an incentive to keep an open mind on how things can stand otherwise.[28] A relevant passage to illustrate this is found in *QC* 5.10, 684F–685A, with regard to the problem of why salt is considered divine. Before starting to inquire into the divine nature of salt itself, Plutarch notes that the problem was complicated by the fact that Egyptian priests made it a point of religion to abstain completely from salt, so that they even eat their bread unsalted. Why then, so Plutarch wonders, if salt is favoured by the gods and divine, is it eschewed on religious grounds by the Egyptians? The question is not a rhetorical one, but requires a proper explanation. Florus, however, Plutarch's Roman patron who hosts the symposium, urges his companions to leave the Egyptians out of the question (ἐᾶν ἐκέλευε τοὺς Αἰγυπτίους) and to find a proper Greek (Ἑλληνιστί) explanation for the problem. This reaction can be taken to imply that the complication of the initial problem may not be appropriate in light of sympotic protocols (which demand that topics of discussion should not become too complex),[29] and that Florus, in his role of συμποσίαρχος, feels obliged to steer the discussion in another direction. Perhaps, Florus (as a Roman in Greek compa-

---

25 That is, with regard to Numa's alleged communion with the goddess Egeria. For reasons of space I cannot discuss this parallel in detail here.

26 Teodorsson (1996) 160.

27 See, e.g., Donini (1986) 205.

28 See, e.g., Van der Stockt (2000b) 94; König (2007) 57–58.

29 See Vamvouri Ruffy (2012) 67–75; Meeusen (2017).

ny) attempts to call on the Greek chauvinism of his guests, but he may have hoped for another reaction than Plutarch's, who tries to defend the case of the Egyptians by arguing that these are actually *not* in conflict (μάχεσθαι) with the Greeks. The use of combat imagery adds a certain jesting tone to the politically-laden discourse, and, in what follows, Plutarch will, indeed, try to make metaphorical peace between Greeks and Egyptians.

In short, Plutarch explains that Egyptian priests abstain from salt perhaps for reasons of purity, because it has aphrodisiac properties owing to its heat, or because it is delicious as a seasoning, some even calling it *charites*, since it makes needful food enjoyable. A little later, Plutarch notes that salt is also considered divine, because it has preservative powers, seeing that it conserves bodies uncorrupted, just like the divine fire of lightning does, when a person is struck by it. At the end of the discussion, Philinus, a friend and fellow citizen of Plutarch, elaborates on the aphrodisiac properties of salt, noting that salt has a generative property, for which it is considered divine.

What this passage shows, basically, is how Greek theories can be supported with Egyptian knowledge. An Egyptian belief may at first seem to be in conflict (μάχεσθαι) with a Greek one, but on closer inspection the reverse appears to be true, so that a stronger case can be made with the help of it. As Teodorsson notes, Plutarch in this passage 'appears as an expert in matters of religion, and with good reason. He was a priest of Apollo at Delphi, and knew Egyptian religion by study on the spot ...'.[30] Plutarch's openness to Egyptian religion and his attempt to ally it with Greek knowledge is revealing for its place in his philosophical programme. The underlying message seems to be that knowledge about the divine principle of generation is not restricted to the Greek tradition but has transcultural value, finding a parallel also in Egyptian culture.

The custom of Egyptian priests of abstaining from salt is part of a larger hypomnematic cluster with parallel passages in *De Is. et Os.* 352F (they eschew salt for various reasons: it has a superfluous nature and sharpens the appetite), 363E (they are forbidden to set salt on the table), and *QC* 8.8, 729A–C.[31] The latter passage deserves closer examination. The symposiasts there discuss why the Pythagoreans abstain especially from fish. The first who attempts an explanation is Theon (presumably the Egyptian γραμματικός: see n. 8). He explains that Pythagoras associated for a long time with the wise men of Egypt, and that he emulated them in many ways. Pythagoras especially approved the rituals of the priests, such as the abstention from beans (Theon refers to Hdt. 2.37.5). It is added that even in the present, Egyptian priests also abstain from fish, and that they consider it a religious duty to *abstain from salt*. Various reasons are given for this, so Theon says, but only

30 Teodorsson (1990) 227.

31 On the method of cluster analysis and its application to Plutarch's writings, see Van der Stockt (1999) and (2004). See also Meeusen (2012).

one is *true* (ἀληθὴς μία), that is, their hatred for the sea, as an element that is alien and strange to us, or rather because it is completely hostile to human nature.[32]

The Greeks were a sea-faring people, which may explain why Plutarch in his subsequent criticism of Theon's argument takes a defensive stance. Plutarch says that many people, both philosophers and laymen, would defend (μαχεῖσθαι) the sea against the Egyptians, summing up in how many ways it improved their way of life. The use of combat imagery again adds a certain jesting tone to the discourse by lifting it onto a political level: in the context of the discussion it denotes a playful rivalry in the formulation of plausible arguments. Presumably, Plutarch did not appreciate Theon's dogmatic position in the debate (ἀληθὴς μία)—though he has (to a debatable extent) created it himself, and made Theon a scapegoat. Plutarch goes on to explain at length that Theon's argument can, in fact, be reversed (διὰ τοὐναντίον), the Pythagoreans actually sparing sea-creatures out of regard for justice, since they do us no harm. Plutarch concludes that the abstinence from fish has been an element of religious duty not only among Egyptians (and Syrians), but among Greeks as well (meaning the Pythagoreans), and that they do this out of righteousness and to free themselves from the luxury of its consumption. This implies that there is a shared sense of justice and a transcultural morality among Greeks and Egyptians.[33] Teodorsson is right, moreover, that the ethical pathos of Plutarch's argument betrays 'a serious commitment to the Pythagorean outlook. One of his sources of inspiration was presumably Ammonius'.[34] The link between Pythagoreanism and Egyptian religion is, indeed, very substantial in this passage, considering the shared abstention from certain foods and reverence for certain animals.

The Egyptians were, in fact, noted for their animal cults, as is illustrated by two other passages in the *Table Talk*. In *QC* 7.4, 703A, Lucius, one of Florus' sons, notes that some Egyptians worship and honour the entire race of dogs, others that of wolves or crocodiles but feed only a single *one* of them (some a dog, some a wolf or a crocodile), because it is impossible to feed them all. The context of this account is again religious in kind: Lucius is explaining the Roman custom to take care for the holy fire in the temple of Vesta as a symbol of their reverence for *all* fire.

A more elaborate account of the religious practice of animal worship is found in *QC* 4.5, 669E–671C, where the symposiasts discuss why Jews abstain from pork. Polycrates wonders whether they do this either out of reverence for the animal or because they abhor it (cf. the *disputatio in utramque partem* concerning the abstention from fish above). Callistratus advocates the first option and argues that the pig enjoys a certain level of respect among the Jews. Even if this creature is misshapen and

32 Moreover, the father and saviour of their country, whom they call an emanation of Osiris (i.e. the Nile), perishes in it. Cf. *De Is. et Os.* 365B, 366A, 371B.

33 This idea may have specific Stoic overtones, as is also attested in Plutarch's rejection of luxury. Cf. also *De soll. an.* 964E–F with Porph. *Abst.* 4.14, p. 251.12 Nauck. See also the anti-Stoic feature in Theon's contrary argument, and Teodorsson (1996) 252.

34 Teodorsson (1996) 729. See also Opsomer (2009).

foul, it has no more absurd appearance or a more distasteful nature than a dung-beetle, field-mouse, crocodile or cat, which are each treated as most sacred animals by a different group of Egyptian priests. They say that the pig is honoured for reasons of usefulness (in agriculture). We need not be surprised, therefore, if some people do not eat pork, so Callistratus adds, since other animals receive even greater honours among the barbarians (παρὰ τοῖς βαρβάροις) for slight and sometimes completely ridiculous reasons. The Egyptians deify the field-mouse for its blindness, since they regard darkness—quite incompatible with Greek preconceptions—as superior to light. They also associate the lion with the sun and honour the ibis for similar dubious reasons. Callistratus wonders, however, how anyone could blame the Egyptians for such absurdity (ἀλογίας), when people say that the Pythagoreans and the Magi (the followers of Zoroaster) also honour such animals. So the Jews would kill pigs if they hated them, but in reality it is forbidden for them to slay or eat pigs.

The fact that the Pythagoreans are also mentioned among the βάρβαροι indicates that this reverence for animals is not that 'barbaric' after all, but, even so, it still remains an absurd practice in Callistratus' opinion. It follows that the general category of 'the barbaric' overlaps with that of 'the absurd' here, rather than that any barbaric people in specific is targeted by Callistratus. In his study on Plutarch and the barbarians, Schmidt has demonstrated that the concept of 'the barbaric' often serves as a rhetorical category in Plutarch's writings (especially aiming to produce a specific moralising effect), but that within this category no real differentiation is made between the barbarian ethnicities themselves.[35] It remains to be seen, therefore, to what extent the Egyptians can really be called barbaric according to Plutarch himself, and what he personally thought, more precisely, of their zoolatry.

With regard to the fact that Jews also abstain from the hare, Lamprias in his ensuing argument mentions that they consider the animal's swiftness and the sharpness of its senses divine, thus perhaps 'thinking like Egyptians do' (αἰγυπτιάζοντες) with regard to the animal's qualities.[36] The verb αἰγυπτιάζειν literally means 'to be like an Egyptian' and has the pejorative connotation of cunning, 'i. e. to be sly and crafty'.[37] In the present context, the word probably has a similar negative meaning, implying that the practice of relating animal qualities to divine categories is specifically considered to be an Egyptian procedure, by which Lamprias especially aims to underline its exotic, non-Greek character.[38]

---

35 Schmidt (1999) 332. He correctly interprets this passage in light of Plutarch's notion of barbaric φαυλότης, showing that barbaric animal cults provoke superstition amongst the common people (ibid. 232). See *De Is. et Os.* 379D–E, where atheism is mentioned as the other religious extreme.
36 As Teodorsson (1990) 100 notes: 'The focusing on Egyptian religion in this talk shows that Jewish and Egyptian religion were regarded as closely related'.
37 See LSJ, s.v.
38 He explains that the hare's eyes are untiring, since they remain open while asleep, and that the Egyptians also admire the hare's unrivalled sharpness of hearing, because an image of its ear serves as a character in their hieroglyphs to represent the notion of hearing.

According to Teodorsson, 'Plut. shows that he tried, not quite successfully, to understand the specific type of deification of animals in Egyptian religion. He thought that they represented certain abstract characteristics or qualities, cf. *De Is. et Os.* 355B ...'.[39] Indeed, in the *On Isis and Osiris* Plutarch on a more philosophical-religious basis shows some sympathy for Egyptian animal cults. In *De Is. et Os.* 382A–B, he reports that animals held in honour in Egypt are actually mirrors (ἐσόπτρων) of the divine. They are a natural instrument or medium for God who orders all things: we should not honour these animals in themselves, but *through* them the divine (θεῖον).[40] This mirror metaphor ties in closely with Plutarch's providential world view, where nature in general is considered an inferior, material reflection of the intelligible realm.[41] This certainly relativises Callistratus' negative evaluation of Egyptian animal cults. In the end, Plutarch is not expressing his own opinion but that of his fellow symposiast.

Conclusions

An important *caveat* that should be kept in mind while drawing our conclusions from the above analysis is that most of the Egyptian passages in the *Table Talk* substantiate very specific lines of thought, where the aspects of plausibility and rhetorical persuasion are considered seminal criteria in solving particular problems. Plutarch is not necessarily putting forward his own opinion in each case but that which suits the discussion best (cf. the apparent absurdity of Egyptian zoolatry). This explains the ambivalence at times towards Egyptian culture: in some cases the symposiasts advocate Egyptian habits or beliefs, whereas in other cases they make a stand against it (cf. the use of combat imagery). However, the incorporation of Egyptian knowledge in the *Table Talk* is not *just* for the sake of contriving ingenious and plausible arguments for complex problems; on the contrary, it often directly contributes to the sympotic speculation about the philosophical truth, which is the eventual goal of such events for Plutarch (his main interest in Egyptian culture goes to broadly philosophical-religious topics).

39 Teodorsson (1990) 114. On the allegorical-symbolic value of Egyptian religion more generally for Plutarch, see Bernard (1990) 208–211. Cf. also Barata Dias (2005) 175.

40 For further discussion, see Boulogne (2005) 203–204. See also Hirsch-Luipold (2002) 211–222. A parallel is found in *De soll. an.* 975B, where Aristotimus hyperbolically claims that he can produce thousands of signs and portents manifested by the gods (παρὰ τῶν θεῶν) through creatures of land and air. For Egyptian priests worshipping crocodiles, cf. also *De soll. an.* 976B–C.

41 See Hirsch-Luipold (2002) 285. I argue elsewhere (regarding Plutarch's writings on 'animal psychology') that '[i]f it is true that Plutarch 'loved' animals almost as much as he did humans, then he loved in them what was rational and virtuous, and therefore could be considered divine.' Meeusen (2013) 132. For further discussion see also Newmyer (2006) 17 ff.

That the kind of truth Plutarch is looking for will always be heavily influenced by Greek preconceptions is less important here than his own personal conviction that this truth will eventually transgress and prevail over issues of cultural identity. Plutarch did, indeed, experience the itch for auto-definition and construction of Greek identity that is so central to the cultural politics of his time (see n. 7), but he was not blinded by it. By consequence, we should not look at him as a navel-gazing Delphic priest, but rather as a genuine enthusiast for the truth, which shines through several human cultures. Egypt had, of course, become a very hellenised country ever since its occupation by Alexander, and this Hellenisation process was further developed by the subsequent rule of the Ptolemaic dynasty and the country's eventual annexation to the Roman empire, but its domestic, Pharaonic culture to a considerable degree retained its own exotic and mystical appeal even in Plutarch's days, providing the necessary food for thought at the Chaeronean's table (and desk).[42] Egypt served as some kind of a 'barbaric' paradigm or mirror culture for Plutarch (a proto-Hellenic society perhaps?), in which he could see the reflection, not only of his own Greek identity but also—and more importantly—that of a divine, transcultural entity.

42 On the persistence of Pharaonic traditions in Ptolemaic Egypt, cf. Stephens (2003).

7 Space, time and notions of community

Evangelos Alexiou

Divisions in Greek culture: *Cultural topoi* in Plutarch's biographical practice

Abstract: Greek culture plays an important role in the evaluation of passions in Plutarch's *Lives*, but at a number of places Plutarch puts more weight not on Greek character generally but on local character: in interaction or diversity (space) and between past and present (time). Such conceptions can be best explained as *cultural topoi*, which Plutarch uses like *rhetorical topoi*, established in Plutarch's era of the so-called Second Sophistic. Local *cultural topoi*, such as Spartan simplicity and honesty or Athenian philanthropy and changeability, are integrated into the themes and the lesson of the *Lives* and receive much attention in biographical practice as means of revealing the character of the heroes. Their introduction allows Plutarch to include an evaluation, which illustrates his biographical techniques, while simultaneously holding up an image of a collective identity, which is familiar to his contemporaries. The more the reader can identify the collective features, the more he will be inclined to understand the similarities or differences of the individuals' characters, such as those of Cimon, Lysander, Callicratidas, Alcibiades or Agis and Cleomenes.

In Plutarch's *Political Precepts* 799C the character of the Athenians, Carthaginians, Thebans and Spartans becomes canonically opposed to each other:[1] The Athenians are easily moved to anger, but they are least inclined to be angry with those who make fun of them. The Carthaginians are bitter and most savage when enraged. The Thebans would not have refrained from reading the letters of the enemies as did the Athenians, and the Spartans would not have endured the insolence of the Athenian orator Stratocles, who cheated the Athenians in making sacrifices on the grounds that they had won a victory.

Such stereotypical conceptions can be best explained as *cultural topoi*, repeated points which Plutarch uses like *rhetorical topoi*, established in Plutarch's era of the so-called Second Sophistic. I wish in this chapter to focus on Plutarch's attitudes to local characteristics and investigate how such *topoi* figure in Plutarch's biographical methods: as collective attitudes (which have an application beyond individual attitudes and are familiar to Plutarch's audience) and as personal attributes which are in harmony or in contradiction to these communal attitudes, and thus serve as moral indicators.

1 For the comparison between the cities cf. Pelling (2010) 220; Pelling (2014) 156. For the characterisation of the Athenian demos in the *Lives*, cf. Pelling (1992) 10–40, esp. 21 ff.

https://doi.org/10.1515/9783110539479-021

I

Cimon was not always in control of his passions. In his earlier years he had the reputation of being dissolute and bibulous (*Cim.* 4.4: ἄτακτος καὶ πολυπότης). The reader might expect here an explanation for Cimon's inability or reluctance to control these passions, based on his character. However, Plutarch's treatment integrates into the structure of the biography a fragment by Stesimbrotus:

> Στησίμβροτος δ' ὁ Θάσιος περὶ τὸν αὐτὸν ὁμοῦ τι χρόνον τῷ Κίμωνι γεγονώς, φησὶν αὐτὸν οὔτε μουσικὴν οὔτ' ἄλλο τι μάθημα τῶν ἐλευθερίων καὶ τοῖς Ἕλλησιν ἐπιχωριαζόντων ἐκδιδαχθῆναι, δεινότητός τε καὶ στωμυλίας Ἀττικῆς ὅλως ἀπηλλάχθαι, καὶ τῷ τρόπῳ πολὺ τὸ γενναῖον καὶ ἀληθὲς ἐνυπάρχειν, καὶ μᾶλλον εἶναι Πελοποννήσιον τὸ σχῆμα τῆς ψυχῆς τοῦ ἀνδρός, φαῦλον, ἄκομψον, τὰ μέγιστ' ἀγαθόν, κατὰ τὸν Εὐριπίδειον Ἡρακλέα· ταῦτα γὰρ ἔστι τοῖς ὑπὸ τοῦ Στησιμβρότου γεγραμμένοις ἐπειπεῖν.
>
> And Stesimbrotus the Thasian, who was of about Cimon's time, says that he acquired no literary education, nor any other liberal and distinctively Hellenic accomplishment; that he lacked entirely the Attic cleverness and fluency of speech; that in his outward bearing there was much nobility and truthfulness; that the fashion of the man's spirit was rather Peloponnesian, 'plain, unadorned, in a great crisis brave and true', as Euripides says of Heracles, a citation which we may add to what Stesimbrotus wrote (*Cim.* 4.5 = *FGrH* 107 F4, transl. B. Perrin, Loeb).

The opening sentences from the Stesimbrotus citation seem to suggest that the lack of Greek education was responsible for the deficits of Cimon. But it is curious that the quotation attributes Cimon's flaws of character to local culture and distinguishes Attic and Peloponnesian behaviour: Cimon was an Athenian, but he had a Peloponnesian spirit (Πελοποννήσιον τὸ σχῆμα τῆς ψυχῆς). More surprising is that Plutarch adds his own words to the quotation of Stesimbrotus, so we don't know exactly where the end of Stesimbrotus' words and the beginning of Plutarch's own words lie. Cimon entirely lacked Attic *deinotēs*, which is the technical term used for cleverness, especially in speech, can be explained with *stōmylia*, an ambiguous term: fluency of speech, but also wordiness.[2] So if the lack of Attic education entails a lack of Attic fluency of speech, this lack also testifies to Cimon's nobility and truthfulness. Plutarch justifies this attitude by suggesting that Cimon's spirit was rather Peloponnesian, as is shown by his reference to the Dorian hero Heracles. Jacoby[3] sees here carelessness and awkwardness on the part of Plutarch, but the text after *stōmylia* and the prolonged stress on the Peloponnesian spirit, which are very likely Plutarch's own invention, offer a brief summary of Cimon's qualities, culminating in a diffe-

---

2 See LSJ s.v. στωμυλία. The word is rare in Plutarch: positive in *De fort. Rom.* 318E, but negative in *De aud.* 42E combined with τερθρεία (see Isocr. *Hel.* 4; Suda s.v.) as 'nonsense'. The word is ambiguous also in Lucian; see Wälchli (2003) 42 n. 71.

3 In his commentary to *FGrH* 107 F4 346, but rightly Gomme (1945) 36 n. 2; cf. Carena et al. (1990) 211. Engels, in Bollansée et al. (1998) 64–65, suggests erroneously that Plutarch's additions to the text of Stesimbrotus begin with the quotation from Euripides.

rence between Athenian and Peloponnesian attitudes. This local *cultural topos* helps to bring out Cimon's ethical stance.

Not only does Plutarch make a subjective comment on Cimon, he also notes a basic link between the gentleness (πρᾳότης) and the artlessness (ἀφέλεια) of his nature (5.5). This is especially striking considering his recurrent emphasis on two basic elements of Greek culture in the *Lives*, *praotēs* and education,[4] but on this occasion Plutarch has nothing to say about education: the simple Peloponnesian nature of Cimon is entirely in accord with his political virtues (5.1). That is why in this *Life* Plutarch does not exploit the opportunity to build a contrast between Cimon as an aristocratic individual and the Athenian *polis*. His larger theme is that the pre-eminently virtuous disposition of his hero can be explained in ethical, not in political terms. Plutarch remarks that the generosity of Cimon surpassed even the hospitality and philanthropy of the Athenians of old (τὴν παλαιὰν τῶν Ἀθηναίων φιλοξενίαν καὶ φιλανθρωπίαν, 10.6). This supplement 'of old' indicates either that the Athenians in the time of Plutarch do not excel in generosity any more or, as is more likely, that this is an idealised *cultural topos* of the past, which continues to prevail among contemporaries of Plutarch.[5] This image of the Athenians, based on a famous statement of Pericles in Thucydides (2.40.4), became a *topos* in the orators of the 4th century BCE.[6] Plutarch similarly mentions it in the *Pelopidas* (6.5; cf. *Demetr.* 22.1). Plutarch's explanation of Cimon's generosity is not pragmatic and political, but ethical:[7]

> οἱ δὲ ταῦτα κολακείαν ὄχλου καὶ δημαγωγίαν εἶναι διαβάλλοντες ὑπὸ τῆς ἄλλης ἐξηλέγχοντο τοῦ ἀνδρὸς προαιρέσεως, ἀριστοκρατικῆς καὶ Λακωνικῆς οὔσης, ... λημμάτων δὲ δημοσίων τοὺς ἄλλους πλὴν Ἀριστείδου καὶ Ἐφιάλτου πάντας ἀναπιμπλαμένους ὁρῶν, αὑτὸν ἀδέκαστον καὶ ἄθικτον ἐν τῇ πολιτείᾳ δωροδοκίας καὶ πάντα προῖκα καὶ καθαρῶς πράττοντα καὶ λέγοντα διὰ τέλους παρέσχε.
>
> Those who slanderously said that this was flattery of the rabble and demagogic art in him, were refuted by the man's political policy, which was aristocratic and Laconian ... and though he saw all the rest except Aristides and Ephialtes filling their purses with gains from their public services, he remained unbought and unapproached by bribes, devoting all his powers to the state, without recompense and in all purity, through to the end (*Cim.* 10.8, transl. B. Perrin, Loeb).

The interweaving of the topics of generosity and Laconian spirit is coherent, and must be read in the context of the consistent disinterestedness of Cimon. And this Laconian spirit is at work in *Cim.* 14.4, too: according to Plutarch, when Cimon was accused of having been bribed, he said before his judges that he was no prox-

4 See *Cor.* 1.5; 15.4; *Brut.* 1.3; *Num.* 3.7; *Them.* 2.7.

5 Cf. *De soll. an.* 970A; *Arist.* 27.7; Becchi (2009) 268 f.

6 See Isocr. *Paneg.* 29; Demosth. 20.109; 21.12; 25.87; Dover (1974) 201–203; Ziolkowski (1981) 102–110.

7 For a political interpretation of Cimon's generosity see Arist. *Athen. Pol.* 27.3; Theop. *FGrH* 115 F89; Stein-Hölkeskamp (1999) 145–164, esp. 152–155.

enus of rich Ionians and Thessalians, but rather of Lacedaemonians, whose temperate simplicity he lovingly imitated (Λακεδαιμονίων ... εὐτέλειαν καὶ σωφροσύνην).[8] No argument is needed to interpret the *cultural topos* of these passages, which make an obvious point. Whatever actually happened, Plutarch was impressed by the honesty of Cimon and, by contrasting his Laconian features to those of other people, he amplifies this impression.

II

The *cultural topoi* of Cimon re-appear in the *Life of Lysander*, but Plutarch is now concerned to analyse not a coherent, unchanging hero, but, as Stadter rightly points out, a paradoxical paradigm.[9] Let us first consider Plutarch's evaluation of Lysander's statue at Delphi at the beginning of the *Life* (1.1–3). In classical literature the terms κομᾶν ('to let the hair grow long') and κομήτης ('long-haired') are recognisably Spartan features and have political connotations: they were favourite habits of aristocrats, and Athenian laconizers were satirized in Aristophanes' comedies.[10] When Plutarch says 'after the ancient custom' (ἔθει τῷ παλαιῷ), he underscores a difference in time and a recognition of a traditional *topos*, which his contemporary readers were expected to spot. Lycurgus is reported to have said that long hair makes the handsome more comely to look upon, and the ugly more terrible (1.3). In other words, the ambiguity surrounding Lysander's handsomeness or ugliness (καλός or αἰσχρός), is not based on the approval or disapproval of the ancient Spartan appearance, namely long hair and beard, but on the way these Spartan characteristics were in accordance or not with his individual features.

The same tension comes to the fore even more in the field of Spartan education (*Lys.* 2.2–5). Lysander conformed to the customs of his people, was of a manly spirit, and superior to every pleasure. Athenaeus (12.543b–c; cf. Nep. *Lys.* 4.2) offers us a tradition about Lysander (in accordance with the image he has from most historians), which put the stress on his luxurious lifestyle. Its negative criticism is reinforced through Lysander's depiction as a parallel to Pausanias. In contrast, Plutarch follows Theopompus' positive evaluation of a temperate Lysander. The divergence demands attention. Plutarch does not force the reader's attention to an alternative ideal. Lysander is not an un-Spartan Spartan, but a Spartan with strong contradictions. It is no coincidence that, for Plutarch, the two main passions of Lysander, his ambition and the spirit of emulation, τὸ φιλότιμον καὶ φιλόνικον, were implanted in him by his Laconian education, whereas Lysander was not by nature ambitious (*Lys.* 2.2).

---

8 For Lacedaemonians and *sōphrosynē* see Thuc. 1.69–70; 1.83–84; 1.120.4; North (1966) 102–104; Rademaker (2005) 208–216.

9 Stadter (1992) 41–55. Cf. Alexiou (2007) 135–169.

10 See Arist. *Rhet.* 1367a30–31. Cf. Aristoph. *Nu.* 14–15; *Equ.* 580; *Lysistr.* 561; *Vesp.* 463–466. See also Hdt. 5.71; Lys. 16.18; Donlan (1980) 160–162.

Both terms are often connected with Plato's τὸ θυμοειδές and are represented by the Laconian State.[11] Plutarch indeed maintains in *Agesilaus* (5.3) that the Spartan lawgiver seems to have introduced the spirit of ambition and rivalry into the state as an incentive to virtue. But ambition is a very common feature in the *Lives*, as politicians and military persons are normally ambitious.[12] Plutarch does not expect his readers to see here a specific Spartan *topos*. The fact that he speaks a lot about Spartan education reveals his keen interest in the power of language to offer a persuasive account of Lysander's motives and to make the audience agree with his analysis of ambition as a Spartan feature of his protagonist.[13]

The situation is essentially different, when Plutarch introduces a natural disposition unusual in a Spartan:

> θεραπευτικὸς δὲ τῶν δυνατῶν μᾶλλον ἢ κατὰ Σπαρτιάτην φύσει δοκεῖ γενέσθαι, καὶ βάρος ἐξουσίας διὰ χρείαν ἐνεγκεῖν εὔκολος· ὃ πολιτικῆς δεινότητος οὐ μικρὸν ἔνιοι ποιοῦνται μέρος.
>
> But he seems to have been naturally subservient to men of power and influence, beyond what was usual in a Spartan, and content to endure an arrogant authority for the sake of gaining his ends, a trait which some hold to be no small part of political ability. (*Lys.* 2.3, transl. B. Perrin, Loeb)

In the case of flattery, Plutarch offers information the author shares with his readers, namely, that the Spartans are not flatterers. And the use of the phrase, without any comment, underlines a Spartan quality Lysander does not share. In this respect, this statement is *ex negativo* comparable to the charges of flattery against Cimon. Plutarch rejected the charge of flattery of the demos, attributing to Cimon an aristocratic and Laconian policy (*Cim.* 10.8: ἀριστοκρατικῆς καὶ Λακωνικῆς). In Lysander, his Spartan ambition is combined with un-Spartan attitudes towards men of power and arrogant authorities. For some, these characteristics illustrate political ability (πολιτικὴ δεινότης),[14] but the term ἔνιοι provides an interpretation from a distance, as it were; the assessment is not made by the application of standards preferred and advocated by the biographer.

This observation is supported by the comparison between Lysander and the Spartan admiral Callicratidas, where *cultural topoi* serve as moral indicators. Callicratidas reflects not only a general Spartan ideal, but more specifically Plutarch's own moral beliefs, so he receives a more laudatory treatment by Plutarch than by Xenophon in the *Hellenica*.[15] Callicratidas was the justest and noblest of men and the manner of his leadership had a certain Dorian simplicity and sincerity (*Lys.* 5.5:

---

11 So Plat. *Resp.* 545a; 545b; cf. 548c; 549a; 550b; 553c; 581a-b; 583a; Arist. *Pol.* 1271a14–16; Frazier (1988) 117.

12 Roskam et al. (2012); Frazier (2014) 488–502.

13 The theme is further developed in the proem of *Agis/Cleomenes-Gracchi* (1–2.8). Cf. Nikolaidis (2012) 48–50; Alexiou (2013) 63–65.

14 Δεινότης is the quality Cimon lacked (*Cim.* 4.5). But Themistocles, Lysander and Alcibiades possess it: *Them.* 15.4; *Lys.* 2.3; 3.2; 11.7; *Alc.* 10.4; 14.10; 16.1; 23.4; 24.5. Cf. Alexiou (2010) 349–350.

15 See Moles (1994) 70–84, esp. 71.

ἁπλοῦν τι καὶ Δώριον ἐχούσης καὶ ἀληθινόν). The three words indicate the point: Plutarch represents Callicratidas as an archetypal Dorian character, whose methods and policy are in accordance with the ideal of Spartan simplicity and honesty, whatever the outcome. Lysander knows how to flatter Cyrus and gain his favour (4.3); he can gratify his friends and allies (5.4). Callicratidas is quite different; he dismissed flattery as a principle of choice and thought any defeat of the Greeks more becoming to them than flattering visits to the houses of barbarians (6.4). Plutarch exploits the ideological connotations the words Sparta, Dorian, Lacedaemon carry for his Greek and Roman audience. When Plutarch says 'Callicratidas' purposes were worthy of Lacedaemon' (7.1: ἄξια τῆς Λακεδαίμονος), the Spartan admiral is used as a model for a local ideal that in this case embodies long established Greek qualities. Plutarch mentions that Callicratidas was able to compete with the most eminent of the Greeks in terms of his δικαιοσύνη, μεγαλοψυχία and ἀνδρεία.

III

In the *Political Precepts* 799C–D Plutarch contrasts the changeable and kindly character of the Athenians with the stubbornness of the Carthaginians. He recounts two stories of the Athenians Cleon and Alcibiades, which are intended to support this contrast. In the *Alcibiades* the protagonist is extraordinary in a number of ways. An essential point to grasp is that Alcibiades does not have the Dorian simplicity of Cimon or Pelopidas or even Coriolanus. In comparison with the simple, straightforward Coriolanus (*Cor.* 15.5: ἁπλοῦς), Alcibiades was, like Lysander,[16] unscrupulous in his public acts and treacherous (*Comp. Alc. et Cor.* 2.1). It is never entirely clear which individual dominating force is most prominent in Alcibiades. In the opening chapters of his *Life,* Plutarch emphasises the many inconsistencies and marked changes of Alcibiades' character (*Alc.* 2.1). Alcibiades' relationship with the people is very impressive in his *Life*. Strictly speaking, we never see a common local *cultural topos* shared between Alcibiades and the Athenians, except changeability. That the *demos* could be assimilated to Alcibiades is in itself significant. As Pelling has argued persuasively, '... it will later be the combination of the two, changeable people and changeable Alcibiades, that will produce such a catastrophic mix'.[17]

If we try, however, to extend the parallel between the Athenians and Alcibiades with respect to changeability, we will find that Alcibiades could never be a paradigmatic Athenian, as his changeability goes beyond that of his polis and works on different levels. The Persian Tissaphernes admires Alcibiades' versatility and surpassing cleverness (*Alc.* 24.5–6), because he too was not straightforward, but malicious and fond of evil

16 See Alexiou (2010) 336 f., 341 f.
17 Pelling (1992) 24.

company. The reciprocal flatteries fit into 'the transformation of Alcibiades *polytropos* into Alcibiades *kolax*'[18] and, in this case, it seems clear that, whatever the similarities between Alcibiades and the Athenian demos, his chameleon mentality makes Alcibiades capable of imitating and practicing everything, good or bad:

> Ἀλκιβιάδῃ δὲ διὰ χρηστῶν ἰόντι καὶ πονηρῶν ὁμοίως οὐδὲν ἦν ἀμίμητον οὐδ' ἀνεπιτήδευτον, ἀλλ' ἐν Σπάρτῃ γυμναστικός, εὐτελής, σκυθρωπός, ἐν Ἰωνίᾳ χλιδανός, ἐπιτερπής, ῥᾴθυμος, ἐν Θρᾴκῃ μεθυστικός, ἐν Θετταλίᾳ [cod. C; ἐν Θετταλοῖς: Loeb] ἱππαστικός, Τισσαφέρνῃ δὲ τῷ σατράπῃ συνὼν ὑπερέβαλεν ὄγκῳ καὶ πολυτελείᾳ τὴν Περσικὴν μεγαλοπρέπειαν, οὐχ αὑτὸν ἐξιστὰς οὕτω ῥᾳδίως εἰς ἕτερον ἐξ ἑτέρου τρόπον, οὐδὲ πᾶσαν δεχόμενος τῷ ἤθει μεταβολήν, ἀλλ' ὅτι τῇ φύσει χρώμενος ἔμελλε λυπεῖν τοὺς ἐντυγχάνοντας, εἰς πᾶν ἀεὶ τὸ πρόσφορον ἐκείνοις σχῆμα καὶ πλάσμα κατεδύετο καὶ κατέφευγεν.
>
> But Alcibiades could associate with good and bad alike, and found naught that he could not imitate and practice. In Sparta, he was all for bodily training, simplicity of life, and severity of countenance; in Ionia, for luxurious ease and pleasure; in Thrace, for drinking deep; in Thessaly, for riding hard; and when he was thrown with Tissaphernes the satrap, he outdid even Persian magnificence in his pomp and lavishness. It was not that he could so easily pass entirely from one manner of man to another, nor that he actually underwent in every case a change in his real character; but when he saw that his natural manners were likely to be annoying to his associates, he was quick to assume any counterfeit exterior which might in each case be suitable for them. (*Alc.* 23.5, transl. B. Perrin, Loeb)

Plutarch here presents several *cultural topoi*, which are indeed more applicable than the problematic similarities between Alcibiades and the Athenians. Thus, he presents not an Alcibiades changeable by nature but an Alcibiades who chose to be changeable, a flatterer,[19] who understood how to manipulate those who met him. By examining together some features of Sparta, Thrace, Thessaly, Ionia, Persia, which are not limited to these *poleis* or nations, nor focused on the relationship between Greek and non-Greek cultures, Plutarch suggests that the chameleon mentality of Alcibiades runs in parallel with his cosmopolitanism. These characteristics form a kind of collective identity shared by a smaller or wider group and they unify the group and the individual who is part of it.

On the other hand, the use of local *cultural topoi* maps out a distinction between past and present. A good example is Agis and Cleomenes and Sparta's decline in their time, contrasted with the idealisation of the fifth-century Spartan past. The two *Lives* reflect a decidedly patriotic colouring of the stories which contributed to the collective memory of Spartans and their contemporaries in the third century BCE, and of Plutarch's readers in the second century CE. These memories of the past, shared by the community, serve to unify the group and constitute a kind of collective identity, based on persons, symbols, and history.[20] Plutarch attributes the dif-

18 Gribble (1999) 271.

19 Cf. *De ad. et am.* 52B. Because of his chameleon mentality Alcibiades is represented as the greatest flatterer and demagogue (52E). Cf. Russell (1966a) 147 n. 1.

20 See Assmann (2008) 109–118. Cf. Assmann (1992); Flashar (1996) 63–85.

ferences between past and present to the decline of Sparta when Agis and Cleomenes were kings: wealth, luxury, effeminacy (*Ag.* 3.1; *Cleom.* 10.4). Throughout the *Lives*, the two heroes function as models and exceptions, identified with the ancient dignity of Sparta, and various phrases and persons illustrate the mastery with which Plutarch adapted the idealisation of the past to his present needs.[21] Most remarkable is the comparison between Aratus and Cleomenes (*Cleom.* 16.6–8), in terms of the important role that local cultural characteristics can play at the panhellenic level. Aratus is presented as a true Greek and a great one (Ἑλληνικὸς γέγονε καὶ μέγας), but he was not free of blame, as, out of personal motives, he proceeded to an act which Plutarch regards as unworthy of a Greek (*Cleom.* 16.3): 'He cast himself and all Achaea down before a diadem, a purple robe, Macedonians and oriental behests' (*Cleom.* 16.5, transl. B. Perrin), because he did not want to obey Cleomenes, a descendant of Heracles, who was seeking to bring back again that restrained and Doric way of life (σώφρονα καὶ Δώριον) which Lycurgus had instituted. This statement serves to link the local Spartan culture with the wider Greek tradition, by preserving the sharp distinction between Greek and oriental habits.[22]

In this respect, Plutarch's arsenal of various *cultural topoi* complements his construction of a *polis*-identity and cultural memory. Shared features of past and present bring into relief his description of an individual's character and distinguish between conditions of individuation and forms of collectivity. They are not an expression of an official ideology, but they must contain a good deal of accepted 'truth' in order to perform their function, that is, to create specific impressions of the protagonists among Plutarch's contemporary readers.

---

21 I mention some of them: departure from the traditions (*Ag.* 3.9: τῇ διαφθορᾷ τοῦ πολιτεύματος; *Cleom.* 18.4: ἴχνος ἐκείνης τῆς ἀγωγῆς), ancient dignity of Sparta (*Ag.* 6.2: τὸ παλαιὸν ἀξίωμα τῆς Σπάρτης; 9.1; *Cleom.* 30.1), worthy of Sparta (*Ag.* 3.1: ταπεινὰ πράττουσα παρ' ἀξίαν; 10.1; 21.2; *Cleom.* 36.6: ἀξίως τῆς Σπάρτης), truly Spartan (*Ag.* 13.1: κάλλιστον διανόημα καὶ Λακωνικώτατον; *Cleom.* 13.3; 16.6: σώφρονα καὶ Δώριον), Laconian simplicity (*Cleom.* 32.3: τὸ Λακωνικὸν καὶ ἀφελές), Comparison with Agesilaus, Lysander, Leonidas of old (*Ag.* 14.3), Lycurgus (*Ag.* 19.7; *Cleom.* 16.6).
22 See Nikolaidis (1986) 229–244.

Maria Vamvouri Ruffy

The construction of a cosmopolitan space in Plutarch's *On Exile*

Abstract: This contribution examines the construction and promotion of a cosmopolitan space in Plutarch's *On Exile*, the only Plutarchan treatise where the term *kosmios* means 'citizen of the world'. The speaker rejects a restricted conception of space by challenging historical, legendary, and ideological constructions through which the Athenians defined themselves in the space of their city-state. He gives the Athenian philosopher Socrates, known for his attachment to his city, a cosmopolitan profile and he presents Theseus as a universal hero. At the same time, he challenges preconceived ideas linked to the notions of homeland and exile by questioning the Athenian foundation myth of autochthony. The speaker proposes to his addressee a vertical look towards the sky, our true homeland. The contemplation of the celestial landscape allows men to realise the immensity of this landscape over the earth. Such a nature-based cosmopolitanism renders the notions of exile and borders invalid, since the true homeland of man and of his soul is the sky.

Plutarch's treatise *On Exile* yields rich insights into Plutarch's perceptions and representations of space, as space forms its central theme. The work is a consolation speech written in the form of a letter which addresses a young man exiled from Sardis, most likely the young Menemachus of Sardis, whom Plutarch also addresses with political advice in the *Political Precepts*. The speaker invites the man in exile to reconsider the idea of homeland and encourages him to feel at home everywhere.

After a broad consideration of how an effective consolation might be achieved, the speaker asserts in the prologue (599A–600E) that exile is not a genuine hardship but the imaginary outcome of an opinion, δόξα, without any real foundation. Like an official currency, νόμισμα δόκιμον, it can be useful to one person and useless to another (599F). In other words, exile is not an 'essence' but rather the outcome of our vision of it. The speaker then highlights the idea that our homeland is the world (600E–602D). He does not reject the notion of homeland but stresses the fact that when we are forced to leave it, we can feel at home at any other location as well. With the help of several examples (602D–604D), he shows to what extent exile is not a cause of poverty or disgracebut, rather, an occasion to accomplish remarkable deeds and lead a quiet life. He then presents various islands as places which are peaceful and far from political troubles, and which liberate from restlessness while offering various pastimes along with commodities that are necessary to life. The speaker goes on (604D–607A) naming poets, historians, and heroes that distinguished themselves in their lives of exile and then disproves unjustified accusations against exile. At the end of the treatise (607A–607F) he says that we are all ex-

https://doi.org/10.1515/9783110539479-022

iled from the sky and that our virtue along with our wisdom cannot be changed by any location. The wise man who feeds on philosophy can live and be happy anywhere. To the wise man, the world is open, without any borders or limits.

The treatise *On Exile* is both recognizably a consolation speech and a rhetorical speech that seeks to convince the imagined interlocutor by using arguments based on examples and quotations.[1] It is similar in subject and content to a *diatribe*, the mainstream philosophical discourse that aims to teach or instruct an audience.[2] Besides, many of the themes found in Plutarch's treatise are also developed in letters by different Stoics and Cynics dealing with the same topic. Exile seems to have been a favourite theme in rhetorical schools.[3]

Even though the principal ideas in *On Exile* are similar to those developed in treatises and orations by writers such as Teles, Musonius Rufus, Dio Chrysostom and Favorinus,[4] the speaker in *On Exile* makes use of some unique themes. Heinz-Günther Nesselrath remarks that, in Plutarch, exile loses all importance, because the speaker offers a prospect of eternity, recommending a mystical ascent.[5] What I would like to show in this article is that the originality of Plutarch's treatise also lies in the way the speaker constructs and promotes a cosmopolitan space. This is achieved by keeping a critical distance from the legendary and ideological past of city-states, notably Athens. As we will see, he gives the Athenian philosopher Socrates, known for his attachment to the city of Athens, a cosmopolitan profile.[6] At the same time, he challenges preconceived ideas linked to the notion of homeland and exile by questioning Athenian foundation myths.[7] His revisionism goes even further, since he suggests to the addressee a vertical look towards the sky, our true homeland, and offers thus a nature-based cosmopolitanism.

---

1 The treatise is most likely posterior to 100 CE. Cf. Ziegler (1949) col. 77; Hani (1980) 134–136; Caballero and Viansino (1995) 8.

2 On the literary genre of the *diatribe*, see Fuentes González (1998) 44–66.

3 According to Cicero, *Tusc.* 3.34.81, there were set outlines on different themes such as exile that were to be developed.

4 On the differences and similarities between Plutarch's treatise *On Exile* and those of Musonius Rufus, Teles and Favorinus, cf. Giesecke (1891) 94 ff., and Nesselrath (2007). These ancient authors question the idea that exile is a wrong and a dishonour. See also Dio Chrysostom's *Or.* 13 (*On Exile*).

5 Nesselrath (2007) 98–99.

6 Socrates' cosmopolitan profile is also found in Epictetus, 1.9.1. Cf. Gourinat (2001) 159–161.

7 Whitmarsh (2001b) argued that in the discourse on exile during the period that we call the Second Sophistic, the 'process of self-definition against the classical past extends from literary fashioning to political revisionism'.

From panhellenism to cosmopolitanism: from Heracles to Socrates

The speaker initiates his argument that a man's homeland is the world by citing words that were attributed to Heracles and to Socrates. Thus, he mentions the following:

> Thus Heracles spoke well when he said
> *an Argive I*
> *or Theban, for I boast no single city;*
> *There is no fort in Greece but is my country;*[8]
> whereas the saying of Socrates is still better (βέλτιον), that he was no Athenian or Greek, but a 'citizen of the world',[9] (as one might say 'Rhodian' or 'Corinthian') ... (*On Exile* 600F)

There is no clear indication within the treatise as to the circumstances in which these words were said. When we read these quotations, coordinated by the μέν and δέ, the lingering impression is that Heracles and Socrates are discussing with each other. However, Socrates lived in the 5th century BCE, whereas we assign Heracles to a legendary and undefined era that we now call mythological. Plutarch himself, in the *Life of Lycurgus*, situates Heracles in a legendary time as the poets do, *mythologousi.*[10] In *On Exile*, then, there is no distinction between historical actions and legendary events. What the speaker is interested in is rather to incorporate and adapt into his speech and reasoning the words of both major figures of Greek culture, so as to give credibility to the perspective on the world and space that he presents.

Heracles appears here explicitly in his panhellenic dimension, since he denies all ties to any particular city:[11] ἅπας μοι πύργος Ἑλλήνων πατρίς. In contrast to the claim of some city-states or royal dynasties who proclaimed themselves to be descendants of the hero or at least to have a special link to him,[12] Heracles specifies that every Greek city is his homeland.

The panhellenic consciousness that is discernible in the above passage is not new. This consciousness had already shaped itself in the 8th century BCE, with the propagation of the alphabet, the Homeric poems, and the foundation of great sanctuaries and panhellenic celebrations.[13] But the speaker seems to want to promote a

---

8 The translations in this article are those of de Lacy and Einarson, in the Loeb.

9 'Citizen of the world' is my translation: the Loeb gives 'Cosmian'.

10 Plu. *Lyc.* 30.1. On myth as a concept of modern anthropology and on the illusions of mythology, see Calame (1996) 9–55.

11 The words of Heracles are, most likely, drawn from a lost tragedy. Cf. *Adesp.* 392 Nauck *TGrFr.*

12 Cf. Lafond (2005).

13 Cf. Nagy (1994) 29–34; Strauss Clay (1989) 9–11; Vamvouri Ruffy (2004) 34–35. Mitchell (2007) 1–36 specifies that the panhellenic consciousness mostly developed from the opposition between Greeks and barbarians in the 5th century. Panhellenism then developed during the next centuries. On this subject, see Flower (2000); Osborne (2009) 231–232; Richter (2011) 90–134.

cosmopolitan outlook rather than a panhellenic one. The comparative βέλτιον, 'better', clearly expresses his preference for the cosmopolitan words of Socrates. The philosopher appears like a κόσμιος, a 'citizen of the world'. The adjective is found in this specific sense only in *On Exile*. This sentence, most likely a quotation from Epictetus,[14] is surprising, because the cosmopolitan portrait of Socrates that emerges here is completely different from the one found in the writings of the 4th century BCE. In Plato's *Apology* 37c–e, for instance, Socrates refuses to suggest exile as his punishment. In *Crito* 52b–c, he explains the reasons why he decided against exile. He specifies that, if the laws were to address him, they would remind him how deeply attached he was to the Athenian laws and to Athens. In a surprising way, in *On Exile* Socrates becomes the symbol of a cosmopolitan viewpoint that wins over confinement and exile, and even over panhellenism.

Promoting cosmopolitanism by challenging Athenian myths

In his attempt to reject the restricted view of space, the speaker challenges historical, legendary, and ideological constructions through which the Athenians defined themselves in the space of their city-state.[15] He takes a critical distance from them, he questions them and puts them on trial in order to favour a much larger principle of extended space, namely, that of cosmopolitanism, over the logic of integration in a specific city-state. He thus sheds new light on Athenian indigenousness, by focussing on the hero Theseus and his relation to the city-state itself, as well as on some monuments and Athenian celebrations.

Questioning autochthony

The speaker keeps a critical distance from the myth of autochthony. The Athenians believed that they were the children of Athena and the Athenian land since Erichthonius, their legendary ancestor, was born of Hephaestus' sperm that fell on the Athenian land when the god tried to rape Athena.[16] Erichthonius as well as the legendary kings Erechtheus and Cecrops were presented in literary texts and in iconography as autochthonous, *sprung from the earth*, and this was also the case for all the Atheni-

14 Epict. 1.9.1. Cf. Sen. *Ad Helviam* 11.7; Musonius Rufus p. 42.1–2 Hense; Cic. *Tusc.* 5.37.108.
15 As Favorinus does. See Whitmarsh (2001b) 301–302.
16 Already in Hom. *Il.* 2.547– 554, there is a reference to king Erechtheus who once (ποτε) was born out of the rich soil of Athens and was nourished by Athena. In Hdt. 8.44 and the Parian marble, *FGrH.* 239 A 10, it is said that it was Erechtheus or Erichthonius who gave Athens its name. On the similarities and the differences between Erichthonius and Erechtheus, see Brulé (1987) 13–22.

ans, their descendants.[17] The tale of autochthony was a unifying ideological construction, a tool elaborated in Athens that made the Athenians seem permanently established on their land. Autochthony gave them an argument to justify and legitimise equality among citizens, together with a certain feeling of superiority in relation to other people and non-citizens.

Yet, in *On Exile*, the speaker shows that the Athenians' autochthonous identity is not enough to keep them in Athens:

> On this account you will find that few men of the greatest good sense and wisdom have been buried in their own country, and that most of them, under compulsion from no one, weighed anchor of their own accord and found a new haven for their lives, removing some to Athens, some from Athens. For who has pronounced such an encomium on his native land as Euripides?
>
> *Where, first, the people are no immigrants*
> *But native to the soil (αὐτόχθονες); All other cities,*
> *Disrupted once, as in the game, have been*
> *Pieced out by importation from abroad.*
>
> Yet the writer of these lines went off to Macedonia and spent his remaining years at the court of Archelaus. (*On Exile* 604D–E)

The work from which all these verses are drawn is the Euripidean tragedy *Erechtheus* that has only reached us in a fragmentary state.[18] The play is set on the Acropolis, in front of the royal palace. Athens is attacked by the Thracian king Eumolpus, the son of Poseidon. Erechtheus, the king of Athens, asks the Delphic oracle whether or not he will defeat the enemy and the god predicts victory as long as he agrees to sacrifice his daughter. Erechtheus and his wife Praxithea agree to perform the sacrifice for the good of the city. After that, Athens wins the battle, and Eumolpus dies. Erechtheus, in turn, is stabbed by Poseidon, who is enraged because of his son's loss. Poseidon shakes Athens with an earthquake that threatens to knock down the royal palace. Athena then appears as *dea ex machina* and asks Poseidon to put an end to his

---

17 For Rosivach (1987) 297–301, *autochthonous* means '*always having the same land* from immemorial time'. Shapiro (1995) studies the iconographical tradition and shows that the central point in the narrative of autochthony is the birth of Erichthonius from the soil in the heart of the city. Cf. also Shapiro (1992). Montanari (1981) situates the belief of Athenian autochthony in Cleisthenes' time. For Rosivach (1987) 296–297, autochthony became prominent after 470 BCE, when the rivalry between the Athenians and the Dorian Spartans started. On Athenian tales of Erichthonius' birth, cf. Loraux (1979), (1981) and (1996). For Loraux, the myth of autochthony legitimises women's exclusion from Athenian politics. Many cities claimed an autochthonous origin but Athens especially took advantage of this origin, thus forging its superiority and a strong shared identity. Calame (2011) questions Loraux's (1981) structural interpretation and her assumption that the tale of autochthony legitimised women's exclusion from politics.

18 One of the main sources is Lyc. *Leoc.* 98–104, along with the Papyrus of Sorbonne inv. 2328. For the reconstructed plot of the play and bibliographical references, see Collard et al. (1995) and Jouan and Van Looy (2002). On the different versions of the war between Erechtheus and Eumolpus, cf. Parker (1987) 200–205.

vengence. She also asks Praxithea to introduce new annual rituals in honour of her dead daughters, who would from then on be celebrated as the Hyacinthides, and in honour of Erechtheus, who would receive a sanctuary in the city center where he would be invoked as Poseidon-Erechtheus.[19]

The passage of *Erechtheus* quoted in *On Exile* is the part where Praxithea speaks and advises Erechtheus to obey Apollo's oracle and sacrifice their eldest daughter.[20] Praxithea heroically makes her choice between her city and her family, between her private ties and the public interest of her fatherland. Praxithea refers here to the autochthonous origin of the Athenians, which is the true pillar of Athenian identity and pride.

Euripides' play initially tells of Eumolpus' defeat by the Athenians and thus, indirectly, of Poseidon's defeat by Athena and ends with the reconciliation of the two gods that had previously fought over Athens and who will be made the recipients of a ritual on the Acropolis. This reconciliation and the pacifism that comes out of it makes perfect sense when we are reminded that the play was most likely staged between 423 and 422 BCE, at the end of the first phase of the Peloponnesian war, when the Athenians were called to continue defending the territory of Attica.[21] During those years, the Erechtheion was under construction on the Acropolis and is it very likely that the dramatic representation supported the resumption of the construction. The play is thus closely linked to the religious and architectural projects of the Athenians.[22] Later, in the 4th century BCE, Lycurgus—who is one of our main sources for this tragedy—in his speech *Against Leocrates* 100, quotes verses from Praxithea's speech in order to arouse and awaken Athenian patriotism after the battle of Chaeronea.[23] So, both Euripides and Lycurgus use Praxithea's speech of civil engagement in order to stir up Athenian patriotism at a critical time in the history of the city, thus legitimating the Athenian claim to autochthony.

The myth of autochthony in *On Exile*, just like any legendary tale, is anchored in a specific political and cultural context.[24] The Athenian autochthony myth is mentioned in *On Exile* as an argument in favour of a cosmopolitan vision of the world. The myth is actualised with a particular, pragmatic goal in view. The speaker in *On Exile* challenges the validity of this legend by decontextualising the Euripidean passage, since he does not mention the name of the play's characters. By not mentioning the speaker in the quotation, and by simply saying that its author is Euripi-

19 On the cult of Poseidon-Erechtheus, see Lacore (1983). Darthou (2005) stresses the importance of the new integration of Erechtheus into Attic soil through his death and his burial.
20 Eur. *Erechtheus*, fr. 14, vv. 7–10 Jouan and Van Looy.
21 According to Carrara (1977) 13, *Erechtheus* must be placed between 425–422 BCE. Cf. also Cropp and Fick (1985) 78–80; Collard et al. (1995) 155. Jouan and Van Looy (2002) 98–99 summarise the problem of this tragedy's date.
22 On the temple of Athena Polias, known as the Erechtheion, cf. Hdt. 8.55 and Paus. 1.26.5; 1.27.1–2. On its construction, Philochorus *FGrH.* 328 F 67. Cf. Hurwit (1999) 200–209 and 316.
23 See Tsagalis (2008).
24 On the pragmatic aspects, the cultic value and the social relevance of the heroic tale actualised in the *Erechtheus*, cf. Calame (2011).

des, he attributes to the poet the words of a character within the play, and thus the theatrical fiction coincides with the historical persona of Euripides. This alteration allows the speaker to assert that the autochthony myth is not quite enough to keep the Athenians in Athens since even Euripides, who was so proud of his homeland, emigrated to Macedonia and spent the rest of his days there. As far as Athens is concerned, it is presented as a city where many foreign people live, while many Athenians have left to live elsewhere. Athens, in Plutarch's treatise, is presented as a cosmopolitan place and not just as a place of the privileged autochthonous Athenians.

Theseus and the Theseion: universal references?

In order to promote the idea of a unified space where there are neither autochthonous nor exiled people, the speaker next discusses Theseus, the hero who was believed to have created the city-state. The speaker emphasises that although he may have been poor, foreign or exiled, Theseus always remained an object of admiration for his virtue. The speaker also mentions an Athenian monument, the Theseion, which, like other important monuments, is the object of honour for everyone, despite the exile that was imposed on Theseus:

> But those who are not carried away by such considerations admire good men, even if they are poor or foreigners or exiles (φυγάδες). Nay, do we not observe that like the Parthenon and the Eleusinium, so the Theseum is saluted with reverence by all (ἅπαντας)? Yet Theseus was banished (ἔφυγε) from Athens, though it is because of him that Athens is now inhabited; and that city was lost to him which he did not possess, but himself created. (*On Exile* 607A)

Among all of Theseus' heroic deeds, the ones that are briefly mentioned here are his unification of Attic *demoi*, his exile, and the construction of a temple in his honour. These three facts are undoubtedly discussed here because they are useful to the argument. Indeed, the unification in a single city-state is a demonstration of the idea that the speaker defends in the treatise. In the *Life of Theseus*, it is specified that, in order to unite and create a democratic city-state, Theseus destroyed every village and every building of political power in various villages, so as to construct common buildings, accessible to everyone (κοινόν).[25] He also established the sacrifice of the *Panathenaea* along with that of the *Metoikia*.[26] The names of these two religious celebrations refer to moving (μετά – οἰκῶ) and to the reunion of all (Πᾶν – Ἀθήναια). Theseus is presented as the legendary figure that brought about the abolition of Attic inner borders and established celebrations glorifying the union of places that were previously separated.

---

25 Cf. Thuc. 2.15.2, from whom Plutarch's extract is most likely inspired. According to Thucydides, Theseus performed the synoecism by getting rid of the councils, τὰ βουλευτήρια, and magistrates, τὰς ἀρχὰς, of villages. On the synoecism, cf. Ampolo and Manfredini (1988) 235–237.

26 The details of this tradition developed in the 5th century BCE.

The mention of the hero's exile and of the Theseion also serves the purpose of Plutarch's argument, since, despite the departure of Theseus from the Attic land, everyone worships the monument that bears his name. Theseus' exile was the consequence of Helen's abduction by Theseus and Pirithous, when they kidnapped (ἁρπάσαντες) her from Sparta as a young girl.[27] During his absence from the city, the demagogue Menestheus jeopardised Theseus' reforms by starting a rebellion in Athens in order to take over power.[28] Moreover, the Tyndarids attacked the city so as to avenge the abduction of their sister. When Theseus came back to the city in the midst of a riot, he attempted to take the helm of political life but he quickly realised that his enemies no longer feared him and that people were corrupted. Theseus was then forced to flee Attica and it is most likely to this departure that the speaker refers in *On Exile*, when he mentions Theseus' exile.

Despite his departure, the hero was not forgotten. On the contrary, the speaker specifies that the Theseion, the monument which was attributed to him, was worshipped to his day. The speaker does not give more details about the Theseion. In Plutarch's *Life of Theseus* it is said that the bones of the hero were repatriated from Scyros by Cimon, so that they could be deposited in a funeral monument—probably the Theseion mentioned in *On Exile*—which was located in the centre of the city-state close to the gymnasium.[29] The bones were welcomed with excitement and were honoured by the Athenians.

Referring to the Theseion, right after mentioning Theseus' exile, the speaker of *On Exile* shows that exile does not make one forget the value or the deeds of an important hero. On the contrary, the admiration for what the hero has accomplished is still alive given the fact that all people without qualification (ἅπαντες) honour the monument bearing his name. In this way, Theseus not only appears as an Athenian hero but also as a universal hero. Through this universal admiration of the Theseion, the Athenian land and its monuments are configured as cosmopolitan places.

Philosophical contemplation and celestial space

In order further to promote cosmopolitanism and his picture of a unified space, the speaker focuses on the natural world, suggesting that exile is the outcome of our judgement, unlike natural elements, which are determined by their own laws:[30]

27 Plu. *Thes.* 29.2. On this abduction, cf. *Cypria* fr. 13 Bernabé; Stesichorus, fr. 191 Page; Alcman, fr. 21 Page; Pindar, frr. 243, 258 Snell-Maehler; Hdt. 9.73. Cf. Calame (1996) 262–264.

28 Plu. *Thes.* 30–33.

29 Plu. *Thes.* 36; *Cim.* 8. On the recovery of Theseus' bones, see Zaccarini (2015). On the celebration of *Theseia*, the Theseion and the excavations, cf. Calame (1996) 153–156 and 181 n. 32.

30 The natural world is a starting point of philosophical reflection; cf. the introduction of Meeusen and Van der Stockt (2015).

> It is by nature (φύσει), that stone is hard, it is by nature that ice is cold; it is not from outside themselves, fortuitously, that they convey the sensation of rigidity and freezing; but banishment, loss of fame, and loss of honours, like their opposites, crowns, public office, and front-seat privileges, whose measure of causing sorrow and joy is not their own nature, but our judgement (κρίσιν). (*On Exile* 599D)

According to this conception, the speaker asserts that our homeland is not something that has been bestowed to us by nature (600E: φύσει γὰρ οὐκ ἔστι πατρίς), but a construction, νομιζομένης πατρίδος, in the same way as the houses, workshops and doctors' offices that man makes. He underscores the fundamental difference between what is constructed on the one hand, that is, homeland and exile, and the natural world along with its laws, on the other.

That being said, if our homeland and exile are our own constructions, which places in the world are not? The speaker does not ask the question in these terms. It is not so much a particular place that is so important as our relationship to it. For him, all areas and spaces can become the homeland of a man whose desire is to take root somewhere:

> For while loss of wealth cannot easily and quickly be repaired, every city at once becomes a native city (πατρὶς δὲ γίνεται πᾶσα πόλις εὐθὺς ἀνθρώπῳ) to the man who has learned to make use of it and has roots which can live and thrive everywhere and take hold in any region … . (*On Exile* 601F)

In order to argue that there are numerous possibilities of residence in the vastness of the world, the speaker specifies that a man only has to look up to the sky and contemplate (ὁρᾷς) the endless ether that surrounds the earth. He says with emphasis:

> This is the boundary of our native land, and here no one is either exile (φυγάς), or foreigner (ξένος), or alien (ἀλλοδαπός). (*On Exile* 601A)

The contemplation of the skyscape allows men to realise the immensity of the landscape over the earth. Such a perception of the space overrides any notion of exile and borders, since the true homeland of a man and of his soul is the sky. Man, the speaker specified shortly before, referring to Plato, is *no earthly*, or *immovable plant*, but *points to heaven:* the head, like a root, keeps the body erect, but it is inverted in order to point to heaven (600F).[31]

From this point of view, man can be seen to act upon his own environment and not the other way around. As the speaker eloquently says:

31 It is interesting that here Plutarch slightly changes Pl. *Ti.* 90A, since man is presented as a plant that contemplates the sky but at the same time sticks to the earth. For Plato, the plant is suspended, with the higher parts up and the lower parts down. This difference could be explained by the fact that Plutarch does not completely condemn the fact of taking roots in a place. After all, he was so attached to Chaeronea that he never abandoned it, so that his city-state would not become empty. This distinguishes him from the Cynics.

> ... from a man no place can take away happiness, as none can take away virtue or wisdom; nay, Anaxagoras in prison was busied with squaring the circle, and Socrates, when he drank the hemlock, engaged in philosophy and invited his companions to do the same, and was by them deemed happy. (*On Exile* 607E–F)

According to this passage, the last paragraph of the treatise, only the wise man who is nurtured by philosophy, who does not let his passions or excessiveness get the better of him, can live anywhere and be a cosmopolitan person.[32] But as it has already been shown in *On Exile,* a critical outlook on tales and ideological constructions linked to the legendary past of city-states, especially Athens, can contribute to the construction of a cosmopolitan space as well.

It is clear that the speaker promotes a unified space, a cosmopolitan world without borders, a space that resembles that of the Roman Empire. During the Roman era and Trajan's rule specifically, space was unified and the *oikoumenē* was constructed not only thanks to military supremacy, but also through the political integration of provinces and through language, the Attic dialect that became the *koinē* of the elite of the *pepaideumenoi.* In *On Exile*, philosophy meets historical reality and redefines it in its own way.

It should be noted, as a conclusion, that the model of cosmopolitanism constructed by *On Exile* does not apply to all men in general but chiefly to those who have a philosophical way of living and an intellectual heritage. Let us not forget that the men who are mentioned as examples of exiled figures are all well-known from literary tradition. This is for example the case with Euripides, Simonides, Herodotus, Thucydides, and Xenophon, who were, for Plutarch's audience, more linked to literary tradition than to the common practices of everyday life. Should we then speak of a cosmopolitanism of an intellectual elite? The question is certainly worth asking.

---

32 On the importance of a philosophical way of life, see Vamvouri Ruffy (2012) 75–78.

Paola Volpe Cacciatore

Il significato del termine ξένος in Plutarco: lo straniero nella realtà dell'Impero cosmopolita

Abstract: In this chapter I analyse the various occurrences of the term *xenos* in Plutarch, and I examine the values that it assumes (guest/host, foreigner, enemy) in Plutarch's complex cosmopolitan context. The examination of the occurrences shows that in Plutarch, similarly to what happens in Plato, it is not possible to find a single notion of *xenos*. What is rather the case is that in the majority of the term's occurrences, we can distinguish between discursive statements, prescriptive statements, and reflective statements which illustrate its meaning.

Il lessico dello straniero: una breve premessa

> Aesch. *Ch.* 680 – 685:
> ἐπείπερ ἄλλως, ὦ ξέν', εἰς Ἄργος κίεις,
> πρὸς τοὺς τεκόντας πανδίκως μεμνημένος
> τεθνεῶτ' Ὀρέστην εἰπέ, μηδαμῶς λάθῃ
> εἴτ' οὖν κομίζειν δόξα νικήσει φίλων
> εἴτ' οὖν μέτοικον, εἰς τὸ πᾶν ἀεὶ ξένον
> θάπτειν, ἐφετμὰς τάσδε πόρθμευσον πάλιν ...
>
> *Visto che comunque vai ad Argo, straniero, ai suoi genitori -tienilo bene a mente- di' che Oreste è morto, e non dimenticarlo. Se vincerà il parere dei suoi cari di riportarlo o se quello di seppellirlo da meteco, straniero per l'eternità, riporta indietro questi loro ordini* ... (trad. it. di L. Battezzato)

Nei versi eschilei riportati, nei quali è proposta la sepoltura da dare a Oreste, sono adombrati le caratteristiche e lo *status* dell'uomo greco, che può essere ἰσόγονος, μέτοικος e ξένος. La differenza naturalmente non è di poco conto perché, a seconda del suo *status*, questi potrà essere considerato πολίτης—nel caso di Oreste figlio e discendente di re—nato dalla terra, alla quale egli appartiene per diritto di nascita, parentela, familiarità e fratellanza. In quanto tale, egli è εὐγενής e, dunque, a lui sarà data—se ciò è riconosciuto—sepoltura in quella stessa terra che lo ha generato.[1] Oppure sarà μέτοικος, il giunto *da altrove* (ἄλλοθεν) e, pertanto, non autoctono, ma pur sempre greco, stabilendo così una forte antitesi tra i Greci ateniesi e i Greci in quanto Ateniesi.[2] Ma Oreste—e così il non greco—sarà ξένος, straniero/ospite/esule, cioè colui che è legato da ξενία, la quale, posta sotto la protezione di Giove Xenio, 'comporta scambio di doni tra i contraenti che dichiarano la loro intenzione di legare

1 '... Ora che hanno ormai cessato di vivere giacciono nei luoghi familiari di colei che li ha generati, nutriti, accolti ...'. Pl. *Mnx.* 237c.
2 Cf. Pl. *Mnx.* 245c–d e Plu. *De aud.* 37F.

https://doi.org/10.1515/9783110539479-023

i loro discendenti con questo patto'.[3] Un patto di ospitalità che bene è espresso nel libro 6 dell'*Iliade*: è l'episodio di Glauco e Diomede che, trovandosi di fronte, scoprono che i loro padri sono stati legati da vincoli di ospitalità. Diomede non esita a chiamare Glauco 'ospite (ξένος) ereditario e da lungo tempo' (v. 215), così come non tarda a dirsi suo ospite. Questo basta a interrompere ogni conflitto. 'Evitiamo allora entrambi il giavellotto l'uno dell'altro ... scambiamoci piuttosto le armi, così che tutti sappiano qui che ci gloriamo di essere degli ospiti ereditari' (vv. 224–231). Ξένος è dunque in Omero l'ospite con cui si stringe un patto, ma—al pari di *hostis*—assume il significato di straniero,[4] di escluso.[5] Oreste stesso allora è ospite ma allo stesso tempo straniero e, quindi, escluso dalla sepoltura nella terra che gli aveva dato la vita. Omero e Eschilo (nelle *Coefore*), cantavano un mondo greco o 'federato' contro il nemico comune (i Troiani) o frazionato in una moltitudine di città—stato; quando la società antica va trasformandosi in *natio* ('kin group')[6] 'le relazioni tra uomo e uomo, tra clan e clan si aboliscono, sussiste solo la distinzione tra ciò che è interno e esterno alla *civitas*. Per un cambiamento, di cui non conosciamo le condizioni precise, la parola *hostis* ha assunto un'accezione *ostile* e non si applica che al *nemico*.[7] Allo stesso modo ξένος ... è divenuto semplicemente lo straniero, il non nazionale'.[8] Chi è posto sotto la protezione degli dei e di un personaggio politicamente importante diveniva ospite, e, talvolta, ospite illustre.

Il 'caso Plutarco': uomo greco, cittadino romano o straniero?

Ospite illustre e cittadino romano fu Plutarco di Cheronea, ovvero Μέστριος Πλούταρχος per la protezione di Lucio, un senatore che aveva combattuto nelle file di Otone e che poi verso il 72/75 era diventato console ad opera di Vespasiano e forse nell' 83/84 proconsole d'Asia sotto Domiziano. Era, dunque, cittadino romano ma era Romano? A tale domanda non è difficile rispondere che Plutarco si sentì soprattutto

3 Benveniste (1976) 71. Cf. Plu. *fr.* 46 Sandbach. Oreste era ξένος e δορύξενος delle case di Strofio, padre di Pilade. Strofio era a sua volta δορύξενος di Agamennone, aveva cioè stretto con lui vincoli d'arme in guerra. Cf. Aesch. *Ch.* 560.

4 Cf. Vegetti (1989) 21.

5 Benveniste (1976) 71.

6 Così Richter (2011) 6.

7 In Tacito, ad esempio, su 209 occorrenze, *hostis* è usato nel significato di nemico della patria 207 volte, due volte nel significato di nemico personale; *hospes* ricorre invece otto volte, cinque volte nel significato di ospite/ospitato e due nel significato di straniero equivalente a *peregrinus*.

8 Benveniste (1976) 71–72. Differente dallo ξένος era evidentemente il barbaro che è 'straniero sia sul piano etnico-culturale sia su quello politico ... <vivendo così> un'esperienza dello stato radicalmente antinomica rispetto a quella dei Greci, da schiavo e da suddito e non da libero cittadino': Bearzot (2009) 82. 'Straniero due volte' è quindi il barbaro secondo la definizione di Moggi (1992) 51–76.

un Greco senza mai, però, assumere atteggiamenti antiromani anche quando ironicamente commenta che un grande demone è stato accanto a Roma,

> inviando un vento favorevole, né per un solo giorno né, in pieno, per breve tempo ... né soltanto sulla terra ... né solamente sul mare ... né tardò a spirare ... né rapido nel cessare ... bensì fin dalle prime origini esso crebbe, si potenziò e partecipò alle vicende politiche insieme con la città e rimase costante per terra e per mare, in guerra e in pace, contro barbarie contro i Greci. (*De Fort. Rom.* 324B[9])

Contro quei barbari che Roma conquistò rendendoli suoi sudditi—il testo plutarcheo enumera le battaglie che avevano reso grande la città—Fortuna e Virtù insieme combatterono per portare a termine 'la più meravigliosa tra le opere umane' (*De Fort. Rom.* 316E), ma solo la Fortuna fu artefice della morte prematura di Alessandro Magno.

Plutarco e l'Impero: un rapporto complesso

La stessa fortuna guidò i Romani contro i Greci—questi ultimi, come è possibile notare, sono ben distinti dai barbari nel citato *De Fort.* 324B (cf. *supra*)—la 'schiavitù' del popolo greco durò però poco perché prima con Tito Quinzio Flaminino e poi con Nerone nella città di Corinto fu proclamata la libertà e l'indipendenza degli Elleni: 'Tito ... per mezzo di un araldo e Nerone con un discorso che tenne egli stesso su una tribuna nella piazza del mercato' (*Flam.* 12.13). Un omaggio dunque all'Imperatore al quale il sacerdote di Delfi riconosceva questo merito tanto che nel *De Sera Numinis Vindicta* egli immaginava che l'anima di Nerone, già condannata a vivere nel corpo di una vipera, avesse avuto destino migliore 'perché qualcosa di buono essa merita dagli dei avendo liberato la stirpe più insigne e più cara agli dei, l'Ellade' (*De Sera Num.* 567E–F[10]). Tale elogio a Nerone fu dettato forse anche dalla scelta, dolorosa per Plutarco, di Vespasiano che ristabilì in Grecia il governo provinciale dei proconsoli: la Grecia diventava così una provincia senatoria, l'Acaia. Ma non per questo egli si

---

9 L'espressione *barbari e Greci* non comporta un atteggiamento ostile nei confronti dei barbari ma di certo il senso preciso della distinzione. A differenza di Dione—cf. Desideri, (1978) 453 n. 24—, che talvolta usa il termine βάρβαρος con valore spregiativo (così e. g., in *Or.* 32.56, 36.43 e 21.4), in Plutarco l'immagine dei barbari 's'insère parfaitement dans ce courant de pensée, doublée néanmoins d'une préoccupation d'ordre moral': così Schmidt (1999) 332. Sull'argomento cf. inoltre Swain (1996) 86–87.

10 'Plutarco non dubitava che Nerone fosse uno sciagurato, sia per la sua tirannide personale, sia per il malgoverno da lui consentito ai suoi procuratori e liberti ... Ma Plutarco ha anche delle attenuanti: era stata l'altrui adulazione a spingere l'imperatore a calcare le scene <e così> rappresenta un Nerone ... non fondamentalmente malvagio, ma il cui debole lato positivo viene duramente represso ... <Sono queste le premesse graduali grazie alle quali Plutarco trasforma Nerone in una ranocchia che> è altrettanto appropriata per accogliere l'anima di un imperatore dal canto naturalmente esile e rauco ... Tutti i suoi crimini sono cancellati dal suo grande atto di generosità filellenica: la liberazione di Acaia dalle tasse': Champlin (2005) 33–34.

mostrò antiromano, un nemico, anzi fu proprio nel periodo flavio che maggiormente si avvicinò a Roma. Fu il suo un atteggiamento contraddittorio? Forse sì, perché mentre diventava cittadino romano esaltava la libertà della Grecia o piuttosto egli, greco, era diventato cittadino romano 'convinto come era della necessità di collaborare con l'Impero dei Flavi e in genere con le autorità romane senza insistere su opposizione di principio'.[11]

Il primato 'morale' dei Greci

Un cosmopolitismo, quello plutarcheo, dettato dalla necessità, ma non di meno egli esalta le doti dei Greci nel tentativo, da uomo greco, di inglobare i Romani nel loro sentire e nel loro ἦθος. 'Le but de Plutarque n'est donc pas une définition de l'identité grecque, mais une glorification des vertus qui en sont la base, de manière à pouvoir mettre en évidence l'hellénisme de ses héros et leur adhésion à un système de valeurs jugé supérieur'.[12] Proprio l'affermazione di tali virtù renderà i Greci diversi dai barbari anche alla luce della Seconda Sofistica che riaffermava con forza l'identità greca. Plutarco stesso si fa portavoce di tale superiorità, pur consapevole, però, che 'al luminoso passato ... si contrapponeva un presente immiserito dalla mancanza di autonomia: di qui la necessità per i Greci di adattarsi alla nuova situazione facendosi comunque valere nel campo della *politikē* su cui essi vantavano il diritto di primogenitura'.[13] Va detto, tuttavia, che tale 'diritto di primogenitura' non era riconosciuto da tutti; basti ricordare Tacito[14] che di certo ebbe un atteggiamento critico nei confronti dei Greci soprattutto dopo i viaggi di Germanico. In *Ann.* 2.53.3 egli, ricordando il viaggio di costui ad Atene, città antica e federata, così commenta:

> *Excepere Graeci quaesitissimis honoribus, vetera suorum facta dictaque praeferentes, quo plus dignationis adulatio haberet.*
>
> I Greci lo accolsero con onori straordinari, rammentando gli antichi detti e fatti dei suoi, per conferire maggiore autorità all'adulazione. (trad. it. di A. Arici)

---

11 Mazzarino (1990) 142. Contrario ad ogni *antitypos* aggressivo, Plutarco 'ha ottime relazioni con i signori Romani; ma li considera dominatori e stranieri. Persino quando ci sono delle disposizioni amichevoli verso i Romani queste sono verso stranieri non concittadini'.

12 Schmidt (1999) 327. Le virtù greche fondamentali per Plutarco sono φρόνησις, σωφροσύνη, δικαιοσύνη, ἀνδρεία, alle quali si aggiungono πραότης, φιλανθρωπία, σύνεσις, πρόνοια, μεγαλοψυχία, χάρις, φιλία. Per una lista completa delle virtù greche, cf. Bucher-Isler (1972) e Frazier (1996) 173–271.

13 Gastaldi (1998) 346.

14 Plutarco fu di circa un decennio più giovane di Tacito, ma può considerarsi contemporaneo. Plutarco era vicino alla 'cerchia' di Sosio Senecione, che dovette costituire un *trait d'union* tra la letteratura greca e latina del II sec. d.C.

E ancora in *Ann.* 2.55.1, citando Pisone che rimproverava a Germanico di avere troppo onorato, con vergogna del popolo romano, ... *non Athenienses tot cladibus extinctos, sed conluviem illam nationum* ... ('... non già gli Ateniesi, tutti estinti dopo tanti disastri, ma quella gentaglia piovuta da ogni paese ...', trad. it. di A. Arici).[15]

La vita, forse, doveva essere difficile qualche volta per un greco-romano, per un romano-greco, anche se concetti come ἰσότης ed ἐλευθερία si diffondevano velocemente diventando più tardi i motivi profondi alla base dell'impero di Marco Aurelio, un impero 'umanistico' come lo definisce Mazzarino.[16] Insieme con l'ἰσότης e l'ἐλευθερία si diffondeva quell'unità di cultura in lingua greca.

Plutarco con le *Vite* aveva dato esempio di tale unità culturale ma possiamo dire con certezza che egli, pur leale cittadino romano, si sia sentito sempre un greco e non mancano in lui accenni, sia pure garbati, di critica all'egemonia (imperialismo) dei Romani, come è possibile ricavare dalla *Synkrisis* fra Licurgo e Numa con la seguente domanda retorica:

> Forse, dirà qualcuno, Roma non progredì ed avanzò grazie alla guerra? La domanda richiederebbe una lunga risposta per certa gente, che pone il progresso nel denaro, nel lusso, nell'egemonia, anziché nella sicurezza, nella mitezza, nell'*autarkeia* con giustizia (*Comp. Lyc. et Num* 4.7).

Ma aggiunge subito dopo che nella vita di Numa resta qualcosa di veramente grande e diverso: l'essere stato chiamato al trono, lui straniero, e di lì con la sola persuasione aver rinnovato completamente lo Stato e dominato una città che non condivideva ancora le sue aspirazioni, e aver fatto tutto ciò senza ricorrere alle armi o violenza alcuna.[17]

Straniero (ξένος), dunque, ma capace di persuadere, straniero, dunque, pronto a condividere le sue aspirazioni, il suo ἦθος con un popolo che da piccolissimo diventava sempre più potente.

L'accento nel testo plutarcheo è posto su due elementi: la ξενία di Numa e la necessità di condividere una linea di condotta ispirata a quelle virtù che Plutarco

---

15 Cf. anche Iuv. 3.60 ss.: *Non possum ferre, Quirites, Graecam urbem. Quamvis quota portio faecis Achaei?... Ingenium velox, audacia perdita, sermo promptus, et Isaeo torrentior: ede, quid illum esse putes? Quemvis hominem secum attulit ad nos: grammaticus rhetor geometres pictor ... omnia novit Graeculus esuriens! Quid, quod adulandi gens prudentissima laudat sermonem indocti, faciem deformis amici; Et longum invalidi collum cervicibus aequat Herculis.* ('Non posso sopportare, Quiriti, una Roma greca; per quanto quale parte della feccia sono mai i veri Greci? ... Intelligenza viva, sfrontata audacia, parola pronta e più torrenziale di quella di Iseo. Dimmi un po', che cosa credi che sia costui? Egli ha portato con sé ogni tipo di uomo: grammatico, retore, geometra, pittore...sa fare di tutto quel grechetto affamato ... E che dire del fatto che questa gente espertissima nell'adulare, loda lo stile di un illetterato, la linea di un amico deforme e paragona il sottile collo di un invalido al collo di Ercole ... .', trad. P. Frassinetti).

16 Giannelli e Mazzarino (1962) 208.

17 Cf. per Numa e, in genere per il rapporto ellenico-barbarico, Nikolaidis (1986) 239–240.

considera fondamentali (e greche): πραότης, δικαιοσύνη, αὐτάρκεια. Una necessità di condivisione che è anche accettazione degli altri come si legge nella *Vita* di Temistocle (*Them.* 27.3–6) dove si narra l'incontro di questi con il chiliarco Artabano:

> 'ὦ ξένε, νόμοι διαφέρουσιν ἀνθρώπων: ἄλλα δ' ἄλλοις καλά: καλὸν δὲ πᾶσι τὰ οἰκεῖα κοσμεῖν καὶ σῴζειν. ὑμᾶς μὲν οὖν ἐλευθερίαν μάλιστα θαυμάζειν καὶ ἰσότητα λόγος. ἡμῖν δὲ πολλῶν νόμων καὶ καλῶν ὄντων κάλλιστος οὗτός ἐστι, τιμᾶν βασιλέα, καὶ προσκυνεῖν ... εἰ μὲν οὖν ἐπαινῶν τὰ ἡμέτερα προσκυνήσεις, ἔστι σοι καὶ θεάσασθαι βασιλέα καὶ προσειπεῖν ...'. ταῦθ' ὁ Θεμιστοκλῆς ἀκούσας λέγει πρὸς αὐτόν: ... 'ὦ Ἀρτάβανε ... καὶ αὐτός τε πείσομαι τοῖς ὑμετέροις νόμοις, ἐπεὶ θεῷ τῷ μεγαλύνοντι Πέρσας οὕτω δοκεῖ, ...'
>
> 'Straniero (sc. Temistocle), diversi sono i costumi degli uomini. A chi piacciono alcuni, a chi piacciono altri, ma a tutti piace aspettare e conservare le proprie usanze. Si dice che voi ammiriate soprattutto la libertà e l'uguaglianza dei cittadini, per noi, fra le molte e belle consuetudini che abbiamo questa è la più bella: onorare il Re e prostrarsi davanti a lui ... Se tu dunque approvi questa nostra usanza e ti prostrerai davanti a lui, è possibile che tu veda il Re e gli parli ...'. A queste parole così rispose Temistocle: '... o Artabano, ... io stesso mi assoggetterò alle vostre usanze, dato che così piace al dio che rende grandi i Persiani, ...'

Rispetto reciproco ed ospitalità ma pur sempre consapevoli della propria grandezza che è anche μετριότης; è il caso di un passo del *De Genio Socratis* (583F–584A) ove Epaminonda riferisce di un incontro avuto con Giasone principe di Tessaglia che gli offriva danaro:

> Gli ho risposto che ingiustamente dispiegava le sue forze dal momento che, essendo fautore della monarchia, con le sue ricchezze metteva alla prova il cittadino di una città libera ed autonoma. Da te, ospite (ξένε), accetto il pensiero che è bello, segno di sapienza e lo tengo caro in modo particolare; ma tu sei venuto a portare medicine a chi non è malato. (trad. A. Aloni)

All'ospite così si contrappone la ricchezza, al lusso la povertà segno di saggezza, perché la povertà delle cose non è disonorevole così come non è disonorevole una ricchezza acquistata con giustizia. Giustizia che è al tempo stesso continenza, resistenza a qualsivoglia piacere illecito. L'ospite, dunque,—continua Epaminonda—riferisca 'ai nobili amici che certo usano le loro ricchezze nel modo migliore che qui hanno amici che vivono bene nella povertà' (*De Genio Socr.* 584B) e che essa non è molesta e che anzi può considerarsi un'amata compagna: sembra qui che φίλος e ξένος assumano il medesimo significato.

Il messaggio del *De Exilio:* l'esule cosmopolita

Ma ospite in terra straniera è l'esule che non è né ἄπολις οὔτ' ἀνέστιος οὔτε ξένος (nel significato di straniero e neppure ἀλλοδαπός né ἀλλότριος). Sono i termini che ricorrono in *De Exilio* 601A–602E dove, oltre che all'influsso stoico, è possibile enucleare motivi propri delle *Consolationes* risalenti a Crantore e al sofista Anti-fonte.[18] Il discorso sullo ξένος in terra straniera assume nel *De Exilio* un significato politico e filosofico. Anzi, rovesciando un'opinione corrente, Plutarco giunge ad affermare che soltanto chi abita in città è da considerarsi ξένος ('straniero') e ἀλλότριος ('estraneo') soprattutto quando ritiene che non è bello né giusto, lasciando la sua città, abitarne un'altra. La patria così non è un concetto astratto ma è luogo dove si stabiliscono rapporti di ospitalità perché sarà la patria il luogo in cui si agirà e si vivrà. Anche un'isola potrà essere patria seppure 'dura, infruttuosa, cattiva per la vegetazione ...' (*De Exilio* 602C)[19] perché anche lì si potrà trovare l'ἡσυχία: non vi è isola dove non vi sia un luogo 'ove abitare, ove passeggiare, ove lavarsi ove andare a pescare e a cacciare' (*De Exilio* 603E). Nell'isola, lontano dalla città, dalla piazza, dai tribunali, dai mallevadori, dai sostenitori di candidature, l'esule potrà dedicarsi ai veri piaceri dell'anima. Il tempo in essa sembra fermarsi alla vista di un mare increspato e di un cielo stellato; gli unici beni che sono stati dati all'uomo e mai gli saranno tolti (*Cons. Helv.* 8.4). Nell'affermare questo Plutarco, riprendendo il pensiero platonico di *Timeo* 90a, considera l'uomo 'non una pianta fissa a terra', né immobile, 'ma celeste e volta verso il cielo'. L'uomo appartiene al mondo, non è 'argivo o tebano: io—dice Eracle[20]—non mi vanto di una sola patria, ogni terra è la mia patria'. L'affermazione di Eracle è quella stessa di Socrate che dice di non sentirsi né ateniese né greco ma di appartenere al mondo.[21] Di certo la frase socratica era divenuta un *topos* letterario e Cicerone stesso testimonia che ogni corrente filosofica poteva far sua la massima di Teucro: 'la mia patria è ovunque mi trovi bene',[22] ma tale massima in Plutarco va interpretata non solo dal punto di vista filosofico ma anche politico. È un riferimento a Roma, alla κοινὴ πατρίς, espressione dello stato universale? Oppure è una ripresa del concetto di cosmopolitismo proprio del I sec. d.C. 'quando rapporti di cultura e ospitalità rendono abitabile ogni terra popolata <quando> gli intellettuali si trasferiscono da un luogo a un altro?'[23] Oppure in nome di questo cosmopolitismo Plutarco rivendicava per sé il diritto di sentirsi un ospite greco nella città di Roma? Sì, un ospite, per quanto illustre, che difendeva il suo essere in un contesto storico nel quale 'la Grecia non è più l'ombelico del mondo e dei suoi disastri, e la pace romana aveva fatto cessare—momentaneamente anche questi ... A Plutarco la filantropia appare il destino dell'uomo Greco ...', una filantropia (ovvero un cosmopolitismo) che 'si realizza nell'atteggiamento dell'individuo, nella normalità di ogni atto: è urbanità di modi e disponibilità a collaborare con gli

18 Cf. Babut (1969b) 121 e n.

19 Questo verso è tratto o da una tragedia (*TrGF* Adesp. 393 Kannicht-Snell) o da una commedia *Com. Adesp.* 1238 Kock (manca nella raccolta Kassel-Austin). Vedi anche Maria Vamvouri Ruffy in questo volume.

20 *TrGF* Adesp. 392 Kannicht-Snell.

21 In Platone Socrate non è 'cosmopolita', ma tale lo rende l'interpretazione cinica e stoica: cf. Caballero e Viansino (1995) 92 n. 71. Cf. inoltre Richter (2011) 86.

22 Cic. *Tusc.* 5.37.108.

23 Caballero e Viansino (1995) 13.

altri, rispetto delle esigenze e delle opinioni altrui ... solidarietà civica, senso dei doveri intellettuali e pratici ...'.[24] Ma da Greco, quale egli sempre si sentì, Plutarco volle rievocare i fasti della patria dove era nato e, in un tempo in cui Roma aveva fuso civiltà e pensiero in un unico dominio, egli ebbe 'la capacità di decantare e sublimare fuori dal contingente, il significato dell'autentica esperienza greca'.[25]

24 Del Corno (1982) 15.

25 Del Corno (1982) 14. Non va dimenticato che in ambito politico-giuridico e linguistico il termine *latinitas* non possedeva una valenza univoca, bensì era legato a concetti opposti quali cittadinanza e alterità: esso rispecchia due aspetti compresenti e apparentemente contraddittori ossia la tendenza universalistica della *urbs caput mundi et orbis terrarum* e l'orgoglio quasi esclusivista del *civis Romanus sum*. Non a caso Elio Aristide nell'*Elogio in gloria di Roma* (26.96 = p. 119, 28 Keil) raccomandava ai Romani di prendersi cura dei Greci 'come padri adottivi'. D'altra parte Cicerone (*Ad Quint, fratr.* 1.1.27) riconosceva alla Grecia il ruolo di 'una nazione dove risiede l'*humanitas* e dove si considera che l'*humanitas* sia arrivata ad altri uomini ancora'.

8 Sympotic spaces: forging links between past and present

Anastasios G. Nikolaidis

Past and present in Plutarch's *Table Talk*[1]

Abstract: One of the best witnesses reflecting Plutarch's intellectual world as well as the trends and broader framework of the contemporary elite is his *Table Talk*. In the first instance, through the Roman participants (and high officials at that) in these learned conversations, the work brings together Greek and Roman intellectuals. In addition, through the participation of thinkers from various philosophical schools, it provides a vivid picture of current issues, interests, and trends as well as contemporary intellectual differences and conflicts over them. Further, in the context of the dinner-party, it reveals social mores, habits, practices and customs, matters of etiquette, and cultural issues. Moreover, it sheds light on Plutarch's family and friends, and bears important witness to stages of his life and career, but also to daily life in Chaeronea and other Greek cities. Finally, through the discussion-topics involved, it mirrors the main intellectual interests of this elite at the end of the first and the beginning of the second century of our era. This chapter will show how Plutarch managed to balance within the Graeco-Roman world of his era, namely, between the realities of the Roman present and his own Hellenic past; in other words, between a present that came from without and a past that emanated from within.

As is known, Plutarch lacks his own biographer and so what we know of Plutarch as a man, his public and private life, his family, his habits and his overall mentality is primarily established by piecing together the information which Plutarch himself scattered in his numerous writings. His political, social and intellectual stand is not difficult to ascertain, and in broad lines we can safely speak of a man born into an aristocratic family of considerably wealthy landowners, who acquired an excellent education in Athens, travelled widely and created friendly relations with several Roman officials, senators and consuls included. But this general picture does not answer certain questions that one might raise: for instance, could Plutarch have made a career either in the Roman administration or as an influential philosopher in the circles of the Roman intellectual elite, had he so wished?[2] Another question: Does Plutarch's decision to remain in Greece, and his obvious reluctance to learn Latin properly, although he did recognise the universality of the Latin language in his time (cf. *Quaest. Plat.* 1010D), tell us anything about his feelings and attitude

1 Many thanks to the anonymous referee for his/her meticulous reading of my article and for the corrections, observations and style improvements he/she kindly suggested.

2 Russell (1972) 7, 10, 29 suggests that Plutarch did try to make his mark in Rome, but, being an unsuccessful orator [on what evidence?], he failed; cf. Bowersock's similar conjecture in Zadorojnyi (2006) 102, who apparently accepts it. *Contra:* Barrow (1967) 43; Sirinelli (2000) 101.

https://doi.org/10.1515/9783110539479-024

to Rome?[3] We know, of course, thanks to a unique Delphic inscription (*CID* 4.150; cf. Dittenberger, *Syll.*³ 843/825 A), that he was a Roman citizen, but we do not know for certain if he obtained the Roman citizenship himself or inherited it from his father, exactly as his own sons became Roman citizens through him.[4] In any case, Plutarch nowhere speaks of his Roman citizenship and no one in antiquity knows him by his epigraphically attested name L. Mestrius Plutarchus; and whereas more than half of his works are addressed or dedicated to various friends and acquaintances, his consular friend Mestrius Florus, who presumably procured the Roman citizenship for Plutarch, gets not even one treatise as a dedicatee.[5] So, to put it differently, how did Plutarch strike a balance within the Graeco-Roman world of his epoch, namely, between the realities of the Roman present and his own Hellenic past?

To cope with questions such as these, Plutarch's *Table Talk* is one of our most illuminating guides, no matter whether we regard this work as a completely rhetorical/literary construct or as an accurate record, to some degree at least, of actual discussions.[6] To this problem I will return, but, as Klotz and Oikonomopoulou rightly point out, the *Table Talk* 'is delicately and inextricably bound up in Plutarch's life', and so 'it would seem a pity', ignoring the abundant autobiographical information contained in this work, 'to maintain that it tells us nothing about Plutarch or his world'.[7]

---

3 According to Flacelière (1964) 14, Plutarch did not take the trouble to learn Latin properly on account of his staunch Hellenism (cf. n. 60 below). See also Titchener (2002) 136 ('his decision to live his life in Chaeronea may provide a key to his underlying feelings'), who further believes that Plutarch's inadequate Latin was another disincentive for living in Rome (pp. 137, 140). However, Stadter (2015) argues that Plutarch's knowledge of Latin was more solid and dependable than is usually thought; see esp. pp. 133–138 and n. 13 ibid. for further bibliography, to which one may add Zadorojnyi (1997), Geiger (2000) and Glucker's (1978) 386–389 judicious remarks. As to the reasons behind Plutarch's decision to settle in Chaeronea, speculation can be limitless; Wilamowitz-Möllendorf (1922) 252, Boulogne (1994) 41–42, Sirinelli (2000) 102–103, 126–127 and Titchener (2002) 138–141 suggest the more plausible ones, but see also Zadorojnyi (2006) 102 and 120.

4 Cf. Wilamowitz-Möllendorf (1922) 251 and Ziegler (1949) col. 14.43. For the Roman citizenship of Plutarch's sons, see Jones (1971) 22 n. 15 (Ditt., *Syll.*³ 844 A), although the identity of the Soclarus in the inscription is disputed; see Teodorsson (1989) 245–246; Puech (1992) 4879–4883.

5 Ziegler's (1949) 51 explanation (Florus had died before Plutarch became a famous author) is not very convincing; for Florus was still alive until after 105, since he entertained the young philosopher Favorinus (cf. 734E), who was born after 80; cf. also Puech (1992) 4860 n. 107. Another possibility is that Plutarch might have dedicated to Florus one or two of his non-extant essays.

6 For a brief overview of this problem see Teodorsson (1989) 13–14, and cf. also Fuhrmann (1972) vii-xiii; Frazier in Frazier and Sirinelli (1996) 192–207; Titchener (2009).

7 Klotz and Oikonomopoulou (2011) 7.

Table Talk: structure, participants, topics

Plutarch's *Table Talk* consists of 95 questions discussed in nine books (10 questions in each of the first eight books and 15 questions in the ninth),[8] of which 9 have not survived, although we know their titles.[9] However, the remaining 86 questions are not distributed to 86 separate occasions, because quite often in one and the same drinking-party two, three or even up to fifteen questions (see 736C) were discussed. Our calculations, then, yield that the 86 questions of the *Table Talk* were discussed in 55 different sympotic occasions.

What sort of people took part in those wine-parties? Appendix 1 shows the different categories of participants, but those of particular interest for our study are: (a) four groups of Greek participants, namely philosophers, teachers of literature, teachers of rhetoric, and physicians, and (b) the sub-category of Roman participants in which we include the hellenised rhetor and philosopher Favorinus of Arelate as well as the Carthaginian erudite Sextius Sulla (Appendix 1b); finally Appendix 2 shows the places where the particular wine-parties were hosted and, whenever Plutarch tells us or we can fairly surmise it from the context, the name of the host is also there.

Before I proceed, I would like to make two brief observations concerning the categories of participants. The first observation is that local magistrates, and politicians in general, are conspicuous by their absence in Plutarch's *Table Talk*; and so are topics of political nature, both theoretical and practical, with the exception of the two last *Talks* of book 7 which were held in Athens: several statesmen are present, the topics are predictably political,[10] but Plutarch takes no part in the discussions and only reports them. No wonder, since he holds the view that political subjects, legal controversies, and rhetorical or sophistic harangues have no place in a drinking-party, as they are liable to generate annoyance or quarrel and thus ruin the purpose of the sympotic gathering, which is to promote friendship.[11]

The second observation concerns the philosophical doctrines represented in the *Table Talk* (see Appendix 1a). The preponderance of Epicureans, Stoics, and Pytha-

---

8 For the significance of number 10 see Klotz and Oikonomopoulou (2011) 25 n. 105; for the 15 questions of the last book see Teodorsson (1996) 300.

9 The headings of the *Table Talk* were most probably added by a later hand. See Hubert (1938) 1; Teodorsson (1989) 116.

10 Topic a: 'That political deliberation over wine was no less a Greek than a Persian custom', and topic b: 'Whether it was a good custom to deliberate over wine'. Teodorsson (1995) 434 justly remarks that these topics are not 'propriamente politiche', but 'piuttosto ... conviviali'.

11 621C: συμποτικὸν τέλος ... ἦν φιλίας ἐπίτασιν ἢ γένεσιν δι' ἡδονῆς ἐνεργάσασθαι τοῖς παροῦσιν· διαγωγὴ γάρ ἐστιν ἐν οἴνῳ τὸ συμπόσιον εἰς φιλίαν ὑπὸ χάριτος τελευτῶσα. See earlier at 621B on the duties of the symposiarch. Cf. also 713B and see Stadter (2015) 111–115. Teodorsson (1995) 436 proposes one more explanation for the absence of politics from the *Table Talk:* the literary tradition, as Plato's and Xenophon's *Symposia* manifest; to which he adds Athenaeus' *Deipnosophists*, and might have also added Lucian's *Symposium*.

goreans or Neopythagoreans accurately reflects the philosophical trends of the time and, therefore, should cause no surprise; but given Plutarch's philosophical position, on the one hand, and the nature of numerous topics that emanate from the so-called *problemata* literature,[12] on the other, one would expect more Platonic and Aristotelian or Peripatetic voices. However, the picture emerging from the interlocutors' philosophical allegiances may be somewhat misleading; because, irrespective of the schools these interlocutors represent, the philosophical subject-matter of the *Table Talk* does come, for the most part, from Plato and Aristotle, as the content of many discussions would reveal.[13]

This chapter, then, will mainly focus (a) on the Roman participants of the *Table Talk*, and the few cases where the banquets took place in Rome, (b) on the capacities of the interlocutors, and (c) on the topics of the discussions. Item (a) will reveal the extent of cross-cultural interactions between Greeks and Romans during the first two centuries of our era, while items (b) and (c) will, *inter alia*, shed light on Plutarch's intellectual world: either through his responses to contemporary trends and ideas or through his dialogue with the past by means of examining, reconsidering or confirming philosophical doctrines, old traditions or literary problems; in other words, items (b) and (c) will explore Plutarch's vision of Greek culture in both its diachronic trajectory and its synchronic aspect.

The present in the *Table Talk* and the issue of authenticity

The present in Plutarch's *Table Talk* manifests itself partly through the participants in those wine-parties, who were all contemporary, partly through allusions to contemporary events (*QC* 4.3, below), and partly through several discussion-topics which refer or are related, in some way or another, to contemporary happenings, issues and problematisations. In the first place, the main Roman participants and interlocutors in the *Table Talk* are Sosius Senecio, the dedicatee of the work (and, incidentally, of the whole series of the *Parallel Lives* and the treatise *On Making Progress in Virtue*), and Lucius Mestrius Florus, who obtained the Roman citizenship for Plutarch's family. Both men ranked highly in the Roman administration of their day, especially Senecio, who was honoured with two ordinary consulates by Trajan, and their relationship with Plutarch was intimate, as we may gather from the manner

12 For this literary genre see Klotz and Oikonomopoulou (2011) 18.

13 For Plato's presence in the *Table Talk*, esp. for the Platonic background of the proems, see Kechagia (2011) 81–87, 99–104; for the other philosophical creeds see p. 92 n. 27 and p. 98 with n. 36; and for Peripatetic knowledge in the *Table Talk* see Oikonomopoulou (2011) 105 ff.

they address and treat one another.[14] Florus, who was considerably older than Plutarch, and eventually retired to Greece,[15] participates in 10 drinking-parties and 13 *Talks*, all but two on Greek soil (see Appendix 2), in some with members of his family, and in about half of them he is also the host.[16] On the other hand, Sosius Senecio, who was much younger than Plutarch,[17] takes part in 6 wine-parties, of which one is occasioned by the wedding of Plutarch's son Autobulus (4.3, 666D), and hosts three of these, one in Patras (2.1, 629F), one in Chaeronea (2.3, 635E),[18] and one in Rome (1.5, 622C).[19]

As for discussion-topics indicating the present, one could adduce, for instance, such *Talks* as 5 and 6 of the fourth book which inquire about the Jewish god and why the Jews abstain from pork. But 'present' pointers *par excellence* are all sympotic topics inquiring 'What sort of man the symposiarch should be' (620A) or 'Whether philosophy is a fitting subject for conversation at a drinking-party' (612E), or 'How the guests will be placed' (615C) and so on, regardless of the fact that these and similar subjects were a legacy of the past.[20] Further, descriptions of the locale,[21] as well as references to the character or manners of some interlocutors, are also linked to the present and seem to constitute a strong evidence in favour of the authenticity of the *Table Talk*.

One can also find similar strong evidence of authenticity in the proems of the *Table Talk*.[22] Some, of course, may contend that these descriptions and portrayals

14 Regarding Senecio, cf. Klotz (2011) 163 and n. 11 ibid. Regarding Florus, our best evidence is his retirement at Chaeronea and his frequent (and fervent at times) participation in Plutarch's banquets. For more details about the career of the two Romans and their relationship with Plutarch, see Jones (1971) 11, 22, 48–49, 54–57, 63; Stadter (2015) 34–42.

15 According to Fuhrmann (1972) 103, Florus may have been born even before 30, while his retirement to Greece is deduced from his frequent participation in the sympotic gatherings there (see Appendix 2). His main residence was probably in Chaeronea, but we know that he also had a house at Thermopylae (see 734D). Cf. also Ziegler (1949) 51; Jones (1971) 49.

16 See Appendix 2 and *QC* 1.9, 5.7, 10, 7.4, 6, 8.10.

17 Perhaps, even twenty years younger, according to Jones (1971) 55; cf. also Puech (1992) 4883. Indeed, Senecio could have been Plutarch's son, as 734E might suggest (Plutarch addressing himself to Senecio): τοῖς μὲν οὖν σοῖς ἑτέροις ἐμοῖς δ' υἱοῖς ... . Cf. Teodorsson (1996) 281 and 283.

18 Cf. also Fuhrmann (1972) 3. Teodorsson (1996) 228 places this *Talk* in Rome, but without adducing any evidence either here or in his commentary on 2.3. However, the Roman location cannot definitely be excluded.

19 For the location of *QC* 1.5 see more below (n. 35 and p. 265). Besides the four *Talks* above, Senecio is also present at two wine-parties that took place in Athens (1.1, 612E and 5.1, 673C), but he apparently takes no part in their discussions.

20 On the good symposiarch and proper sympotic topics see Vamvouri Ruffy (2011) 144–149 and (2012), 29–61, 166–179; see also n. 32 below, and cf. Stadter (2015) 110–115.

21 For the significance of the locale see König (2007) 62–68, and cf. Klotz and Oikonomopoulou (2011) 15–16.

22 See esp. the proems to the first (612E), second (629D), eighth (717 A) and ninth (736C) books; also *QC* 5.5, 678C; 6.4–6, 690B ff.; 8.1–2, 717B ff.; 8.10, 734D, and cf. Teodorsson (1989) 36, 170, 184, (1990) 184 and (1996) 148, 162, 281, 299–300, and particularly Abramowicz (1962).

prove the very opposite, namely the artificiality of the *Table Talk*, something that could perhaps be conceded, if Plutarch offered such characterisations for all participants or speakers; in which case we might discern a motif or mannerism and thus plausibly argue for some kind of artificiality. But this is not the case here, as not only are several participants or interlocutors in the *Table Talk* neither described nor characterised, but quite often they are not introduced at all.[23] Similarly, it has also been remarked that Plutarch's apology, as it were, for the random and unsystematic presentation of his sympotic memories (629D: σποράδην δ' ἀναγέγραπται καὶ οὐ διακεκριμένως ἀλλ' ὡς ἕκαστον εἰς μνήμην ἦλθεν) is a sham one, and in fact Plutarch 'perhaps composed in this way in order to increase the impression of authenticity'.[24] Needless to say that such a rationale can render questionable all kinds of literary evidence regardless of author or work concerned.

I see no reason to question Plutarch's general descriptions of the sympotic events or his particular characterisations of certain guests.[25] The weight of proof, after all, devolves upon those who deny rather than accept Plutarch's sincerity. To accept at face value what our author explicitly states does not, of course, prevent us from assuming or suspecting other things that he might have had in mind but chose not to express; but even if one manages to somehow substantiate Plutarch's hidden thoughts, purposes or motives, his explicit statements do not become invalid, unless it is also proven that they are false, which would mean that our author is trying to mislead his readers. Nor does the dialogic fiction-technique suggest that we should necessarily doubt what the author says or believes. Further, the discussion-topics must be also authentic, more or less, although the meticulous argumentation we often see in the *Table Talk* is certainly a product of literary elaboration that took place at a much later time.[26] This product, however, based on Plutarch's notes, library and memory, is not fictitious but rather a follow-up and a refined advancement of views and arguments that were actually put forward sometime. The numerous quotations, for instance, are good witnesses to such an elaboration.[27] As Teodorsson puts it, in his *Table Talk* 'presumably Plutarch depended to a large extent upon reminiscences of real talks, which he complemented with material from literary sources or, in some talks, inversely'.[28] Thus, Titchener rightly sees Plutarch of the *Table*

---

23 See, e.g., *QC* 1.5–8; 2.7–10; 5.2–3, 5.6, 5.9; 6.1, 6.4–6, 6.8–10; 7.2–3, 7.9; 8.5; also 639B, 643C, 655F, 656A, 658A, 669C, 671C, 677D, 684E, 722E, 728D, 730D, 744F.

24 Cf. Teodorsson (1989) 37.

25 See, e.g., 615D, 617E, 622E, 646A, 678C, 704D–E, 710B, 726A; cf. also the telling participles in 613A (ἀνακραγών), 617F, 657B (ἀναβοήσας), 692E (γελάσας), 738F.

26 As is known, the *Table Talk* is one of the latest works of Plutarch (see Jones [1971] 56, 137 and Fuhrmann [1972] xxvi), but it contains material going as back as to the time of his studies in Athens under Ammonius (see book 9).

27 Cf. also Teodorsson (1989) 296.

28 Teodorsson (1989) 170. Stadter (2015) 34 n. 50 also takes the *Table Talk* 'as Plutarch's reconstruction and literary refinement of actual conversations, and the speakers as real participants'; and Sirinelli (2000) 381–382 affirms, *inter alia*, that 'Plutarque n'invente pas; au mieux, il brode'.

Talk as both insider (participant) and outsider (editor),[29] and Klotz and Oikonomopoulou aptly point out that the reality of the *Table Talk* 'is doubly filtered through authorial memory and narratorial voice'.[30] With his literary refinement, Titchener (n. 29) observes, Plutarch gives us 'a greater, or enhanced reality than that of the actual banquets', and in this sense 'the *QC* do not need to be authentic [I would say 'literally authentic'] to be real and true'.

The past in *Table Talk*

On the other hand, the past in Plutarch's *Table Talk* is mainly represented by discussion-topics which either investigate various literary matters (mostly in connection with the Homeric poems and problems pertaining to the *exegesis* of Homer)[31] or bring forward and revive philosophical doctrines going back to Pythagoras and the Presocratics (especially Democritus) or review issues of the *problemata* literary tradition (such as old customs, strange practices, and proverbial sayings, for example) that are totally irrelevant to present interests and concerns. Such topics exclusively pertaining to the past, are, for instance, the inquiry into why the chorus of the phylê Aiantis in Athens was never judged last (1.10), or 'Why the pine was held sacred to Poseidon and Dionysus' (5.3), or 'Why alpha stands first in the alphabet?' (9.2), or 'What is the hidden meaning of Poseidon's defeat' (9.6), and so on. Besides, Plutarch's prologues to each book of the *Table Talk* refer mostly to the past, even though the starting-point of some of them is linked to the present. The prologue to book 3, for instance, features Simonides, Aesop and Plato; the prologue to book 6 exalts the simplicity of the Platonic dinners in the Academy; and the prologue to book 7 refers to some sympotic sayings of Homer and the elegiac poet Euēnos as well as to sympotic practices of the ancient Spartans, although the opening for Plutarch's remarks is provided by a popular witticism of a contemporary, presumably, Roman (see 697C).

Further, the past is recalled through the special capacity of some interlocutors. Contemporary philosophers and men of letters, such as grammarians and rhetors, were naturally expected to participate in those learned discussions; but why so many physicians, who appear to take part in almost half of those sympotic gatherings, even in cases where the topic of inquiry is not a medical one?[32] One can hardly

---

29 Titchener (2009) 400; so also Klotz (2011) 166.

30 Klotz and Oikonomopoulou (2011) 4; see also 26. Surprisingly, however, they seem to doubt (p. 19) whether the *QC* is clearly a more literary work than the other *Quaestiones* treatises (*Quaest. Graec., Quaest. Rom., Quaest. Nat., Quaest. Plat.*).

31 See, e.g., *QC* 2.5: 'Why Homer always arranges boxing first, then wrestling, and last racing'; 5.10: 'Why Homer calls salt divine'; 9.4: 'Which of Aphrodite's hands did Diomedes wound?'

32 See, for example, *QC* 2.6, 3.1, 5.8, 9.14. Physicians constitute the second, *pace* Teodorsson (1995) 433, most numerous category of guests after the philosophers (see Appendix 1), and this in all like-

help recalling here, regardless of the popularity of medical themes at the dinner-parties of the time, the model of all sympotic literature where, importantly enough, it is the physician Eryximachus who proposes the topic of the discussion.[33] In this way, through such characters, the present appears strongly linked with the past.

Table talks at Rome

Although Plutarch's prologue to book 1, practically a prologue to the whole *Table Talk*, speaks of several sympotic occasions in Rome and Greece where Senecio was present,[34] there is only one occasion in the whole *Table Talk* which we know for sure that it took place in Rome (8.7–8); and in this Senecio was not present![35]The host of that dinner-party was Plutarch's Carthaginian friend Sextius Sulla,[36] and the occasion was a welcome-dinner to celebrate Plutarch's coming to Rome again after a considerable lapse of time.[37] The participants, we are told, were a small num-

lihood suggests both the status of doctors in intellectual society and the popularity of medical themes at the drinking-parties of the time (cf. also Teodorsson [1989] 297); a popularity that is further confirmed by the participation of several physicians also in Lucian's *Symposium* and Athenaeus' *Deipnosophists*. cf. Hirzel (1895) 362.

33 Cf. Pl. *Smp.* 177a–d. For bibliography on the role of medicine in Plutarch's work, see Vamvouri Ruffy (2011) 131 n. 1 and (2012) 29–124. For the relationship of the *QC* (esp. the first *Talk*) with Plato's *Symposium* in general, see Klotz (2011) 167–171, and cf. Klotz and Oikonomopoulou (2011) 4: 'The philosopher reshapes his memories into a Platonic sympotic form'.

34 Cf. 612E: ... ᾠήθης τε δεῖν ἡμᾶς τῶν σποράδην *πολλάκις ἔν τε Ῥώμῃ μεθ' ὑμῶν καὶ παρ' ἡμῖν ἐν τῇ Ἑλλάδι* παρούσης ἅμα τραπέζης καὶ κύλικος φιλολογηθέντων συναγαγεῖν τὰ ἐπιτήδεια ... (Clement's transl. in Loeb: '... you thought that I ought to collect such talk as suits our purpose from among the learned discussions *in which I have often participated in various places both at Rome in your company and among us in Greece*, with table and goblet before us').

35 The other three occasions registered in Appendix 2 are located in Rome by conjecture; see below pp. 265-266 with n. 41, and p. 267 with n. 48. According to Teodorsson (1989) 37, other *Talks* 'that can with some probability be located in Rome' are: 1.9, 5.7 and 10, 7.4 (Mestrius Florus stated as the host), and 1.5, 2.3 which took place in Senecio's home'. Teodorsson (1996) 228 suggests that *QC* 3.3 may also be located in Rome, but in discussing 1.9 (p. 145) he places (together with Ziegler [1949] 51) all *Talks* featuring Mestrius Florus in Greece. Fuhrmann (1972) 3 places *QC* 2.3 at Chaeronea, and most probably Chaeronea is again the place of *QC* 3.3, because it is rather hard for one to imagine a local military trainer (Apollonides) and a physician from Thasos (Athryitus), interlocutors in 3.4 and 5 (belonging to one and the same banquet with 3.3), as visitors in Rome; and the same holds true for *QC* 2.3 (see n. 41).

36 Cherniss in Cherniss and Helmbold (1957) 3 would distinguish this Sulla (also an interlocutor in the *De coh. ira* and *De facie*, and mentioned by Plutarch in *Rom.* 15.3) from the Sulla in *QC* 2.3 (636A) and 3.3 (650A), but there is no cogent evidence against their identification; see Ziegler (1949) 53–54, and cf. Puech (1992) 4878–4879.

37 Cf. 727B: Σύλλας ὁ Καρχηδόνιος εἰς Ῥώμην ἀφικομένῳ μοι διὰ χρόνου τὸ ὑποδεκτικόν, ὡς Ῥωμαῖοι καλοῦσιν, καταγγείλας δεῖπνον ... . In all likelihood, that was Plutarch's third and last visit to Rome in the beginning of the second century, many years indeed after his previous one in 92/93. See Puech (1992) 4878–4879 with n. 183, and cf. Teodorsson (1996) 228.

ber of close friends including the invitee of the host Nestor, the grammarian Theon who came from Egypt,[38] a certain Lucius of Etruria, who was a Pythagorean, Plutarch's intimate friend Philinus, also a follower of Pythagoras, and a certain Empedocles, another Pythagorean. Not surprisingly, therefore, the discussion-topics in this gathering were, first, an inquiry into the meaning and explanation of some Pythagorean precepts, and then the examination of the question why the Pythagoreans abstained from fish more strictly than from any other living creature (*QC* 8. 7–8).

The occasion is located, as we saw, in Rome, but Romans do not participate in it, and the discussion is a journey to the remote past of the epoch of Homer, Hesiod and Anaximander (cf. 730C–E). The only contact with the present might perhaps emerge from considering this *Talk* as evidence of the revival of Pythagoreanism in Plutarch's times, of the thriving Neopythagoreanism, to be more precise.[39] So, this appears to be all that Plutarch remembered or thought fit to take down from this sympotic experience and from this visit to Rome that took place a decade or so after his previous one. No description, no indication whatever of the surroundings, as is often the case with sympotic occasions in Greece;[40] no suspicion of a hint that could somehow point to Rome, its landscape, its buildings, its sights, its people.

Insofar as the *Table Talk* indeed includes sympotic occasions that took place in Rome (see n. 35 above), which is very probable given Plutarch's visits and long sojourns there, and if Senecio, on the basis of the general proem to the work, was present in at least one of them, that one must of necessity be the fifth *Talk* of the first book where Senecio is also the host.[41] The topic of the discussion was the meaning of the saying 'love teaches a poet, even if he had no skill before',[42] the participants are all anonymous, which is perfectly understandable if the guests were unknown to Plutarch, and Senecio makes an important contribution based on Theophrastus' book *On music*. Once again (as in the banquet which Sextius Sulla hosted), we have a wine-party in Rome, the interlocutors are probably Romans this time, but indications of or references to the locale are again completely missing, while the whole *Talk* is in essence a learned browsing through landmarks of Greek literature: from Sappho to Aeschylus, Sophocles and Pindar, and from Herodotus to Plato and Theophrastus. One may plausibly conclude, I think, that this *Talk*, featuring Senecio as the only

---

38 So Cherniss (1957) 7 note b. On whether or not this Theon is the same as Plutarch's friend Θέων ὁ ἑταῖρος opinions differ. See Teodorsson (1989) 92 (Ziegler and Flacelière see only one Theon, but Babut two) and cf. Fuhrmann (1972) 6–8 (one Theon). By contrast, Puech (1992) 4886, and Frazier (1996) 291 distinguish two persons.

39 Cf. also Teodorsson (1989) 211.

40 See, for instance, 640B or 667C.

41 The location of the other five *Talks* where Senecio was present is known (see above n. 19): 1.1 and 5.1: Athens, 2.1: Patras, 4.3: Chaeronea, and 2.3 Chaeronea again, mainly on the basis of some of the participants who appear to be locals (Alexander and Plutarch's γαμβρός Firmus); cf. Puech (1992) 4833–4835, and Fuhrmann in nn. 18 and 35 above.

42 Cf. Eur. *fr.* 663N/Collard and Cropp (Loeb).

named speaker (unique case in the entire *Table Talk*), is meant as a compliment to the host.[43]

The seventh book of the *Table Talk* contains two dinner-parties (7.4 and 6) which were probably hosted by Mestrius Florus in Rome.[44] For, besides Florus and Plutarch, the other interlocutors are the host's son Lucius, his son-in-law Caesernius,[45] and the Athenian Eustrophus, Plutarch's fellow student under Ammonius.[46] Yet, unlike the *Talk* with Senecio, the two *Table Talks* at Florus' house contain Roman elements. The first one is the topic itself: 'Why the ancient Romans had the custom not to allow a table to be removed empty nor to let a lamp be extinguished?'. Florus, who loved traditions and old things (φιλάρχαιος ὤν, 702D), Plutarch tells us, observed this custom scrupulously and found fault with his contemporaries who would put the light out directly after eating, so as not to waste oil. But being of a philosophical nature as well (734D), Florus suggests that the company should investigate the origin of or the reason lying behind these customs.[47] Interestingly, the views that are subsequently put forward seem to be coloured by the nationality of the speakers. The Romans Caesernius and Lucius connect the light of the lamp with the unquenchable sacred fire kept by the Vestal Virgins (702E), whereas Plutarch and Eustrophus attempt some symbolic explanations for both customs by appealing *inter alia*, on the one hand, to Pythagoras and Homer and, on the other, to contemporary Boeotian sayings (703E–F).

The second *Talk* (7.6), which can be placed in Rome with even greater probability, examines the problem of the so-called 'shadows', namely those who go to a drinking-party without a personal invitation by the host, but invited by another invitee whom they, as a rule, accompany. And the question to be considered is whether and under what circumstances one should or may accept such a secondary invitation. Expectedly enough, Plutarch associates the origin of this practice with Socrates, who went to Agathon's banquet at the invitation of Aristodemus, but Florus' son-in-law Caesernius strongly disapproves of this custom, which apparently had been common, giving his preference to Hesiod's advice instead, which was to invite only friends and not unknown people to one's wine-party (*QC* 707C = *WD* 342). His line of argument betrays a remarkable discretion: A man of breeding should never go

---

43 At 623A Sosius says that he had just read Theophrastus' book *On music* (καὶ γὰρ ἔναγχος ... τὸ βιβλίον ἀνέγνων). And Abramowicz with justice finds it hard to think that Plutarch could have put this statement in Sosius' mouth, if it was not true (see Teodorsson [1989] 112, and my own remarks on p. 262 above).

44 Cf. Teodorsson (1996) 51–52. The phrase οὕτως ἐνταῦθα in Lucius' speech (703A) and the ἰσχυρὰ γὰρ ἡ τῆς πόλεως συνήθεια in Caesernius' speech (708A) convincingly point to Rome; see also below, n. 48.

45 Perhaps the same person as the Gaius in 5.7 (682F); see below, and Teodorsson (1996) 53: C(aius) Caesernius.

46 Wealth was indispensable for higher education (cf. Luc. *Somn.* 1). So, since Eustrophus had the means to study, he could also have had the means to travel abroad.

47 Searching for causes is typical of Florus: see 650A, 651F, 680C–D, 684E.

to a dinner-party as a shadow of someone else. If the host does not know him, as is more often the case, the shadow is simply rude; but if the host knows him, is on friendly terms with him and still has not invited him, it is much more disgraceful for him to go and thus greatly embarrass the host. Besides, the shadow-invitee cannot speak as freely as the other legitimate guests, and this may easily make him a target of criticism, which is even more disgraceful. 'This is why', Caesernius goes on, 'when I invite my friends, I occasionally allow a place for shadows, recognizing that this is a firmly established custom of the city, yet I myself never go to another man's house at a secondary invitation'.[48] This passage decisively points to Rome, I believe, as the place of this banquet. In any case, Florus appears not so uncompromising as his son-in-law on this matter, and Plutarch is even more tolerant of this practice, since, according to his theory, wine-parties should give, *inter alia*, everyone the opportunity to acquire new friends (621C).[49]

Conclusions

Time to conclude. How then does the evidence from the *Table Talk*, and especially the *Talks* we have discussed, colour the standard picture of Plutarch, as this picture emerges from his writings and the historical data? Where exactly does he stand within the Graeco-Roman world of his epoch and how does he strike a balance between the Roman realities of today and the heavy Greek legacy of yesterday? In other words, between a present that came from without and a past that emanated from within? The evidence from the *Table Talk* suggests that this work 'necessarily ... embeds itself in second-century Graeco-Roman society'.[50] But in this context, it was mainly the past, I think, that offered Plutarch intellectual vitality and enjoyment as well as emotional relaxation and repose, something that can further be confirmed by his prologue to the *Aemilius–Timoleon* pair (1.5). Doubtless, he recognises the Roman power and achievement, and accepts the political reality of his times,[51] but again it is Rome and the Romans of the past that he primarily admires and respects. It is the Romans of the Republic that he compares with the most eminent Greek worthies in his *Parallel Lives*, and whom he often regards as better and superior to their

48 Cf. 708A: Διὸ καλῶν μὲν ἑταίρους ἔδωκα τόπον σκιαῖς, αὐτὸς δὲ κληθεὶς ὑφ' ἑτέρου πρὸς ἕτερον ἄχρι γε νῦν ἀντέχω μὴ ὑπακοῦσαι.

49 Cf. 621C (above, p. 259 with n. 11), and the proem to book 4. However, it may be worth noting here that in his *De amicorum multitudine* Plutarch advises against having many friends (93E ff.).

50 Klotz and Oikonomopoulou (2011) 7; cf. also Bowie (1970) 4.

51 See notably *Praec ger. reip.* 824C–F. Holford-Strevens (2003) 6 aptly remarks that Plutarch's re-creation of his era expresses 'the contradiction between the reviving fortunes of the Greek elite and its constant awareness of subjection'. Cf. also Bowie (1970) 4, who speaks of 'the Greeks' preoccupation with their past' in connection with 'their dissatisfaction with the political situation of the present'.

Greek counterparts.[52] With the exception of the *Roman Questions*, the purpose of which is narrow and specific,[53] and the Roman section of the *Apophthegmata*, which in one way or another is connected with the *Lives*,[54] there is little of Rome and Romans in the *Moralia*, which, unlike the *Parallel Lives*, are mostly related to the present. To say nothing of his rhetorical essay *On the Fortune of the Romans*, which is hardly complimentary to the ruling power.[55] He probably visited (and stayed in) Rome at least three times,[56] but, unlike Strabo or Dionysius of Halicarnassus, for instance, he appears rather unimpressed by the world capital of his day.[57] True, at some point of his treatise *On the Cleverness of Animals* he calls Rome beautiful, but the context and expression allow the suspicion that Plutarch is being somewhat ironic there;[58] for seldom in his writings as a whole—and certainly nowhere in the *Table Talk*—does he approvingly mention or describe things that he saw and particularly liked or admired in the city; on the contrary, he often mentions and describes things that he detests in Rome: the gladiatorial shows, for instance, the incredible costliness and luxury of the Domitian palace, the desecration of Plato's dialogues through their performance over dinner by slaves, the monster-market where all kinds of deformity and crippledom were exhibited, and so on.[59]

Thus, in performing his duties as a local magistrate or ambassador or priest or teacher of philosophy or lecturer, Plutarch was the practical man of affairs and as such he lived in the present. But his huge literary production suggests that contentment and fulfilment he mainly found not in the tasks of the present, but in his immersion in the intellectual and moral achievements of the Greek past. It was this immersion that supported his dignity and gave him even an air of superiority over his Roman acquaintances, and made him, as Wilamowitz put it, not only remain a Hellene, but also insist on showing this off.[60] Having said that, I close with a slight mo-

52 Cf. Nikolaidis (forthcoming); see also Geiger in this volume.

53 See on this Boulogne (2002) 95–96.

54 The *Reg. et imp. apophth.* is certainly linked to the *Lives* whether these anecdote collections are prior (so Stadter [2014b] 675 n. 31 and 685: a separate Plutarchan work that was used also for his biographies) or subsequent to the *Lives* (so Pelling [2002] 70: garnered from the biographies). On this problem, see also Fuhrmann (1988) 3–13.

55 See esp. 318B–319F and cf. Babbitt (1936a) 320: 'The thesis that Fortune was responsible for the great Roman Empire would hardly be pleasing to Romans'. See also Frazier in Frazier and Froidefond (1990) 14–15, with the footnotes there.

56 The third time is convincingly upheld by Puech (1992) 4878–4879 (cf. n. 37 above); but see also Jones (1971) 20–25, and Ziegler (1949) 19–20.

57 Contrast Strabo 5.3.7–8, 8.6.23 and Dion. Hal. 3.67.5, 68.2–4, 4.13.5. According to Titchener (2002) 138, Plutarch had an 'antipathy toward Roman and/or urban life'.

58 Cf. *De soll. an.* 963C: ... ἡ καλὴ Ῥώμη.

59 For the gladiatorial shows, cf. Jones (1971) 123 with n. 8 ibid.; for the Domitian palace: *Publ.* 15.4–5; for Plato's desecration: *QC* 711C–D; for the monster-market: *De cur.* 520C. Cf. also Titchener's remark in n. 57.

60 Wilamowitz (1922) 251: 'Er blieb Hellene und hielt darauf, es auch zu scheinen'. Cf. Ziegler (1949) 14–37, and see also the broader context set down by Swain (1996) 88–89.

dification of Ewen Bowie's conclusion in a seminal article: 'Most often the past was resorted to as an alternative to rather than an explicit reflection on the present, for most Greeks were not in a real sense *anti*-Roman, and their absorption in the Greek past *complemented* their acquiescence in the politically defective Roman present'.[61] For the sophists of the time this conclusion may be perfectly valid; yet Plutarch's absorption in his Greek past, did not, in my view, so much complement, but rather *compensated for* his acquiescence in a Roman present, which he probably deemed unworthy of his race.

Appendices

1 Categories of participants (except Plutarch)

Members of Plutarch's family: 10
Plutarch's Greek friends and other acquaintances: 44
Philosophers: 12 by name and some unnamed
Physicians: 10 by name and some unnamed
Scholars/Teachers of literature (γραμματικοί): 7 by name and some unnamed
Teachers of rhetoric (ῥήτορες): 5 by name and one unnamed
Sophists: 2 (one unnamed)
Poets: 2
Priests: 2
Directors (ἐπιμεληταί) of the Amphictyons: 2.
One gymnastic trainer (παιδοτρίβης), one military trainer/trainer of the *ephēbes* (τακτικός), one land-measurer (γεωμέτρης), one director of games (ἀγωνοθέτης), one guide/geographer (περιηγητής), one musician/music theorist (ἁρμονικός), one farmer

1a The Philosophers
Epicureans: 4 (Alexander 635E, Boëthus 673C; cf. *De Pyth. or.* 398B, Xenocles 635A), Zōpyrus 653C, E), and some unnamed.
Stoics: 3 (Philip 710D, Sarapion 628A; cf. *De Pyth. or.* 400B–C, Themistocles 626E), and one unnamed.
Pythagoreans: 3/4 (Alexicrates 728D, Lucius 727B–C [see below 1b], Philinus 727B, and
possibly Sextius Sulla of 1b; cf. 728C, 729C)
Platonists: 2 (Ammonius, Tyndares 717E–718C)
Aristotelian/Peripatetics: 2 (Favorinus 734F, Menephyllus 741A, 745C)

61 Bowie (1970) 41.

1b Roman Participants (including the hellenised Favorinus and Sextius Sulla of Carthage): 8/9 (Sosius Senecio, Mestrius Florus, Lucius [Florus' son], Gaius [Florus' son-in-law], Caesernius [another son-in-law], Lucius, a Pythagorean from Etruria), Aufidius Modestus [the Roman grammarian?]

2 Places and Hosts

Chaeronea: **31** 1.2–3 (Plutarch's brother Timon), 1.4, 1.6–8, 1.9 (Florus); 2.3 (Senecio), 2.7, 2.8, 2.9, 2.10 (Plutarch as ἄρχων); 3.3–5 (Florus) 3.6, 3.7–9 (Plutarch or Plutarch's father), 3.10 (Euthydemus, Plutarch's fellow-priest)*; 4.3 (Plutarch); 5.4, 5.5–6 (Onēsicrates the physician), 5.7 (Florus), 8–9 (Plutarch?), 5.10 (Florus); 6.1–3 (a physician of Philōn's circle: cf. 687B and 689A)**, 6.4–6 (Plutarch?), 6.7 (Aristion, Plutarch's fellow-citizen), 6.8 (Plutarch as ἄρχων), 6.9 (Plutarch?), 6.10 (Aristion); 7.1 (Plutarch or Florus), 7.3 (Plutarch), 7.7–8 (Plutarch), 8.1–2 (Plutarch), 8.5 (Plutarch?), 8.6 (Plutarch?), 8.9**

* alternative place: Delphi ** alternative place: Hyampolis

Athens: **7** 1.1, 1.10 (Sarapion the poet); 3.1–2 (Erato the musician/ἁρμονικός); 5.1 (Boëthus the Epicurean); 7.9–10 (Nicostratus, an Athenian acquaintance of Plutarch); 8.3 (Ammonius); 9.1–15 (Ammonius)
Delphi: **4** 2.4–5 (Plutarch); 5.2 (Petraeus, director of the Pythian Games/ἀγωνοθέτης); 7.2 (Plutarch); 7.5 (Callistratus, director/ἐπιμελητής of the Amphictyons)
Rome: **4** 1.5 (Senecio); 7.4 (Florus); 7.6 (Florus or Florus' γαμβρός Caesernius); 8.7–8 (Sulla the Carthaginian)

* * *

Aidēpsus (late spring): 4.4–6 (Callistratus the sophist)
Corinth (Isthmian Games): 2 5.3 (Lucanius, the chief priest); 8.4 (Sospis, ἀγωνοθέτης)
Eleusis (the Great Mysteries festival, end of September): 2.2 (Glaucias the rhetor)
Ēlis (Olympic Games?): 4.2 (Agemachus)
Hyampolis (festival of *Elaphēbolia*, springtime): 4.1 (Philōn the physician)
Patras: 2.1 (Senecio)
Thermopylae: 8.10 (Florus)
Tithōra (garden party by the river Cēphissus): 2.6 (Sōclarus)

David Driscoll

Sympotic space, hierarchy and Homeric quotation in *Table Talk* 1.2

Abstract: *Table-Talk* 1.2, on whether guests should have reserved seating in the sympotic space, has been recognised as programmatically establishing the political discourse of the *QC* generally, as the question ties in with various larger issues: Roman vs. Greek, democracy/egalitarianism vs. tyranny/hierarchy. Little attention, however, has been paid to the close of Plutarch's speech (617D–E) for his self-represented negotiation between these poles. I argue that his misreading of *Il.* 23 there allows him to have it both ways. I first set out the problem. Plutarch recommends choosing a family member if one cannot choose between two similarly important guests. He justifies this practice with reference to *Il.* 23.534 f., where, he claims, Achilles gives his fellow Thessalian Eumelus second place to forestall a conflict between Menelaus and Antilochus. This misreading, however, reverses the order of events in *Il.* 23. I suggest that a very learned audience may have interpreted this misreading as prioritising Greek *paideia* over Roman political power. According to Σ *Il.* 23.536–537 T, Achilles' gift to Eumelus teaches us to follow the 'natural' hierarchy, as ordered by 'excellence' (ἀρετή), not the actual result, which the scholion calls 'chance' (τύχη). Parallels in diction and thought suggest a connection between the scholion and Plutarch. By comparing his family members and other intimates to Eumelus, Plutarch hints that these figures are in another, more authentic hierarchy superior to the figures who by chance appear to be better. For an unlearned audience Plutarch respects the social hierarchies of the external world, but to the *pepaideumenoi* he asserts the superiority of Greek *paideia*. Homeric quotation and interpretation map out social hierarchies, physically in the sympotic space and verbally in the legitimisation of elite status.

What is at stake when characters quote and interpret Homer in Plutarch's *Table Talk?* Much recent work on quotation in Plutarch has richly explored the breadth and patterns of Plutarch's reading and deployment of quotations, as well as the *Quellenforschung* of individual quotations, but comparatively little attention has been paid to the interpersonal dynamics of quotation of 'the poet'. Such dynamics are particularly important in the special environment of the sympotic space, where implicit social hierarchies become plainly visible in seating arrangements and where symposiasts legitimise these hierarchies by offering the appropriate quotation and interpretation of canonical poetry.[1] I intend in this chapter to suggest the possible stakes by looking at a single rich moment—the end of Plutarch's own contribution—in *Table Talk* 1.2. I first set the scene by connect-

1 Important recent contributions specifically on Homeric quotations in Plutarch include Cannata Fera (1996); Alexiou (2000); Bréchet (2003); D'Ippolito (2004); Sluiter (2005), and Díaz Lavado (2010).

https://doi.org/10.1515/9783110539479-025

ing the dialogue to some larger questions related to symposia and politics. I then argue that the close of Plutarch's speech can be read ironically, and that the interpretation has consequences for our reading of the *Table Talk*.

As Oikonomopoulou observes, *Table Talk* 1.2 is 'a programmatic chapter of crucial significance for our understanding of the political discourse of the *Table Talk* as a whole'.[2] The dialogue's topic, the proper way to seat guests, asks whether the symposium's own spatial organization should mirror the hierarchy of the external world: should guests choose their own seats, or should hosts seat their guests according to their social status? In the character Timon's own words, should their symposium be one governed like a Persian satrapy (616E: σατραπικόν) or a democracy (616F: δημοκρατικόν)? Following the common Second Sophistic habit of referring to Roman Imperial institutions with Persian names, like 'satrap' for proconsul or 'Great King' for emperor,[3] we might imagine that σατραπικόν here alludes in particular to Roman administration, especially since σατραπικόν in its sense here of 'formal', 'official', 'strict', as Teodorsson glosses it, is unparalleled.[4] Furthermore, this juxtaposition evokes the following 1.3, which distinguishes Roman and Greek places of honor at the symposium and calls the Roman the 'consul's place'.[5] Timon then suggests that the different ways of spatially arranging a symposium map onto an opposition between Roman hierarchical rule and Greek democracy.

The vignette that opens this dialogue makes this political question even more loaded by directly connecting it to the Roman presence in Greece. Plutarch's brother Timon is holding a symposium to which he has invited 'foreigners, citizens, friends, relatives, and altogether all sorts'[6] and has directed them to sit where they want. After everyone is seated, a new guest arrives and is unhappy not to find a seat that he deems he is worthy of:

> πολλῶν οὖν ἤδη παρόντων ξένος τις ὥσπερ εὐπάρυφος ἐκ κωμῳδίας, ἐσθῆτί τε περιττῇ καὶ ἀκολουθίᾳ παίδων ὑποσολοικότερος, ἧκεν ἄχρι τῶν θυρῶν τοῦ ἀνδρῶνος, καὶ κύκλῳ ταῖς ὄψεσιν ἐπελθὼν τοὺς κατακειμένους οὐκ ἠθέλησεν εἰσελθεῖν ἀλλ' ᾤχετ' ἀπιών.
>
> When many were already present, a foreigner, who like an arriviste from comedy was committing a little *faux pas* with his strange clothing and entourage of slaves, came as far as the room's doors. After looking over those reclining, he refused to come in and left. (*QC* 1.2, 615D)

This outlandish figure has no place at Timon's symposium.[7] The comparison to an εὐπάρυφος, which I translate 'arriviste', is particularly suggestive: the term refers originally to an ornate piece of clothing with a purple border, but elsewhere Plutarch uses the term pejoratively to refer to newcomers to elite society who refuse to learn its rules.

---

2 Oikonomopoulou (2007) 175.

3 Almagor (2014) 288 n. 64, ultimately going back to Bowie (1970).

4 Teodorsson (1989) 73.

5 Oikonomopoulou (2007) 190 – 191.

6 615D: ... καὶ ξένους καὶ πολίτας καὶ συνήθεις καὶ οἰκείους καὶ ὅλως παντοδαποὺς ... (All translations are my own).

7 Cf. the wealthy but tactless foreigner at Luc. *Nigr.* 13.

This use is most clear in the *On Praising Oneself*, when Plutarch characterises the obnoxious discourse of soldiers and the *nouveaux riches* as 'pompous and haughty' (εὐπάρυφα καὶ σοβαρὰ).[8] Like them, this character's flaunting of his ignorance of sympotic etiquette is revealed in his social blunders (ὑποσολοικότερος) in his clothing and entourage: he either does not know or purposely ignores the customs on what to wear, who to bring, and generally how to act at a symposium.[9] His reaction to the full room is also suggestive: to judge from his unexpected departure and the symposiasts' subsequent conversation, this character is surprised not to find a space reserved for him. In Alcock's view, this character expects to find a style of reserved seating that Plutarch often connects to Rome, such as in the immediately following conversation when Plutarch's father recalls Aemilius Paulus' highly structured symposia.[10] Different conceptions of the sympotic space are hence connected to Greek and Roman cultural identities, and this elite, boorish outsider, perhaps a Roman himself, but certainly anticipating a Roman-style banquet, is an embodiment of the difficulties Greeks and Romans can have when they dine together.

The opening of the dialogue, then, connects the difficulty Greeks and Romans can have in creating a shared sympotic space with characteristically Greek and Roman styles of government. Four characters take up these questions in turn: Plutarch's father, in favour of a hierarchical symposium; Timon, who supports an egalitarian symposium; Plutarch himself, who tries to walk a fine line between these two extremes; and finally Lamprias, who ends the dialogue arguing for a mixed symposium.

I briefly summarise Plutarch's speech before coming to its close, which I will linger on for its importance for understanding the place of early Greek poetry in relationship to these larger questions. Beginning his speech, Plutarch promises to 'walk a middle path' between the two types of symposia favoured by his father and brother. He distinguishes between symposia for two categories of guests; it is appropriate for the young, citizens, and friends (νέους ... καὶ πολίτας καὶ συνήθεις) to drink in the egalitarian mode, but it is better to use hierarchical seating in the case of foreigners, rulers, and elders (ἐν δὲ ξένοις ἢ ἄρχουσιν ἢ πρεσβυτέροις). He argues for hierarchical seating by providing four examples from Homer and one from Pindar. He then tackles some of the problems with hierarchical seating: if multiple people vie for one seat, how to choose?

It is here that Plutarch advises that if one cannot make a decision between two guests, it is best to honor a family member or somebody with a close connection to the host. He ends his speech by providing an *exemplum* for this point of social etiquette with his claim that he takes this rule from Homer, citing *Iliad* 23. I provide the relevant quotation:

---

8 Plu. *De se ipsum laud.* 547E.

9 For ξένος as 'foreigner', cf. *QC* 7.6, 707B: 'But later, in the case of the entertainment of foreigners, especially those of consular rank ...' (ὕστερον μέντοι περὶ τὰς τῶν ξένων ὑποδοχάς, μάλιστα τῶν ἡγεμονικῶν, ...).

10 Alcock (2003) 602–603.

κατακλίνω γὰρ εἰς τὸν ἔνδοξον μάλιστα τόπον, ἂν μὲν ᾖ πατήρ, τοῦτον ἀράμενος, εἰ δὲ μή, πάππον ἢ πενθερὸν ἢ πατρὸς ἀδελφὸν ἤ τινα τῶν ὁμολογουμένην καὶ ἰδίαν ἐχόντων παρὰ τῷ δεχομένῳ τιμῆς ὑπεροχήν, ἐκ τῶν Ὁμήρου τὸ θεώρημα τοῦτο λαμβάνων καθηκόντων. καὶ γὰρ ἐκεῖ (Ψ 534 ff.) δήπουθεν ὁ Ἀχιλλεὺς τὸν Μενέλεων καὶ τὸν Ἀντίλοχον περὶ τῶν δευτερείων τῆς ἱπποδρομίας ὁρῶν διαφερομένους καὶ δεδοικὼς μὴ πορρωτέρω προέλθωσιν ὀργῆς καὶ φιλονεικίας ἑτέρῳ βούλεται τὸ ἔπαθλον ἀποδιδόναι, λόγῳ μὲν Εὔμηλον οἰκτίρων καὶ τιμῶν, ἔργῳ δὲ τῆς ἐκείνων διαφορᾶς τὴν αἰτίαν ἀφαιρῶν.

For I set my father into the most honored location, elevating him, if he is there; if not, my grandfather, or my father-in-law, or my uncle, or anyone who has an uncontroversial special claim to honor in the eyes of the host. I take this rule of etiquette from the poems of Homer. For there, I believe, Achilles sees Menelaus and Antilochus at odds over second place in the horse race. Since he is afraid that they will proceed further in anger and rivalry, he wishes to give the prize to another, nominally because he pities and honors Eumelus, but in reality to remove the cause of their disagreement. (*QC* 1.2, 617D–E)

Several points here should make us question whether Plutarch is being fully transparent. Not only does Plutarch's version of the Iliadic chariot race greatly differ from *Il.* 23, as I will show; it is also unclear how Plutarch even derives his rule from the situation as he describes it. How does Eumelus have a special claim to honor in Achilles' eyes? Rather, Plutarch's 'I believe' (δήπουθεν) invites the audience to further consideration of his account. I will first summarise the situation in *Il.* 23 and its interpretation in the scholia, before analyzing Plutarch's puzzling reading.

At the conclusion of the chariot race in *Il.* 23, Diomedes is the unquestioned victor who carries off the prize for first place (23.499–513). Antilochus comes in second thanks to tricks, not speed (23.515: κέρδεσιν, οὔ τι τάχει γε), but Menelaus is just a chariot's length behind (23.517); if the race had lasted any longer, Menelaus would have won (23.526–527). Finally, Eumelus comes to the finish line last, though he is the best racer in the eyes of the narrator (23.289: ἱπποσύνῃ ἐκέκαστο) and Achilles (23.536: ἀνὴρ ὤριστος),[11] and his early lead is only lost because of the intervention of Athena, who had supported Diomedes by breaking Eumelus' chariot yoke (23.388–397). In recognition of Eumelus' inherent excellence Achilles proposes to give him the prize for second place (23.537: δεύτερ'). All agree except (naturally) Antilochus, who vehemently objects (23.539–554). Achilles in reply agrees to give Antilochus the prize for second place, giving Eumelus instead an additional prize (23.555–565).

The most relevant scholion for the scene in Plutarch is Σ *Il.* 23.536–537 T, which I provide here with the appropriate Iliadic passage, where Achilles speaks immediately after the conclusion of the race:

11 Narrator: *Il.* 23.289; Achilles: *Il.* 23.536.

λοῖσθος ἀνὴρ ὤριστος ἐλαύνει μώνυχας ἵππους· ἀλλ' ἄγε δή οἱ δῶμεν ἀέθλιον ὡς ἐπιεικὲς δεύτερ'· δεύτερ'· ἀτὰρ τὰ πρῶτα φερέσθω Τυδέος υἱός. (*Il.* 23.536–538)	The best man is driving his single-foot horses in last. Come then, we must give some kind of prize, and well he deserves it; Second place; let first place go to the son of Tydeus.
<λοῖσθος ἀνὴρ ὤριστος ... δῶμεν ἀέθλιον>: διδάσκει τοὺς παρὰ τὴν ἀξίαν ἀτυχοῦντας ἐλεεῖν καὶ μὴ τῆς ἀρετῆς ἐᾶν ὑπερτερεῖν τὴν τύχην· b(BCE³E⁴)T ἄλλως τε καὶ Θεσσαλὸς ὁ Εὔμηλος. Σ *Il.* 23.536–537 T	[Homer] teaches us to take pity on those suffering unworthily and not to allow chance to surpass excellence, especially since Eumelus is a Thessalian.

The scholion sees the text as providing a moral exemplar for readers to learn from. The scholion's view is that one should follow inherent virtue, as ordered by 'excellence' (ἀρετή), not the actual result, which it calls 'chance' (τύχη). The scholion, as preserved in the generally fuller T, then emphasizes one feature of that inherent virtue: Eumelus' ἀρετή arises especially from his status as a Thessalian.

While this additional detail may hint at a special connection between Achilles and Eumelus thanks to their common homeland of Thessaly—perhaps the special connection Plutarch alludes to in his treatment of this scene—it is more likely that the scholiast at least has in mind Thessaly's reputation for horses, a theme the scholia often return to.[12] In particular, the scholion probably refers to Eumelus' Thessalian horses and his Thessalian expertise in horses, which the scholia repeatedly claim is part of the reason for his excellence.[13] For example, commenting on the narrator's assertion at the beginning of the Catalogue of Ships that Eumelus' horses are the best, the scholion attributes their excellence to their Thessalian origin:

Ἵπποι μὲν μέγ' ἄρισται ἔσαν Φηρητιάδαο, τὰς Εὔμηλος ἔλαυνε ποδώκεας ὄρνιθας ὣς ὄτριχας οἰέτεας σταφύλῃ ἐπὶ νῶτον ἐΐσας· (*Il.* 2.763–765)	Best by far among the horses were the mares of Eumelos, Pheres' son, that he drove, swift-moving like birds, alike in texture of coat, in age, both backs drawn level like a plumb-line. (trans. Lattimore)
... Εὐμήλου δὲ ἵππους προκρίνει ὡς Θεσσαλικάς. καὶ εἰ προὔχουσι δὲ αἱ Ἀχιλλέως ὡς ἐκ θεῶν, ἀλλ' οὖν γε καὶ αὐταὶ Θεσσαλὸν ἡνίοχον ἔχουσιν. σπέρματα δὲ προκαταβάλλει τῇ ἱπποδρομίᾳ καὶ τῇ ἀριστείᾳ τούτων τῶν ἵππων. Σ *Il.* 2.761–765 b	... And he prefers Eumelus' horses, on the grounds that they are Thessalian. And even if Achilles' horses are superior because they come from the gods, they [Eumelus' horses] still also have a Thessalian charioteer. And he plants seeds in advance of the chariot race and the *aristeia* of these horses.

12 Thessalians are horse-breeders: Σ *Il.* 9.123a bT. For the scholia 'horse-rearing Argos' (Ἄργος ἐς ἱππόβοτον) generally refers to the Thessalian Argos: Σ *Il.* 3.75a T, Σ *Il.* 3.258 bT. The Myrmidons are 'swift-horsed' (ταχύπωλοι) because they are Thessalian: Σ *Il.* 23.6 T.

13 Cf. Σ *Il.* 2.763 D, Σ *Il.* 23.288 T.

This scholion highlights Eumelus' Thessalian origin and his Thessalian horses as being key to his excellence in horse-racing, as well as seeing the reference to Eumelus in the Catalogue as foreshadowing the chariot race in *Il.* 23.

Let us return our attention to Plutarch's exploitation of this scene. As Teodorsson has previously noted, Plutarch's reading of *Il.* 23 does not match the Homeric text.[14] In the *Iliad*, Achilles' decision to give the prize to Eumelus *starts*, not ends the quarrel between Antilochus and Menelaus. The conflict is in fact resolved not by Achilles' action, but by Antilochus' willingness to yield to Menelaus. Plutarch's reading reverses the temporal order of these events—intentionally, as I will argue.

On a superficial reading that does not consider the consequences of the intertext and sees Plutarch's citation as simply a misreading of *Il.* 23, Plutarch's interpretation supports his advice to choose a family member or other intimate when it is difficult to determine who is socially superior. In Teodorsson's words, 'Plutarch distorts the story to make it suit his argument'.[15] Assimilating athletic performance in Homer to social standing in Roman Greece, Plutarch reads Achilles as a model for sympotic hosts.

But we may wish to take this misreading of *Il.* 23 seriously, as Plutarch appears to signal with his 'I believe' (δήπουθεν), which is occasionally used elsewhere in the *Table Talk* to mark a significant engagement with a literary text and Homer in particular. For example, *Talk* 5.4 concerns itself with the problem of why at *Il.* 9.204 Achilles orders Patroclus to mix stronger (ζωρότερον) wine. Responding to 'childish' (μειρακιώδη) readings of *Il.* 9 which deny ζωρότερον its conventional meaning, Plutarch fills in the story by imagining motivations not explicit in the Homeric text:

> ἔπειτα Χείρωνος ὢν μαθητὴς καὶ τῆς περὶ τὸ σῶμα διαίτης οὐκ ἄπειρος ἐλογίζετο δήπουθεν, ὅτι τοῖς ἀργοῦσι καὶ σχολάζουσι παρὰ τὸ εἰωθὸς σώμασιν ἀνειμένη καὶ μαλακωτέρα κρᾶσις ἁρμόζει·
>
> Second, since he was a student of Cheiron and not ignorant of proper diet, he was reasoning, I imagine, that for bodies that were unaccustomedly idle and at leisure a relaxed, gentle mixture was appropriate. (*QC* 5.4, 677F)

Plutarch's daring and inventive solution stands in contrast to the more pedestrian, 'childish' solutions of the earlier interlocutors. While Plutarch purports to be merely reading what is implicit in the *Iliad*, he acknowledges the tendentiousness of his reading with the particle δήπουθεν. This particle, also found in *Table Talk* 1.2, can thus be taken as a marker for readings that engage seriously and creatively with the text.

Furthermore, Plutarch's reading, while not matching the scholion's interpretation, nonetheless suggests a knowledge of the scholion or perhaps its ancestral commentary. In Plutarch's reading, the gift to Eumelus is a way out of determining who is superior: instead of choosing the person who is genuinely more worthy of the gift, Achilles gives the prize to someone who has a special connection to him. The scholion, however, adopts the opposite reading: Achilles gives the prize to Eumelus in

---

14 Teodorsson (1989) 78.
15 Teodorsson (1989) 78.

recognition of his inherent excellence. Despite these differences, both Plutarch and the scholion see this episode of Achilles and Eumelus as a teaching moment: as if following the scholion's advice to learn from Achilles, Plutarch does create a general rule from this episode, but this rule is not what the scholion advises.

In light of these facts, we might be prompted to attempt an ironic reading of Plutarch's account of *Il.* 23. Achilles' choice of Eumelus is in fact the correct choice (even if that is not what happens in the *Iliad*), but not for the reasons explicitly given. Rather, it is a choice to reward ἀρετή over τύχη, to respect inherent excellence over chance results. In the terms of the dialogue, which thus constructs an analogy between racing on the race-course and racing at the symposium, it is a choice to give priority in seating to those who deserve it over those who happen to be quicker. This reading lines up well with Plutarch's slightly earlier criticism of Timon's support of free seating, when he compares that kind of seating to theft:

> ἀφαιρεῖται γὰρ ὁ κοινὸν ποιῶν τὸ ἴδιον (ἴδιον δὲ τὸ κατ' ἀξίαν ἑκάστου) καὶ ποιεῖ δρόμου καὶ σπουδῆς τὸ πρωτεῖον ἀρετῇ καὶ εὐγενείᾳ καὶ ἀρχῇ καὶ τοῖς τοιούτοις ὀφειλόμενον.
>
> For in making shared what is each man's own (i.e. what belongs to him according to his merit) Timon deprives him of it, and Timon gives first place to the footrace and speed, though it is owed to virtue, good birth, and sovereignty. (*QC* 1.2, 617C)

Perhaps laying a foundation for the reading of *Il.* 23 that ends his speech, Plutarch here likewise prefers inherent ἀρετή to the unreliable results of the footrace. The similarity strengthens the possibility of the ironic reading of the close of Plutarch's speech.

Who, however, has this ἀρετή? The nature of this coded message suggests one possible answer: the *pepaideumenoi*, steeped in, among other things, Homer and Homeric interpretation. In some ways this is no surprise; for Plutarch, the moral excellence that παιδεία can provide is necessary for successful political leadership, as can be seen in the emphasis often given to education in the *Lives*.[16] In this case, Plutarch subtly sends two messages to two different groups, one to the *pepaideumenoi*, one to boorish figures like the elite outsider whose entrance begins the talk. In the case of reserved seating, Plutarch explicitly suggests assigning seats according to 'what belongs to him according to his merit', making use of criteria that the boor would recognise—'virtue, good birth, and sovereignty'. At the close of Plutarch's speech, when he recommends choosing a person with a close connection to the host over others, this message is intended for the boor and figures like him at this eclectic symposium, and allows them to save face: they might be passed over not out of disrespect but because the person preferred has a special tie to the host. To the *pepaideumenoi* Plutarch has another message, however: ἀρετή can derive from and can include an education that the boor fails to recognize, including a knowledge of sym-

16 Stadter (2014a) 27 n. 42.

potic etiquette and Homer.[17] Therefore, just as in Plutarch's reading Achilles rewards Eumelus for his inherent ἀρετή, but with the pretence that he is rewarding him for some special connection to the host, Plutarch recommends elevating those who have acquired ἀρετή through their παιδεία, even if the boor would not recognise it.

In light of the broader stakes of the dialogue, however,—hierarchical rule vs. democracy, Greece vs. Rome—it is tempting to push this reading just a little further. The elite boor whose entrance began the narrative of the symposium, according to the opposition that has structured our reading so far, must possess not ἀρετή but τύχη; his status arises from chance or providence, not any inherent virtue. Successful political leadership, however, requires both, just as at *On the Fortune of the Romans* 316E it is a combination of ἀρετή and τύχη that leads to the rise of the Roman Empire.[18] If ἀρετή can arise from Greek παιδεία, and some foreigners fail to possess παιδεία, they are unlikely to be effective leaders. Swain has shown that in the *Lives* Romans are more likely than Greeks to be scrutinized as to their acquisition of παιδεία, and that in the *Table Talk* 'Plutarch expects his Roman dining companions to use Hellenic culture in the conversations while paradoxically presenting them as being not fully and absolutely at ease with it'.[19] For Plutarch rulers need not just τύχη but also an ἀρετή that can come from παιδεία, but Romans are less likely to possess that education than Greeks. We see this view again in this passage in the *Table Talk*. Here, the Greeks and Romans who can recall their Homer—presumably much of the readership of the *Table Talk*—can flatteringly see that according to Plutarch's interpretation they would be better leaders than those without παιδεία, while figures like the outlandish, uneducated foreigner who began this dialogue may possess a high social status and can attempt to demand privileges, but are unlikely to be as effective leaders.

Such a reading in the programmatic 1.2 suggests the importance of the use of early Greek poetry and its interpretation in the *Table Talk*. Characters' quotation and interpretation of Homer at table is not always simply an innocent attempt to add colour to one's speech: it can carry further ideological significance in the Roman Greece of the 2nd century CE. When Plutarch proves himself a master symposiast through his ability to offer a proper Homeric interpretation that ambiguously caters to multiple audiences, he demonstrates how the social hierarchies of his world are mapped out in the sympotic space both physically, in the seating arrangement of the guests, and verbally, as correct knowledge of poetry legitimises one's elite status and right to be present at the symposium.

---

17 Cf. *QC* 1.1, 613D–E for a contrast by Plutarch between the *pepaideumenoi* and *idiōtai*, where the former are characterized by their ability to combine Dionysus (sympotic etiquette) with the Muses (intellectual activity, including engagement with the Muses' domain of poetry).

18 Typically of Plutarch's thought; see Swain (1989c), esp. 509–510, for characters relying on both ἀρετή and τύχη for success in both the *Lives* and the *Moralia*.

19 Swain (1990b) 131.

Johann Goeken

Plutarque et la tradition rhétorique du banquet

Abstract: L'article étudie comment la présence de la rhétorique (prise au sens propre d'art de la persuasion) dans les œuvres conviviales de Plutarque s'articule avec la référence à Platon. Conformément aux règles de la *mimèsis*, le *Banquet* de Platon est présenté comme le modèle principal qui permet de comprendre la démarche littéraire de Plutarque dans le *Banquet des sept sages* et les *Propos de table*. Pourtant, du point de vue strictement rhétorique, ce modèle est problématique: les convives d'Agathon sont tous influencés par la sophistique que dénoncent Socrate et Platon, et cet emploi de la rhétorique empêche en réalité l'éclosion d'un vrai dialogue. En revanche, dans les banquets de Plutarque, la parole est mieux répartie et, même si les excès de la rhétorique sont aussi condamnés, les convives utilisent des techniques répertoriées par les rhétoriciens. Or, si Platon ne donnait pas l'image d'un banquet modèle, Plutarque entend livrer des exemples de conversations où la rhétorique se révèle un réflexe culturel et un outil épistémologique. Que ce soit dans le cadre d'un banquet précis ou dans l'espace global de l'Empire, la rhétorique joue, pour Plutarque, un rôle important dans la recherche philosophique et dans le dialogue entre Grecs et Romains.

La place du *Banquet des sept sages* et des *Propos de table* dans la littérature conviviale a fait l'objet de plusieurs contributions importantes consacrées notamment aux aspects philosophiques et littéraires de la question.[1] Mais la présence de la rhétorique dans les œuvres conviviales de Plutarque est un thème qui reste relativement peu exploré et dont le traitement se justifie par l'importance que les Anciens accordent, dans le cadre du *symposion*, à la parole.[2] Par 'rhétorique', il faut entendre la théorie et la pratique du discours persuasif. À l'époque de Plutarque, il s'agit d'une discipline reconnue, avec des représentants et des techniques bien identifiés.

Les *Propos de table* et le *Banquet des sept sages* se réclament de plusieurs modèles. Mais, pour ce qui concerne le thème de la rhétorique, la référence est surtout platonicienne, puisque c'est Platon qui, dans son *Banquet*, problématise explicitement la présence de la rhétorique parmi les convives en mettant en scène un concours d'éloge. La démarche de Plutarque se comprend dans le contexte de la Seconde Sophistique, dont les productions sont gouvernées par les lois de la *mimèsis*. Toutefois Platon s'avère un modèle problématique et Plutarque prend parfois

1 Martin (1931); Mossman (1997); Montes Cala et al. (1999); Vetta (2000); Teodorsson (2009); Titchener (2009); König (2012) 30–32, 35–41, 60–89; Klotz et Oikonomopoulou (2011).
2 Romeri (2002); González Julia (2009).

https://doi.org/10.1515/9783110539479-026

ses distances par rapport au modèle platonicien, en ne critiquant pas systématiquement la rhétorique. Il convient alors de se demander dans quelle mesure Plutarque témoigne ainsi des évolutions de l'histoire et du contexte sophistique dans lequel il vit.

Dans cette perspective rhétorique, il est utile d'étudier comment Platon est pris comme modèle dans des œuvres qui entendent retranscrire des λόγοι dignes de mémoire, qui font de la rhétorique un thème de réflexion et qui attestent elles aussi l'usage de la rhétorique dans la conversation. Une telle étude, qui privilégiera ici les *Propos de table*, présente l'intérêt de souligner une tension dynamique, propre à la Seconde Sophistique, entre classicisme et réalité romaine, c'est-à-dire entre passé et présent. Il s'agit de montrer que, pour Plutarque et ses amis, le banquet est un espace ouvert qui permet aux Grecs et aux Romains de communiquer, grâce à l'emploi réfléchi de la rhétorique, laquelle est devenue une *lingua franca* dans les cercles lettrés de l'Empire.

Plutarque et la tradition des discours sympotiques

En composant le *Banquet des sept sages* et les *Propos de table*, Plutarque s'inscrit dans une tradition littéraire particulière: celle des récits de discours conviviaux qui sont dignes de mémoire. Dans les *Propos*, il se réfère ainsi à

> Πλάτωνα καὶ Ξενοφῶντα καὶ Ἀριστοτέλη καὶ Σπεύσιππον Ἐπίκουρόν τε καὶ Πρύτανιν καὶ Ἱερώνυμον καὶ Δίωνα τὸν ἐξ Ἀκαδημίας, ὡς ἄξιόν τινος σπουδῆς πεποιημένους ἔργον ἀναγράψασθαι λόγους παρὰ πότον γενομένους.
>
> Platon, Xénophon, Aristote, Speusippe, Épicure, Prytanis, Hiéronymos et Dion l'Académicien, qui ont regardé comme une œuvre digne de quelque intérêt de consigner par écrit des propos tenus dans des banquets. (*QC* 1, 612D–E, trad. Fuhrmann)

Platon, cité le premier, constitue l'archétype de la tradition dans laquelle le médio-platonicien Plutarque veut s'inscrire, étant entendu que seuls les *Banquets* de Platon et Xénophon nous sont parvenus. Plutarque s'intéresse aux seuls λόγοι et donne peu de détails sur le cadre des réunions ou sur les actions des convives. D'autre part, dans les *Propos*, Plutarque ne retranscrit pas des paroles échangées lors d'une seule occasion, mais lors de plusieurs banquets organisés en divers endroits, surtout en Grèce mais aussi à Rome.[3] La multiplication des banquets va de pair avec un élargissement de l'espace: on passe du cadre de la cité athénienne à l'étendue de l'Empire gréco-romain, où résonnent partout des prises de parole conviviales.

Si Plutarque se réclame de Platon ou de Xénophon, il faut bien voir que les conversations chez Callias et Agathon posent quelques problèmes. Dans un cas,

3 *QC* 1, 612E. Voir Nikolaidis dans ce volume. Pour une approche spatiale des *Propos de table*, cf. Andurand (2015).

Socrate doit lutter contre l'impact des spectacles qui tend à réduire les invités de Callias au silence. Dans l'autre, les convives prononcent surtout des éloges rhétoriques. En tant que lieu de parole, le banquet s'avère donc, chez Plutarque, une situation rhétorique, comme c'était déjà le cas chez Xénophon et surtout chez Platon, dont les personnages s'apparentent souvent à des sophistes. Les modèles choisis ont donc leurs limites et, dans les banquets de Plutarque, la parole est clairement plus vivante et mieux répartie entre les convives.

La rhétorique comme sujet de réflexion

Dans les *Propos de table* et même dans le *Banquet des sept sages*, la rhétorique devient un sujet de réflexion, comme elle l'était déjà dans les banquets socratiques et en particulier chez Platon. On retrouve tout d'abord chez Plutarque les critiques classiques lancées contre les sophistes. Dans cette optique, Isocrate a bien fait de s'abstenir de parler au banquet, car les périodes de ses discours ne sont pas adaptées à une telle occasion et, pour Craton, bannir le λόγος ῥητορικός, ce n'est pas la même chose que de bannir le λόγος φιλόσοφος: la philosophie ne doit être tenue à l'écart d'aucun divertissement, car elle apporte le μέτρον et le καιρός, c'est-à-dire la σωφροσύνη et la δικαιοσύνη.[4] On remarque que l'exemple d'Isocrate donne lieu à une attaque de type stylistique. Or la philosophie n'est pas liée aux mots: elle est un mode de vie qui est associé aux vertus cardinales de la tempérance et de la justice, ce qui signifie que la rhétorique est assimilée aux valeurs contraires: démesure, inopportunité, intempérance, injustice. En clair, il ne s'agit pas d'exclure les rhéteurs du banquet, mais d'éviter les discours trop travaillés et les prestiges de la sophistique.

Cette critique se retrouve dans les propos de Théon sur le symposiarque dessiné par Craton:

> <εὔκρα>τον δέ μοι δοκεῖ, τοι<οῦτο>ς ὤν, τὸ συμπόσιον <διαφυλ>άξειν ἡμῖν καὶ μὴ περιόψεσθαι νῦν μὲν ἐκκλησίαν δημοκρατικήν, νῦν δὲ σχολὴν σοφιστοῦ γινομένην, αὖθις δὲ κυβευτήριον, εἶτά που σκηνὴν καὶ θυμέλην. Ἦ γὰρ οὐχ ὁρᾶτε τοὺς μὲν δημαγωγοῦντας καὶ δικαζομένους παρὰ δεῖπνον, τοὺς δὲ μελετῶντας καὶ ἀναγινώσκοντας αὑτῶν τινα συγγράμματα, τοὺς δὲ μίμοις καὶ ὀρχησταῖς ἀγωνοθετοῦντας;
>
> Tel qu'il est, je crois qu'il maintiendra une atmosphère agréable dans notre banquet et qu'il ne laissera point celui-ci se transformer tantôt en assemblée populaire, tantôt en école de sophiste, ou même en tripot, voire en scène et en orchestre. Ne voyez-vous pas, en effet, certains prononcer à table des harangues politiques et des plaidoiries, d'autres s'y exercer à la déclamation et donner lecture de leurs ouvrages, d'autres encore y instituer des concours de bouffons et de danseurs? (*QC* 1.4, 621B–C)

Il s'agit de refuser la rhétorique spécialisée et de ne pas transposer au banquet ce qui se dit à l'ἐκκλησία et à l'école, voire en mauvais lieu ou au théâtre. Sont exclus les

4 *QC* 1.1, 613A–B. Sur ces notions, voir Vamvouri Ruffy (2012) 78–122.

débats houleux, les discours en bonne et due forme, les controverses et les suasoires, qui risqueraient de créer des conflits. Il ne faut tolérer, en matière de discours, de spectacles et de plaisanteries, que ce qui peut accroître ou faire naître l'amitié.

Si, en théorie, Plutarque reprend l'essentiel de la critique platonicienne de la rhétorique, force est de constater que, dans la pratique, son jugement est plus mesuré. Chez Agathon, tous les personnages sont influencés par les sophistes et tous leurs discours sont critiqués par Socrate. Pour Plutarque, le terme 'sophiste' sert à moquer plutôt des philosophes et l'auteur ou le narrateur a tendance à réintégrer la rhétorique et les rhéteurs sans (trop) se moquer d'eux, tandis que la rhétorique est reconnue comme un des champs du savoir.[5]

On touche là à l'utilité de la rhétorique dans la conversation—une question qui est traitée plus en détail au livre 9 des *Propos de table*. Le rhéteur Hérode parle en effet des Muses qui patronnent les orateurs: Calliope (pour l'éloquence politique), Clio (pour le genre encomiastique), Polymnie (pour l'histoire). Mais Hérode revendique aussi en partie le patronage d'Euterpe:

> ὥς φησι Χρύσιππος, αὕτη τὸ περὶ τὰς ὁμιλίας ἐπιτερπὲς εἴληχε καὶ κεχαρισμένον· ὁμιλητικὸς γὰρ οὐδὲν ἧττον ἢ δικανικὸς ὁ ῥήτωρ καὶ συμβουλευτικός· αἱ γὰρ <διαλ>έξεις ἔχουσι καὶ συμβουλίας καὶ συνηγορίας καὶ ἀπολογίας· πλείστῳ δὲ τῷ ἐπαινεῖν χρώμεθα καὶ τῷ ψέγειν ἐν τούτοις, οὐ φαύλων οὐδὲ μικρῶν τυγχάνοντες, ἂν τεχνικῶς τοῦτο πράττωμεν, ἂν δ' ἀπείρως καὶ ἀτέχνως, ἀστοχοῦντες.
>
> Comme le dit Chrysippe, elle a reçu en partage le charme et la grâce de la conversation, car le rhéteur n'est pas moins l'homme de la conversation que des plaidoiries et des harangues; et, de fait, les entretiens comportent tout à la fois conseils, plaidoyers et apologies; l'éloge et le blâme aussi nous y sont du plus grand usage, et nous en tirons des résultats qui ne sont ni minces ni négligeables pour peu que nous procédions avec art, tandis que l'inexpérience et l'ignorance de l'art nous font manquer notre but. (*QC* 9.14, 743D–E, trad. Frazier et Sirinelli)

Dans ces conditions, la rhétorique annexe le domaine des ὁμιλίαι et Hérode défend l'idée que la conversation de banquet obéit aussi, du moins en partie, aux lois de la rhétorique dont elle reprend les trois genres canoniques. Puis Hérode continue:

> τὸ γὰρ 'ὢ πόποι, ὡς ὅδε πᾶσι φίλος καὶ τίμιός ἐστιν ἀνθρώποις' * * δὲ μᾶλλον, ὡς τὸ περὶ τὰς ὁμιλίας εὐάρμοστον ἔχουσι, πειθὼ καὶ χάριν, οἶμαι προσήκειν.
>
> L'exclamation: *«Misère! en voilà un que tous aiment et respectent»*... c'est plus encore, je pense, <aux rhéteurs> qu'elle s'applique, parce qu'ils possèdent ce qui s'accorde à la conversation, à savoir la persuasion et la grâce.[6] (*QC* 9.14, 743E)

La citation, qui provient de l'*Odyssée* (10.38–39), reprend des propos de l'équipage d'Ulysse, l'orateur par excellence. Elle permet à Hérode d'associer deux notions-clés: la χάρις (qui est de règle au banquet) et la πειθώ (qui sert à définir la rhétorique). Sur

5 Sirinelli (2000) 73 n. 1; Frazier (2000) 189–191; Goeken (2013); Oikonomopoulou (2015).
6 Le texte est abîmé: Teodorsson (1996) 349–350.

ces mots, Ammonios taquine un peu Hérode, mais il ne contredit pas les propos du rhéteur et il admet que tous ont besoin du discours.[7] Peu après, le frère du narrateur intervient pour dire que les Muses patronnent la philosophie, la rhétorique et la mathématique. Ensuite d'autres théories sur le rôle des Muses sont proposées, où l'on retrouve une mise en vedette de la notion (rhétorique) de 'persuasion'.[8] Tous ces propos ne sont pas condamnés en bloc et c'est ce qui était annoncé dans le prologue du livre 9 où Plutarque, justifiant le nombre de 'questions' plus élevé que dans les livres précédents, déclarait: 'c'est qu'il fallait rendre aux Muses tout ce qui appartient aux Muses sans rien y retrancher, non plus qu'à des choses sacrées, tant ce que nous leur devons dépasse en nombre et en beauté ce faible hommage'.[9] Ce sont là des propos qui incluent les considérations sur la rhétorique de la conversation au banquet et qui tranchent avec le rejet ironique et catégorique de la rhétorique qu'on observe chez Platon.

La rhétorique des banqueteurs

Platon donne une image spéciale du banquet et des propos qui y sont tenus. Les invités d'Agathon se mettent d'accord pour prononcer chacun à tour de rôle un éloge rhétorique qu'ils écoutent en silence et qu'ils ne commentent (presque) pas. En vertu de ce protocole rigide, il est évident que Platon ne fournit pas un modèle de discours à tenir. Or il en va tout différemment des banquets de Plutarque, comme en témoigne par exemple la préface du livre 7 des *Propos de table*:

> ἡμεῖς δ' ἑαυτοὺς χρῆσθαι λόγοις συνεθίζωμεν ὧν πᾶσίν ἐστιν καὶ πρὸς πάντας ἐξαγωγή, διὰ τὰς ὑποθέσεις μηδὲν ἀκόλαστον μηδὲ βλάσφημον μηδὲ κακόηθες ἐχούσας μηδ' ἀνελεύθερον. Ἔξεστι δὲ κρίνειν τοῖς παραδείγμασιν ὧν τὴν ἑβδόμην τουτὶ περιέχει τὸ βιβλίον.
>
> Nous, accoutumons-nous à ne tenir que des propos que tous aient la possibilité de divulguer à l'adresse de tous, en raison de leurs sujets qui ne comportent rien de licencieux, de diffamatoire, de vicieux, ni de bas. On peut en juger d'après les exemples dont ce livre contient la septième douzaine. (*QC* 7, 697E)

Plutarque indique clairement que les *Propos de table* constituent des 'modèles' de 'sujets' et de 'discours' à tenir. Et une lecture attentive révèle que ces παραδείγματα impliquent un recours à la rhétorique, comme le suggère le rhéteur Hérode et comme le démontre l'attitude du narrateur.

---

7 *QC* 9.14, 743E–744C.
8 *QC* 9.14, 745C–D.
9 *QC* 9, 736C: ἔδει γὰρ πάντα ταῖς Μούσαις ἀποδοῦναι τὰ τῶν Μουσῶν καὶ μηδὲν ἀφελεῖν ὥσπερ ἀφ' ἱερῶν, πλείονα καὶ καλλίονα τούτων ὀφείλοντας αὐταῖς.

En effet, la rhétorique gouverne souvent les modalités des échanges. C'est pourquoi l'εὑρησιλογία des convives est louée.[10] Au banquet, il faut toujours improviser en quelque sorte et trouver des idées et des arguments originaux pour construire son discours, ce qui n'est pas sans rappeler l'εὕρεσις (la première tâche de l'orateur).

Semblablement, les convives de Plutarque recourent à des formes répertoriées par les rhétoriciens, qu'il s'agisse de *progymnasmata*[11] ou de discours qui ressemblent à des antilogies, sans parler des nombreuses prises de parole qui sont assimilées à des discours d'accusation ou de défense, à des discours délibératifs ou encore à des éloges.[12]

Chez Platon, ces schémas rhétoriques étaient condamnés. Chez Plutarque, ils font partie du bagage culturel des convives. Dans ces conditions, les convives se comportent en orateurs: ils ont des réflexes langagiers qui leur sont dictés par l'éducation rhétorique qu'ils ont reçue. Dans le *Banquet des sept sages*, quand Ésope prend la défense de Périandre qui vient d'être attaqué par les autres convives, il se réfère aux trois genres oratoires en disant qu'il fallait 'n'aborder ce sujet qu'entre nous, et ne pas, en nous prétendant leurs conseillers et leurs amis, nous ériger en accusateurs de ceux qui exercent le pouvoir'.[13] Quant aux amis de Plutarque, ils sont d'accord pour 'poser des questions dont les réponses ne provoqueront pas le blâme, mais l'éloge, ni la haine ou l'envie, mais la bienveillance et la gratitude de la part des auditeurs',[14] ce qui implique une moralisation des notions propres au domaine rhétorique. De même, est recommandé un usage élogieux de la plaisanterie et, si l'éloge indirect est prisé pour sa grâce plus piquante, il est fréquent de voir les convives se louer les uns les autres pour ce qu'ils ont dit.[15]

De manière plus fondamentale, chaque convive doit prouver ce qu'il avance—et le nombre d'occurrences du champ lexical de la preuve et du témoignage confirme qu'il s'agit là d'une préoccupation majeure des banqueteurs.[16] Ainsi les interventions doivent respecter deux critères essentiels: le 'naturel' (εἰκός) et le 'vraisemblable' (πιθανόν), deux notions héritées elles aussi de la rhétorique.[17]

Les convives ont à cœur, pour chaque question, de discuter une réponse commune ou de réfléchir à la validité d'une opinion première, et cette recherche de type philosophique se sert de certains outils de la rhétorique. En fait, l'enjeu de la con-

---

10 *QC* 1.8, 625C; 3.7, 656A; 5.7, 682B. Cf. Oikonomopoulou (2011) 120 – 123.

11 Schenkeveld (1996); Vicente Sánchez (2009); González Equihua (2009); Fernández Delgado et Pordomingo Pardo dans ce volume.

12 Schenkeveld (1997); Goeken (2013).

13 *Sept. sap. conv.* 152B–C, trad. Defradas, Hani et Klaerr modifiée: τοῦτο καθ' ἑαυτοὺς περαίνειν καὶ μὴ συμβούλους φάσκοντας εἶναι καὶ φίλους κατηγόρους τῶν ἀρχόντων.

14 *QC* 2.1, 631B, trad. Furhmann modifiée: ἐρωτήσεις προφέρονται ὧν ταῖς ἀποκρίσεσιν οὐ ψόγος, ἀλλ' ἔπαινος, οὐδὲ μῖσος ἢ νέμεσις, ἀλλ' εὔνοια καὶ χάρις ἕπεται παρὰ τῶν ἀκουσάντων.

15 *QC* 2.1, 633A; 6.4, 690D.

16 Par exemple *QC* 1.6, 623E (ἀπεδείκνυεν); 4.6, 671C (τεκμηρίων); 6.9, 696C (τεκμήριον).

17 Goeken (2013) 185.

versation sympotique reste la persuasion, même s'il s'agit peut-être parfois d'une persuasion provisoire. Au banquet, il faut se montrer persuasif et la persuasion bienveillante (πειθὼ φιλάνθρωπος), associée à la χάρις, joue un rôle dans le processus de socialisation, même dans le cas d'amitiés utiles et intéressées—c'est l'idée que Plutarque défend dans la préface du livre 4 des *Propos de table.*[18]

L'importance de la persuasion est mise en vedette tout au long des conversations et l'idée apparaît même quand il s'agit de discuter de la place de la philosophie au banquet:

> Ἂν δὲ πλῆθος ᾖ τοιούτων ἀνθρώπων οἳ παντὸς μὲν ὀρνέου, παντὸς δὲ νεύρου καὶ ξύλου μᾶλλον ἢ φιλοσόφου φωνὴν ὑπομένουσιν, τὸ τοῦ Πεισιστράτου χρήσιμον· ἐκεῖνος γὰρ ἐν διαφορᾷ τινι πρὸς τοὺς υἱοὺς γενόμενος, ὡς ᾔσθετο τοὺς ἐχθροὺς χαίροντας, ἐκκλησίαν συναγαγὼν ἔφη βούλεσθαι μὲν αὐτὸς πεῖσαι τοὺς παῖδας, ἐπεὶ δὲ δυσκόλως ἔχουσιν, αὐτὸς ἐκείνοις πείσεσθαι καὶ ἀκολουθήσειν. Οὕτω δὴ καὶ φιλόσοφος ἀνὴρ ἐν συμπόταις μὴ δεχομένοις τοὺς λόγους αὐτοῦ μεταθέμενος ἕψεται καὶ ἀγαπήσει τὴν ἐκείνων διατριβήν, ἐφ' ὅσον μὴ ἐκβαίνει τὸ εὔσχημον.
>
> S'il y a un grand nombre de ces hommes qui préfèrent la voix de n'importe quel oiseau ou celle de n'importe quel morceau de bois muni d'une corde à celle du philosophe, l'histoire de Pisistrate est utile: celui-ci, s'étant brouillé avec ses fils, s'aperçut que ses ennemis s'en réjouissaient; il convoqua alors une assemblée et annonça que, malgré son désir de persuader (πεῖσαι) ses enfants, ce serait lui qui, vu leur mauvaise humeur, se laisserait persuader de suivre leur avis (πείσεσθαι καὶ ἀκολουθήσειν). Ainsi un homme philosophe, parmi des buveurs que ses discours n'intéressent pas, changera de ton pour suivre les autres et prendra plaisir à leur manière de passer le temps, pour autant que celle-ci reste dans la décence. (*QC* 1.1, 613E–F, trad. Fuhrmann modifiée)

Évoquant ensuite l'action dissimulée de la philosophie qui s'adapte aux circonstances, le narrateur ajoute:

> Ὡς γὰρ αἱ παρ' Εὐριπίδῃ μαινάδες ἄνοπλοι καὶ ἀσίδηροι τοῖς θυρσαρίοις παίουσαι τοὺς ἐπιτιθεμένους τραυματίζουσιν, οὕτω τῶν ἀληθινῶν φιλοσόφων καὶ τὰ σκώμματα καὶ οἱ γέλωτες τοὺς μὴ παντελῶς ἀτρώτους κινοῦσιν ἁμωσγέπως καὶ συνεπιστρέφουσιν.
>
> De même que, chez Euripide, les Ménades sans arme et sans épée, blessent leurs adversaires des seuls coups de leur thyrse, de même les railleries et les rires des vrais philosophes ne laissent pas d'émouvoir (κινοῦσιν) et d'ébranler (συνεπιστρέφουσιν) ceux qui ne sont pas totalement invulnérables. (*QC* 1.1, 614A)

L'histoire de Pisistrate qui renonce à persuader pour se laisser persuader indique les implications rhétoriques de la conversation et, de manière plus générale, du comportement des convives. En outre, l'image dionysiaque des Ménades armées de thyrses suggère que la philosophie, même déguisée, reste une arme (comme la rhétorique) qui vise à émouvoir—le *movere* étant bien une des fonctions de la rhétorique, avec le *docere* et le *delectare.* Mais l'importance de la rhétorique apparaît aussi au début du passage, quand il est question du chant des oiseaux et de la

18 *QC* 4, 660A.

cithare. En effet, cette phrase rappelle l'opposition établie par Socrate, dans le *Protagoras*, entre les banquets vulgaires (où l'on fait appel à la poésie et à la musique) et les banquets civilisés (où les convives discutent sans faire appel à une voix étrangère).[19] Elle rappelle aussi qu'au début du *Banquet* de Platon, l'αὐλητρίς est expulsée de la salle des convives.[20]

La situation où les philosophes ne sont pas majoritaires est plutôt fréquente. Or Plutarque et ses amis acceptent les pratiques que Platon et Socrate rejetaient comme dignes de Protagoras et des sophistes. Même si la réflexion personnelle et les arguments tirés de sa propre expérience sont valorisés, il est possible de recourir aux voix étrangères. De fait, les amis de Plutarque, quelle que soit leur spécialité, citent beaucoup d'auteurs anciens et le recours aux citations est même justifié au début du livre 9 des *Propos*.[21] Tout au plus faut-il éviter l'excès de citations agressives.[22]

De même, le recours à l'αὐλητρίς est admis par les amis de Plutarque, qui s'opposent ainsi à un stoïcien rigide. Comme pour les citations, la question fait l'objet d'une discussion particulière[23] et, de manière plus générale, le recours à la musique se défend en ce qu'il permet de calmer les convives en cas de querelle ou d'ἀγὼν σοφιστικός.[24] Dans le *Protagoras*, le banquet jugé positif était celui des πεπαιδευμένοι συμπόται, capables de trouver en eux-mêmes la matière de leurs entretiens. Chez Plutarque, la notion de παιδεία a évolué: les πεπαιδευμένοι de l'Empire réfléchissent par eux-mêmes, mais ils vouent aussi un culte au passé; c'est pourquoi ils tiennent aux pratiques ancestrales et à la voix des poètes ou des prosateurs. Les amis de Plutarque opèrent donc une sorte de synthèse entre les deux types de banquet que Socrate opposait dans le *Protagoras*.

Mais il y a plus. Plutarque parle aussi de sa propre rhétorique en se réclamant des Anciens:

> Οἱ δὲ χαρίεντες, κἂν ἀπ' εὐθείας φιλοσοφῶσιν, τηνικαῦτα διὰ τοῦ πιθανοῦ μᾶλλον ἢ βιαστικοῦ τῶν ἀποδείξεων ἄγουσι τὸν λόγον. ... Πλάτων ἐν τῷ Συμποσίῳ περὶ τέλους διαλεγόμενος καὶ τοῦ πρώτου ἀγαθοῦ καὶ ὅλως θεολογῶν οὐκ ἐντείνει τὴν ἀπόδειξιν οὐδ' ὑποκονίεται, τὴν λαβὴν ὥσπερ εἴωθεν εὔτονον ποιῶν καὶ ἄφυκτον, ἀλλ' ὑγροτέροις λήμμασι καὶ παραδείγμασι καὶ μυθολογίαις προσάγεται τοὺς ἄνδρας.

> Les hommes d'esprit (χαρίεντες), même lorsqu'ils font directement de la philosophie, conduisent leurs propos par les voies de la persuasion (διὰ τοῦ πιθανοῦ), plutôt que par celles des démonstrations péremptoires. Platon lui-même ... , dans son dialogue du *Banquet* portant sur la fin dernière et le souverain bien, et traitant de la divinité en général, ne bande pas sa démonstration ni ne se couvre de poussière pour faire, selon son habitude, une prise solide et

19 *Prot.* 347c3–348a6.
20 *Smp.* 176e6–8.
21 *QC* 9.1, 736C–737C.
22 *QC* 5.3, 676E.
23 *QC* 7.7, 710A–711A.
24 *QC* 7.8, 713F.

irrésistible, mais il presse les hommes par des arguments moins arides, des exemples et des récits mythiques. (*QC* 1.1, 614C–D, trad. Fuhrmann modifiée)

Les χαρίεντες sont les philosophes qui se comportent comme s'ils étaient au banquet (dont la χάρις est une valeur fondatrice). Ils font preuve de souplesse en n'imposant pas de démonstration austère, mais en s'immisçant dans l'esprit des interlocuteurs, c'est-à-dire en recourant au πιθανόν. L'exemple qui vient directement à l'esprit du locuteur est précisément le *Banquet* de Platon, qui est pensé ici comme un ouvrage composé par le philosophe à destination des 'hommes', comme un dialogue convivial entre l'auteur et les lecteurs. Un parallèle s'instaure par conséquent entre d'un côté Platon et ses lecteurs, de l'autre Plutarque et Sossius Sénécion, voire Plutarque et les autres lecteurs. Et en introduisant la notion rhétorique du πιθανόν pour décrire la démarche de Platon et se référer à son propre ouvrage, Plutarque fait donc de Platon un rhéteur qui, comme lui et ses amis, se conforme aux règles de l'homilétique conviviale.

Conclusion

La présence de la rhétorique dans les œuvres conviviales de Plutarque, et en particulier dans les *Propos de table*, révèle une attitude complexe à l'égard des modèles du passé. Malgré la multiplicité des patronages revendiqués, Platon se révèle sans aucun doute le modèle principal, celui qui sert de référence première pour transcrire des λόγοι et pour faire de la rhétorique un sujet central aussi bien de la conversation que de l'œuvre publiée.

Pourtant Plutarque se démarque sensiblement du modèle platonicien: pour lui, la rhétorique est, sous certaines conditions, une pratique acceptable et même nécessaire au banquet. Si Platon présente en réalité une version ironique du *symposion*, Plutarque propose des modèles de discours qui sont inspirés en partie des préceptes rhétoriques: à la vision négative de la rhétorique des sophistes succède donc la vision positive d'une rhétorique qui s'adapte aux exigences de la philosophie. Sans aller jusqu'à critiquer frontalement son modèle, Plutarque souligne la complexité du problème traité par Platon en proposant d'autres réponses.

De telles considérations révèlent que les œuvres conviviales de Plutarque constituent une source importante et encore inexploitée sur l'histoire de la rhétorique et donc de la Seconde Sophistique. Elles montrent ainsi que le contexte culturel a changé. En mettant en scène la bonne société d'Agathon, Platon écrivait pour son monde. Plutarque, quant à lui, écrit en tant que Grec pour des πεπαιδευμένοι qui peuvent être aussi Romains; et s'il réfléchit à la manière de mener la conversation au banquet, c'est dans le cadre d'une réflexion plus vaste sur les réseaux de sociabilité de l'Empire.[25] En s'adressant à Sossius Sénécion, Plutarque instaure un dialogue entre la Grèce et Rome et il se conforme ainsi aux représentations de son époque. Cette adaptation au contexte socioculturel de l'Empire implique de réserver une place à la rhétorique, laquelle constitue un réflexe, mais aussi une valeur cardinale pour les élites gréco-romaines. La rhétorique n'est plus une nouveauté qui fait peur. Elle est au contraire bien ancrée dans la vie quotidienne de Plutarque et de ses amis. Mais surtout elle participe d'une vision normative et consensuelle, qui est en accord avec l'éthique traditionnelle du banquet, sans empêcher *a priori* l'éclosion d'un vrai dialogue aussi bien entre convives, à l'occasion d'un banquet circonscrit, qu'entre Grecs et Romains dans l'espace global de l'Empire.

25 Sur ces réseaux, cf. Andurand and Bonnet (2016). Voir aussi, sur la figure de Platon dans les cercles de Plutarque, Andurand and Bonnet (à paraître).

José Antonio Fernández Delgado and Francisca Pordomingo

Theseis rather than *quaestiones convivales*[1]

Abstract: A high proportion of the *quaestiones convivales* in Plutarch's *Table Talk* do not follow the traditional interrogative format of 'Why ...', 'What is the cause/reason ...', 'Who ...', 'What ...', 'How ...', but are formulated as 'Whether ...' questions. This format is characteristic of the school exercise called *thesis*, one of the exercises preparatory for rhetorical education. Based on progymnasmatic theory, this chapter analyses other *thesis* features present in these 'Whether ...' questions, such as the possibility of classifying them as theoretical/practical or as simple/complex, their composition by means of an introductory *ephodos*, and a number of characteristic principles of argumentation, their capacity to admit refutation and confirmation, or their eventual presentation in the form of a *synkrisis*. We thus aim to clarify the influence this important phase of Graeco-Roman education had on the author and on his literary work, taking the *Table Talk* as our main case study.

One of the few studies dedicated to the structure of the different *quaestiones* in Plutarch's *Table Talk* notes that the dialogue, or sometimes monologue, involving the participants of the various symposia described within the work is presented in the form of successive argumentative sequences.[2] In twos and threes, and sometimes fours, these sequences tend to present the theoretically weakest positions first, and conclude with the argument with which the author himself most closely identifies, with the author often being one of the members of the discussion group. A related study by Teodorsson has considered the possible criteria involved in the composition, distribution and ordering of the different *quaestiones*, with a view to addressing the question of how much in these conversations was real and how much rhetorical-literary elaboration. The latter interpretation tends to receive more credence.[3] However, no attention has been paid to the fact that a considerable proportion of the *quaestiones convivales*—all of which are in reality called προβλήματα (or sometimes ζητήματα) by Plutarch—[4] offer a structure that can be clearly differentiated from the rest: that is, they not only contribute arguments on a given topic, but also, when a topic is posed as a dilemma, they contribute arguments in favour of one stance and implicitly against the other, or else in favour of both sides in cases of dis-

1 This chapter was written within the framework of Research Project FFI2010–21125, financed by Spain's Ministry of Science and Innovation (MICINN).

2 Martín García (1983); cf. also Martín García (1987) 10–14.

3 Teodorsson (1996); cf. also Teodorsson (1989–1996). See also Nikolaidis in this volume.

4 These *problemata* and *zetemata* are not essentially different from the other type of Plutarchan *zetemata*, such as the *Platonic Questions*, the *Roman Questions*, and the *Natural Questions*. Cf. Opsomer (1996).

https://doi.org/10.1515/9783110539479-027

junctive comparisons. This suggests that they have the same form as the type of questions known as *theseis* in school rhetoric. In this context, they were used as part of the *progymnasmata*, which served as a basis for the first stages of rhetorical education from the Graeco-Roman period of antiquity through to the Renaissance. This chapter seeks to explore the features and functions of this type of question, together with its presence in the privileged domain for intellectual and literary discussion that is the symposium, as reflected in Plutarch's *Table Talk*.

The following are the *Table-Talk*'s questions or προβλήματα that fit the *thesis* structural model:

1.1 Εἰ δεῖ φιλοσοφεῖν παρὰ πότον ('Whether philosophy is a fitting topic for conversation at a drinking-party')

1.2 Πότερον αὐτὸν δεῖ κατακλίνειν τοὺς ἑστιωμένους τὸν ὑποδεχόμενον ἢ ἐπ'αὐτοῖς ἐκείνοις ποιεῖσθαι ('Whether the host should arrange the placing of his guests or leave it to the guests themselves')

2.3 Πότερον ἡ ὄρνις πρότερον ἢ τὸ ᾠὸν ἐγένετο ('Whether the hen or the egg came first')

2.4 Εἰ πρεσβύτατον ἡ πάλη τῶν ἀγωνισμάτων ('Whether wrestling is the oldest of the sports')

2.10 Πότερον οἱ παλαιοὶ βέλτιον ἐποίουν πρὸς μερίδας ἢ οἱ νῦν ἐκ κοινοῦ δειπνοῦντες ('Whether people of old did better with portions served to each, or people of today, who dine from a common supply')

3.1 Εἰ χρηστέον ἀνθίνοις στεφάνοις παρὰ πότον ('Whether flower-garlands should be used at drinking-parties')

3.2 Περὶ τοῦ κιττοῦ πότερον τῇ φύσει θερμὸς ἢ ψυχρός ἐστιν ('Concerning ivy, whether its nature is hot or cold')

3.4 Πότερον ψυχρότεραι τῇ κράσει τῶν ἀνδρῶν ἢ θερμότεραί εἰσιν αἱ γυναῖκες ('Whether women are colder in temperament than men or hotter')

3.5 Εἰ ψυχρότερος τῇ δυνάμει ὁ οἶνος ('Whether wine is on the cold side in its power')

3.6 Περὶ καιροῦ συνουσίας ('Concerning the suitable time for coition')

4.1 Εἰ ἡ ποικίλη τροφὴ τῆς ἁπλῆς εὐπεπτοτέρα ('Whether a variety of food is more easily digested than one kind alone')

4.4 Εἰ ἡ θάλασσα τῆς γῆς εὐοψοτέρα ('Whether the sea is richer in delicacies than the land')

4.5 Πότερον οἱ Ἰουδαῖοι σεβόμενοι τὴν ὗν ἢ δυσχεραίνοντες ἀπέχονται τῶν κρεῶν ('Whether the Jews abstain from pork because of reverence or aversion for the pig')

4.9 Εἰ δεῖ θεῶν εἰκόνας ἐν ταῖς σφραγίσιν ἢ σοφῶν ἀνδρῶν φορεῖν (lost) ('Whether it is more proper to wear images of the gods or of wise men on seal rings')

6.2 Πότερον ἔνδεια ποιεῖ τὸ πεινῆν καὶ διψῆν ἢ πόρων μετασχηματισμός ('Whether hunger and thirst are caused by deficiency or by a change in shape of the passages')

6.7 Εἰ δεῖ τὸν οἶνον ἐνδιηθεῖν ('Whether it is right to strain wine')

7.6 Περὶ τῶν λεγομένων σκιῶν, καὶ εἰ δεῖ βαδίζειν καλούμενον πρὸς ἑτέρους ὑφ' ἑτέρων ἐπὶ δεῖπνον, καὶ πότε, καὶ παρὰ τίνας ('On 'shadows', so-called, and whether one should go to one man's dinner at another man's invitation, and on what occasions, and to what kind of host')

7.7 Εἰ δεῖ παρὰ πότον αὐλητρίσι χρῆσθαι ('Whether the music of flute-girls is proper after-dinner entertainment')

7.10 Εἰ καλῶς ἐποίουν βουλευόμενοι παρὰ πότον ('Whether it was a good custom to deliberate over wine')

8.9 Εἰ δυνατόν ἐστι συστῆναι νοσήματα καινὰ καὶ δι' ἃς αἰτίας ('Whether it is possible for new diseases to come into being, and from what causes')

9.12 Πότερόν ἐστι πιθανώτερον τὸ ἀρτίους εἶναι τοὺς σύμπαντας ἀστέρας ἢ περιττούση [too much of it is lost to be able to judge it] ('Whether it is more plausible that the total number of the stars is even than it is odd')

The first feature that betrays these *quaestiones*' peculiar structure is the titles themselves, since they do not fit the type of question the title usually introduces in the case of all the other 'questions': Διὰ τί / Διὰ τίν' αἰτίαν, 'Why/For what cause ...' or Τίς αἰτία, 'What is the cause/reason ...', Τίς / Τίνες, 'Who ...', Ποῖόν τινα, 'Whom ...', Ὅτι, 'That ...', Πῶς, 'How', or Περὶ, 'About ...'. Rather, they are questions introduced by the interrogative particle 'Whether ...', either in its simple (εἰ) or disjunctive (πότερον ... ἢ ...) form.[5] We therefore have 21 questions (two of which, 4.9 and 9.12, are identified only by the title) out of a total of 95 (among them, the last three in Book 4 and questions 7–11, in Book 9, are missing, except for the title). The ratio of the former to the latter is thus far from negligible, namely, 22%.

Some two-thirds of these 'Whether ...' questions address topics inherent to the symposium itself: namely, συμποτικά-questions pertaining to wine, food or drink, for example. By contrast, the other third address other kinds of matters characteristic of the occasion, or συμποσιακά, according to the classification made by Plutarch himself in the preface to Book 2 (*Mor.* 629D). The ratio of the former to the latter is much lower in questions other than the 'Whether ...' type, especially after Book 1, and with the exception of Book 7. In addition, some of these questions, which address aspects particularly emblematic of the symposium, occupy significant positions in the general ordering and distribution of the *Table Talk*. Thus, the complete series opens with question 1.1, which deals with the issue of whether the symposiasts should philosophise during their meal, and continues with question 1.2, dealing with a no less pertinent and relevant topic in this respect, namely, whether the host should personally seat the guests or whether the guests should seat themselves. Book 2 closes with the question of whether the ancient practice of eating with por-

5 In the case of 'question' 3.6, the Περί-type enquiry ('Concerning the suitable time for coition') quickly takes the πότερον ... ἢ ...form in the dilemma about whether it is better after going to bed or during the day, in a parallel way to what happens with 'questions' 3.2 and 7.6.

tions served to each one was better than the current practice of dining from a common dish. Book 3 begins with the issue of whether garlands of flowers should be used during drinking, while Book 4 opens with the question whether a variety of food is more digestible than a simple diet. Leaving aside Books 8 and 9, which leave room only for one sympotic question (question 8.6),[6] Book 7 includes three sympotic issues of the type mentioned above: chapter 6, about whether one should go to a dinner when invited by another guest, but not the host himself, chapter 7, about whether flute players should be admitted during drinking, and the tenth and final chapter, no less emblematic than 1.1, about whether it was a good idea to talk about politics when drinking. Some of these questions are in fact introduced by explicit indications as to their importance; thus, 1.1 begins with the following programmatic declaration: 'The question of philosophical talk over the cups I have placed first of all'. [7]

Let us consider an example of each one of the two modalities of the aforesaid compositional structure, and see how it operates both in the questions in which the alternative to the issue posed is simply a negative (which is almost always implicit in the title), and in the ones in which the alternative offered is a disjunctive option. Let these be, respectively, question 1.1 and question 2.3, 'Whether the chicken or the egg came first', the first non-sympotic question in the set.

In πρόβλημα 1.1, by way of *prooemium* (1), a reference to a prior symposium uses an *exemplum* and a comparison as dissuasive elements ending in a *chreia* by Isocrates. Later on (2), Crato argues in favour of philosophical conversation in the symposium, since it is natural that it should accompany pleasure by adding moderation to it. Otherwise, one must recognise that temperance should not be admitted in the symposium, as seriousness is considered inappropriate to it. Besides, it would be senseless to deprive a symposium of the contribution of philosophy. After an exhortation by Sosius Senecio to seek the philosophical topics to be addressed in the symposium, the author (3) says he proposes first to investigate the character of the attendees, since, if the majority of them are given to dialectics, it is legitimate to mix philosophy with wine and water. If, on the contrary, there are many who tolerate the voice of the philosopher less than that of any bird, that person, following the example of Peisistratus when faced with drinkers who pay no attention to his words, will join their conversation himself, as long as it remains decent. Plutarch jokingly opposes what Plato says about seeming to be just when one is not, arguing that it is very intelligent not to appear to philosophise when one is in fact doing so. For his part, Plutarch (4) is of the opinion that there is a genre of convivial topics, some of which are provided by history, others by daily matters, that contain many examples for philosophy, as well as for piety and the emulation of virtuous deeds, in imitation of Homer's Helen. As for people of taste, although they philosophise,

6 Cf. Teodorsson (1996).
7 Cf. Teodorsson (1989) 38 f.

they do so more by way of persuasion than by demonstration, just as Plato uses exempla and myths in his *Symposium*. Enquiries, continues Plutarch (5), should be fluid, the issues should be comprehensible, the questions fitting, as Democritus said, in the same way that drinkers can rock gently to the rhythm of the dance. But it would be dangerous if they were obliged to do violent exercise, the opposite of what happened with the crane and the fox in Aesop's fable or with Phrynichus and Aeschylus who, when tragedy was taken to the terrain of myths for the first time, were asked what this had to do with Dionysus. Singing would be reasonable, but using annoying conversations during the symposium is neither appealing nor appropriate.

In πρόβλημα 2.3, in a *prooemium* (1) Plutarch narrates how, at a dinner-party hosted by Sosius Senecio, his refraining from eating eggs for a long time was thought to be linked to Orphic or Pythagorean beliefs. The controversial problem of whether the chicken or the egg came first, then occupies the centre of the conversation. (2) Sulla says that with a small problem they were rocking loose a great and heavy one, with Alexander ridiculing the enquiry. Firmus invites him to expound the principles of atomic theory, according to which, if small things are the beginning of big ones, it is natural for the egg to come before the chicken, and since progress is something intermediate between natural qualities and perfection, the egg is like a progression of nature towards a living being; just as in living beings the arteries and veins are the first things to form, it makes sense that the egg would come before the living being. In the same way that the arts first model amorphous masses, and then articulate them into figures, it is natural that matter is first subjected slowly to nature, producing amorphous figures such as eggs. These then take shape, and the living being is produced. Similarly, just as the caterpillar comes first and later releases a different being, the butterfly, the egg, too, pre-exists as material for generation, since it is necessary that in every change, the thing that changed was preceded by that from which it changed. Senecio (3) replies that the last of these images was the first that opposed him, and 'instead of the proverbial door, you opened up to the world, to your own despite', since it is perfect and it is logical that what is perfect is so by nature before what is imperfect, just as the whole comes before the part, since it is not logical that a part can be part of what has not come into being. The nature of eggs needs the solidity that exists in any living thing, whereas the earth produces self-sufficient and complete animals, which, upon coupling, are by nature oviparous or viviparous; one who posits how chickens came to be if there were no eggs, differs not at all from the one who asks how could there be men and women before there were penises and wombs. Therefore what has need of something else for its generation must necessarily come last, whereas what can constitute itself is at the beginning of the generation.

According to Aelius Theon of Alexandria (1st c. CE) (120 ff. Sp. = 120, 12 ff. Patillon), the only one of the surviving authors of progymnasmastic theory who may perhaps have been known by Plutarch,[8] the *thesis* is 'a logical examination that admits controversy', without dealing with specific characters and without any other necessity, as, for example, 'Whether it is a good idea to get married'. The *prooemium* of the *thesis* should be based either on a maxim or a *chreia* that confirms it, or on an encomium or vituperation of the thing in question, and then it should move on to the arguments.

Some *theseis*, Theon says, are theoretical: those that are pursued because of a simple desire for knowledge, while others are practical. The principles of argumentation for the practical *theseis*, as well as for the theoretical ones to a great extent, shall be taken from necessity, beauty, suitability and pleasure, and their refutation from their opposites. The *topoi* of argumentation shall be based on the fact that it is possible and in agreement with nature, it can easily occur, we are not the only ones to do it nor the first, and it is suitable, pious, necessary, it enjoys consideration, it is useful, suitable for security, the origin of more important things, agreeable and difficult to repair. As Aelius Theon further states, argumentation can also be built by means of the opposite, or by basing ourselves on the similarity, inferiority, superiority, and relationship of the part to the whole and its purpose. It is therefore based on the content, as well as on the circumstances before, during and after the fact. One can also adduce the testimony of poets and eminences and notable examples. The refutation is to be achieved based on the opposite *topoi*. Amplifications and digressions can be used to the extent that the parts of the *thesis* permit them, presenting many circumstances of life with respectively suitable arguments.

A mere glance at the two samples selected from the *Table Talk* and the summary of Theon's theory of the *thesis* that we have highlighted suffices to show that, just as the theory advocates, the samples correspond, respectively, to the practical and theoretical types of *thesis*. They both contain a *prooemium* built around a *chreia* (the *prooemium* of 2.3 includes a κατὰ χαριεντισμόν *chreia*)[9] with elements of vituperation and, based on it, a series of arguments are developed in order of increasing importance in favour of the topic proposed in the first case; first in favour of one alternative and then in favour of the other, in the form of a *refutatio/confutatio* or an ἀνασκευή/κατασκευή, in the second case. The principles and *topoi* of argumentation used by Plutarch also largely coincide with those proposed by Theon. They take into account that, as Theon says, it is not always possible to offer arguments for all the questions based on any source of argumentation (Theon 121), nor does the order of the arguments have to be the same in the theoretical *theseis* and in the practical ones, but rather the order should be adapted to the question put forward (Theon 125).[10]

---

8 Cf. Kennedy (2003) XII.

9 Cf. Bellu (2007).

10 The practical *thesis* is called *quaestio finita* in Latin and also *causa*, and the theoretical one is called *quaestio infinita*, or simply *quaestio* (in Quintilian's adaptation of progymnasmatic theory: see *Inst. Or.* 3.5.5; 3.5.14). Actually, the theory of the *thesis* is older than the dates we have for the first *progymnasmata*. It appears in Aristotle's *Rhetoric*, and its use goes back as far as Protagoras.

The compositional model of the *thesis* serves as the framework for other pieces of the *Moralia* as well, such as the treatise *Soul Diseases are Worse than Body Diseases* and *Whether Fire or Water is more Useful*, both utilising the theoretical and complex type of *thesis* according to Theon's classification, which sets out arguments on two opposing sides of a question (*synkrisis*). The second of these pieces comprises five sections in favour of water (chap. 2 to 7) and another five in favour of fire (chap. 8 to 13), and is similar to the comparative counterpositions so frequent in Plutarch's work, except that instead of the *antilogia* based on *anaskeue* and *kastaskeue*, it is based on a double *encomion*. It is incomplete, like *Soul Diseases are Worse than Body Diseases*, as well as the works *Whether Vice be Sufficient to cause Unhappiness*, *Can Virtue be Taught?* and *On the Love of Wealth*, often considered a juvenile exercise carried out in a school rhetorical context.[11]

Such a degree of coincidence between Theon's theory about this *progymnasma* and the *problemata*, both practical and theoretical, selected from among the corpus of Plutarch's *Table Talk*, cannot help but lead us to think that, just as in the case of the other *Moralia* examples we have pointed out, the *progymnasma-thesis* served as the compositional matrix for at least part of this work. This suggests once again that school rhetoric played a role in Plutarch's literary production.[12] The *progymnasmata*, in which both the author and many of his educated readers were undoubtedly versed, clearly left their trace beyond his youthful works, also influencing his more mature works, as demonstrated by the case of the *Table Talk*.

It was practiced in schools of rhetoric at least from the time of Hermagoras (c. 150 BCE). Cf. Reichel (1909) 97–104; Throm (1932); Kennedy (1963) 305. Theon (69 Spengel) says that there are many books by Aristotle (and Theophrastus) with titles of *theseis*, and some of them are called 'models of *thesis*' (*thetika kephalaia*) by the rhetors, insofar as they provide means for their argumentation.

11 Cf. Milazzo (1991); Vicente Sánchez (2005).

12 Cf. Fernández Delgado (2013).

9 Space, place, landscape: symbolic and metaphorical aspects

Michael Lipka

Individuated gods and sacred space in Plutarch[1]

Abstract: Despite Plutarch's ubiquitous references to the names of individuated gods, such gods appear in two capacities only: as mythical figures, that is, as a by-product of the poetic imagination; and as objects of worship in traditional cults. On the other hand, when Plutarchan gods are represented as *actively* affecting the reality, they are replaced by abstract notions such as God, Tyche or Daimon. Furthermore, as a short excursus on Herodotus shows, individuated gods of traditional religion defined—and were defined by—the ownership of sacred space (temples, altars, etc.). By contrast, Plutarch's God, Tyche and Daimon were spatially indifferent. Not even Delphi is represented as the sacred ground owned by an individuated Apollo.

Individuated Gods

Divine names of individuated gods are ubiquitous in Plutarch, as a quick glance at the relevant Plutarchan indices shows. But the vast majority of these references do not represent individuated gods as entities *actively* involved in the historical process. Rather, individuated gods appear in Plutarch almost exclusively in two capacities: as mythical figures, a by-product (so to speak) of the poetic imagination; and as recipients of worship in traditional cults.

While Plutarch diligently documents myths and cults where this is expedient for his biographical or philosophical endeavour, he himself clearly does not share the underlying concept of individuated gods. On the contrary, he is rather outspoken about his view that mythical accounts and the concomitant human worship of individuated, human-shaped gods are make-shift conceptualisations in order to make the underlying metaphysical powers treatable in a literary and ritual fashion. In fact, Plutarch explicitly states that human-shaped gods of traditional representations are flawed concepts: 'for the deity does not resemble the human in terms of nature, movement, skill and power ... being different in almost everything, it is most of all dissimilar in its works'.[2] Not surprisingly, then, in Plutarch's view, poets should not be taken at face value when they speak of anthropomorphic gods with a

1 By 'individuated gods' I mean gods whose individual features are not restricted to human-shaped iconography (such as Moira, Tyche, Hygeia) but deities that are endowed with a specific human *individuality*, normally marked by a) a *personal* name (as opposed to personifications), b) relevant myths with connections to other individuated gods, c) 'human behaviour', including 'failures', 'passions', 'ambitions', etc.

I would like to thank Frederick Brenk for helpful suggestions.

2 Plu. *Cor.* 38.4.

https://doi.org/10.1515/9783110539479-028

human temperament and emotions.[3] Language, at any rate, is not to be trusted, according to Plutarch, since the names of individuated gods may be employed as metaphors for their main spheres of competence, without any metaphysical meaning: Hephaestus can stand for fire, Ares for war, and–we may add–Aphrodite for love.[4] Furthermore, Plutarch is much too perspicacious a thinker to ignore the fact that even when a metaphysical entity is denoted by a specific name, this entity does not have to be the same in all contexts. Take, for instance, Plutarch's statement: '... under the name of Zeus (or Zen) they sometimes address the god, sometimes Fortune (τύχη), sometimes Fate (εἱμαρμένη)'.[5] It does not come as a surprise that, unlike his near-contemporary Pausanias, Plutarch is markedly distanced and often caustic about reports of divine epiphanies in the world of man. To quote just one statement concerning such epiphanies: 'these fabulous and ridiculous tales (τὰ μυθώδη καὶ γελοῖα) display the attitude of the humans towards the divine, (an attitude) which convention (ἐθισμός) has forced upon them'.[6]

I hasten to add that Plutarch's disbelief in human-shaped individuated gods is *not* contradicted by the frequent appearances of human-shaped individuated deities in Plutarchan dreams.[7] Plutarch's relation to dreams is complex. In works of his early

3 Plu. *De aud. poet.*, esp. 16D–17F.

4 Plu. *De aud. poet.*, 23A–24C.

5 Plu. *De aud. poet.* 23C–D.

6 Plu. *Num.* 15.6. Numa's capture of two daimons, Picus and Faunus, and his encounter with Zeus on the Aventine Hill is labeled as 'surpassing all absurdity (ἀτοπία)'. The tale is well attested in the early first century BCE (Val. Ant., *FRHist* 25 F 8), and in Augustan poetry (Ov. *Fast.* 3.291–346). In a similarly suspicious vein, Plutarch reports that in 83 BCE, Sulla captured a Satyr 'such as sculptors and painters portray' in a sacred precinct of the Nymphs. Sulla tried to communicate with the Satyr 'through many interpreters' (in what language?), but it merely produced a sound somewhere between neighing and bleating, Plu. *Sull.* 27.1 f. The anecdotal/comic character of this passage is of course suggested by his cautious 'as they say' and by the explicit lack of all communication between Sulla and the Satyr. Another 'epiphany' of Artemis is explicitly said to have been due to the imagination of the citizens of Pellene, who mistook a mortal woman, the 'daughter of Epigethes', for the goddess. And, of course, Aratus' own journals have nothing of that sort, as Plutarch punctiliously observes, Plu. *Arat.* 32.1–3. Like Herodotus and others, Plutarch mentions heroic battle epiphanies, always with a pinch of salt. Some Athenians saw Theseus participating in the battle of Marathon and thus gave the man heroic status, Plu. *Thes.* 35.5. But in what follows, Plutarch makes it quite clear that Theseus' heroization was due not to an epiphany, but to a Delphic oracle which demanded the translation of his bones from Skiros, a deed eventually accomplished by Cimon. It was his new burial place in the centre of Athens that served as the focus of the new hero cult, Plu. *Thes.* 36, cf. *Cim.* 8.3–6. The participation of the Aeacidae during the battle of Salamis was an arbitrary conjecture by some fanciful observers, as Plutarch's wording makes clear. For the only thing 'they thought to have seen' were 'shades of armed men coming from Aigina', Plu. *Them.* 15.1, cf. Hdt. 8.64. Last, consider the following: 'some' witnessed the Dioscuri accompanying Lysander's ship on its way to the battle of Aegospotami in 405 BCE, Plu. *Lys.* 12.1.

7 E.g., Plu. *Cam.* 6.1 [cult statue of Veian Iuno]; *Rom.* 2.5 [Hestia], *Cic.* 44.2–4 [Zeus], *Cor.* 24.2 [Zeus], *Arist.* 11.2–8 [Zeus], *Luc.* 10 [Persephone, Athena], *Luc.* 12.1 f. [Aphrodite]; *Per.* 13.8 [Athena]; *Lys.* 20.5 [Zeus Ammon], *Sull.* 9.4 [Luna, with Brenk (1977) 223 f.], *Tim.* 8.1 [Demeter].

period (such as the *On Superstition*), he ridiculed them, but later on, he connects them with the discussion of the soul and the question of the vision of true forms. What matters here is the observation that throughout his life he considered dreams as essentially a human state of mind, to be explained by science rather than by divine intervention.[8]

In sum, then, Plutarch represents individuated deities as having no impact whatsoever on reality. In Plutarch's thinking, they are venerable remnants of convention and tradition that may appear in poetic or ritual contexts; but in the real world, as Plutarch sees it, divinity acts in a different shape, namely, in the abstract form of God, Tyche or Daimon (to mention only the most prominent names or attributes assigned to them).

Sacred space

What consequences does the absence of active, individuated gods have for the notion of sacred space? To begin with, the notion of sacred space in Plutarch is no less complex than his notion of God, Tyche, and Daimon. However, unlike these terms, it has not received the same scholarly attention, apparently because Plutarch conceals his quite heretical notion of 'sacred space' behind the façade of a very traditional vocabulary.[9]

How is sacred space generally designated? The Greek term par excellence is the ubiquitous ἱερός. Throughout pagan antiquity, the word denotes divine possession of all sorts of objects. Already the poet of the *Iliad* calls a temple ἱερὸς δόμος.[10] The name of the deity whose possession is indicated by ἱερός is either implied or added in the genitive: the Homeric ἱερὸν Ἀθηναίης is to be rendered as 'sacred to Athena'.[11] By contrast, both Judaism and Christianity avoid ἱερός, apparently because of its markedly pagan/polytheistic connotations: instead, the *Septuagint* and the *New Testament* consistently employ the term ἅγιος, where the Greek Gentile would have used ἱερός and the Jew *qodäš*.[12]

In short, then, sacred space in a pagan context is ἱερός, that means, the property of a deity. Plutarch is by no means an exception. Time and again, temples are called ἱερά, and the act of dedication, that is to say, the transfer from human to divine ownership, is predictably called καθιερόω.[13] If we then take the employment of ἱερός as

8 Brenk (1977) 16–27, 214–235; Brenk (1987) 260 f. The belief in dreams characterised the superstitious rather than the pious. This view is expressed not only in Plutarch's theoretical writings, but also in his *Lives*, when Cassius explains the apparition of Brutus' daimon on merely scientific grounds (Plu. *Brut.* 36 f. with Brenk (1977) 124 f., 152 f.). From an artistic point of view, dreams offered a welcome literary device to introduce the notion of individuated gods and their intervention in the human world, without compromising one's philosophical/scientific convictions.

9 For example, Beck (2012) does not mention sacred space.

10 *Il.* 6.88 f. For ἱερόν as sacrifice, see *Il.* 1.147. A temple was ἱερόν since Herodotus, Hdt. 1.183.

11 *Od.* 6.322, cf. 13.104.

12 G. Schrenk, in: *ThWNT* III (1938) 225 f., 229.

13 Plu. *Sol.* 4.3.

the basic parameter for the definition of sacred space, Plutarch's world is full of it. But what does ownership of sacred space mean in practice?

In the world of traditional polytheism, it means first of all that the owner deity is both entitled and able to ward off unwanted trespassers. To demonstrate Plutarch's conceptual distance from traditional polytheism, I will briefly turn to the notion of sacred space in Herodotus.

The best known passage is Herodotus' account of the repulsion from Delphi of a Persian detachment by divine intervention in 480 BCE. Herodotus indulges in a rather detailed—and apparently exemplary—description of the chain of events: when the Persians drew near Delphi, the Delphians consulted the god, but 'the god ordered them to do nothing, saying that he was able to protect his own property'.[14] And so he did: Herodotus suggests (without saying so explicitly) that the god moved his sacred and untouchable weapons out of the inner sanctum to the open space in front of the sanctuary, and then stopped the Persian advance by means of thunderbolts and landslides. Some say that two local heroes (Phylacus and Autonous) actually engaged in a fight with the aggressors. Thus the Persians abandoned their attack.[15]

To take another very similar Herodotean account: when the Assyrians attacked the temple of Ptah in Memphis in the first half of the seventh century, the priest bewailed the looming disaster to the god's image. The god comforted his distressed servant in a dream, announcing that he would send his champions. He sent a swarm of field mice to destroy the equipment of the invaders.[16]

Even when sacred space is trespassed upon, Herodotus is keen to detect some punishment of the trespassers among later events. Thus, at Plataea, no Persian corpses are found inside the sanctuary of Demeter, because—according to Herodotus' own verdict—'the goddess herself denied them entry because they had turned to ashes her sanctuary (ἱερόν), the shrine at Eleusis'.[17] In fact, the Greek–Persian hostilities had been triggered by the incineration of the temple of Cybele in Sardis by the Greeks in 499 BCE. The Persians retaliated against the Greeks by destroying their sanctuaries, which in turn led to the wrath of the Greek gods and ultimately the defeat of the Persians.[18]

One could heap up the evidence in order to show that, in Herodotus, the gods are fully in charge of their sacred ground, punishing relentlessly either at the moment of trespassing on their land or after a short respite.[19] It should also be noted that in all

14 Hdt. 8.36.
15 Hdt. 8.37–39.
16 Hdt. 2.141.
17 Hdt. 9.65.
18 Hdt. 5.102.1 with Mikalson (2002) 187–189.
19 Protesialus is empowered 'by gods' to punish Artayctes for the defilation of his shrine (Hdt. 9.116, 120). The Spartans' atonement for the murder of Darius' heralds was labeled by Herodotus or his source as the 'wrath of Talthybius' (μῆνις Ταλθυβίου), because it was in Sparta that Talthybius had a shrine and was worshipped as the founding hero of all matters concerning heralds (Hdt. 7.134).

the Herodotean passages just referred to, the divine actors are individuated gods, identified by their names and specific characteristics and commanding their own sacred ground.

Not so in Plutarch. As I have shown in the first part of this article, his individuated gods are mere ciphers. As the driving force of Plutarch's world, individuated gods are normally replaced by God (θεός, sometimes with an article), Tyche (τύχη = Fortune, in opposition to τὸ αὐτόματον = Chance)[20] or Daimon (δαίμων) or another of the approximately twenty terms collected by Swain.[21] Fortunately, for the scope of this chapter, there is no need for us to discuss these terms in detail here.[22] It is enough to stress that Plutarch's God and Daimon and their like were fully detached from the ritual geography of the world of man. As for Tyche, although Plutarch once refers to the dedication of sacred space to her and the αὐτόματον in his *Timoleon*, Frederick Brenk is certainly correct when he remarks: 'Plutarch avoids conceiving *tyche* as the Hellenistic goddess personifying the favoring circumstance, the *eutychia*, of an individual or city'.[23] Moreover, Tyche in *Timoleon* is a special case anyway and has repeatedly been studied as such in recent times.[24] Where then did the Plutarchan concept of divinity, which was 'basically monotheistic',[25] leave the traditional notion of sacred space? In fact, there was no room for sacred space in Plutarch's world. Like his individuated gods, his sacred space was a nostalgic reminiscence of a polytheistic past.

As a consequence, in marked contrast to Herodotus, the Plutarchan Apollo is never represented as defending his precinct against trespassers: neither the mercenaries under Philomelus and Onomarchus, who sacked the shrine in 356 BCE,[26] nor Sulla, who plundered the god's treasures during the Mithridatic War,[27] are ever really represented as being punished for their sacrilege by Apollo; in the case of Sulla, this

---

And the flood that drowned part of Artabazus' troops was (according to the Potidaeans, with whom Herodotus agrees) sent by Poseidon, because some Persians had desecrated the temple and the image of the god in the suburb of their city (Hdt. 8.129). During the battle of Plataea, Pausanias directed his prayer towards the temple of Hera and was promptly rewarded (Hdt. 9.62), apparently by Hera herself.

20 For the distinction, see Swain (1989a) 273.

21 Swain (1989a) 298–302.

22 For God in Plutarch, Brenk (1987) 262–275; Brenk (2005); Van Nuffelen (2011) 167–171; Hirsch-Luipold (2014) 168 f.; for Plutarch's daimonology, Brenk (1987) 275–294; for Tyche and Daimon, ibid. 305–316; Swain (1989a); various contributions in Frazier and Leão (2010); Van Nuffelen (2011) 164–167; Titchener (2014).

23 Brenk (1987) 315.

24 Plu. *Tim.* 36.4: Timoleon dedicates a shrine to Automatia *in* his house, while he dedicates his entire house to the 'sacred Daimon' (ἱερῷ δαίμονι). One should note that in the previous sentence (36.3) Plutarch refers to Timoleon's letters as his source. For Tyche and Daimon in *Timoleon*, see Brenk (1987) 311 f.; Swain (1989b), esp. 327–334; Tatum (2010); Piettre (2012).

25 Brenk (1987) 341.

26 Plu. *Tim.* 30.3 f. (shunned by all mankind 'because they had put themselves under a curse') for the occupation of Delphi during the Sacred War; see also Plu. *Mul. virt.*, 249E–F. Cf. Paus. 5.3.2–4.

27 Plu. *Sull.* 12.4–9, cf. Diod. 38.7.

may be because he offered amends later on.[28] In fact, Plutarch would have had plenty of opportunities to refer to Herodotus' version of Apollo's defence of his Delphic sanctuary, especially in his *Themistocles* and *Aristides*, but also elsewhere in his Delphic writings. After all, in his extant work he refers to some fifty Herodotean passages.[29] Plutarch's near contemporary Pausanias, who took his information from Herodotus or local tourist guides (who in turn may have drawn on Herodotus), refers to the story. So it is hard to see why it should not have been known to Plutarch.[30] In fact, Plutarch more than any other writer would have been expected to mention Apollo's defence of his property against the Persians (even if only in order to praise his beloved Delphi), had he believed that there was a germ of truth in it. After all, it is Plutarch, not Herodotus, who makes Delphi responsible for the Plataean victory.[31] The fact is that Plutarch did *not* believe that the god had protected his sanctuary, but that the temple had been burnt to the ground by the Persians.[32]

Plutarch's silence about Apollo's intervention in 480 is not due to inadvertence, as can be shown by another omission of the same kind: when Brennus invaded Greece in 279 BCE, he made a bid for Delphi. Eventually, he was repelled by the Delphians and—most importantly—by the decisive intervention of the god himself, which led ultimately to Brennus' painful death. The story is mentioned by Pausanias and Diodorus, but it appears in its most explicit form in Justin's epitome of the *Philippic Histories* of Pompeius Trogus, dating to the Augustan period. All these accounts apparently go back to one or more Hellenistic sources, and a version of the god's intervention is epigraphically attested shortly after the event.[33] Remarkably, *nothing of this* is found in Plutarch, who *must* have known the story, given his position as a Delphic priest and the apparently wide circulation of the story in Roman times. We should also note that Pausanias enumerates in passing a number of pillages of the Delphic treasury. At least in one case, namely, the attack of the Phlegyians, he also reports the god's personal intervention to protect his property.[34] Plutarch, for his part, is silent about this tradition.

Plutarch's God, Tyche and Daimon may act *at* meaningful places, without however laying any claim to their ownership. To mention just one case in point, Caesar is

---

28 Some Sicilian mercenaries died after a number of illustrious victories. This is attributed to Dike by Plutarch (Plu. *Tim.* 30.5) and offers only a very general—and far-fetched—sense of retribution. Brenk (1977) 252 argued that the slaughter of the thousand mercenaries at Plu. *Tim.* 30.2 was due to their earlier sack of Delphi, while in fact these soldiers were punished for their betrayal of Timoleon in Syracuse, as reported earlier (Plu. *Tim.* 25.3 f.). Sulla offered amends at a later time, cf. Plu. *Sull.* 19.6 with Brenk (1977) 240 f.
29 Schettino (2014) 419.
30 Paus. 10.8.7; Hdt. 8.39.
31 Brenk (1987) 331.
32 Plu. *Num.* 9.6.
33 D.S. 22.9.1; Paus. 1.4.4; 8.10.9; 10.3.4; 10.7.1; 10.8.3; 10.23.1; Just. *epit. Pomp. Trog.* 24.7 f.; *Syll.*[3] 398 (transl. in Bagnall and Derow (2004) 34 f.; Austin (2006) 129 f.).
34 Paus. 9.36.2 f.; 10.6.6 – 10.7.1.

murdered under the statue of Pompey, a fact that demonstrated, according to Plutarch, 'that it was the work of some Daimon which was calling and guiding the action to this point'.[35] The Daimon intervenes at a specific meaningful place that is intrinsically *un*connected to himself. It would be impossible to identify the Daimon with a specific deity and thus turn him into an individuated god. The ultimate purpose of the intervention of God, Tyche or Daimon is never to preclude or punish trespassing, but to guarantee the unfolding of some (fully or partially preordained) divine order.

Of course, Plutarch is not the kind of author to completely break with traditional concepts. There are faint traces of the notion of divine possession of sacred space beyond mere lip service in his work. The most conspicuous passage I could locate is the following: in Plutarch's *Aristides,* the protagonist received instructions from Delphi before the battle of Plataea, to the effect that the precondition for a Greek victory was vows to numerous individuated gods, namely, Zeus, Hera, Pan and the nymphs, as well as sacrifices to a number of Plataean heroes at their shrines, apart from the injunction to pitch the battle in the plain of Eleusinian Demeter and Kore. Aristides followed the instructions to the letter, and the Greeks were victorious.[36] Underlying this section is clearly the notion of individuated gods bound to sacred ground and potentially intervening as beneficent agents in the course of the battle. But this passage, for whatever reasons, remains an exception.[37]

What about the Pythian Apollo? Is he not a clearly identifiable individuated god, spatially bound to his sanctuary at Delphi? I suggest that he is not, despite the fact that Plutarch is anything but shy about Delphi's contribution to Greek history over the centuries.[38] At the very end of his treatise *On the Pythian Oracles,* Plutarch remarks that Delphi would not be what it is in his day, unless God had been present at Delphi and allowed the oracle to share in his divine inspiration (συνεπιθειάζοντος).[39] By writing θεός instead of ὁ θεός, Plutarch is decidedly *not* pointing to Apollo in particular, but to God in general. In what manner, then, is God present at Delphi, and what is the meaning of συνεπιθειάζειν here? In order to find an answer, we have

---

35 Plu. *Caes.* 66.1 f.; Brenk (1987) 313 f., 329.

36 Plu. *Arist.* 11.

37 Furthermore, in the same *Life,* as well as in the *Life of Lucullus,* a god appears in a dream vision to a protagonist sleeping in or close to the deity's sanctuary. This suggests a connection between the appearance of the god and the space sacred to him: 1. When Lucullus camped in Aphrodite's sanctuary in the Troad, the goddess addresses him in a dream in nice hexameters, encouraging him to get up and seize a number of hostile ships; Lucullus immediately does so (Plu. *Luc.* 12.1 f.). 2. A Lydian was sent by Mardonius to the sanctuary of Amphiaraus, where he slept and in his sleep saw an attendant of the god instructing him leave; when he did not do so, the attendant crushed his skull with a stone (which was actually a foreboding of Mardonius' death), Plu. *Arist.* 19.1 f.

38 Brenk (1987) 330 – 336.

39 Plu. *De Pyth. or.* 409C: ἀλλ' οὐκ ἔστιν ἄλλως ποτὲ τηλικαύτην καὶ τοσαύτην μεταβολὴν ἐν ὀλίγῳ χρόνῳ γενέσθαι δι' ἀνθρωπίνης ἐπιμελείας, μὴ θεοῦ παρόντος ἐνταῦθα καὶ συνεπιθειάζοντος τὸ χρηστήριον.

to turn to the more common cognate compound ἐπιθειάζειν.[40] In two passages of his treatise *On the Sign of Socrates*, Plutarch employs the verb ἐπιθειάζειν in the sense of 'to inspire' or 'to give divine insight'.[41] In both cases, the subject of the verb is the Daimonion, not a specific or identifiable individuated god. If we are allowed to extrapolate from these two passages, we may suggest that also at the end of the treatise *On the Pythian Oracles,* the faceless God—in the Greek text termed θεός—is actually to be construed as an equivalent to the Daimonion. To corroborate this, we may point to Frederick Brenk's observation that the Plutarchan Daimonion shows a particular propensity to appear in revelatory or prophetic contexts: time and again, it is the Daimonion that reveals, predicts, forebodes.[42] On the other hand, both passages in *On the Sign of Socrates* show that ἐπιθειάζειν is spatially indifferent. 'Inspiration' or 'knowledge of the future' is granted by the Daimonion to anyone graced with its favour and predisposed to listen to it, be it Socrates or the Pythia, be it in Athens or in Delphi, that is, anywhere and at any time.

---

40 The verb συνεπιθειάζειν appears only once again in Plutarch, in a very different context and with a human, not a god, as the subject of the verb. This passage can thus be ignored without much ado, cf. Plu. *Sull.* 6.4: Σύλλας ... συνεπιθειάζων τὰ πραττόμενα. Other usages of ἐπιθειάζειν: ἐπιθειάζειν occurs in Plutarch as an equivalent to the Latin *inauguro* = to inaugurate, Plu. *Cam.* 31.3. In Thucydides (2.75) it can also mean 'to call on the gods/to summon the gods to turn to the gods' when it refers to a prayer offered by Archidamus.

41 Plu. *De genio Socr.* 580D, 589D.

42 This includes the δαιμόνιον which warned the Spartan through a (Delphic?) oracle of the ascent of a lame king = Agesilaus (here, τὸ δαιμόνιον presumably stands for 'Apollo') (Plu. *Ages.* 30.1 with ibid. 3.3–5, also Fontenrose (1978) 322 no. Q163), Socrates' δαιμόνιον predicting the Sicilian disaster (Plu. *Alc.* 17.4 f., *Nic.* 13.6), the δαιμόνιον which prepares great things for Marius after sacrifices (Plu. *Mar.* 8.4 f.), and a dozen or more passages collected already by Brenk (1977) 272 f. n. 13.

Carlos Alcalde-Martín

Espacio monumental y *autopsia* en las *Vidas Paralelas* de Plutarco

Abstract: Plutarch mentions in his *Lives* many places, monuments, or events which he must have witnessed, but there is rarely any explicit indication of his witnessing them. This chapter deals with such rare instances and analyses Plutarch's eyewitness testimony and the different functions that it represents, especially when it comes to the depiction of his protagonists' moral character. Moreover, the monuments and places the author claims to have seen serve to validate the link between the past and the present.

Al tratar de la *autopsía* en las *Vidas paralelas*, se puede partir de un conocido pasaje de la *Vida de Nicias* (1.5):

> ἃς γοῦν Θουκυδίδης ἐξήνεγκε πράξεις καὶ Φίλιστος ἐπεὶ παρελθεῖν οὐκ ἔστι ... ἐπιδραμὼν βραχέως καὶ διὰ τῶν ἀναγκαίων ... τὰ διαφεύγοντα τοὺς πολλούς, ὑφ' ἑτέρων δ' εἰρημένα σποράδην ἢ πρὸς ἀναθήμασιν ἢ ψηφίσμασιν εὑρημένα παλαιοῖς πεπείραμαι συναγαγεῖν, οὐ τὴν ἄχρηστον ἀθροίζων ἱστορίαν, ἀλλὰ τὴν πρὸς κατανόησιν ἤθους καὶ τρόπου παραδιδούς.
>
> Las acciones que han narrado Tucídides y Filisto, ya que no es posible omitirlas ... las he recorrido brevemente y con los datos indispensables ... los datos que pasan inadvertidos a la mayoría y han sido mencionados por otros esporádicamente o se han hallado en ofrendas votivas y antiguos decretos, he tratado de reunirlos, pues no pretendo componer una historia inútil sino proporcionar la que sirva para la comprensión del carácter y la forma de ser.

Plutarco, pues, selecciona los materiales literarios de que dispone con el propósito de mostrar el carácter del personaje, y completa su retrato moral utilizando documentos distintos de los propios del discurso historiográfico, visibles y a menudo también vigentes en la época del autor, como decretos y ofrendas votivas (ἀναθήματα).[1] Gran parte de los casos de *autopsia* de las *Vidas* ofrecen datos de este tipo u otros semejantes, como estatuas, tumbas o monumentos en general.

En un trabajo de referencia sobre la *autopsia* en Plutarco, Buckler cita la mayoría de los pasajes de *Moralia* y *Vidas* en los que el autor dice expresamente haber visto algo o, al menos, se puede deducir con facilidad. A pesar de que Buckler reconoce que el interés primordial en las *Vidas* es el retrato del carácter y esto condiciona la selección de materiales, lamenta que Plutarco no transmita más información sobre las cosas que vio y de que no empleara sus viajes para examinar los escenarios de las acciones que narra. También considera que su interés por los vestigios del pasado es propio de un amante de las antigüedades pero que no realiza un gran esfuerzo en

1 Desideri (1992b) 4543.

https://doi.org/10.1515/9783110539479-029

este sentido, y que sus testimonios de *autopsia* sirven para confirmar los datos que ya conoce de fuentes literarias. Concluye Buckler afirmando que la *autopsia* no tenía para Plutarco un interés particular ni central.[2] Pensamos, sin embargo, que Plutarco sí tiene especial interés en dejar constancia de su testimonio visual, que no significa una mera confirmación de fuentes escritas consultadas. Por eso, aunque haya visto muchas cosas, pocas veces lo dice expresamente ya que su interés primordial no es ratificar la veracidad de un hecho. El valor que suele tener la indicación expresa de Plutarco de que ha visto algo es su incidencia en el aspecto moral de un hecho o aportar una prueba del carácter del personaje biografiado completando así su retrato moral. Además Plutarco destaca con su testimonio visual que determinados monumentos del pasado no solo tienen visibilidad en el espacio que ocupan sino, sobre todo, que los hechos que representan y sus protagonistas tienen continuidad en el presente en el que vive.

Centramos el estudio en algunos tipos de monumentos especialmente significativos en las *Vidas paralelas*.

Estatuas y santuarios

En el proemio de la *Vida de Cimón* (2.2), leemos que la intervención de Lúculo a favor de Queronea tras unos disturbios hizo que fuera considerado el salvador de la ciudad y los habitantes, agradecidos, erigieron una estatua suya en el ágora. A propósito de la estatua no se menciona ningún rasgo físico ni del carácter del personaje, pues la intención del autor es otra: justifica la inclusión de Lúculo en las *Vidas paralelas* pues el favor que hizo a los habitantes de Queronea perdura hasta la época de Plutarco, y este, también agradecido como sus paisanos del pasado, le dedica la narración de sus acciones: un retrato de su carácter y costumbres más bello que el de mármol, que solo representa el cuerpo y el rostro.[3]

No podemos saber si Plutarco conocía la estatua de Lúculo por la tradición o porque aún existía en su tiempo; probablemente la vio, porque la localiza junto a la de Dioniso, pero no consideró relevante decirlo porque la estatua pierde importancia, aunque siga visible en el ágora de Queronea, como testigo del nexo entre el pasado y el presente: esa función la cumple mejor el retrato biográfico.

Las estatuas pueden confirmar el aspecto físico o el carácter, apoyar una información o desmentirla.[4] Pero Plutarco suele ir más lejos e introduce con su testimonio visual una valiosa prueba que no consiste en la descripción de la imagen que contempla y que puede derivar de su contexto o del significado que le atribuye. Podemos citar los ejemplos de Filopemén y Flaminino antes de tratar otros más detenidamente:

2 Buckler (1992) 4821, 4828–4829.

3 Kaesser (2004) 364–367.

4 Esto es lo único que destaca Buckler (1992) 4829–4830, entre otros ejemplos en las estatuas de Filopemén y Flaminino, que analizamos en el texto.

Ἦν δὲ τὸ μὲν εἶδος οὐκ αἰσχρός, ὡς ἔνιοι νομίζουσιν· εἰκόνα γὰρ αὐτοῦ διαμένουσαν ἐν Δελφοῖς ὁρῶμεν.

No tenía una figura deforme como piensan algunos, pues vemos una imagen suya que todavía se conserva en Delfos. (*Phil.* 2.1)

Con esta concisión cuenta Plutarco que vio un retrato de Filopemén en Delfos. No se plantea si era realista o idealizado ni lo describe, pues su propósito no es tanto desmentir lo que ha leído sobre la fealdad física del personaje como ofrecer de él un aspecto externo más acorde con el elogio que acaba de hacer de sus virtudes, por las que fue llamado 'el último de los griegos' (*Phil.* 1.7). Del aspecto físico de Flaminino, dice Plutarco que se puede ver en una estatua en Roma (1.1). Ni lo describe ni lo relaciona con el carácter, pero añade un detalle: la estatua tiene una inscripción en griego (γράμμασιν Ἑλληνικοῖς ἐπιγεγραμμένη). Sin revelar el contenido de la inscripción, tal vez Plutarco quiera anticipar ya al filoheleno que proclamó en Corinto la libertad de las ciudades griegas (10.4–5).

Podemos extraer ya, y anticipar, algunas conclusiones. En primer lugar, que no siempre es fácil determinar si lo que Plutarco menciona (lugares, monumentos, ritos, celebraciones, etc.) lo ha visto con sus propios ojos o lo ha conocido por fuentes literarias o de otro tipo. Ni cuenta todo lo que ve ni siempre que cuenta algo aclara si lo ha visto o no, pues deja constancia de su *autopsia* normalmente con un propósito determinado. En algunas ocasiones, la descripción o el comentario de Plutarco sobre algo hacen evidente la *autopsia*. En otros casos, el léxico, con expresiones que, por su frecuencia, adquieren carácter formular, la indica directamente (ὁρῶμεν, 'vemos') o conecta el pasado con la época de Plutarco (ἔτι καθ' ἡμᾶς, 'todavía en nuestros días')[5] o alude a la permanencia de algo, a través de los siglos, en su espacio original (δείκνυται, 'se muestra'; διαμένει, 'permanece'; κεῖται, 'se encuentra').

En la *Vida de Nicias*, Plutarco hace una relación de algunos gastos públicos realizados por el personaje (3.1–8) y refiere sus ofrendas votivas (ἀναθήματα) conservadas todavía en su tiempo: en la Acrópolis, la estatua de Palas que ya ha perdido su revestimiento de oro, y en el santuario de Dioniso, el templete para los trípodes que había ganado como corego.[6] Menciona además la estela que dejó en el santuario de Delos, en la que mandó grabar sus donativos.[7] Concluye calificando estos actos de vulgar ostentación 'para conseguir gloria y honores' (4.1): πρὸς δόξαν καὶ φιλοτιμίαν.[8]

Ofrendas votivas y *philotimia* van asociadas con frecuencia en las *Vidas paralelas*, como veremos a continuación. En la *Vida de Lisandro* 18.1, la ambición o *philotimia* es un rasgo fundamental del carácter del protagonista. Del ostentoso monumento consagrado en Delfos para conmemorar su victoria de Egospótamos,

---

5 Ver Frazier, Oikonomopoulou y Roskam en este volumen.

6 Marín-Valdés (2008) 53.

7 Liddel (2008) 132–136.

8 Sobre *philodoxia*, *philotimia* y términos relacionados, Frazier (2014) 489–443. Ver Aloumpi en este volumen.

Plutarco apenas hace mención, pero muestra su disgusto señalando que fue erigido con el botín conseguido (ἀπὸ τῶν λαφύρων).[9] En cambio, otra estatua atrae más su atención: la que había dentro del tesoro de los acantios. Con su descripción comienza la biografía (1.1–2). Se creía que era de Brásidas pero Plutarco afirma que, en realidad, es la imagen de Lisandro, con barba y con el pelo al estilo antiguo, largo según había dispuesto Licurgo. Plutarco no argumenta su afirmación pues su objetivo no es corregir una tradición equivocada ni hablar del aspecto del personaje, sino indicar su carácter. Se ha considerado la dificultad para identificar los rasgos del personaje en la estatua como un símbolo de la dificultad para definir su carácter debido a su ambigüedad y contradicciones.[10] Como se comprueba en la biografía, el aspecto tradicional espartano representado en la estatua, acorde con algunas virtudes de Lisandro, contradice sus comportamientos opuestos a la tradición, como los desmedidos honores que busca(18), la introducción de dinero en Esparta (17) o la conspiración para cambiar el régimen político (*Lys.* 24.2–5, *Ages.* 8.3; 20.3–5).

El hecho de que la estatua ya no fuera reconocida contrasta con la *philotimia* del personaje, rasgo fundamental de su carácter, y el testimonio visual de Plutarco constata lo que queda en su época de los desmedidos honores buscados por Lisandro en el pasado. También muestra lo que piensa Plutarco sobre el inconsistente honor que significa una estatua.[11] Podemos relacionar esto con otro personaje importante en esta biografía, Agesilao, cuya envidia (φθόνος), causada por la incontinente *philotimia* de Lisandro, provocará el declive de este.[12] Plutarco refiere la actitud de Agesilao ante las estatuas: prohibió que se hicieran esculturas que lo representaran (*Ages.* 2.4).[13] El motivo lo expresa en *Reg. et imp. Apophth.* 191D: εἰ γάρ τι καλὸν ἔργον πεποίηκα, τοῦτό μου μνημεῖον ἔσται· εἰ δὲ μηδέν, οὐδ' οἱ πάντες ἀνδριάντες ('Si he hecho algo noble, esto es mi mejor recuerdo; pero si no hice nada, de nada valdrían todas las estatuas'). Los hechos y su recuerdo son, como vimos a propósito de Lúculo, un retrato superior al que ofrecen pinturas y esculturas, puede ser identificado con seguridad y es inconfundible.

Cuando se trata sistemáticamente de la austeridad de Agesilao (19.5–11), el biógrafo ofrece un testimonio de *autopsia* que completa su caracterización (19.6): ἔστι δὲ καὶ λόγχην ἰδεῖν αὐτοῦ κειμένην ἄχρι νῦν ἐν Λακεδαίμονι, μηδὲν τῶν ἄλλων διαφέρουσαν ('Se puede ver también una lanza suya conservada hasta hoy en La-

9 Plutarco censura este tipo de monumentos: *Nic.* 4.1; *De Pyth. or.* 401A–E. Buckler (1992) 4810 dice solo que Plutarco se refiere al monumento para demostrar que Lisandro era el hombre más poderoso de Grecia después de Egospótamos.

10 Stadter (1992); Duff (1999) 162–168; Candau-Morón (2000).

11 Mossman (1991) 98 ss., al estudiar el uso de las estatuas en Plutarco, no centra su atención en la *autopsia*. Ilustra con un poema de Shelley el fenómeno general del olvido de las glorias pretéritas y muestra que el desprecio de Plutarco por los honores tributados con estatuas cuenta con larga tradición en la literatura griega, destacando especialmente el poeta Simónides.

12 Bearzot (2005); Frazier (2014) 494.

13 Mossman (1991) 113.

cedemonia, que no se diferencia en nada de las demás'). Plutarco deja constancia de que la vio porque la considera un objeto importante como testimonio visible todavía en su tiempo del carácter del héroe. La modesta lanza simboliza la austeridad y el igualitarismo espartano, proyecta sobre el carácter de Agesilao los valores tradicionales, suple a la estatua inexistente y ofrece una imagen segura del personaje, algo que no logra la equívoca estatua de Lisandro del tesoro de los acantios.

En la *Vida de Temístocles* encontramos también notorios ejemplos de *autopsia* como un importante elemento descriptivo de la *philotimia* del personaje, que 'en ambición los superó a todos': Τῇ δὲ φιλοτιμίᾳ πάντας ὑπερέβαλεν (5.3). Tenía su casa en el barrio de Mélite, donde, según el biógrafo, mandó construir un templo a Ártemis Aristobula[14] dando a entender, con este apelativo de la diosa, que él había tomado las mejores decisiones para la ciudad y los griegos. Pero esa muestra de *philotimia* disgustó a los atenienses (22.2–3).

En el templo había una pequeña estatua de Temístocles[15] que vio Plutarco, y dice que era un personaje no solo de espíritu sino también de aspecto heroico.[16]

> ἔκειτο δὲ καὶ τοῦ Θεμιστοκλέους εἰκόνιον ἐν τῷ ναῷ τῆς Ἀριστοβούλης ἔτι καθ' ἡμᾶς, καὶ φαίνεταί τις οὐ τὴν ψυχὴν μόνον, ἀλλὰ καὶ τὴν ὄψιν ἡρωικὸς γενόμενος.
>
> Había todavía en nuestros días una pequeña estatua de Temístocles en el santuario de Aristobula, y parece que fue alguien no solo de espíritu sino también de aspecto heroico. (*Them.* 22.3)

De la expresión ἔκειτο ... ἔτι καθ' ἡμᾶς ('había todavía en nuestro tiempo') se puede deducir que Plutarco vio la estatua. La constatación de la *autopsia* se hace normalmente en presente, pero aquí el pretérito imperfecto puede indicar que el autor se refiere al momento pasado en que la vio; sin embargo, el aspecto durativo del imperfecto subraya la continuidad del pasado relatado en el tiempo del autor.[17]En cambio, el presente φαίνεται ('parece') podría considerarse un presente general con validez en todo tiempo. Con su testimonio, Plutarco pretende ilustrar, en nuestra opinión, dos aspectos del personaje: su carácter y aspecto heroico, sin cuestionar la fidelidad del retrato, y la *philotimia* (por la presencia de la estatua en el templo), rasgo fundamental del carácter de Temístocles e hilo conductor de toda la biografía. Por eso Plutarco no menciona la estatua al comienzo de la obra, como en otras *Vidas* cuando hace la presentación del personaje. Aquí la encontramos en el contexto de la envidia y el rechazo que provoca en los ciudadanos la *philotimia* de Temístocles y que le ocasionarán el ostracismo y el exilio.[18]

---

14 Podlecki (1975) 174–176; Piccirilli (1983) 263–264; Marín-Valdés (2008) 186–193.

15 Podlecki (1975) 143–146; Frost (1980): 185; Marín-Valdés (2008) 193–204.

16 Sobre la conexión entre carácter y aspecto físico, Mossman (1991) 111; Georgiadou (1992) 4621–4623; Pérez-Jiménez (2013) 195–196.

17 Según observación de Mme. Françoise Frazier en la 10th International Plutarch Conference. Le agradezco su generosidad al enviarme su comunicación. Ver Frazier en este volumen.

18 Podlecki (1975) 138; Piccirilli (1983) XV-XVI.

La edificación de este templo por Temístocles para gloria suya responde en la estructura de la biografía al de Ártemis Proseoa en el cabo Artemision,[19] mencionado en el capítulo 8; Plutarco lo visitó y reproduce una elegía que se encontraba allí, grabada en una estela, y elogia una gran acción colectiva de los atenienses, en contraste con el autoelogio personal que significaría el templo de Mélite y el retrato de Temístocles.

En suma, Plutarco usa a menudo las estatuas como metáfora, dando más importancia a su significado que a su apariencia.[20] Deja constancia de su autopsía para dar validez a su argumentación y como un testimonio en su época del carácter de los personajes biografiados y, también a veces, como un símbolo visible de la vigencia de los hechos del pasado en su propia época.

Sepulcros

También en el barrio de Mélite, Plutarco comprobó la εὐτέλεια ('sencillez de vida') de Foción (18.8): ἡ δ' οἰκία τοῦ Φωκίωνος ἔτι νῦν ἐν Μελίτῃ δείκνυται, χαλκαῖς λεπίσι κεκοσμημένη, τὰ δ' ἄλλα λιτὴ καὶ ἀφελής. ('La casa de Foción se enseña todavía en Mélite, decorada con placas de bronce pero en lo demás sencilla y simple'). El capítulo 18 de su biografía ilustra esa sencillez de vida del protagonista y el testimonio visual de Plutarco aporta en su época la prueba evidente del carácter del héroe.

Junto con las estatuas, las tumbas son también testimonios del pasado y unas y otras contribuyen al retrato del héroe. Tampoco de ellas ofrece Plutarco descripciones, sino que selecciona los aspectos que le interesan o se limita a ofrecer una interpretación moral.[21] De la tumba de Foción no habla Plutarco, pues no se conservaría en su tiempo pero, a cambio, habla de la tumba de la hetera Pitonice, que todavía existía en su época. El contexto es (igual que al hablar de su casa) el de la sencillez de vida de Foción y su incorruptibilidad: tras rechazar los regalos de Alejandro, resiste los intentos de Hárpalo para corromperlo (*Phoc.* 21.3–5). Por el contrario, su yerno Caricles se dejó corromper y además aceptó el innoble encargo de Hárpalo de erigirle una tumba a su amante, la hetera Pitonice, cuando murió.

Algunos autores admiran la grandiosidad del sepulcro, que parecería el monumento de un importante personaje del pasado.[22] Plutarco, tras contar que Hárpalo quiso erigir una tumba muy costosa (μνημεῖον ἀπὸ χρημάτων πολλῶν) y encomendó el encargo a Caricles, continúa:

---

19 Frost (1980) 110; Marín Valdés (2008) 172–186.

20 Como apunta Mossman (1991) 98–99.

21 Podemos ver algunos ejemplos en *Sul.* 38.4, *Oth.* 18.1–2, *Lys.* 29.3, *Cim.* 4.3, 19.5. *Ca. Ma.* 5.4.

22 Paus. 1.37.5; Ath. 13.594–595.

οὖσαν δὲ τὴν ὑπουργίαν ταύτην ἀγεννῆ, προσκατῄσχυνεν ὁ τάφος συντελεσθείς—διαμένει γὰρ ἔτι νῦν ἐν Ἑρμείῳ, ᾗ βαδίζομεν ἐξ ἄστεος εἰς Ἐλευσῖνα—μηδὲν ἔχων τῶν τριάκοντα ταλάντων ἄξιον, ὅσα τῷ Ἁρπάλῳ λογισθῆναί φασιν εἰς τὸ ἔργον ὑπὸ τοῦ Χαρικλέους.

A este servicio, ya de por sí innoble, se le añadió aún más vergüenza por la ejecución del sepulcro. Pues se conserva todavía hoy en Hermeo, en el camino que va desde la ciudad a Eleusis, y no vale los treinta talentos que, según cuentan, presupuestó Caricles a Hárpalo para la obra. (*Phoc.* 22.2)

Lo único que Plutarco desea destacar con su testimonio visual es la incorruptibilidad y austeridad de Foción por el contraste con la corrupción de Caricles y el lujo ostentoso de otros. Por eso no describe la tumba y, aparte del emplazamiento, solo dice que fue muy costosa y que Caricles, además de aceptar un encargo indigno, estafó a Hárpalo. No le importa su esplendor, como a otros autores, sino el aspecto moral. Plutarco desea presentarnos no una bella antigüedad sino un símbolo del carácter de Foción visible todavía en sus días. Hay un gran contraste entre esa tumba ostentosa ubicada en un lugar muy conocido y a la vista de todos y la narración, al final de la biografía, del exilio del cadáver de Foción, que fue transportado más allá de Eleusis (por tanto quizás pasó por delante del sepulcro de Pitonice) y tuvo un triste entierro fuera de las fronteras del Ática; su esposa levantó allí un cenotafio y se llevó los huesos en secreto para enterrarlos junto al hogar de la casa (37.3–5). Sin decirlo espresamente, Plutarco vuelve a señalar que el recuerdo de las nobles acciones constituye un monumento superior a uno de piedra muy costoso.

En el testimonio visual de Plutarco de vestigios de épocas más antiguas, podemos apreciar a veces un interés por resaltar la vigencia del pasado en su propia época[23] superior incluso a la intención de apuntar a una enseñanza moral. Hay dos casos muy notorios. Uno está en la *Vida de Pericles* (12.1): la descripción de los monumentos de Atenas, que en el pasado fueron el mejor ornato de la ciudad y, en el tiempo de Plutarco, su permanencia son 'lo único que atestigua a favor de Grecia' su pasada grandeza (Plutarco desea equiparar esa grandeza de Atenas y Grecia en el pasado a la grandeza de Roma en su época). El otro, del que nos ocupamos a continuación, lo encontramos en la *Vida de Teseo.*

Leemos lo siguiente sobre la tumba de Teseo:

κεῖται μὲν ἐν μέσῃ τῇ πόλει παρὰ τὸ νῦν γυμνάσιον, ἔστι δὲ φύξιμον οἰκέταις καὶ πᾶσι τοῖς ταπεινοτέροις καὶ δεδιόσι κρείττονας, ὡς καὶ τοῦ Θησέως προστατικοῦ τινος καὶ βοηθητικοῦ γενομένου καὶ προσδεχομένου φιλανθρώπως τὰς τῶν ταπεινοτέρων δεήσεις.

Se encuentra en el centro de la ciudad junto al actual gimnasio, y es un refugio para los esclavos, para todos los débiles y cuantos tienen miedo de los más poderosos, puesto que también Teseo fue amparo y defensor y acogía con humanidad las súplicas de los más débiles. (*Thes.* 36.4)

23 Desideri (1992b) 4562–4566; Payen (2014) 237–241.

Tampoco en este caso se describe la tumba; solo se menciona el lugar que sigue ocupando en Atenas con una referencia concreta al presente y a los cambios en el entorno urbano: 'junto al actual gimnasio'. No se pone en duda que los restos de Teseo se encontraban allí, y el hecho de que sirva de refugio para los débiles confirma su autenticidad pues concuerda con el carácter que se le atribuye al héroe. Por tanto, dos cosas han permanecido a través de los siglos: el emplazamiento del sepulcro y la beneficiosa influencia de Teseo en la ciudad. Además de la tumba, Plutarco da muestras de haber visto otros lugares relacionados con Teseo, como el sitio donde Egeo, al reconocer a su hijo, dejó caer la copa del veneno (12.6); y recuerdos de la lucha con las amazonas, como las tumbas de los atenienses caídos en combate, la estela que señala dónde murió Antíope, y tumbas de amazonas en Mégara, Queronea y Tesalia (27.4–9).

El valor de estos testimonios se entiende plenamente a la luz del proemio de la *Vida de Teseo*, donde afirma el autor que en la composición de las *Vidas paralelas* ha ido haciendo un relato verosímil y conforme a la historia que se atiene a los hechos pero que, al ocuparse de épocas más antiguas, encuentra sucesos fantásticos y propios de la tragedia (*Thes*. 1.3).[24] Algo más adelante, encontramos el procedimiento metodológico que seguirá:

> εἴη μὲν οὖν ἡμῖν ἐκκαθαιρόμενον λόγῳ τὸ μυθῶδες ὑπακοῦσαι καὶ λαβεῖν ἱστορίας ὄψιν.
>
> Ojalá podamos hacer que el mito, depurado, se someta a la razón y tome el aspecto de historia. (*Thes*. 1.5)

Es decir, depurará de elementos fantásticos las narraciones míticas aportadas por mitógrafos y poetas para lograr un relato biográfico verosímil y con aspecto de historia.[25] Teseo, fundador de la polis ateniense, constituye el fundamento del ordenamiento político posterior de la ciudad (24.1–25–3), de instituciones y costumbres (como la acuñación de moneda: 25.3), de festividades religiosas (22.3–22.4). La presencia del héroe en estos campos es tomada por Plutarco como una verificación de los inseguros relatos sobre su vida.[26] Otra aportación fundamental al aspecto de historia que toma esta biografía lo constituye el testimonio ocular de Plutarco de los lugares de Atenas en los que sigue presente Teseo o sus hechos. De esta forma, la vida del héroe no concluye con su muerte en un tiempo remoto; se proyecta en la historia de Atenas y sus instituciones hasta la propia época de Plutarco no solo por sus acciones pasadas sino también por su efecto en el presente, como muestra su carácter humanitario, que adquiere visibilidad, para Plutarco y sus contemporáneos, en su tumba, situada en el centro de la ciudad.

---

24 Stadter (1988) 284.

25 Pelling (1999a) = (2002) 173–177; Pérez-Jiménez (2005) 343–347; Casanova (2013).

26 Desideri (1992b) 4562–4564.

Conclusión

De este recorrido por pasajes de las *Vidas paralelas* podemos extraer algunas conclusiones válidas para otros muchos casos. Plutarco no siempre que ha visto algo lo dice, y si lo hace es con un propósito determinado que hace relevante la constancia de la *autopsia*. Esta constituye a veces la única fuente documental del tipo de datos que precisa el biógrafo para su obra, pero incluso en estos casos no se trata de un elemento aislado en la biografía, sino que encuentra correspondencias en otros y se inserta en el entramado de todos los que que configuran el retrato moral del protagonista. Estatuas, tumbas y otros monumentos, que permanecen en su lugar original aunque haya cambiado el entorno, no son citados como una curiosidad de tiempos antiguos o un mero recuerdo del pasado. La *autopsia* de Plutarco es el puente que, en la biografía, enlaza ese pasado con el presente, y le permite presentar tales monumentos conservados de tiempos anteriores como testimonio fehaciente de la continuidad y vigencia, en la época en que vive, de los hechos del pasado y del carácter de sus protagonistas.

Sophia Xenophontos

Military space and *paideia* in the *Lives of Pyrrhus and Marius*

Abstract: In this chapter, I explore the dangers and limitations of an exclusively military education in the light of the *Pyrrhus and Marius*. As I shall argue below, for Plutarch the military sphere is not merely a background setting in which the characters exhibit their valour, but rather a vital environment for the construction and interpretation of the biographical account; it helps to cast light on how the hero behaves in other contexts, e. g., in the family, in politics, philosophy, and rhetoric, which in turn has implications for the hero's morality and cultural identity.

At the beginning of the *Life of Pyrrhus* Plutarch provides his readers with the mythical narrative explaining the foundation and settlement of Epirus. According to tradition, Neoptolemus, son of Achilles, was the country's first ruler, but the line of kings descended from him, the Pyrrhidae, soon sank into barbarism in terms both of their power and way of life (ἐκβαρβαρωθέντων ... τῇ τε δυνάμει καὶ τοῖς βίοις, *Pyrrh.* 1.4). The situation was ameliorated when Tharrhypas, Pyrrhus' great-great-grandfather, introduced Greek customs and letters, and administered the cities with humane laws (*Pyrrh.* 1.4).

This mythical account anticipates a key theme arising from the subsequent narrative, namely the contradiction between barbarism and acculturation, and–as so often in Plutarch—the value of Greek education as a means to combat moral depravity. On another level, it is also linked to an incident from Pyrrhus' infancy, which describes the agonising escape of some native fugitives in an attempt to rescue the baby Pyrrhus from some ill-disposed pursuers. The decisive factor in this episode that ensures the hero's physical survival is the crossing of a turbulent river. Interestingly, the spatial description of the river as a location where the action takes place involves such terms as 'forbidding', 'savage', 'violent' and 'dark' (*Pyrrh.* 2.3–5), while crossing it is only made possible by means of written messages of support that were sent across (ἐνέγραψεν ... γράμματα, ἀνέγνωσαν ... τὰ γράμματα, *Pyrrh.* 2.6–7). This exciting story surely reflects Plutarch's creative imagination,[1] and the distanced —and in all likelihood fictitious—geography of the river in full flood becomes a cul-

1 The closest parallel is found in Justin, *Epit. Pomp. Trog.* 17.3.17–20, who merely refers to the persecution of the child Pyrrhus by his father's enemies until he is handed over to the wife of King Glaucias.

https://doi.org/10.1515/9783110539479-030

turally charged 'frame'[2] that shapes the story in a suggestive way: the preservation of the Epirotes and their king is predicated upon the use of letters, an index of civilisation and education.

This central pattern permeates the rest of the *Life* in inverted versions, in which negligence of cultural training brings about the hero's plights. In this chapter, I wish to look above all at the passages in which Pyrrhus encounters wise advisors and intellectual men, and fails to grasp their deeper lessons. My aim is to examine the dangers and constraints arising from Pyrrhus' warlike arrogance (*pleonexia*), rather than identify its causes, as has hitherto been the general tendency.[3] In particular, through a comparison with *Marius*, I want to show that whenever the military sphere is untouched by Greek philosophy, rhetoric, and politics, various flaws occur in private and public life, such as neglect of parental duties, political calamities, social isolation, and blind acceptance of divine omens verging on superstition. In line with the aims of the present volume, my analysis of the 'military space' in this pair of *Lives* will elucidate that this is a concept far more complex than initially assumed: it denotes not just a background setting in which biographical subjects traditionally perform grand achievements. Rather it is a textual environment where the construction and interpretation of the biographical account takes on new meanings and connotations by linking the hero's military role to his moral character and cultural identity, both of them central notions in Plutarch's biographical project. Therefore, considerations of 'military space' in Plutarch could cast new light on the purpose and overarching concerns of the *Parallel Lives* more broadly.

Spaces of action and inaction in the *Pyrrhus*

The *Pyrrhus* is unusual in having a prolonged account of the hero's childhood, adolescence, and early adulthood, and a rather awkward postponement of the account of his education. In the first seven and a half chapters, the reader becomes acquainted with Pyrrhus' physical valour, plans for expansion, and the ways he distinguishes himself on campaign. The Epirote king is admired for his bellicosity and depicted as a reflection of Achilles and Alexander.[4] Plutarch uses Homeric language to describe

2 de Jong (2014) 107–108 distinguishes 'setting' from 'frame', in that the latter encompasses distant, hypothetical reports, dreams, or memories that can nevertheless 'expand the space of a story in significant ways'.

3 Swain (1989d), (1990b) and Pelling (1989) have discussed the causes that led to the destruction of Plutarch's Roman subjects, stressing the fact that their uncontrollable passions were hardly ever tamed by Hellenic *paideia*.

4 Mossman (1988), (1992) and Braund (1997) have done much on the tragic and epic texturing of such allusions; also Duff (1999) 123–126. Mossman (2005) 512–515 discusses how Pyrrhus failed to live up to Alexander's example.

his subject's area of action in heroic terms, thus preparing for the military success he encounters later on in his career:

> There was a sharp and terrible conflict between the soldiers who engaged (in the battle), and especially also between the leaders. ... and Pyrrhus, who yielded to none of the kings in daring and prowess, and wished that the glory of Achilles should belong to him by right of valour rather than of blood alone, advanced through the foremost fighters to confront Pantauchus. At first they hurled their spears, then, coming to close quarters, they plied their swords with might and skill. Pyrrhus got one wound, but gave Pantauchus two, one in the thigh, and one along the neck, and put him to flight and overthrew him; he did not kill him, however, for his friends hauled him away. Then the Epirotes, exalted by the victory of their king and admiring his valour, overwhelmed and cut to pieces the phalanx of the Macedonians, pursued them as they fled, slew many of them, and took five thousand of them alive. (*Pyrrh.* 7.6 – 10, transl. B. Perrin[5])

This description of Pyrrhus' military excellence accounts for the stabilisation in his reign and his expedition against Macedon, but it also explains his long-standing commitment to military studies (*Pyrrh.* 8.3 – 4). According to Plutarch's account, Pyrrhus had produced writings on military tactics and leadership (8.6). He seems to have been forever studying and meditating upon military affairs, and made his subjects think only about weapons (16.1 – 3).

The presentation of successful incidents from his military career, however, is counterbalanced by reports of other activities from which Pyrrhus consciously withdraws. A good example that features early in the *Life* is the anecdote with the flute-players whom Pyrrhus openly despised at a drinking-party (*Pyrrh.* 8.7). What stands out in this incident is not merely Pyrrhus' championing of the public status of a general over that of a flute-player, but rather his failure to conform to the spirit of conviviality.[6] It is nevertheless true that there are also some more positive examples of his opposition to the symposium, as, for instance, when he magnanimously excuses some young fellows who were castigating him under the influence of wine (8.11 – 12). Soon after this, Plutarch presents another incident, which reflects other people's response to Pyrrhus' one-sided lifestyle. The implication here is that, by accepting Pyrrhus as their leader, the people of Tarentum will themselves have to abandon frivolity and parties in favour of an austere and military life, which is nicely encapsulated in Meton's warning to the people of Tarentum (*Pyrrh.* 13.8 – 10):

> Men of Tarentum, you do well not to frown upon those who wish to sport and revel (παίζειν καὶ κωμάζειν), while they can. And if you are wise, you will all also get some enjoyment still out of your freedom (καὶ πάντες ἀπολαύσετε ἔτι τῆς ἐλευθερίας), assured that you will have other busi-

---

5 All translations of Plutarch's texts are taken from B. Perrin (1920) with minor alterations.

6 *Pyrrh.* 8.7: λέγεται γάρ ὡς ἐρωτηθεὶς ἔν τινι πότῳ, πότερον αὐτῷ φαίνεται Πύθων αὐλητὴς ἀμείνων ἢ Καφισίας, εἰπεῖν ὅτι Πολυπέρχων στρατηγός, ὡς ταῦτα τῷ βασιλεῖ ζητεῖν μόνα καὶ γινώσκειν προσῆκον. ('For instance, we are told that when he was asked at a drinking-party whether he thought Python or Caphisias the better flute-player, he replied that Polyperchon was a good general, implying that it became a king to investigate and understand such matters only').

ness and a different life and diet (ὡς ἕτερα πράγματα καὶ βίον καὶ δίαιταν ἕξοντες) when Pyrrhus has come into the city.

Again the description of the sympotic space encompasses forces opposed to those of Pyrrhus' space of action. In fact, the reference to the withered garland and the torch, the dancing and singing by drunken symposiasts and Meton in particular, and the presence of a flute-girl create a scene characteristic of a *kōmos* or drunken procession of revellers,[7] something which Pyrrhus prohibits as soon as he assumes power. We learn that he closed the gymnasia and banned drinking bouts (16.2). Plutarch is careful here to present the impressions of both sides: to Pyrrhus' mind withdrawal from social festivities was a matter of prudent constraint; the Tarentines called it servitude (16.3). This balanced approach is meant to encourage Plutarch's readers to engage in critical reflection, especially as regards the effects of strict adherence to notions of generalship. The narrator's description of the sympotic space relates in this instance to the so-called psychologising function of space, meant to elucidate the hero's character and disposition.[8]

Furthermore, Pyrrhus' commitment to military affairs undermines his approach to his parental responsibilities, as it produces a distorted image of how a father is supposed to behave. It is interesting that Plutarch glosses over in silence any non-military rapport between Pyrrhus and his sons, and the only lesson that Pyrrhus appears to be transmitting to them is that of the importance of military prowess with a view to the imposition of crude authority (*Pyrrh.* 9.4–6). Such military *paideia* puts at risk the morality of his male children: he is going to bequeath his kingdom to the son who demonstrates that he has the most powerful weapon (*Pyrrh.* 9.3–6, *Reg. et imp. apophth.* 184C). Pyrrhus impels his successors towards a brutal competition that could potentially lead to fratricide. Indeed that is the outcome of such a contest in Euripides' *Phoenician Women,* a play which provides a sub-text for a reading of the *Pyrrhus* and highlights the connection between power and distorted familial

7 *Pyrrh.* 13.6–7: εἷς δέ τις ἀνήρ ἐπιεικής, Μέτων ὄνομα, τῆς ἡμέρας ἐκείνης ἐν ᾗ τὸ δόγμα κυροῦν ἔμελλον ἐνστάσης καὶ τοῦ δήμου καθεζομένου, λαβὼν στέφανον τῶν ἑώλων καὶ λαμπάδιον, ὥσπερ οἱ μεθύοντες, αὐλητρίδος ὑφηγουμένης αὐτῷ πρὸς τὴν ἐκκλησίαν ἐκώμαζεν. οἷα δὲ ἐν ὄχλῳ δημοκρατίας κόσμον οὐκ ἐχούσης οἱ μὲν ἐκρότουν ἰδόντες, οἱ δὲ ἐγέλων, ἐκώλυε δὲ οὐδείς, ἀλλὰ καὶ τὸ γύναιον αὐλεῖν κἀκεῖνον ᾄδειν ἐκέλευον εἰς μέσον προελθόντα: καὶ τοῦτο ποιήσων ἐπίδοξος ἦν. ('But there was a certain worthy man, Meton by name, who, when the day on which the decree was to be ratified was at hand and the people were taking their seats in the assembly, took a withered garland and a torch, after the way of revellers, and came dancing in behind a flute-girl who led the way for him. Then, as will happen in a throng of free people not given to decorum, some clapped their hands at sight of him, and others laughed, but none tried to stop him; nay, they bade the woman play on her flute and called upon Meton to come forward and give them a song; and it was expected that he would do so').
8 de Jong (2014) 127–128.

ties.[9] Braund justifiably stresses Pyrrhus' 'very limited concern for his sons', suggesting that 'their education serves merely as a ploy in the king's *pleonexia*'.[10]

The case of Marius resembles that of Pyrrhus in that the former is once again the type of the fierce general who has been well equipped with a purely military training, and whose identity as a man of power outweighs his role as a father. At one point Marius had even abandoned his son to save himself (*Mar.* 35.9 – 11). The ineffectual parenting of powerful fathers is compensated for in the life-pair by the presence of Antigonus in chapter 34 of *Pyrrhus:* the moralising programme that Antigonus wished to transmit to his son was founded on the virtues of humanity (*philanthrōpia*) and mildness (*praotēs*), and not on self-assertive authority. Antigonus rebukes his son, Alcyoneus, for bringing him the head of Pyrrhus, considering this an impious and barbaric act, while he praised him for treating Pyrrhus' son, Helenus, with compassion. The story of the hero's death survives in several versions, but Plutarch follows the basic core found in Strabo (8.6.18), and augments it with dramatic details and emotional effect:[11] the mighty king is wounded by a tile thrown by a distressed mother in her attempt to save her son (*Pyrrh.* 34.1 – 3). Although Pausanias (1.13.8) too follows the traditional line of the story, he refers only vaguely to a woman but not specifically to a distraught mother. Plutarch's version of the story might simply be the result of his creative imagination as influenced by other versions he consulted, but in case it is not, could it have been informed at least to some extent by Plutarch's general emphasis on Pyrrhus' inadequate performance in the parental sphere, suggestively portraying how this brought about his final catastrophe?

There are further insights into the weaknesses resulting from Pyrrhus' short-sighted military agenda, which, as we have seen, ignores vital areas of human life and activity. In the *Lives* of dedicated generals, Plutarch tends to contrast the military hero with other

---

9 On the *Phoenician Women* as a sub-text for the *Pyrrhus*, see Braund (1997) 1 – 10.

10 Braund (1997) 6.

11 Polyaenus, *Strat.* 8.68, refers to a group of Argive women—and not specifically to a mother—who were throwing all sorts of solid objects: Πύρρος Ἠπειρώτης ἐνέβαλεν εἰς Ἄργος καλέσαντος αὐτὸν Ἀριστέως Ἀργείου. οἱ μὲν Ἀργεῖοι συνέδραμον ἐς τὴν ἀγορὰν μετὰ τῶν ὅπλων, αἱ δὲ γυναῖκες αὐτῶν προκαταλαβόμεναι τὰ τέγη τοὺς Ἠπειρώτας ἄνωθεν βάλλουσαι ἀναχωρῆσαι βιάζονται, ὥστε καὶ Πύρρος, ὁ δεινότατος τῶν στρατηγῶν, αὐτὸς ἔπεσεν ἐμπεσούσης ἐπὶ τὴν κεφαλὴν αὐτοῦ κεραμίδος. αἱ δὲ Ἀργολίδες μέγιστον κλέος ἐν τοῖς Ἕλλησιν ἤραντο Πύρρου τοῦ πολεμικωτάτου μηδὲ ὑπ' ἀνδρῶν, ἀλλὰ ὑπὸ γυναικῶν Ἀργολίδων πεσόντος. Justin, *Epit. Pomp. Trog.* 25.5.1 – 2, reports that Pyrrhus was killed outside the wall of Argos by stones thrown by the army of Antigonus: *Repulsus ab Spartanis Pyrrhus Argos petit; ibi dum Antigonum in urbe clausum expugnare conatur, inter confertissimos uiolentissime dimicans saxo de muris ictus occiditur. Caput eius Antigono refertur, qui uictoria mitius usus filium eius Helenum cum Epirotis sibi deditum in regnum remisit eique insepulti patris ossa in patriam referenda tradidit.* Finally, Zonaras (*Hist. Epit.* 8.VI, vol. II, p. 90, ll. 22 – 28 Dindorf 1869), reports that Pyrrhus was killed by a woman who lost her balance and fell onto his head: Πύρρος ... ἐπὶ τὴν 'Ελλάδα στρατεύσας οὐ πολλῷ ὕστερον ἐν Ἄργει ἀπέθανε. γυνὴ γάρ τις, ὡς λόγος ἔχει, παριόντα αὐτὸν ἰδεῖν ἀπὸ τοῦ τέγους ἐπιθυμήσασα ἐσφάλη καὶ ἐμπεσοῦσα διέφθειρεν αὐτόν.

figures, who normally exhibit a more varied range of *paideia*.[12] These have been called 'foil figures' (*Folienfiguren*), and we get two apt examples of that confrontational technique in Pyrrhus' interaction with Cineas and Fabricius in chapters 14 and 20–21, respectively. Although Cineas is sketched as Pyrrhus' political counsellor, his contribution is not just aimed at strengthening Pyrrhus' kingship, but more at improving his moral condition.[13] For that reason, he employs his rhetorical eloquence—otherwise a vital tool of political success—in order to lead Pyrrhus to some degree of self-awareness through a series of dialectical questions. This Socratic scene is omitted from both Dio Cassius' corresponding account (*Hist. Rom.*, 40.5) and Appian's *History of Rome* ('The Samnite War', 17–18, 22–30), presumably because the moralising strands of the confrontation would have been of less interest to the two historians. Cineas exposes the vanity of Pyrrhus' *pleonexia*, trying to rectify his political aspirations through a better regulated moral outlook, yet to no avail; in the end, rather than changing his mind, Pyrrhus is simply disquieted.

Fabricius is a very similar case. He is a Roman ambassador whom Pyrrhus approaches with compliments and attempts to bribe (*Pyrrh.* 20.2–4). Here Pyrrhus is naive in his political actions that might have caused some mild laughter among spectators: despite his failure to get Fabricius to accept gold, the very next day he offers him an elephant, which Fabricius once again rejects. Pyrrhus does not seem to understand the priorities of a politically-informed person, and mistakenly confuses the things that might have attracted a general with those potentially more appropriate to a politician. The remarks which conclude Fabricius' conversation with Pyrrhus (20.9) also have a bearing on the latter's lack of philosophical awareness. In this episode a discussion on Epicureanism takes place, which, interestingly, is launched and articulated by Fabricius and Cineas, not Pyrrhus, who is uninterested in intellectual topics. This discussion—which is omitted in Dio Cassius—dwells, *inter alia*, on the Epicurean view of public retirement that Pyrrhus and his followers appear to have espoused, despite the fact that they seem to have devoted little attention to the true essence of this doctrine. Plutarch is the only source to refer explicitly to the name of Epicurus in relation to this episode (cf. Cic., *De Sen.* 43, Val. Max. 4.3.6), and this reflects the biographer's concern to stress Pyrrhus' limitations in philosophical contemplation, an area which might have allowed the hero's natural abilities to flourish.

Plutarch is also interested in looking at Pyrrhus' behaviour in the religious sphere, and at how this might have affected his military choices. A recurrent motif throughout the hero's polemical exploits is his loyalty to the gods (*Pyrrh.* 3.4, 12.7, 29.3, 30.5, 32.10) and fortune (*tychē*) (26.7), the attention he pays to dreams (29.2–4),[14] omens (31.7), and prophecies (30.5, 32.8), and his decision to dedicate

12 Bucher-Isler (1972) 62–68; Beck H. (2002) 468–469. Cf. Xenophontos (2012) on Fabius Maximus.

13 Buszard (2008) 203–205 compares Pyrrhus' dialogue with Cineas to Socrates' dialogue with Alcibiades in *Alcibiades I*.

14 For dreams in Tacitus and Plutarch, see Pelling (1997b).

his military spoils to the gods (26.9). One has to reflect upon the implications of such commitment to the divine, especially in instances when Pyrrhus trusts to fate rather than his reason in military campaigns (καὶ τῇ τύχῃ μᾶλλον ἢ τοῖς λογισμοῖς χρώμενος, 26.7).[15] In Plutarch the limits between reverence and superstition are often blurred and I would suggest that Pyrrhus shows the signs of a superstitious man that are probably consistent with his lack of proper *paideia*.

The most famous description of the superstitious man can be found in Theophrastus' *Characters* (Δεισιδαιμονίας).[16] Yet where Theophrastus mocks,[17] Plutarch analyses and ponders.[18] Plutarch classifies *deisidaimonia* as a barbarian and not a Greek characteristic, considering it fitted only to uncivilised men,[19] and in his *On superstition* he offers practical instructions for distinguishing the superstitious man from the atheist.[20] Pyrrhus shares a number of common features with the superstitious figure of the essay. To begin with, ignorance stimulates superstition in feeble and pliable characters (*De superst.* 164E). Pyrrhus' lack of political and philosophical knowledge leads him to vacillate between his military toughness on the one hand and his mildness and humanity on the other hand. Secondly, if emotion (*pathos*) is added to that psychological state, then something very bad happens (164E). Pyrrhus' *pathos* is his *pleonexia*, which eventually causes him some serious problems on a number of levels, as we have observed. Thirdly, the *deisidaimōn* is agitated by dreams (165E ff.),[21] and is constantly preoccupied with seers and omens (168F); so too, as we have seen, is Pyrrhus. Furthermore, in 167B Plutarch cites Plato in support of the claim that music can control the 'troubled' part of the soul that lacks sophistication and refinement; Pyrrhus' aversion to both music and leisure could explain his deficiencies in this respect. Finally, the *deisidaimōn* dislikes politicians and philosophers (167E); one might recall Pyrrhus' encounter with Cineas and Fabricius and his poor grasp of Epicurean doctrines.

---

15 This is reminiscent of the portrayal of Marius in Sallust's *Jugurtha*, esp. 91–95, where similar issues arise; in 92.6–7 Marius' success is presented as the result of chance rather than of skill. In 94.6 Marius has some good fortune and gains glory through an error of judgment. In 92.2 Marius' circle believed that he was either possessed of divine insight or everything was revealed to him by the favours of the gods.

16 On Plutarch's predecessors and the use of the superstitious character, see Moellering (1963) 80–88.

17 Despite the moralising purpose of Theophrastus' *Characters* (Diggle [2004] 12, obvious also from the *Preface* of the work), the comic strand in the delineation of characters cannot easily be overlooked. Diggle (2004) 349 observes that whereas *deisidaimōn* used to be a neutral or even approbative term, with Theophrastus it acquired negative connotations.

18 On superstition as a feature that Plutarch's politician should moderate in his behaviour, Wardman (1974) 86–93.

19 Nikolaidis (1986) contrasts Greek and barbarian elements. Cf. Schmidt (1999).

20 On superstition in Plutarch, Pérez-Jiménez (1996), esp. 204–205 n. 42 for further references.

21 Theophrastus 16.11 with Diggle (2004) 368.

True, there is more to Pyrrhus' portrayal than this. He neither hates nor blames the gods (168C; 170E), nor does he remain idle for fear of divine retribution (168C). Still, we do find substantial indications in his conduct of superstition,[22] which are consistent with the emphasis usually given to the superstitious conduct of other military heroes. Even Nicias (*Nic.* 23–24), who is not an exclusively military figure, is confounded by his superstitious nature when on campaign. The anecdote in *Pericles* 6, on the other hand, about the educated response to popular *deisidaimonia*, reflects how reasoned explanation can eliminate unfounded impressions, and thus set the limits within which consulting the divine is supposed to operate in the military and political sphere.

Military space and *paideia* in the *Marius*

In turning to *Marius*, I wish to explore how Plutarch adjusts some common elements in the spaces of action of the two heroes, and comment on the significance of this readjustment for Plutarchan notions of culture. Unlike Pyrrhus, from the very beginning of his *Life* Marius' education is the focus of attention, with forceful references to his rejection of Hellenic culture in favour of its Roman, military counterpart (2.1–2): it would be completely absurd, Marius thinks, to be educated by teachers who were the slaves of the Romans. His resistance to Greek *paideia* is also manifested in a dictum ascribed to Plato that highlights his opposition to the Muses and the Graces,[23] thus explaining the many hasty changes in his behaviour, quite unlike what happens in the *Pyrrhus*. That trait becomes critical in the ensuing narrative and is recurrently stressed throughout: in 3.2 we encounter Marius' change and adaptability to a new way of life; in 4.7 and 5.3 we get two instances of quick changes of a political resolution; in 28.1–2 there is a superficial change in his character, when he pretends to be mild and democratically inclined in order to please the people; in 29.1 he is detested by the patricians for changing his political position; finally, in 34.3–7 we see Marius succumbing to luxury in his old age, despite his hitherto unsophisticated lifestyle.

---

22 *Alex.* 75.1–2; cf. *Fab.* 4.4.

23 *Mar.* 2.3–4: ὥσπερ οὖν Ξενοκράτει τῷ φιλοσόφῳ σκυθρωποτέρῳ δοκοῦντι τὸ ἦθος εἶναι πολλάκις εἰώθει λέγειν ὁ Πλάτων· 'ὦ μακάριε Ξενόκρατες, θῦε ταῖς Χάρισιν', οὕτως εἴ τις ἔπεισε Μάριον, θύειν ταῖς Ἑλληνικαῖς Μούσαις καὶ Χάρισιν, οὐκ ἂν ἐκπρεπεστάταις στρατηγίαις καὶ πολιτείαις ἀμορφοτάτην ἐπέθηκε, ὑπὸ θυμοῦ καὶ φιλαρχίας ἀώρου καὶ πλεονεξιῶν ἀπαρηγορήτων εἰς ὠμότατον καὶ ἀγριώτατον γῆρας ἐξοκείλας. ταῦτα μὲν οὖν ἐπὶ τῶν πράξεων αὐτῶν εὐθὺς θεωρείσθω. ('Accordingly, just as Plato was wont to say often to Xenocrates the philosopher, who had the reputation of being rather morose in his disposition, "My good Xenocrates, sacrifice to the Graces," so if Marius could have been persuaded to sacrifice to the Greek Muses and Graces", he would not have put the ugliest possible crown upon a most illustrious career in field and forum, nor have been driven by the blasts of passion, ill-timed ambition, and insatiable greed upon the shore of a most cruel and savage old age. However, his actual career shall at once bring this into clear view').

In chapter 29 the presence of Metellus is designed to highlight a contrast between the political practices of the two men, to shed light on Marius' lack of proper *paideia* and ultimately demonstrate Marius' perverted form of statecraft: the general considers public dishonesty an element of political virtue, whereas Metellus embraces the Pindaric axiom which is in line with the principle of political morality.[24] Metellus is a typical 'foil figure' and it is interesting that he is philosophically educated (*Mar.* 29.12), which is why Plutarch thinks very highly of him.[25]

The Platonic quotation mentioned above, demonstrating Plutarch's view of Marius' rejection of the Muses and the Graces, recalls its deployment in *On superstition*, where helps classify Pyrrhus in the category of superstitious men. This makes it tempting to start searching for similar elements in *Marius* as well. Indeed, one finds numerous references there too to the role of the gods (4.1, 8.5), the significance of *tychē* (7.2, 14.1, 23.1, 19.10), of prophets, consultation of omens and celestial signs, and sacrifices (17.2, 22, 26.2–4, 36.8–10, 38.7–10, 40.13).[26] Moreover, in 42.7 Plutarch suggests that prophecies are a disease and considers it a paradox that whereas Marius was saved thanks to using them, Octavius was destroyed by them. What is more, in 45.5 Marius sleeps badly and is agitated by nightmares. Just like Pyrrhus, Marius too crosses the boundary between reverence and superstition, though in this case it does not contribute directly to his downfall. In general, his rejection of culture keeps him focused on war, and his natural inclination towards peace and political participation (31.3) fails to develop due to lack of suitable didactic influences. In *Marius* one can also see how the same qualities may be shown to have positive effects in warfare, but turn out disastrously in politics.

---

24 *Mar.* 29.5: αὐτὸς μὲν γὰρ εἰς ἀρετῆς καὶ δεινότητος μερίδα τὸ ψεύσασθαι τιθέμενος, λόγον οὐδένα τῶν πρὸς τὴν σύγκλητον ὡμολογημένων ἕξειν ἔμελλε, τὸν δὲ Μέτελλον εἰδὼς βέβαιον ἄνδρα καὶ τὴν 'ἀλήθειαν ἀρχὴν μεγάλης ἀρετῆς' κατὰ Πίνδαρον ἡγούμενον, ἐβούλετο τῇ πρὸς τὴν σύγκλητον ἀρνήσει προληφθέντα καὶ μὴ δεξάμενον τὸν ὅρκον εἰς ἀνήκεστον ἐμβαλεῖν πρὸς τὸν δῆμον ἔχθραν. ('For he himself regarded lying as part of a man's excellence and ability, made no account of his agreements with the senators, and did not intend to keep them; whereas he knew that Metellus was a steadfast man, who thought with Pindar that "truth is the foundation of great excellence", and he therefore wished to bind him beforehand by a statement to the senate that he would not take the oath, and then have his refusal to do so plunge him into a hatred on the part of the people that could never be removed. And this was what came to pass').

25 Consider Buszard's apt distinction between Metellus' personal virtue and his political effectiveness. It is true that Metellus, despite his 'admirable figure' and 'noble behaviour', does succumb to political mistakes. This, however, does not undermine his political feasibility: his mistakes, unlike Marius', do not derive from his uncultured character, (2005) 491. Metellus is a foil for Marius in Sallust too, but education does not feature in the discussion so much. A compare-and-contrast with Sallust would be interesting, though Plutarch might not have known the *Jugurtha* directly (unlike the *Histories*). On Metellus and Marius in *Jugurtha*, see also Syme (1964) 142–151.

26 On Marius and the gods, McDonnell (2006) 267–271.

Conclusions: Military space and cultural identity

In his *Lives of Pyrrhus and Marius* Plutarch emphasises the heroes' engagement with military affairs, and their lack of interest in intellectual, social, and familial activities. However, their devotion to military affairs is justified for different reasons. Pyrrhus acquires a very complex cultural identity in the text; he is certainly not a Roman, but then again not a pure Greek either. And this remains unresolved partly because Plutarch does not explicitly associate his military greed specifically with his unfamiliarity with Greek culture. The biographer is keen to play down any direct links with Hellenism. For Marius, by contrast, the hero's sense of *Romanitas* is clearly in play from the start of the biography and is explicitly related to his aversion to Greek *paideia*. This might have to do with Plutarch's general tendency to evaluate and comment upon the Hellenic affiliations of his Roman heroes in particular; yet one of the consequences of his references to cultural categories is that Plutarch is not interested in determining the identity of his heroes simply on the basis of origin, but more in the light of their ethical behaviour. As we have seen, in Plutarch's biographical project, the heroes' successes are heavily dependent on their performance in the cultural and moral sphere.[27]

27 See also discussion in Xenophontos (2016).

Andrea Catanzaro

Astronomical and political space: The sun's course and the statesman's power in Plutarch and Dio

Abstract: One of the most relevant topics of Roman and Greek political thought between the 1st and the 2nd century CE is the problem of the *princeps*' unlimited power and how it could be restricted institutionally; so the authors were deeply concerned, paradoxically, about limiting what was unlimited (ἀνυπεύθυνος) by definition. To many of them the metaphor of the sun was the best answer to their question. The course of the sun made the *princeps*' space of action as clear as possible: he was both absolute and limited, able to give life or death, essential for all creatures that were indispensable to justify his power. Astronomical space invests the space of politics and the metaphor of the sun with a common political language. In this chapter I focus on the transposition of this idea of space from the metaphorical dimension to the pragmatic field of the *princeps*' power as it appears in the *To An Uneducated Ruler* of Plutarch and in the *Third Discourse on Kingship* of Dio Chrysostom, in order to highlight analogies and differences, according to their respective political visions.

The limits of the *princeps*' absolute power (ἀνυπεύθυνος[1] ἀρχή) was one of the thorniest political issues for imperial authors during the 1st and 2nd centuries CE. Was it possible to limit what was unlimited by nature? What kind of boundaries, if any, could prevent a *princeps* from becoming a tyrant? The evolution of the *principatus* had rendered these questions crucial.[2] The issue was important to Plutarch and Dio Chrysostom as well.[3] These authors addressed this problem which they had seen arise dramatically during Nero's and Domitian's *principatus*. In Plutarch's *To An Uneducated Ruler* and in the *Third Discourse on Kingship* by Dio, for example, the problem of the limits to the *princeps*' power is widely discussed. Both authors stress it by using the sun's metaphor, which allowed them to make a theoretical

1 On the significances of the adjective ἀνυπεύθυνος, see LSJ s.v. On the uses of ἀνυπεύθυνος in ancient political works, see for example Plato, *Laws* (3, 691c; 6, 761e; 9, 875b) and Aristotle, *Politics* (2, 1272a; 4, 1292a, 1292b, 1295a). Two uses in Plutarch are worth pointing out, because of their importance from the perspective of political thought. In *De unius* (826E), he uses ἀνυπεύθυνος in order to describe the Persian regime; in *Fab.* 3.7, this adjective characterises the Roman dictatorship.
2 Cf. Mazzarino (1982) 805–875.
3 On the contextualisation of these authors, cf. Stadter and Van Der Stockt (2002); Whitmarsh (2001a); Swain (1996); Swain (2000); Hidalgo De La Vega (1998) 1015–1058; Salmeri (2000) 60–63; Aalders (1982); Desideri (1978); Ziegler (1965); Von Arnim (1898); cf. Desideri (2000) 93: 'Dio of Prusa was, together with Plutarch, the first author, and one of the most important ones, of the so-called Greek Renaissance in the time of the High Roman Empire'.

https://doi.org/10.1515/9783110539479-031

idea concrete, in order to show to what extent the *princeps* could use his power without becoming a tyrant.

Two concepts need to be especially highlighted in terms of their political connotations, with a view to analysing how Plutarch and Dio deal with the limits of absolute power through the image of the sun: namely, space and time as they appear in the metaphor in question.[4] In the treatise *To An Uneducated Ruler* and in the *Third Discourse*, Plutarch and Dio Chrysostom, respectively, summarise the problem, albeit in different ways: the former rhetorically asks himself Τίς οὖν ἄρξει τοῦ ἄρχοντος; ('Who, then, shall rule the ruler?', 780C),[5] the latter highlights that, in an absolutist regime, the νόμος corresponds to the βασιλέως δόγμα (3.43).[6] Both pose the same question: the core of their analysis deals with the issue of limits. Where there are no institutional means capable of allowing men to choose their rulers, to remove a *princeps* become tyrant or to compel him to pursue the common good, it is essential to establish whether other limits to his power might exist and, therefore, what these limits are. Plutarch and Dio suggest managing the problem through education *to* and *in* virtue, which are considered as an effective means to reach the target of the common good.[7] Their main difference lies in the emphasis exclusively laid by Dio on the *princeps*, whereas Plutarch's view seems related to other figures as well.[8]

The Roman transition to monocratic power changed the way through which philosophers could perceive themselves within the political arena. Preceptors and *princeps*' mentors started to take a remarkable position in the political game, where they sometimes became eminent actors.[9] Although they easily solved the problem of finding *who* would be able to train the *princeps* with a view to making him a virtuous ruler and *how*, two questions were still far from being answered. What kind of virtue were they talking about? And, given that the purpose was to teach this virtue, how was it possible to apply something that was theoretical by nature? Since both in Plutarch's and in Dio's conception the model of this virtue was Zeus,[10] how could it be possible to make

4 On the political metaphor, see Euchner et al. (1993); Rigotti (1989).

5 From this point forward, all the Plutarchan excerpts and their translations that I mention in this contribution come from the Loeb; about this Plutarchan work see Tirelli (2005); Whitmarsh (2001a) 186.

6 From this point forward, all the Dionian excerpts and their translations that I mention in this essay come from the Loeb.

7 Roskam (2002) 181: 'it is the task of a ruler to take care of his subjects, and the best way to do this consists in leading them towards moral virtue'.

8 In Plutarch's *To An Uneducated Ruler*, for example, we find a mention of both kings and the rulers (779F) and a reference to the ruler ἐν πόλεσι (780F), from which we might deduce that he is not exclusively speaking about the *princeps*.

9 Cf. Dillon (2002) 29–40.

10 On Plutarch's idea about kingship, cf. Aalders (1982) 34; on Dio's ideas concerning monarchy, cf. Swain (1996) 195–206; on Zeus as a model in Dio's *Discourses on Kingship*, cf. Desideri (2015) 322 and (2012b) 14, Gangloff (2009) 17–29, Swain (1996) 196; cf. also Whitmarsh (2001a) 214: 'In certain passages of the *Kingships*, the divine represents an unattainable limit of power, and a warning to the em-

that example clear to the *princeps?* In other words, if kingship 'is a high duty, even a service of divinity, of whom the (perfect) king is an image',[11] how could this idea be shifted from the metaphysical sphere to the physical one?

In *To An Uneducated Ruler* and in the *Third Discourse*, Plutarch and Dio choose to use the sun metaphor with a view to illustrating the idea of Zeus as the best model of the virtuous *princeps.*[12] Unfortunately, the mere comparison between the king and the sun was not enough to highlight the idea of limited power. What was also needed was to make concrete what was theoretically difficult to be demonstrated. That is why the concepts of space and time that these authors employ, albeit to a different degree, seem worth stressing: in political terms they suggest that the *princeps*' power has some limits—in spite of its absolute nature—and illustrate what these limits are. Plutarch's and Dio's ultimate objective is the common good (*To An Uneducated Ruler*, 780D–E; *Third Discourse*, 73). The metaphysical model is the father of the gods (780D; 3.82) and its physical translation is the image of the sun (780F and 781F; 3.73–81). The sun's elevation from the earth finds its parallel in the notion of *vertical* power, while the sun's course points to the unremitting *duration* of the *princeps*' virtuous activity (metaphorically, from his rise to power to his sunset). The sun's force, finally, suggests that absolute power can both *give life* and *kill*. According to Plutarch and Dio, the sum of all these elements—*verticality*, *duration* and *authority over life and death*, or, in other words, *space*, *time* and *purpose*—should make clear to the ruler or the rulers how carefully they have to manage their power.[13] There are evident differences in detail: Dio's description appears richer than Plutarch's. Some of the divergences concern spatial and temporal dimensions. But the two authors share a crucial conceptual analogy which is about the purposes of political power: the sun's existence and role are linked to a political model whose roots go back to the Homeric poems:[14] rulers have to take care of the common good; their power is strictly linked to their subjects, since it exists only in their presence.[15]

In *To An Uneducated Ruler* the aim of the ruler's activities is spelled out:

peror not to overreach mortal boundaries ... Divinity ... in the *Kingships* functions as a transcendent signifier of kingship'.

11 Aalders (1982) 34.

12 On this metaphor in Plutarch's *Ad princ. iner.*, cf. Roskam (2002) 179–180; on this image in Dio, cf. Desideri (2015) 322.

13 On the spatial metaphors of politics and on the correspondences between natural order of universe and the hierarchical structure of societies, see Rigotti (1989) 85–102.

14 Cf. for example Hom., *Il.* 2.204–206; on the significance of this model in Dio's thought, cf. Whitmarsh (2001a) 212 and 244–245.

15 Cf. Aalders (1982) 34: '[The king] shall devote himself fully to his subjects'.

ἀληθέστερον δ᾽ ἄν τις εἴποι τοὺς ἄρχοντας ὑπηρετεῖν θεῷ πρὸς ἀνθρώπων ἐπιμέλειαν καὶ σωτηρίαν, ὅπως ὧν θεὸς δίδωσιν ἀνθρώποις καλῶν καὶ ἀγαθῶν τὰ μὲν νέμωσι τὰ δὲ φυλάττωσιν.	One might more truly say that rulers serve god for the care and preservation of men, in order that of the glorious gifts which the god gives to men they may distribute some and safeguard others.
ὁρᾷς τὸν ὑψοῦ τόνδ᾽ ἄπειρον αἰθέρα, *καὶ γῆν πέριξ ἔχονθ᾽ ὑγραῖς ἐν ἀγκάλαις;*	*Dost thou behold this lofty, boundless sky* *Which holds the earth enwrapped in soft embrace?*
ὁ μὲν καθίησιν ἀρχὰς σπερμάτων προσηκόντων γῆ δ᾽ ἀναδίδωσιν, αὔξεται δὲ τὰ μὲν ὄμβροις τὰ δ᾽ ἀνέμοις τὰ δ᾽ ἄστροις ἐπιθαλπόμενα καὶ σελήνῃ, κοσμεῖ δ᾽ ἥλιος ἅπαντα καὶ πᾶσι τοῦτο δὴ τὸ παρ᾽ αὑτοῦ φίλτρον ἐγκεράννυσιν.	The sky sends down the beginnings of the appropriate seeds, and the earth causes them to sprout up: some are made to grow by showers and some by winds, and some by the warmth of stars and moon; but it is the sun which adorns all things and mingles in all things that men call the 'love charm' which is derived from himself. (780D–E)

The sun has an enormous and essential power: it has to put God's gift in order, that is, to take care of the common good. Accordingly, 'the ruler should ... have correct opinions about his task: he should know that a true ruler serves God for the care and preservation of men, and that he should either distribute or safeguard the beautiful and good things which God gives to men ..., thus functioning in a sense as an intermediary between the divine and the human world'.[16]

In the *Third Discourse* of Dio (3.73–83) we find a very elaborate comparison between the sun and the ruler. It is worth discussing it, since it clarifies the point of the common good, suggests something remarkable about the theme of time and introduces the question of unremitting work that will be taken into account later.

Through a detailed description of the role of the sun in the universe and, particularly, of its influence on seasonal changes, Dio shows the importance of the *princeps*' temperance in handling his absolute power (3.74–80). He lays great emphasis on space and time as limits in these sections: if the sun were not careful in spreading its force in accordance with the processes indispensable for creating and growing natural creatures, the universe would fall into disorder, and would be completely unfit for life. Furthermore, as it will be shown in the next pages, the sun is not allowed to stop doing so even for a moment. Because of this strong and paradoxical dichotomy between this absolute power and the virtuous behaviour requested of its holder, Dio compares the sun's tasks to slavery (3.75).

According to both authors, the best example of how to take care of one's subjects is offered by Zeus' government of the universe. But in Dio's *Third Discourse*, the emphasis is on the spatial and temporal aspects of the sun which serve to stress the limits of power. Whereas Plutarch simply says that the sun creates order in all things (κοσμεῖ δ᾽ ἥλιος ἅπαντα, 780E), Dio lays more emphasis on the nature of this task. Since, as shown above, the sun is a symbol of absolute power, the metaphor

16 Roskam (2002) 180.

makes it evident that this force is not to be used outside certain boundaries. Within the *Third Discourse*, this is articulated by stressing the spatial dimension:

οὕτω δὲ πάνυ ἀσφαλῶς καὶ ἀραρότως τηρεῖ τοὺς ὅρους πρὸς τὸ ἡμῖν συμφέρον, ὥστε εἴτε προσιὼν ὀλίγον ἐγγυτέρω γένοιτο, πάντα ἂν συμφλέξειεν, εἴτε ἀπιὼν μικρὸν ὑπερβάλοι, πάντα ἂν ἀποψυγείη τῷ κρύει.	And with such perfect nicety of adjustment does he [the sun] observe his bounds with respect to our advantage that, if in his approach he got a little nearer, he would set everything on fire, and if he went a little too far in his departure, everything would be stiffened with frost. (3.79)

The sun's course becomes a powerful metaphor for the spatial and temporal limits/ unlimits of the *princeps'* absolute power.

Although Plutarch's metaphor is less detailed than Dio's, he stresses the same theme in *To An Uneducated Ruler:*

φοβεῖσθαι δὲ δεῖ τὸν ἄρχοντα τοῦ παθεῖν κακῶς μᾶλλον τὸ ποιῆσαι: τοῦτο γὰρ αἴτιόν ἐστιν ἐκείνου καὶ οὗτός ἐστιν ὁ φόβος τοῦ ἄρχοντος φιλάνθρωπος καὶ οὐκ ἀγεννής, ὑπὲρ τῶν ἀρχομένων δεδιέναι μὴ λάθωσι βλαβέντες.	But the ruler should have more fear of doing than suffering evil; for the former is the cause of the latter; and that kind of fear on the part of the ruler is humane and not ignoble to be afraid on behalf of his subjects lest they may without his knowledge suffer harm. (781C)

Common to both authors is the idea that the common good depends on a careful ruler's government that has to be continuously adjusted so as not to allow absolute power to escape its boundaries and weaken the ruled. Furthermore, both Plutarch and Dio stress the existence of a superior equilibrium whose observance makes the *princeps*' power not completely absolute. Moreover, the possession of political power is linked to a responsibility towards a higher power:

ἀλλ' οἱ πολλοὶ κακῶς φρονοῦντες οἴονται πρῶτον ἐν τῷ ἄρχειν ἀγαθὸν εἶναι τὸ μὴ ἄρχεσθαι.	But most people foolishly believe that the first advantage of ruling is freedom from being ruled. (780C)

The warning in *Political Precepts* 813E, 'You who rule are a subject, ruling a State controlled by proconsuls, the agents of Caesar' (ἀρχόμενος ἄρχεις, ὑποτεταγμένης πόλεως ἀνθυπάτοις, ἐπιτρόποις Καίσαρος), can be considered as a clear example of the ruler's limited power. Obviously Plutarch is writing here about a local political situation and he is clearly not speaking about an absolute *princeps*. However, the idea of *controlling and being controlled* applies to all men who are in power, whether we consider it in its astronomic or human dimensions. The *princeps* has nothing similar to Roman proconsuls in terms of his power's limits, but shares with them a more powerful bond: he has to guarantee his subjects the common good through a careful handling of his absolute power. If the fragile balance between the *princeps*' absolute power and the need to keep it within clear-cut boundaries were to be upset, the *principatus* would run the risk of becoming a tyrannical regime. Plutarch effectively summarizes the issue as follows: 'There is indeed great danger that he who can do what he wishes may wish what he ought not to do' (μέγας οὖν ὁ κίνδυνος βούλεσθαι ἃ μὴ δεῖ τὸν ἃ βούλεται ποιεῖν δυνάμενον, 782C).

As previously outlined, there is another point which is worth highlighting. In order to achieve the objective of the common good, it is sometimes not enough to adopt the behaviour requested to a good king: a ruler has to operate unremittingly in accordance with virtue. According to Plutarch and Dio, the subjection of absolute power to the subjects' care and pursuit of the common good renders this power dependent on a continual virtuous behaviour on the part of its holder. This emphasis on continuity represents another essential means through which the *princeps*' absolute power can be kept under control, in spite of the lack of institutional process capable of ensuring it. Dio's lengthy development of the sun's metaphor illustrates this aspect: whoever is in charge has to take care of his subjects *perpetually* (δι' αἰῶνος,[17] 3.73 and 74). The ruler is not allowed to ignore his political duties even for a moment nor to put aside his assignments even for a short period. Due to this, Dio observes that the power relation between ruler and rulers—in metaphorical terms the vertical space between the sun and the earth—[18] seems to be overturned: the king becomes a sort of slave because of his unremitting duties towards his subjects.[19]

Plutarch, in turn, does not explicitly discuss the theme of the constancy of virtue through the sun metaphor, but uses another image to highlight that those who want to rule other people must previously learn to control themselves according to the precepts of virtue.

δεῖ δέ, ὥσπερ ὁ κανὼν αὐτός, ἀστραβὴς γενόμενος καὶ ἀδιάστροφος, οὕτως ἀπευθύνει τὰ λοιπὰ τῇ πρὸς αὐτὸν ἐφαρμογῇ καὶ παραθέσει συνεξομοιῶν, παραπλησίως τὸν ἄρχοντα πρῶτον αὐτὸν τὴν ἀρχὴν κτησάμενον ἐν ἑαυτῷ καὶ κατευθύναντα τὴν ψυχὴν καὶ καταστησάμενον τὸ ἦθος οὕτω συναρμόττειν τὸ ὑπήκοον οὔτε γὰρ πίπτοντός ἐστιν ὀρθοῦν οὔτε διδάσκειν ἀγνοοῦντος οὔτε κοσμεῖν ἀκοσμοῦντος ἢ τάττειν ἀτακτοῦντος ἢ ἄρχειν μὴ ἀρχομένου.	But just as a rule, if it is made rigid and inflexible, makes other things straight when they are fitted to it and laid alongside it, in like manner the sovereign must first gain command on himself, must regulate his own soul and establish his own character, then make his subjects fit his pattern. For one who is falling cannot hold others up, nor can one who is ignorant teach, nor the uncultivated impart culture, nor the disorderly make order, nor can he rule who is under no rule. (780B–C)[20]

17 On the meanings of the expression δι' αἰῶνος, see LSJ s.v.

18 On the vertical space and its significances in the political metaphors, see Rigotti (1989) 93–102.

19 On this topic in Plutarch, see *Ad. princ. iner.* 779E; cf. Roskam (2002) 184.

20 See Roskam (2002) 182: 'A virtuous ruler ... can carry his people with him by means of his own example of moral excellence. A ruler must therefore first put his own soul in order, and then model his subjects after his own pattern, just as a rule should first itself be straight and rigid, before it can make other things as straight as itself'.

This idea of both controlling and unremittingly presenting himself as an archetype of virtue is shown both in *Old Men in Public Affairs* 791C and in *Political Precepts* 800B and 800E–F. In *To An Uneducated Ruler*, the temporal dimension of the political office is implicit in the sun's metaphor (780D–E). Seasons and other atmospheric variations are not clearly mentioned, but it is significant enough that Plutarch, like Dio, highlights the role played by the sun in making the seeds that gods bestow human beings germinate. Albeit in different ways, both authors perceive time as another element capable of limiting the *princeps*' absolute power. Pauses, stops and hesitations are not allowed: avoiding virtuous behaviour even for a while affects his subjects, whose lives a good ruler has to safeguard carefully.

To sum up: the sun's metaphor in Plutarch's *To An Uneducated Ruler* is not as detailed as in Dio, because Plutarch seems mainly focused on the political consequences of imitating Zeus. For this reason, the concepts of time and space—particularly the former, since the latter is expressed as a function of it—have a stronger presence in the *Third Discourse on Kingship* than in the Plutarchan work.[21] However, the outcome seems to be very similar: in order to propose a solution to the issue of the limits of the ἀνυπεύθυνος ἀρχή, both authors choose the same metaphor, according to which a ruler's individual space and time are conceived as being closely linked with those of the ruled people, within the framework of a superior natural order that exists because of Zeus, and the unremitting virtuous behaviour of the good *princeps*.

21 Cf. Plu. *Ad princ. iner.*, 780E–F, 781A, 781F, 782D.

Bibliography

Aalders (1982): Gerhard J.D. Aalders, *Plutarch's Political Thought*, Amsterdam, Oxford and New York.

Abramowicz (1962): Zofia Abramowicz, "Plutarchs 'Tischgespräche'", *Das Altertum* 8, 80–88.

Alcock (1996): Susan E. Alcock, "Landscapes of Memory and the Authority of Pausanias", in: Jean Bingen (ed.), *Pausanias Historien* (Entretiens de la Fondation Hardt, 41, 15–19 Août 1994), Vandœuvres and Geneva, 241–276.

Alcock et al. (2001): Susan E. Alcock, John F. Cherry and Jaś Elsner (eds.), *Pausanias: Travel and Memory in Roman Greece*, Oxford.

Alcock (2003): Susan E. Alcock, "Power Lunches in the Eastern Roman Empire", *Michigan Quarterly Review* 42.4, 591–606.

Alexiou (2000): Evangelos Alexiou, "Die Funktion der Homerzitate in Plutarchs Biographien", in: Andreas Haltenhoff and Fritz-Heiner Mutschler (eds.), *Hortus Litterarum Antiquorum*, Heidelberg, 51–65.

Alexiou (2007): Evangelos Alexiou, *Πλουτάρχου Παράλληλοι Βίοι. Η προβληματική των θετικών και αρνητικών παραδειγμάτων*, Thessaloniki.

Alexiou (2010): Evangelos Alexiou, "Plutarchs Lysander und Alkibiades als Syzygie: Ein Beitrag zum moralischen Programm Plutarchs", *RhM* 153, 323–352.

Alexiou (2013): Evangelos Alexiou, "Philotimia oder Pleonexia als κακίστη δαιμόνων? Dion von Prusa 17,9 und die geistegeschichtlichen Konturen eines euripideischen Verses", *RhM* 156, 47–73.

Allan (2007): Rutger J. Allan, "Sense and sentence complexity: sentence structure, sentence connection, and tense-aspect as indicators of narrative mode in Thucydides' *Histories*", in: Rutger J. Allan and Michel Buijs (eds.), *The Language of Literature: Linguistic Approaches to Classical Texts*, Leiden, 93–121.

Allan (2009): Rutger J. Allan, "Towards a typology of the narrative modes in ancient Greek: text types and narrative structure in Euripidean messenger speeches", in: Stéphanie J. Bakker and Gerry C. Wakker (eds.), *Discourse Cohesion in Ancient Greek* (Amsterdam Studies in Classical Philology 16), Leiden, 171–203.

Allan and Buijs (2007): Rutger J. Allan and Michel Buijs (eds.), *The Language of Literature: Linguistic Approaches to Classical Texts*, Leiden.

Almagor (2013a): Eran Almagor, "Dualism and the Self in Plutarch's Thought", in: Jorg Rüpke and Gregory D. Woolf (eds.), *Religious Dimensions of the Self in the Second Century CE*, Tübingen, 3–22.

Almagor (2013b): Eran Almagor, "But This Belongs to Another Discussion": Ethnographic Digressions in Plutarch' in: Eran Almagor and Joseph Skinner (eds.), *Ancient Ethnography: New Approaches*, London, 153–178.

Almagor (2014): Eran Almagor, "The *Aratus* and the *Artaxerxes*", in: Mark Beck (ed.), *A Companion to Plutarch*, Malden, Mass. and Oxford, 278–291.

Alt (1993): Karin Alt, *Weltflucht und Weltbejahung. Zur Frage des Dualismus bei Plutarch, Numenius, Plotin*, Stuttgart.

Amandry (1984): Pierre Amandry, *Delphi and Its History*, Archaeological Guides, Athens.

Ampolo and Manfredini (1988): Carmine Ampolo and Mario Manfredini (eds.), *Plutarco. Le Vite di Teseo e di Romolo*, Milan.

Andurand (2015): Anthony Andurand, "Le monde plutarquéen des banquets savants: essai d'approche spatiale", *Histoire & Informatique* 18, 46–53.

Andurand and Bonnet (2016): Anthony Andurand and Corinne Bonnet, "'Les coutumes et les lois des nations barbares' (*Quest. conv.* 2, 1). Réseaux savants entre centre et périphérie dans les

https://doi.org/10.1515/9783110539479-032

Propos de Table de Plutarque", in: Sydney H. Aufrère and Frédéric Möri (eds.), *Alexandrie la Divine. Sagesses barbares, échanges et réappropriation dans l'espace culturel gréco-romain*, 109–141.

Andurand and Bonnet (forthcoming): Anthony Andurand and Corinne Bonnet, "'Le divin Platon' à la table des Grecs et des Romains. Dynamique et enjeux de la fabrique d'une mémoire savante dans les *Propos de table* de Plutarque", in: Frédéric Chapot, Johann Goeken and Maud Pfaff (eds.), *Figures mythiques et discours religieux dans l'Empire gréco-romain* (forthcoming).

Arnush (2005): Michael Arnush, "Pilgrimage to the Oracle of Apollo at Delphi: Patterns of Public and Private Consultation", in: Jaś Elsner and Ian Rutherford (eds.), *Pilgrimage in Graeco-Roman and Early Christian Antiquity: Seeing the Gods*, Oxford, 97–110.

Ash (1997): Rhiannon Ash, "Severed heads: individual portraits and irrational forces in Plutarch's *Galba* and *Otho*", in: Judith Mossman (ed.), *Plutarch and his Intellectual World*, London, 189–214.

Asheri et al. (2007): David Asheri, Alan B. Lloyd, and Aldo Corcella, *A Commentary on Herodotus, books I-IV*, edited by O. Murray and A. Moreno, Oxford.

Assmann (1992): Jan Assmann, *Das kulturelle Gedächtnis: Schrift, Erinnerung und politische Identität in frühen Hochkulturen*, Munich.

Assmann (2008): Jan Assmann, "Communicative and Cultural Memory", in: Astrid Erll and Ansgar Nünning (eds.), *Cultural Memory Studies. An International and Interdisciplinary Handbook*, Berlin and New York, 109–118.

Aufrère and Möri (2016): Sydney H. Aufrère and Frédéric Möri (eds.), *Alexandrie la Divine. Sagesses barbares, échanges et réappropriation dans l'espace culturel gréco-romain*, Geneva.

Austin (2006): Michel M. Austin, *The Hellenistic World from Alexander to the Roman Conquest. A Selection of Ancient Sources in Translation*, Cambridge.

Babbitt (1936a): Frank C. Babbitt, *Plutarch's Moralia*, Vol. IV, Loeb Classical Library, Cambridge, Mass. and London.

Babbitt (1936b): Frank C. Babbitt, *Plutarch's Moralia*, Vol. V, Loeb Classical Library, Cambridge, Mass. and London.

Babut (1969a): Daniel Babut, *Plutarque de la Vertu Éthique* (Bibliothèque de la Faculté des Lettres de Lyon 15), Paris.

Babut (1969b): Daniel Babut, *Plutarco e lo stoicismo*, transl. Alberto Bellanti, Milan.

Babut (1992): Daniel Babut, "La composition des *Dialogues Pythiques* de Plutarque et le problème de leur unité", *JS* 27, 189–234.

Badian (1967): Ernest Badian, "Agis III", *Hermes* 95, 170–192.

Bagnall and Derow (2004): Roger S. Bagnall and Peter Derow, *The Hellenistic Period. Historical Sources in Translation*, Malden, Mass. and Oxford.

Bakhtin (1981): Mikhail M. Bakhtin, *The Dialogic Imagination: Four Essays*, ed. Michael Holquist, transl. Caryl Emerson and Michael Holquist, Austin.

Bakker (1997): Egbert J. Bakker, "Verbal aspect and mimetic description in Thucydides", in: Egbert J. Bakker (ed.), *Grammar as Interpretation: Greek Literature in its Linguistic Contexts* (*Mnemosyne* Supplement 171), Leiden, 7–54.

Bakker (2007): Egbert J. Bakker, "Time, tense and Thucydides", *CW* 100, 113–122.

Bakker and Wakker (2009): Stéphanie J. Bakker and Gerry C. Wakker (eds.), *Discourse Cohesion in Ancient Greek* (Amsterdam Studies in Classical Philology 16), Leiden.

Bal (2009): Mieke Bal: *Narratology: Introduction to the Theory of Narrative*, Toronto and Buffalo.

Balsdon (1979): John P.V.D. Balsdon, *Romans and Aliens*, London.

Banta (2006): Jason Banta, *Imperium cum Finibus: Plutarch's Archaic Rome*, Diss., State University of New York at Buffalo.

Banta (2007a): Jason Banta, "'Who Gives a Fig (Tree a Name)?': Chronotopic Conflicts in Plutarch's *Romulus*", *Intertexts* 11.1, 25–41.
Banta (2007b): Jason Banta, "The Gates of Janus: Bakhtin and Plutarch's Roman *meta-chronotope*", in: David H.J. Larmour and Diane Spencer (eds.), *The Sites of Rome: Time Space, Memory*, Oxford, 238–270.
Barata Dias (2005): Paula Barata Dias, "La religion égyptienne chez Plutarque et chez Athanase d'Alexandrie: Les animaux, des signes d'une continuité", in: Jacques Boulogne (ed.), *Les Grecs de l'antiquité et les animaux. Le cas remarquable de Plutarque*, Lille, 173–187.
Barker et al. (2010): Elton Barker, Stefan Bouzarovski, Christopher B.R. Pelling and Leif Isaksen, "Mapping an ancient historian in a digital age: the Herodotus Encoded Space-Text-Image Archive (HESTIA)", *Leeds International Classical Studies* 9, article no.1 (http://www.leeds.ac.uk/classics/lics/).
Barker et al. (2013): Elton Barker, Leif Isaksen, Nancy Rabinowitz, Christopher B.R. Pelling and Stefan Bouzarovski, "On using digital resources for the study of an ancient text: the case of Herodotus' *Histories*", *Digital Classicist* 2013 (*BICS* Supplements 122), 45–62.
Barker et al. (2016): Elton Barker, Stefan Bouzarovski, Leif Isaksen and Christopher B.R. Pelling (eds.), *New Worlds from Old Texts: Revisiting Ancient Space and Place*, Oxford.
Barker and Pelling (2016): Elton Barker and Christopher B.R. Pelling, "Space-travelling in Herodotus 5", in: Elton Barker, Stefan Bouzarovski, Leif Isaksen and Christopher B.R. Pelling (eds.), *New Worlds from Old Texts: Revisiting Ancient Space and Place*, Oxford, 225–251.
Barrow (1967): Reginald H. Barrow, *Plutarch and his Times*, London.
Basset (2009): Louis Basset, "The use of the imperfect to express completed states of affairs: the imperfect as a marker of narrative cohesion", in: Stéphanie J. Bakker and Gerry C. Wakker (eds.), *Discourse Cohesion in Ancient Greek* (Amsterdam Studies in Classical Philology 16), Leiden, 205–219.
Bearzot (2005): Cinzia Bearzot, "Philotimia, tradizione e innovazione. Lisandro e Agesilao a confronto in Plutarco", in: Aurelio Pérez-Jiménez and Frances Titchener (eds.), *Historical and Biographical Values of Plutarch's Works: Studies devoted to Professor Philip A. Stadter by the International Plutarch Society*, Málaga and Logan, Utah, 31–50.
Bearzot (2009): Cinzia Bearzot, *La polis greca*, Bologna.
Becchi (2009): Francesco Becchi, "La notion de philanthropia chez Plutarque. Contexte social et sources philosophiques", in: José Ribeiro Ferreira, Delfim Leão, Manuel Tröster and Paula Barata Dias (eds.), *Symposion and Philanthropia in Plutarch*, Coimbra, 263-273.
Beck H. (2002): Hans Beck, "Interne *synkrisis* bei Plutarch", *Hermes* 130.4, 467–489.
Beck (2004): Mark Beck, "Plutarch on the statesman's independence of action", in: Lukas de Blois, Jeroen Bons, Ton Kessels and Dirk M. Schenkeveld (eds.), *The Statesman in Plutarch's Works, Volume I: Plutarch's Statesman and His Aftermath – Political, Philosophical, and Literary Aspects* (*Mnemosyne* Supplements 250.1), Leiden, 105–114.
Beck (2007): Mark Beck, "Plutarch", in: Irene J.F. de Jong and René Nünlist (eds.), *Time in Ancient Greek Narratives: Studies in Ancient Greek Narrative*, Vol. 2, Leiden and Boston, 397–411.
Beck (2012): Mark Beck, "Plutarch", in: Irene J.F. de Jong (ed.), *Space in Ancient Greek Literature: Studies in Ancient Greek Narrative*, Vol. 3, Leiden and Boston, 441–462.
Beck (2014): Mark Beck (ed.), *A Companion to Plutarch*, Malden, Mass. and Oxford.
Beetham (2002): Frank Beetham, "The aorist indicative", *G&R* 49, 227–236.
Bellu (2007): Mariangela A. Bellu, "Las chreiai κατὰ χαριεντισμόν del *Conjugalia praecepta*", in: José A. Fernández Delgado, Frascisca Pordomingo Pardo and Antonio Stramaglia (eds.), *Escuela y literatura en Grecia Antigua. Actas del Simposio Internacional Universidad de Salamanca, 17–19 de Noviembre de 2004, Università degli Studi di Cassino*, Frosinone, 363–372.

Bemong et al. (2010): Nele Bemong, Pieter Borghart, Michel De Dobbeleer and Kristoffel Demoen, *Bakhtin's Theory of the Literary Chronotope: Reflections, Applications, Perspectives*, Ghent.
Beneker (2005): Jeffrey Beneker, "Thematic Correspondences in Plutarch's *Lives* of Caesar, Pompey, and Crassus", in: Lukas de Blois, Jeroen Bons, Ton Kessels and Dirk M. Schenkeveld (eds.), *The Statesman in Plutarch's Works, Volume II: The Statesman in Plutarch's Greek and Roman Lives* (*Mnemosyne* Supplements 250.2), Leiden and Boston, 315–325.
Benveniste (1976): Émile Benveniste, *Il vocabolario delle istituzioni indoeuropee*, vol. I, Turin.
Bernard (1990): Wolfgang Bernard, *Spätantike Dichtungstheorien: Untersuchungen zu Proklos, Herakleitos und Plutarch*, Stuttgart.
Bernardakis (2009): Panagiotis D. Bernardakis and Heinz G. Ingenkamp (eds.), *Plutarchi Chaeronensis Moralia, recognovit Gregorius N. Bernardakis. Editio Maior Vol.* II, Athens.
Bianchi (1987): Ugo Bianchi, "Plutarch und der Dualismus", *ANRW* II.36.1, 350–365.
Blanchard (1997): Alain Blanchard, "Destins de Ménandre", *Ktèma* 22, 213–225.
Bolkestein (1964): Hendrik Bolkestein, "Some notes on Plutarch's *De Pythiae Oraculis*", *Mnemosyne* 4.17, 367–374.
Bollansée et al. (1998): Jan Bollansée, Johannes Engels, Guido Schepens and Jan Radicke (eds.), *Die Fragmente der griechischen Historiker* IV A: Biography, Fasc. 1: The Pre-Hellenistic Period (ed. Guido Schepens), Leiden, Boston and Cologne.
Bommelaer (2011): Jean-François Bommelaer, "Delphica 3: le monument des 'Navarques'", *BCH* 135, 199–235.
Bonnechère (2007): Pierre Bonnechère, "Divination", in: Daniel Ogden (ed.), *A Companion to Greek Religion*, Malden, Mass. and Oxford, 145–159.
Boulogne (1992): Jacques Boulogne, "Les 'Questions Romaines' de Plutarque", *ANRW* II.33.6, 4682–4708.
Boulogne (1994): Jacques Boulogne, *Plutarque. Un Aristocrate Grec sous l'Occupation Romaine*, Lille.
Boulogne (2002): Jacques Boulogne, *Plutarque, Œuvres morales*, Tome IV, Budé, Paris.
Boulogne (2005): Jacques Boulogne, "Le culte égyptien des animaux vu par Plutarque. Une étiologie égyptienne (*Isis et Osiris*, 71–76, 379D-382C)", in: Jacques Boulogne (ed.), *Les Grecs de l'antiquité et les animaux. Le cas remarquable de Plutarque*, Lille, 197–205.
Bowen (1992): Anthony J. Bowen, *The Malice of Herodotus (de Malignitate Herodoti). Translated with an Introduction and Commentary*, Warminster.
Bowie (1970): Ewen Bowie, "Greeks and Their Past in the Second Sophistic", *P&P* 46, 3–41, repr. in: Moses I. Finley (ed.), *Studies in Ancient Society*, London, 166–209.
Bowie (1996): Ewen Bowie, "Past and Present in Pausanias", in: Jean Bingen (ed.), *Pausanias Historien*, (Entretiens de la Fondation Hardt, 41, 15–19 Août 1994), Vandœuvres and Geneva, 207–239.
Bowie (2014): Ewen Bowie, "Poetry and Education", in: Mark Beck (ed.), *A Companion to Plutarch*, Malden, Mass. and Oxford, 177–190.
Bowra (1940): Cecil M. Bowra, "Sophocles on his own Development", *AJPh* 61, 385–401.
Boys-Stones (2001): George R. Boys-Stones, *Post-Hellenistic Philosophy. A Study of its Development from the Stoics to Origen*, Oxford.
Braund (1988a): David Braund, "Introduction: The Growth of the Roman Empire (241 BC–AD 193)", in: David Braund (ed.), *The Administration of the Roman Empire: 241 BC–AD 193*, Exeter, 1–14.
Braund (1988b): David Braund (ed.), *The Administration of the Roman Empire: 241 BC–AD 193*, Exeter.
Braund (1997): David Braund, "Plutarch's *Pyrrhus* and Euripides' *Phoenician Women:* Biography and Tragedy on Pleonectic Parenting", *Histos* 1, 1–10.

Bravo García (1973): Antonio Bravo García, "El pensamiento de Plutarco acerca de la paz y la guerra", *CFC* 5, 141–191.
Bréchet (2003): Christophe Bréchet, *Homère dans l'œuvre de Plutarque: la référence homérique dans les 'Œuvres Morales'*, Thèse de doctorat, Université Paul Valéry–Montpellier III.
Bremmer (2005): Jan M. Bremmer, "Plutarch and the 'liberation of Greece'", in: Lukas de Blois, Jeroen Bons, Ton Kessels and Dirk M. Schenkeveld (eds.), *The Statesman in Plutarch's Works, Volume II: The Statesman in Plutarch's Greek and Roman Lives* (*Mnemosyne* Supplements 250.2), Leiden and Boston, 257–267.
Brenk (1977): Frederick E. Brenk, *In Mist Apparelled. Religious Themes in Plutarch's Moralia and Lives*, Leiden.
Brenk (1987): Frederick E. Brenk, "An Imperial Heritage: The Religious Spirit of Plutarch of Chaironeia", *ANRW* II.36.1, 248–349.
Brenk (1999): Frederick E. Brenk, "'Isis is a Greek Word'. Plutarch's Allegorization of Egyptian Religion", in: Aurelio Pérez Jiménez, José García López and Rosa M. Aguilar (eds.), *Plutarco, Platón y Aristóteles. Actas del V Congreso Internacional de la I.P.S., Madrid-Cuenca, 4–7 de mayo de 1999*, Madrid, 227–238.
Brenk (2005): Frederick E. Brenk, "Plutarch's Middle-Platonic God: About to Enter (or Remake) the Academy", in: Rainer Hirsch-Luipold (ed.), *Gott und die Götter bei Plutarch. Götterbilder–Gottesbilder–Weltbilder*, Berlin and New York, 27–49.
Bridgeman (2007): Teresa Bridgeman, 'Time and Space', in: David Herman (ed.), *The Cambridge Companion to Narrative*, 52–65.
Brock (1991): Roger Brock, "The Emergence of Democratic Ideology", *Historia* 40, 160–169.
Broeker (1954): Helmut Broecker, *Animadversiones ad Plutarchi libellum ΠΕΡΙ ΕΥΘΥΜΙΑΣ*, Bonn.
Brooks (1984): Peter L. Brooks, *Reading for the Plot: Design and Intention in Narrative*, Oxford.
Brouillette (2014): Xavier Brouillette, *La philosophie delphique de Plutarque. L'itinéraire des Dialogues pythiques*, Paris.
Brout (2004): Nicolette Brout, "Typhon dans le *De Iside* de Plutarque", *RPhA* 22, 71–94.
Bruce (1960): Iain A.F. Bruce, "Internal Politics and the Outbreak of the Corinthian War", *Emerita* 28, 75–86.
Bruit Zaidman (2003), Louise Bruit Zaidman, "La notion d'archaion dans la Périégèse de Pausanias", in: Guy Lachenaud and Dominique Longrée (eds.), *Grecs et Romains aux prises avec l'histoire*, Rennes I, 21–30.
Brulé (1987): Pierre Brulé, *La fille d'Athènes: la religion des filles à Athènes à l'époque classique: mythes, cultes et société*, Paris.
Bucher-Isler (1972): Barbara Bucher-Isler, *Norm und Individualität in den Biographien Plutarchs*, Bern and Stuttgart.
Buckler (1992): John Buckler, "Plutarch and Autopsy", *ANRW* II.33.6, 4788–4830.
Buijs (2007): Michel Buijs, "Aspectual differences & narrative technique: Xenophon's *Hellenica* and *Agesilaus*", in: Rutger J. Allan and Michel Buijs (eds.), *The Language of Literature: Linguistic Approaches to Classical Texts*, Leiden, 122–153.
Burkert (1996): Walter Burkert, "Plutarco: religiosità personale e teologia filosofica", in: Italo Gallo (ed.), *Plutarco e la religione. Atti del VI convegno plutarcheo, Ravello, 29–31 maggio 1995*, Naples, 11–28.
Burridge (1992): Richard A. Burridge, *What are the Gospels? A Comparison with Graeco-Roman Biography* (Society for New Testament Studies Monograph Series 70), Cambridge.
Burridge (1997): Richard A. Burridge, "Biography", in: Stanley E. Porter (ed.), *Handbook of Classical Rhetoric in the Hellenistic Period: 330 B.C.-A.D. 400*, Leiden, New York and Cologne, 371–391.
Buszard (2005): Bradley B. Buszard, "The decline of the Roman Republic in the *Pyrrhus–Marius*", in: Lukas de Blois, Jeroen Bons, Ton Kessels and Dirk M. Schenkeveld (eds.), *The Statesman*

in Plutarch's Works, Volume II: The Statesman in Plutarch's Greek and Roman Lives (*Mnemosyne* Supplements 250.2), Leiden and Boston, 281–296.
Buszard (2008): Bradley Bruce Buszard, "Caesar's Ambition: a combined reading of Plutarch's *Alexander–Caesar* and *Pyrrhus–Marius*", *TAPA* 138, 185–215.
Caballero and Viansino (1995): Raul Caballero and Giovanni Viansino (eds.), *Plutarco, L'esilio*, Naples.
Calame (1996): Claude Calame, *Thésée et l'imaginaire athénien: Légende et culte en Grèce antique*, Lausanne.
Calame (2011): Claude Calame, "Myth and Performance in the Athenian Stage: Praxithea, Erechtheus, their Daughters, and the Etiology of Autochthony", *CPh* 106, 1–19.
Camassa (1996): Giorgio Camassa, "Leggi orali e leggi scritte. I legislatori", in: Salvatore Settis (ed.), *I Greci. Storia Cultura Arte Società 2. Una storia greca I. Formazione*, Turin, 561–576.
Candau-Morón (2000): José María Candau-Morón, "Plutarch's Lysander and Sulla: integrated characters in roman historical perspective", *AJPh* 121, 453–478.
Cannata Fera (1996): Maria Cannata Fera, "Plutarco e la Parola dei Poeti", in: José A. Fernández Delgado and Francisca Pordomingo Pardo (eds.), *Estudios Sobre Plutarco: Aspectos Formales (IV Simposio Espanol Sobre Plutarco, Salamanca, 26 a 28 de Mayo de 1994)*, Madrid, 415–428.
Carena et al. (1990): Carlo Carena, Mario Manfredini and Luigi Piccirilli, *Plutarco. Le Vite di Cimone e di Lucullo*, Milan.
Carrara (1977): Paolo Carrara (ed.), *Euripide: Eretteo: introduzione, testo e commento*, Florence.
Cartledge and Spawforth (1989): Paul Cartledge and Anthony Spawforth, *Hellenistic and Roman Sparta: A Tale of Two Cities*, London and New York.
Casanova (2012): Angelo Casanova, "Plutarch as Apollo's priest at Delphi", in: Lautaro R. Lanzillotta and Israel Muñoz Gallarte (eds.), *Plutarch in the Religious and Philosophical Discourse of Late Antiquity*, Leiden and Boston, 151–158.
Casanova (2013): Angelo Casanova, "La *Vita di Teseo* e la tradizione letteraria", in: Angelo Casanova (ed.), *Figure d'Atene nelle opere di Plutarco*, Florence, 9–18.
Champion (2009): Jeff Champion, *Pyrrhus of Epirus*, Barnsley.
Champlin (2005): Edward Champlin, *Nerone*, transl. Mario Carpitella, Milan.
Charlesworth (1926): Martin Percival Charlesworth, "The Fear of the Orient in the Roman Empire", *Cambridge Historical Journal* 2, 1–16.
Cherniss and Helmbold (1957): Harold Cherniss and William Helmbold, *Plutarch's Moralia*, Vol. XII, Loeb Classical Library, Cambridge, Mass. and London.
Chirassi Colombo (1996): Ileana Chirassi Colombo, "Pythia e Sybilla. I problemi dell'*atechnos mantike* in Plutarco", in: Italo Gallo (ed.), *Plutarco e la religione. Atti del VI convegno plutarcheo, Ravello, 29–31 maggio 1995*, Naples, 429–450.
Clarke (2008): Katherine Clarke, *Making Time for the Past: Local History and the Polis*, Oxford and New York.
Clay (2011): Jenny Strauss Clay, *Homer's Trojan Theater. Space, Vision, and Memory in the Iliad*, Cambridge.
Clement and Hoffleit (1969): Paul A. Clement and Herbert B. Hoffleit, *Plutarch's Moralia in Sixteen Volumes*, Vol. VIII, Loeb Classical Library, Cambridge, Mass. and London.
Cohoon (1971): James Wilfred Cohoon (ed.), *Dio Chrysostomus*, Vol. I, Loeb Classical Library, Cambridge, Mass. and London.
Collard et al. (1995): Chiristopher Collard, Martin J. Cropp and K. H. Lee (eds.), *Euripides: Selected Fragmentary Plays*, Warminster.
Constantakopoulou (2007): Christy Constantakopoulou, *The Dance of the Islands: Insularity, Networks, the Athenian Empire and the Aegean World*, Oxford.

Cook (2001): Brad L. Cook, "Plutarch's Use of λέγεται: Narrative Design and Source in *Alexander*", *GRBS* 42, 329–360.
Cornell (2013): Tim J. Cornell (ed.), *The Fragments of the Roman Historians*, Vols. I-III, Oxford 2013.
Cropp and Fick (1985): Martin Cropp and Gordon Fick (eds.), *Resolutions and Chronology in Euripides*, London.
Darthou (2005): Sonia Darthou, "Retour à la terre: La fin de la Geste d'Erechthée", *Kernos* 18, 69–83.
Davidson (1976): Donald A. Davidson, "Processes of Tell Formation and Erosion", in: Donald A. Davidson and Myra L. Shackley (eds.), *Geoarchaeology: Earth Science and the Past*, London, 255–266.
Davies (1988): John K. Davies, "Religion and the State", in: *The Cambridge Ancient History*, Vol. 4, Cambridge, 368–388.
de Blois (2004): Lukas de Blois, "Classical and Contemporary Statesmen in Plutarch's *Praecepta*" in: Lukas de Blois, Jeroen Bons, Ton Kessels and Dirk M. Schenkeveld (eds.), *The Statesman in Plutarch's Works, Volume I: Plutarch's Statesman and His Aftermath – Political, Philosophical, and Literary Aspects* (*Mnemosyne* Supplements 250.1), Leiden, 57–63.
de Blois (2008): Lukas de Blois, "The ideal statesman: A commonplace in Plutarch's Political Treatises, his Solon and his Lycurgus", in: Anastasios G. Nikolaidis (ed.), *The Unity of Plutarch's Work: 'Moralia' Themes in the 'Lives', Features of the 'Lives' in the 'Moralia'* (*Millennium-Studien* 19), Berlin and New York, 317–324.
de Jong (2007): Irene J.F. de Jong, "Introduction. Narratological theory on time", in: Irene J.F. de Jong and René Nünlist (eds.), *Time in ancient Greek literature. Studies in Ancient Greek Narrative, Vol. 2* (*Mnemosyne* Supplements 291), Leiden and Boston, 1–14.
de Jong (2012a): Irene J.F. de Jong (ed.), *Space in ancient Greek literature. Studies in Ancient Greek Narrative, Vol. 3* (*Mnemosyne* Supplements 339), Leiden and Boston.
de Jong (2012b): Irene J.F. de Jong, "Introduction. Narratological theory on space", in: Irene J.F. de Jong (ed.), *Space in ancient Greek literature. Studies in Ancient Greek Narrative, Vol. 3* (*Mnemosyne* Supplements 339), Leiden and Boston, 1–18.
de Jong (2014): Irene J.F. de Jong, *Narratology and Classics: A Practical Guide*, Oxford.
de Jong and Nünlist (2007): Irene J.F. de Jong and René Nünlist (eds.), *Time in ancient Greek literature. Studies in Ancient Greek Narrative, Vol. 2* (*Mnemosyne* Supplements 291), Leiden and Boston.
de Lacy (1952): Philip de Lacy, "Biography and Tragedy in Plutarch", *AJP* 73.2, 159–171.
Del Corno (1982): Dario Del Corno, Plutarco, *Il demone di Socrate. I ritardi della punizione divina*, transl. Antonio Aloni and Giulio Guidorizzi, Introduction by Dario Del Corno, Milan.
de Romilly (1979): Jacqueline de Romilly, *La douceur dans la pensée grecque*, Paris.
Desideri (1978): Paolo Desideri, *Dione di Prusa. Un intellettuale greco nell'impero romano*, Messina and Florence.
Desideri (1985): Paolo Desideri, "L'impossibile misura della ricchezza (Plutarco, *Sept. sap. conv.*, 154 f sgg.)", *Annali dell'Istituto di Storia, Fac. Magistero Firenze* 3, 21–32.
Desideri (1992a): Paolo Desideri, "La formazione delle coppie nelle Vite plutarchee", *ANRW* II.33.6, Berlin and New York, 4470–4486; repr. in: Paolo Desideri (2012a), *Saggi su Plutarco e la sua fortuna*, ed. Angelo Casanova, Florence, 229–245.
Desideri (1992b): Paolo Desideri, "I documenti di Plutarco", *ANRW* II.33.6, Berlin and New York, 4536–4567.
Desideri (2000): Paolo Desideri, "City and Country in Dio", in: Simon C.R. Swain (ed.), *Dio Chrysostom. Politics, Letters and Philosophy*, Oxford, 93–107.
Desideri (2012a): Paolo Desideri, *Saggi su Plutarco e la sua fortuna*, ed. Angelo Casanova, Florence.

Desideri (2012b): Paolo Desideri, "Dione di fronte all'imperatore: riflessioni sulla regalità", in: Gustavo Vagnone (ed.), *Dione di Prusa. Orazioni I-II-III-IV ("Sulla regalità"). Orazione LXII ("Sulla regalità e sulla tirannide")*, Accademia Nazionale dei Lincei, Rome, 7–18.

Desideri (2013): Paolo Desideri, "La città di Pericle", in: Angelo Casanova (ed.), *Figure d'Atene nelle opere di Plutarco*, Florence, 19–30.

Desideri (2015): Paolo Desideri, "L'immagine dell'imperatore nei discorsi 'Sulla Regalità' di Dione di Prusa", in: Jean-Louis Ferrary and John Sheid (eds.), *Il princeps romano: autocrate o magistrato? Fattori giuridici e fattori sociali del potere imperiale da Augusto a Commodo*, Pavia, 293–325.

Díaz Lavado (2010): Juan Manuel Díaz Lavado, *Las citas de Homero en Plutarco*, Zaragoza.

Diggle (1994): James Diggle (ed.), *Euripidis Fabulae: Tomus III, Oxford Classical Texts*, Oxford.

Diggle (2004): James Diggle, *Theophrastus Characters. Edited with introduction, translation and commentary*, Cambridge.

Dignas and Smith (2012): Beate Dignas and Reginald R. R. Smith, "Introduction", in: Beate Dignas and Reginald R.R. Smith (eds.), *Historical and Religious Memory in the Ancient World*, Oxford, 1–11.

Dillon (1996): John Dillon, *The Middle Platonists. 80 B.C. to A.D. 220*, Ithaca, NY.

Dillon (1997): John Dillon, "Plutarch and the End of History", in: Judith M. Mossman (ed.), *Plutarch and his Intellectual World: Essays on Plutarch*, London, 233–240.

Dillon (2002): John Dillon, "The Social Role of the Philosopher in the Second Century CE: Some Remarks", in: Philip A. Stadter and Luc Van Der Stockt (eds.), *Sage and Emperor: Plutarch, Greek Intellectuals, and Roman Power in the Time of Trajan, 98–117 A.D.*, Leuven, 29–40.

Dillon M. (1997): Matthew Dillon, *Pilgrims and Pilgrimage in Ancient Greece*, London.

Dimauro (2008): Elisabetta Dimauro, *Re contro. La rivalità dinastica a Sparta fino al regno di Agide II*, Alessandria.

Dindorf (1869): Ludwig August Dindorf, *Ioannis Zonarae epitome historiarum*, Vol. 2, Leipzig.

D'Ippolito (2004): Gennaro D'Ippolito, "L'Omero Di Plutarco", in: Italo Gallo (ed.), *La Biblioteca di Plutarco. Atti dell'IX convegno plutarcheo, Pavia, 13–15 giugno 2002*, Naples, 11–35.

D'Ippolito (2010): Gennaro D'Ippolito, "Norma e variazione nella scrittura plutarchea", in: Giuseppe Zanetto and Stefano Martinelli Tempesta (eds.), *Plutarco: lingua e testo. Atti dell'XI Convegno plutarcheo della IPS, Milano, 18–20 giugno 2009*, Milan, 85–111.

Dognini (2007): Cristiano Dognini, "Il *De Herodoti malignitate* e la fortuna di Erodoto", in: Yves Perrin (ed.), *Neronia VII. Rome, l'Italie et la Grèce. Hellénisme et philhellénisme au premier siècle après J.-C. Actes du VIIe Colloque international de la SIEN (Athènes, 21–23 octobre 2004)*, Brussels, 481–502.

Donini (1986): Pierluigi Donini, "Lo scetticismo academico. Aristotele e l'unità della tradizione platonica secondo Plutarco", in: Giuseppe Cambiano (ed.), *Storiografia e dossografia nella filosofia antica*, Turin, 203–226.

Donlan (1980): Walter Donlan, *The Aristocratic Ideal in Ancient Greece. Attitudes of Superiority from Homer to the End of the Fifth Century B.C.*, Lawrence, Kansas.

Dover (1974): Kenneth Dover, *Popular Morality in the Time of Plato and Aristotle*, Oxford.

Dubel, Favreau-Linder and Oudot (2015): Sandrine Dubel, Anne-Marie Favreau-Linder and Estelle Oudot (eds.), *À l'école d'Homère. La culture des orateurs et des sophistes*, Paris.

Duff (1999): Timothy E. Duff, *Plutarch's Lives. Exploring Virtue and Vice*, Oxford.

Duff (2001): Timothy E. Duff, "The Prologue to the *Lives* of Pericles and Fabius", in: Aurelio Pérez Jiménez and Francesc Casadesús Bordoy (eds.), *Estudios sobre Plutarco: Misticismo y religiones mistéricas en la obra de Plutarco. Actas del VII simposio Español sobre Plutarco, Palma de Mallorca, 2–4 de Noviembre de 2000*, Madrid and Málaga, 351–363.

Duff (2007): Timothy E. Duff, "Plutarch's readers and the moralism of the *Lives*", *Ploutarchos* 5, 3–18.

Duff (2008a): Timothy E. Duff, "How *Lives* begin", in: Anastasios G. Nikolaidis (ed.), *The Unity of Plutarch's Work: "Moralia" Themes in the "Lives", Features of the "Lives" in the "Moralia"* (*Millennium-Studien* 19), Berlin and New York, 187–207.
Duff (2008b): Timothy E. Duff, "Models of Education in Plutarch", *JHS* 128, 1–26.
Duff (2011a): Timothy E. Duff, "The Structure of the Plutarchan Book", *CA* 30.2, 213–278.
Duff (2011b): Timothy E. Duff, "Plutarch's *Lives* and the Critical Reader", in: Geert Roskam and Luc Van der Stockt (eds.), *Virtues for the People: Aspects of Plutarch's Ethics*, Leuven, 59–82.
Duff (2014): Timothy E. Duff, "The Prologues", in: Mark Beck (ed.), *A Companion to Plutarch*, Malden, Mass. and Oxford, 333–349.
Duff (2015): Timothy E. Duff, "Aspect and subordination in Plutarchan narrative", in: Rhiannon Ash, Judith Mossman and Frances Titchener (eds.), *Fame and Infamy: Essays for Christopher Pelling on Characterization in Greek and Roman Biography and History*, Oxford, 129–148.
Eidinow (2013): Esther Eidinow, *Oracles, Curses, and Risk among the Ancient Greeks*, Oxford.
Eckerman (2014): Chris Eckerman, "Pindar's Delphi", in: Kate Gilhuly and Nancy Worman (eds.), *Space, Place, and Landscape in Greek Literature and Culture*, Cambridge, 21–62.
Elsner (2001): Jaś Elsner, "Structuring 'Greece'. Pausanias's *Periegesis* as a Literary Construct", in: Susan E. Alcock, Jon F. Cherry and Jaś Elsner (eds.), *Pausanias: Travel and Memory in Roman Greece*, Oxford, 3–20.
Erbse (1956): Hartmut Erbse, "Die Bedeutung der *Synkrisis* in den Parallelbiographien Plutarchs", *Hermes* 84, 398–424.
Étienne (2004): Roland Étienne, *Athènes, espaces urbains et histoire. Des origines à la fin du IIIe siècle ap. J.-C.*, Paris.
Euchner et al. (1993): Walter Euchner, Frangesca Rigotti and Pierangelo Schiera (eds.), *Il potere delle immagini*, Bologna and Berlin.
Favro (1996): Diana Favro, *The Urban Image of Augustan Rome*, Cambridge.
Fernández Delgado (1990): José A. Fernández Delgado, "Carácter y función de los neologismos de Plutarco en *Moralia*", in: Aurelio Pérez Jiménez and Gonzalo Del Cerro Calderón (eds.), *Estudios sobre Plutarco: obra y tradicion. Actas del I Symposion español sobre Plutarco, Fuengirola 1988*, Málaga, 141–153.
Fernández Delgado (2013): José A. Fernández Delgado, "El arte de la retórica en Plutarco", in: Germán Santana Henríquez (ed.), *Plutarco y las artes. XI Simposio Internacional de la Sociedad Española de Plutarquistas*, Madrid, 13–44.
Fischer (1974): Eitel Fischer, *Die Ekloge des Phrynichos*, Berlin.
Ferrari (2003): Giovanni R.F. Ferrari, *City and Soul in Plato's Republic*, Chicago.
Flacelière (1943): Robert Flacelière, "Plutarque et la Pythie", *REG* 46, 72–111.
Flacelière (1964): Robert Flacelière, *La sagesse de Plutarque*, Paris.
Flacelière (1971): Robert Flacelière, "Hadrien et Delphes", *CRAI* 115.1, 168–185.
Flacelière et al. (1972), Robert Flacelière, Emile Chambry and Marcel Juneaux, *Plutarque. Vies*, Tome VII, Budé, Paris.
Flacelière (1974): Robert Flacelière, *Plutarque. Œuvres Morales*, Tome VI, Budé, Paris.
Flacelière et al. (1987): Robert Flacelière, Jean Irigoin, Jean Sirinelli and André Philippon, *Plutarque. Œuvres Morales*, Tome I.1, Budé, Paris.
Flashar (1996): Martin Flashar, "Die Sieger von Marathon: zwischen Mythisierung und Vorbildlichkeit", in: Martin Flashar, Hans-Joachim Gehrke and Ernst Heinrich (eds.), *Retrospektive: Konzepte von Vergangenheit in der griechisch–römischen Antike*, Munich, 63–85.
Fletcher (1931): G.B.A. Fletcher, "Two Notes on Plutarch *De Herodoti malignitate*", *CPh* 26, 426–427.
Flower (2000): Michael A. Flower, "From Simonides to Isocrates: the Fifth-Century Origins of Fifth-Century Panhellenism", *CA* 19, 65–101.

Flower (2009): Michael A. Flower, "Spartan 'religion' and Greek 'religion'", in: Stephen Hodkinson (ed.), *Sparta: Comparative Approaches*, Swansea, 193–229.

Fontenrose (1978): Joseph Fontenrose, *The Delphic Oracle. Its Responses and Operations with a Catalogue of Responses*, Berkeley.

Fox (1983): Barbara A. Fox, "The discourse function of the participle in Ancient Greek", in: Flora Klein-Andreu (ed.), *Discourse Perspectives on Syntax*, 23–41, New York.

Frazier (1988): Françoise Frazier, "A propos de la 'philotimia' dans les 'Vies': quelques jalons dans l'histoire d'une notion", *RPh* 62, 109–127.

Frazier (1992): Françoise Frazier, "Contribution à l'étude de la composition des *Vies* de Plutarque: l'élaboration des grandes scènes", *ANRW* II.33.6, 4487–4535.

Frazier (1996): Françoise Frazier, *Histoire et Morale dans les Vies Parallèles de Plutarque* (Collection d'études anciennes 124), Paris.

Frazier (2000): Françoise Frazier, "Les visages de la rhétorique contemporaine sous le regard de Plutarque", in: Luc Van der Stockt (ed.), *Rhetorical Theory and Praxis in Plutarch. Acta of the IVth International Congress of the International Plutarch Society, Leuven, July 3–6, 1996*, Leuven and Namur, 184–202.

Frazier (2007/2008): Françoise Frazier, Review of Frédérique Ildefonse, *Plutarque, Dialogues Pythiques. L'E de Delphes, Pourquoi la Pythie ne rend plus ses oracles en vers, La disparition des oracles* (Paris 2006), *Ploutarchos* 5, 114–121.

Frazier (2014): Françoise Frazier, "The Perils of Ambition", in: Mark Beck (ed.), *A Companion to Plutarch*, Malden, Mass. and Oxford, 488–502.

Frazier (2016): Françoise Frazier, "Athènes historique, Athènes éternelle. Le regard de Plutarque sur la ville et ses monuments", *Euphrosyne* 44, 65–82.

Frazier and Froidefond (1990): Françoise Frazier and Christian Froidefond (eds.), *Plutarque. Œuvres Morales*, Tome V.1, Budé, Paris.

Frazier and Leão (2010): Françoise Frazier and Delfim F. Leão (eds.), *Tychè et Pronoia (La marche du monde selon Plutarque)*, Coimbra.

Frazier and Sirinelli (1996): Françoise Frazier and Jean Sirinelli, *Plutarque, Œuvres Morales*, Tome IX.3, Budé, Paris.

Frost (1980): Frank J. Frost, *Plutarch's Themistocles. A Historical Commentary*, Princeton, New Jersey.

Fuentes González (1998): Pedro P. Fuentes González, *Les Diatribes de Telès: Introduction, texte revu, traduction et commentaire des fragments (avec en appendice une traduction espagnole)*, Paris.

Fuhrmann (1972): François Fuhrmann, *Plutarque, Œuvres Morales*, Tome IX.1, Budé, Paris.

Fuhrmann (1988): François Fuhrmann, *Plutarque, Œuvres Morales*, Tome III, Budé, Paris.

Gaertner (2007): Jan F. Gaertner, "The discourse of displacement in Greco-Roman antiquity", in: Jan F. Gaertner (ed.), *Writing Exile: The Discourse of Displacement in Greco-Roman Antiquity and Beyond*, Leiden and Boston, 1–20.

Galli (2005): Marco Galli, "Pilgrimage as elite *Habitus:* educated pilgrims in sacred landscape during the Second Sophistic", in: Jaś Elsner and Ian Rutherford (eds.), *Pilgrimage in Graeco-Roman and Early Christian Antiquity: Seeing the Gods*, Oxford, 253–290.

Gallo (1992): Italo Gallo, "Ecdotica e critica testuale nei 'Moralia' di Plutarco", in: Italo Gallo (ed.), *Ricerche plutarchee* (Università degli Studi di Salerno. Quaderni del Dipartimento di Scienze dell' Antichità 12), Naples, 11–37; repr. in: Italo Gallo (1999), *Parerga Plutarchea* (Strumenti per la Ricerca Plutarchea 4), Naples, 125–155.

Gallo (1998): Italo Gallo, "Forma letteraria nei 'Moralia' di Plutarco: aspetti e problemi", *ANRW* II.34.4, 3511–3540.

Gallo and Mocci (1992): Italo Gallo and Maria Mocci, *Plutarco: La gloria di Atene* (Corpus Plutarchi Moralium 11), Naples.

Gallo and Moreschini (2000): Italo Gallo and Claudio Moreschini (eds.), *I generi letterari in Plutarco. Atti del VIII Convegno plutarcheo, Pisa, 2–4 giugno 1999, Napoli* (*Collectanea* 19), Naples, 177–216.
Gangloff (2009): Anne Gangloff, "Le sophiste Dion de Pruse, le bon roi et l'empereur", *Revue historique* 2009/1 (no. 649), 3–38.
Garnsey and Saller (1982): Peter Garnsey and Robert Saller, *The Early Principate*, Oxford.
Garoufalias (1979): Petros Garoufalias, *Pyrrhus, King of Epirus*, London.
Gascó (1990): Fernando Gascó, "Maraton, Eurimedonte y Platea (*Praec. Ger. Reip.* 814A–C)", in: Aurelio Pérez Jiménez and Gonzalo Del Cerro Calderón (eds.), *Estudios sobre Plutarco: obra y tradicion. Actas del I Symposion español sobre Plutarco, Fuengirola 1988*, Málaga, 211–215.
Gastaldi (1998): Silvia Gastaldi, *Storia del pensiero politico antico*, Rome and Bari.
Gawlinski (2015): Laura Gawlinski, "Greek Calendars", in: Georgia L. Irby (ed.), *A Companion to Science, Technology, and Medicine in Ancient Greece and Rome*, Vol. 2, Malden, Mass. and Oxford, 891–905.
Geiger (1981): Joseph Geiger, "Plutarch's *Parallel Lives*: The Choice of Heroes", *Hermes* 109, 85–104; repr. in: Barbara Scardigli (ed.), *Essays on Plutarch's Lives*, Oxford, 165–190.
Geiger (1985): Joseph Geiger, *Cornelius Nepos and Ancient Political Biography* (*Historia* Einzelschriften 47), Stuttgart.
Geiger (1988): Joseph Geiger, "Nepos and Plutarch: From Latin to Greek Political Biography", *ICS* 13.2, 245–256.
Geiger (2000), Joseph Geiger, "Plutarch on Late Republican Orators and Rhetoric", in: Luc Van der Stockt (ed.), *Rhetorical Theory and Praxis in Plutarch. Acta of the IVth International Congress of the International Plutarch Society, Leuven, July 3–6, 1996*, Leuven and Namur, 211–223.
Geiger (2002): Joseph Geiger, "*Felicitas Temporum* and Plutarch's Choice of Heroes", in: Philip A. Stadter and Luc Van der Stockt (eds.), *Sage and Emperor. Plutarch, Greek Intellectuals, and Roman Power in the Time of Trajan, 98–117 A.D.*, Leuven, 93–102.
Geiger (2005): Joseph Geiger, "Plutarch's Choice of Roman Heroes: Further Considerations", in: Aurelio Pérez Jiménez and Frances Titchener (eds.), *Historical and Biographical Values of Plutarch's Works. Studies Devoted to Professor Philip A. Stadter by the International Plutarch Society*, Málaga and Logan, Utah, 231–242.
Geiger (2008): Joseph Geiger, *The First Hall of Fame. A Study of the Statues in the* Forum Augustum (*Mnemosyne* Supplements 295), Leiden and Boston.
Geiger (2014): Joseph Geiger, "The Project of the *Parallel Lives:* Plutarch's Conception of Biography", in: Mark Beck (ed.), *A Companion to Plutarch*, Malden, Mass. and Oxford, 292–303.
Genette (1980): Gérard Genette, *Narrative Discourse: An Essay in Method*, transl. J.E. Lewin, Ithaca, NY and New York.
Genette (1988): Gérard Genette, *Narrative Discourse Revisited*, transl. J.E. Lewin, Ithaca, NY and New York.
Georgiadou (1992): Aristoula Georgiadou, "Idealistic and realistic portraiture in the *Lives* of Plutarch", *ANRW* II.33.6, Berlin and New York, 4616–4623.
Georgiadou (1997): Aristoula Georgiadou, *Plutarch's Pelopidas: A Historical and Philological Commentary* (Beiträge zur Klassichen Altertumskunde Band 105), Stuttgart and Leipzig.
Georgoudi, Piettre and Schmidt (2012): Stella Georgoudi, Renée K. Piettre and Francis Schmidt (eds.), *La raison des signes: présages, rites, destin dans les sociétés de la Méditerranéé ancienne*, Leiden.
Giangiulio (2010): Maurizio Giangiulio, "Collective Identities, Imagined Past, and Delphi", in: Lin Foxhall, Hans-Joachim Gehrke and Nino Luraghi (eds.), *Intentional History. Spinning Time in Ancient Greece*, Stuttgart, 121–135.

Giangrande (1988): Giuseppe Giangrande, "Problemi di critica testuale nei *Moralia* I. II.", in: Italo Gallo (ed.), *Sulla tradizione manoscritta dei "Moralia" di Plutarco. Atti del Convegno salernitano del 4–5 dicembre 1986*, Salerno, 55–103.
Giangrande (1990): Giuseppe Giangrande, "On the Text of Plutarch's *Non posse suaviter vivi*", in: Italo Gallo (ed.), *Contributi di filologia greca*, Naples, 61–90.
Giangrande (1992): Giuseppe Giangrande, "La lingua dei *Moralia* di Plutarco: normativismo e questioni di metodo", in: Italo Gallo and Renato Laurenti (eds.), *I "Moralia" di Plutarco tra filologia e filosofia. Atti della giornata plutarchea di Napoli, Istituto Suor Orsola Benincasa, 10 aprile 1992*, Naples, 29–46.
Giannelli and Mazzarino (1962): Giulio Giannelli and Santo Mazzarino, *Trattato di storia romana*, Rome.
Giesecke (1891): Christian A. Giesecke, *De philosophorum veterum quae ad exilium spectant sententiis*, Leipzig.
Giesen (1901): Karl Giesen, "Plutarchs *Quaestiones Graecae* und Aristoteles' *Politien*", *Philologus* 60, 446–471.
Gigon (1987): Olof Gigon, *Aristotelis Opera Vol. 3: Librorum Deperditorum Fragmenta*, Berlin.
Gilhuly and Worman (2014): Kate Gilhuly and Nancy Worman, "Introduction", in: Kate Gilhuly and Nancy Worman (eds.), *Space, Place and Landscape in Ancient Greek Literature and Culture*, Cambridge, 1–20.
Gill (1983): Christopher Gill, "The Question of Character-Development: Plutarch and Tacitus", *CQ* 33.2, 469–487.
Gill (1996): Christopher Gill, *Personality in Greek Epic, Tragedy, and Philosophy: the Self in Dialogue*, Oxford.
Glazebrook and Henry (2011): Allison Glazebrook and Madeleine M. Henry, "Introduction: Why Prostitutes, Why Greek? Why Now?", in: Allison Glazebrook and Madeleine M. Henry (eds.), *Greek Prostitutes in the Ancient Mediterranean, 800 BCE–200 CE.*, Madison, Wisconsin, 3–13.
Glucker (1978), John Glucker, *Antiochus and the Late Academy* (*Hypomnemata* 56), Göttingen.
Goeken (2013): Johann Goeken, "La rhétorique aux banquets de Plutarque", in: Giovanna Pace and Paola Volpe Cacciatore (eds.), *Gli scritti di Plutarco: tradizione, traduzione, ricezione, commento. Atti del IX Convegno internazionale della International Plutarch Society, Ravello-Auditorium Oscar Niemeyer, 29 settembre-1° ottobre 2011*, Naples, 173–186.
Goldhill (2001): Simon Goldhill (ed.), *Being Greek under Rome. Cultural Identity, the Second Sophistic and the Development of Empire*, Cambridge.
Gomme (1945): Arnold Gomme, *A Historical Commentary on Thucydides*, Vol. I, Oxford.
González Equihua (2009): Rodolfo González Equihua, "El *Banquete de los Siete Sabios* como *tour de force* escolar", in: José Ribeiro Ferreira, Delfim Leão, Manuel Tröster and Paula Barata Dias (eds.), *Symposion and Philanthropia in Plutarch*, Coimbra, 471–479.
González Julia (2009): Lluis González Julia, "Plutarch's *techne rhetorike* for the symposium in *Quaestiones Convivales*: the importance of speaking well to cultivate friendship", in: José Ribeiro Ferreira, Delfim Leão, Manuel Tröster and Paula Barata Dias (eds.), *Symposion and Philanthropia in Plutarch*, Coimbra, 63–74.
Gorman and Gorman (2014): Robert J. Gorman and Vanessa B. Gorman, *Corrupting Luxury in Ancient Greek Literature*, Ann Arbor.
Gourinat (2001), Jean-Baptiste Gourinat, "Le Socrate d'Epictète", *Philosophie antique* 1, 137–165.
Graf (1996): Fritz Graf, "Plutarco e la religione romana", in: Italo Gallo (ed.), *Plutarco e la religione. Atti del VI Convegno plutarcheo, Ravello, 29–31 maggio 1995*, Naples, 269–284.
Graindor (1934): Paul Graindor, *Athènes sous Hadrien*, Cairo.
Grasmück (1978): Ernst L. Grasmück, *Exilium. Untersuchungen zur Verbannung in der Antike*, Paderborn.
Greenhalgh (1980): Peter A.L. Greenhalgh, *Pompey: the Roman Alexander*, London.

Griffin (1989): Miriam T. Griffin, "Philosophy, Politics, and Politicians at Rome", in: Miriam T. Griffin and Jonathan Barnes (eds.), *Philosophia Togata: Essays on Roman Philosophy and Society*, Oxford, 1–37.
Griffiths (1970): John Gwyn Griffiths, *Plutarch's De Iside et Osiride*, Cardiff.
Gruen (1990): Erich S. Gruen, "Philosophy, Rhetoric, and Roman Anxieties", in: Erich S. Gruen (ed.), *Studies in Greek Culture and Roman Policy*, Leiden, 158–192.
Gruen (1992): Erich S. Gruen, *Culture and National Identity in Republican Rome*, Ithaca, NY.
Gutzwiller (1969): Kurt Gutzwiller, *ψυχρός und ὄγκος. Untersuchungen zur rhetorischen Terminologie*, Zurich.
Haacke (2008): Matthias Haacke, "Philosopher and Priest: The Image of the Intellectual and the Social Practice of the Elites in the Eastern Roman Empire", in: Beate Dignas and Kai Trampedach (eds.), *Practitioners of the Divine. Greek Priests and Reigious Officials from Homer to Heliodorus*, Cambridge, Mass., 145–166.
Hall (1997): Jonathan M. Hall, *Ethnic Identity in Greek Antiquity*, Cambridge.
Halliday (1928): William R. Halliday, *The Greek Questions of Plutarch*, Oxford.
Hamilton (1979): Charles D. Hamilton, *Sparta's Bitter Victories: Politics and Diplomacy in the Corinthian War*, Ithaca, NY and London.
Hamilton (1992): Charles D. Hamilton, "Plutarch's *Life of Agesilaus*", *ANRW* II.33.6, 4201–4221.
Hamilton (1968): James R. Hamilton, *Plutarch, Alexander, A Commentary*, Oxford.
Hani (1976): Jean Hani, *La religion égyptienne dans la pensée de Plutarque*, Paris.
Hani (1980): Jean Hani, *Plutarque. Œuvres Morales*, Tome VIII, Budé, Paris.
Hannah (2015a): Robert Hannah, "Roman Calendars", in: Georgia L. Irby (ed.), *A Companion to Science, Technology, and Medicine in Ancient Greece and Rome*, Vol. 2, Malden, Mass. and Oxford, 906–922.
Hannah (2015b): Robert Hannah, "Time-Telling Devices", in: Georgia L. Irby (ed.), *A Companion to Science, Technology, and Medicine in Ancient Greece and Rome*, Vol. 2, Malden, Mass. and Oxford, 923–940.
Hardie (1992): Philip R. Hardie, "Plutarch and the interpretation of myth", *ANRW* II.33.6, 4743–4787.
Harris (1989): Edward Harris, "Demosthenes' Speech against Meidias", *HSCPh* 92, 117–136.
Harrison (1992): George W.M. Harrison, "The Critical Trends in Scholarship on the Non-Philosophical Works in Plutarch's *Moralia*", *ANRW* II.33.6, Berlin and New York, 4646–4681.
Harrison (1995): George W.M. Harrison, "The Semiotics of Plutarch's Συγκρίσεις: The Hellenistic *Lives* of *Demetrius–Antony* and *Agesilaus–Pompey*", *RBPhil* 73, 91–104.
Harrison (2000): George W.M. Harrison, "Problems with the Genre of *Problems:* Plutarch's Literary Innovations", *CPh* 95.2, 193–199.
Hauvette (1894): A. Hauvette, *Hérodote. Historien des Guerres Médiques*, Paris.
Heftner (1995): Herbert Heftner, *Plutarch und der Aufstieg des Pompeius: Ein historischer Kommentar zu Plutarchs Pompeiusvita. Teil I, Kap. 1–45*, Frankfurt am Main and New York.
Hershbell (1993): Jackson P. Hershbell, "Plutarch and Herodotus – the Beetle in the Rose", *RhM* 136, 143–163.
Hidalgo De La Vega (1998): Maria J. Hidalgo De La Vega, *La teoria monarchica e il culto imperiale*, in: Salvatore Settis (ed.), *I Greci. Storia Cultura Arte e Società*, Turin, 1015–1058.
Hillman (1994): Thomas P. Hillman, "Authorial Statements, Narrative, and Character in Plutarch's *Agesilaus–Pompey*", *GRBS* 35, 255–280.
Hillyard (1981): Brian P. Hillyard, *Plutarch: De audiendo. A Text and Commentary*, New York.
Hirsch-Luipold (2002): Rainer Hirsch-Luipold, *Plutarchs Denken in Bildern: Studien zur literarischen, philosophischen und religiösen Funktion des Bildhaften*, Tübingen.

Hirsch-Luipold (2013): Rainer Hirsch-Luipold, "'The Most Ennobling Gift of the Gods': Religious Traditions as the Basis for Philosophical Interpretation in Plutarch", in: Giovanna Pace and Paola Volpe Cacciatore (eds.), *Gli scritti di Plutarco: tradizione, traduzione, ricezione, commento. Atti del IX Convegno internazionale della International Plutarch Society, Ravello-Auditorium Oscar Niemeyer, 29 settembre-1° ottobre 2011*, Naples, 203–217.
Hirsch-Luipold (2014): Rainer Hirsch-Luipold, "Religion and Myth", in: Mark Beck (ed.), *A Companion to Plutarch*, Malden, Mass. and Oxford, 163–176.
Hirzel (1895): Rudolf Hirzel, *Der Dialog*, 2 vols., Leipzig.
Hodkinson (1983): Stephen Hodkinson, "Social order and the conflict of values in classical Sparta", *Chiron* 13, 239–281.
Hodkinson (2000): Stephen Hodkinson, *Property and Wealth in Classical Sparta*, Swansea.
Hodkinson (2005): Stephen Hodkinson, "The imaginary Spartan *politeia*", in: Mogens H. Hansen (ed.), *The Imaginary Polis* (Symposium, January 7–10, 2004), Copenhagen, 222–281.
Holden (1887): Hubert A. Holden, *Πλουτάρχου Νικίας: Plutarch's Life of Nicias with Introduction Notes and Lexicon*, Cambridge.
Holford-Strevens (2003, 2nd edn.): Leofranc Holford-Strevens, *Aulus Gellius*, London.
Holzapfel (1884): Ludwig Holzapfel, "Über die Echtheit der plutarchischen Schrift *De Herodoti malignitate*", *Philologus* 42, 23–53.
Holzhausen (1993): Jens Holzhausen, "Zur Inspirationslehre Plutarchs in *De Pythiae oraculis*", *Philologus* 137, 72–91.
Homeyer (1967): Helene Homeyer, "Zu Plutarchs *De malignitate Herodoti*", *Klio* 49, 181–187.
Hubert (1938): Kurt Hubert, *Plutarchi Moralia*, Vol. IV, 2nd edn. 1971, Teubner, Leipzig.
Humbert (1982): Jean Humbert, *Syntaxe grecque*, Paris.
Humble (2010): Noreen Humble (ed.), *Plutarch's Lives: Parallelism and Purpose*, Swansea.
Humm (2013): Michel Humm, "Aristote et les Romains: entre hellénisme et barbarie, une vision grecque de Rome du ive siècle avant J.-C.", in: Yves Lehmann (ed.), *Aristoteles Romanus. La réception de la science aristotélicienne dans l'Empire gréco-romain*, Turnhout, 425–462.
Hummel (2003): Pascale Hummel, "La poésie lyrique entre archaïsme et classicisme", in: Jacques Jouanna and Jean Leclant (eds.), *Actes de Colloque 'La poésie grecque antique'*, Paris, 185–201.
Hurwit (1999): Jeffrey M. Hurwit, *The Athenian Acropolis: History, Mythology and Archaeology from the Neolithic Era to the Present*, Cambridge.
Hutton (2005a): William Hutton, *Describing Greece. Landscape and Literature in the Periegesis of Pausanias*, Cambridge.
Hutton (2005b): William Hutton, "The Construction of Religious Space in Pausanias", in: Jaś Elsner and Ian Rutherford (eds.), *Pilgrimage in Graeco-Roman and Early Christian Antiquity: Seeing the Gods*, Oxford, 291–318.
Innes (2002): Doreen C. Innes, "Longinus and Caecilius: Models of the Sublime", *Mnemosyne* 55, 259–284.
Jazdzewska (2013): Katarzyna Jazdzewska, "Reading Plato's 'big letters': the opening of Plutarch's *De audiendo* and Plato's *Republic*", in: Giovanna Pace and Paola Volpe Cacciatore (eds.), *Gli scritti di Plutarco: tradizione, traduzione, ricezione, commento* (IX Convegno Internazionale della *IPS*, Ravello, 29 settembre–1 ottobre 2011), Naples, 245–250.
Jeuckens (1908): Robert Jeuckens, *Plutarch von Chaeronea und die Rhetorik*, Strassburg.
Jones A.H.M. (1966): Arnold H.M. Jones, *The Greek City, from Alexander to Justinian*, Oxford.
Jones (1966): Christopher P. Jones, "Towards a Chronology of Plutarch's Works", *JRS* 56, 61–74; repr. in: Barbara Scardigli (ed.) (1995), *Essays on Plutarch's Lives*, Oxford, 95–123.
Jones (1971): Christopher P. Jones, *Plutarch and Rome*, Oxford.

Jones (2001): Christopher P. Jones, "Pausanias and his Guides", in: Susan E. Alcock, John F. Cherry and Jaś Elsner (eds.), *Pausanias. Travel and Memory in Roman Greece*, Oxford, 33–39.
Jouan and Van Looy (2002): François Jouan and Herman Van Looy, *Euripide, Tragédies*, Vol. 8, *Fragments*, part 2, *De Bellérophon à Protésilas*, Paris.
Kaesser (2004): Christian Kaesser, "Tweaking the real: art theory and the borderline between history and morality in Plutarch's *Lives*", *GRBS* 44, 361–374.
Kagan (1961): Donald Kagan, "The Economic Origins of the Corinthian War (395–387 B.C.)", *La parola del passato* 16, 321–341.
Kannicht (2004): Richard Kannicht, *Tragicorum Graecorum Fragmenta*, Vol. 5, *Euripides*, Göttingen.
Karavas and Vix (2014): Orestis Karavas and Jean-Luc Vix, "On the reception of Menander in the Imperial Period", in: Alan H. Sommerstein (ed.), *Menander in Contexts*, New York and London, 183–198.
Kechagia (2011): Eleni Kechagia, "Philosophy in Plutarch's *Table Talk*: In Jest or in Earnest?", in: Frieda Klotz and Katerina Oikonomopoulou (eds.), *The Philosopher's Banquet: Plutarch's Table Talk in the Intellectual Culture of the Roman Empire*, Oxford, 77–104.
Kennedy (1963): George A. Kennedy, *The Art of Persuasion in Greece*, Princeton.
Kennedy (2003): George A. Kennedy, *Progymnasmata: Greek Textbooks of Prose Composition and Rhetoric*, Atlanta.
Kermode (1966): Frank Kermode, *The Sense of an Ending: Studies in the Theory of Fiction*, Oxford and New York.
Kim (2010): Lawrence Kim, "The Literary Heritage as Language: Atticism and the Second Sophistic", in: Egbert J. Bakker (ed.), *A Companion to the Ancient Greek Language*, Hoboken, New Jersey, 468–482.
Kim (2014): Lawrence Kim, "Archaizing and Classicism in the Literary Historical Thinking of Dionysius of Halicarnassus", in: James Ker and Christoph Pieper (eds.), *Valuing the Past in the Greco-Roman World*, Leiden and Boston, 357–387.
Kindt (2016): Julia Kindt, *Revisiting Delphi. Religion and Storytelling in Ancient Greece*, Cambridge.
Kittel and Friedrich (1933–1973): Gerhard Kittel and Gerhard Friedrich, *Theologisches Wörterbuch zum Neuen Testament*, Vols. I-IX, Stuttgart.
Klotz (2011): Frieda Klotz, "Imagining the Past. Plutarch's Play with Time", in: Frieda Klotz and Katerina Oikonomopoulou (eds.), *The Philosopher's Banquet: Plutarch's Table Talk in the Intellectual Culture of the Roman Empire*, Oxford, 161–178.
Klotz and Oikonomopoulou (2011): Frieda Klotz and Katerina Oikonomopoulou (eds.), *The Philosopher's Banquet: Plutarch's Table Talk in the Intellectual Culture of the Roman Empire*, Oxford.
Klotz and Oikonomopoulou (2011): Frieda Klotz and Katerina Oikonomopoulou, "Introduction", in: Frieda Klotz and Katerina Oikonomopoulou (eds.), *The Philosopher's Banquet: Plutarch's Table Talk in the Intellectual Culture of the Roman Empire*, Oxford, 1–31.
Kokolakis (1960): Minos Kokolakis, "Lucian and the Tragic Performances in his time", *Platon* 12, 67–109.
König (2007): Jason König, "Fragmentation and coherence in Plutarch's *Sympotic Questions*", in: Jason König and Tim Whitmarsh (eds.), *Ordering Knowledge in the Roman Empire*, Cambridge, 43–68.
König (2012): Jason König, *Saints and Symposiasts: The Literature of Food and the Symposium in Greco-Roman and Early Christian Culture*, Cambridge.
Koskenniemi (1956): Heikki Koskenniemi, *Studien zur Idee und Phraseologie des griechischen Briefes bis 400 n. Chr.*, Helsinki.
Lachenaud (1981): Guy Lachenaud, *Plutarque. Œuvres morales*, Tome XII.1, Budé, Paris.

Lacore (1983): Michelle Lacore, "Euripide et le culte de Poséidon-Érechthée", *Revue des Études Anciennes* 85, 215–234.
Lafond (2005): Yves Lafond, "Le mythe, référence identitaire pour les cités grecques d'époque impériale: l'exemple du Péloponnèse", *Kernos* 18, 329–346.
La Matina (2000): Marcello La Matina, "La conferenza in Plutarco", in: Italo Gallo and Claudio Moreschini (eds.), *I generi letterari in Plutarco. Atti del VIII Convegno plutarcheo, Pisa, 2–4 giugno 1999, Napoli* (*Collectanea* 19), Naples, 177–216.
Lamberton (1997): Robert Lamberton, "Plutarch and the Romanization of Athens", in: Michael C. Hoff and Susan I. Rotroff (eds.), *The Romanization of Athens* (Oxbow Monographs 94), Oxford, 150–160.
Lamberton (2001): Robert Lamberton, *Plutarch*, New Haven and London.
Larmour (1992), David H.J. Larmour, "Making Parallels: *Synkrisis* and Plutarch's *Themistocles* and *Camillus*", *ANRW* II.33.6, 4154–4200.
Larmour (2005): David H.J. Larmour, "Statesman and Self in the *Parallel Lives*", in: Lukas de Blois, Jeroen Bons, Ton Kessels and Dirk M. Schenkeveld (eds.), *The Statesman in Plutarch's Works, Volume II: The Statesman in Plutarch's Greek and Roman Lives* (*Mnemosyne* Supplements 250.2), Leiden and Boston, 43–51.
Lau (1967): Otto Lau, *Schuster und Schusterhandwerk in der griechisch-römischen Literatur und Kunst*, Bonn.
Lausberg (1960): Heinrich Lausberg, *Handbuch der literarischen Rhetorik. Eine Grundlegung der Literaturwissenschaft*, Munich.
Leeck (2010): Christian Leeck, *Das Bild Roms in Plutarchs Römerbiographien. Schmeichelei oder ernsthafte Völkerverständigung?* (Wissenschaftliche Beiträge aus dem Tectum Verlag: Geschichtswissenschaft Band 8), Marburg.
Lefkowitz (1991): Mary Lefkowitz, *First Person Fictions: Pindar's Poetic 'I'*, Oxford.
Legrand (1932): Philippe-Ernest Legrand, "De la 'malignité' d'Hérodote", in: Mélanges Gustave Glotz, Vol. II, Paris, 535–547.
Lendon (1989): Jon E. Lendon, "The Oxyrhynchus Historian and the Origins of the Corinthian War", *Historia* 38, 299–313.
Lendon (1997): Jon E. Lendon, *Empire of Honor: The Art of Government in the Roman World*, Oxford.
Liddel (2008): Peter Liddel, "Scholarship and morality: Plutarch's use of inscriptions", in: Anastasios G. Nikolaidis (ed.), *The Unity of Plutarch's Work: 'Moralia' Themes in the 'Lives', Features of the 'Lives' in the 'Moralia'* (*Millennium-Studien* 19), Berlin and New York, 125–137.
Link (2004): Stefan Link, "Die Ehrenrechte der spartanischen Könige", *Philologus* 148, 222–244.
Lintott (1972): Andrew W. Lintott, "Imperial Expansion and Moral Decline in the Roman Republic", *Historia* 21.4, 626–638.
Lock (1972): Robert Lock, "The Date of Agis III's War in Greece", *Antichthon* 6, 15–27.
Loraux (1979): Nicole Loraux, "L'autochthonie: une topique athénienne. Le mythe dans l'espace civique", *Annales, Economie, Sociétés, Civilisations* 34, 3–26.
Loraux (1981): Nicole Loraux, *Les enfants d'Athéna: Idées athéniennes sur la citoyenneté et la division des sexes*, Paris.
Loraux (1996): Nicole Loraux, *Né de la terre: Mythe et politique à Athènes*, Paris.
MacDowell (1990): Douglas M. MacDowell, *Against Meidias (Oration 21)*, Oxford.
MacDowell (2000): Douglas M. MacDowell, *On the False Embassy (Oration 19)*, Oxford.
Mahaffy (1890): John P. Mahaffy, *The Greek World under Roman Sway*, London.
Malkin (1989): Irad Malkin, "Delphoi and the founding of social order in archaic Greece", *Mètis* 4/1, 129–153.

Malkin (1998): Irad Malkin, "Myth, religion, and Spartan 'ideology'", in: Wolfgang Schuller (ed.), *Politische Theorie und Praxis im Altertum*, Darmstadt, 43–49.
Malkin (2011): Irad Malkin, *A Small Greek World: Networks in the Ancient Mediterranean*, Oxford.
Manfredini and Piccirilli (1977): Mario Manfredini and Luigi Piccirilli, *Plutarco. La Vita di Solone*, Milan.
Manfredini and Piccirilli (1980): Mario Manfredini and Luigi Piccirilli, *Plutarco. Le Vite di Licurgo e di Numa*, Milan.
Marín-Valdés (2008): Fernando Marín-Valdés, *Plutarco y el arte de la Atenas hegemónica*, Oviedo.
Marincola (1994): John M. Marincola, "Plutarch's Refutation of Herodotus", *AncW* 25, 191–203.
Marincola (1997): John M. Marincola, *Authority and Tradition in Ancient Historiography*, Cambridge.
Martin H. (1960): Hubert Martin, "The Concept of *Praotes* in Plutarch's *Lives*", *GRBS* 3, 65–73.
Martin (1931): Josef Martin, *Symposion: die Geschichte einer literarischen Form*, Paderborn.
Martín García (1983): Francisco Martín García, "Las *Cuestiones Convivales* de Plutarco: estructura, fuentes y finalidad de la obra", *Revista del Colegio Universitario de Ciudad Real*, 2, 109–134.
Martín García (1987): Francisco Martín García, *Plutarco. Obras Morales y de Costumbres (Moralia) IV, Charlas de Sobremesa*, Madrid.
Mathewson (2008): David Mathewson, "Verbal aspect in the Apocalypse of John: an analysis of Revelation 5", *Novum Testamentum* 50, 58–77.
Mazzarino (1982): Santo Mazzarino, "Il pensiero politico pagano nell'età imperiale", in: Luigi Firpo (ed.), *Storia delle Idee Politiche Economiche e Sociali*, Turin, 805–875.
Mazzarino (1990): Santo Mazzarino, *Il pensiero storico classico*, Bari.
McDonnell (2006): Myles McDonnell, *Roman Manliness: Virtus and the Roman Republic*, Cambridge and New York.
McInerney (2004): Jeremy McInerney, "'Do You See What I See?': Plutarch and Pausanias at Delphi", in: Lukas de Blois, Jeroen Bons, Ton Kessels and Dirk M. Schenkeveld (eds.), *The Statesman in Plutarch's Works, Volume I: Plutarch's Statesman and his Aftermath: Political, Philosophical, and Literary Aspects*, Leiden and Boston, 43–55.
McKay (1953): Kenneth L. McKay, "The Oxyrhynchus Historian and the Outbreak of the 'Corinthian War'", *CR* 3, 6–7
Meeusen (2012): Michiel Meeusen, "Matching in Mind the Sea Beast's Complexion. On the Pragmatics of Plutarch's *Hypomnemata* and Scientific Innovation: the Case of *Q.N.* 19 (916BF)", *Philologus* 156, 234–259.
Meeusen (2013): Michiel Meeusen, "How to Treat a Bee-Sting? On the Higher Cause in Plutarch's *Causes of Natural Phenomena:* the Case of *Q.N.* 35–36", *QUCC* 105, 131–157.
Meeusen (2017): Michiel Meeusen, *Plutarch's Science of Natural Problems. A Study with Commentary on Quaestiones Naturales*, Leuven.
Meeusen and Van der Stockt (2015): Michiel Meeusen and Luc Van der Stockt (eds.), *Natural Spectaculars. Aspects of Plutarch's Philosophy of Nature*, Leuven.
Mendels (1981): Doron Mendels, "The Five Empires. A Note on a Propagandistic *Topos*", *AJP* 102, 330–337.
Mikalson (2002): Jon D. Mikalson, "Religion in Herodotus", in: Egbert J. Bakker, Irene J.F. de Jong and Hans van Wees (eds.), *Brill's Companion to Herodotus*, Leiden and Boston, 187–198.
Milazzo (1991): Antonino M. Milazzo, "Forme e funzioni retoriche dell'opuscolo *Aqua an ignis utilior* attribuito a Plutarco", in: G. D'ippolito and I. Gallo (eds.), *Strutture formali dei "Moralia" di Plutarco. Atti III Convegno plutarcheo, Palermo, 3–5 May 1989*, Naples, 419–433.
Millar (1969): Fergus Millar, "P. Herennius Dexippus: The Greek World and the Third-Century Invasions", *JRS* 59, 12–29; repr. in Hannah M. Cotton and Guy M. Rogers (eds.), *Rome, The Greek World, and the East*, vol. 2, Chapel Hill and London, 265–297.

Millender (2009): Ellen Millender, "The Spartan dyarchy. A comparative perspective", in: Stephen Hodkinson (ed.), *Sparta: comparative approaches and classical tradition*, Swansea, 1–67.
Miller (1998): Dean A. Miller, "The Spartan kingship: some extended notes on complex duality", *Arethusa* 31, 1–17.
Miller Rosen (1986): Arlene Miller Rosen, *Cities of Clay: The Geoarchaeology of Tells*, Chicago.
Minchin (2001): Elizabeth Minchin, *Homer and the Resources of Memory*, Oxford, 84–87.
Mitchell (2007): Lynette G. Mitcell, *Panhellenism and the Barbarian in Archaic and Classical Greece*, Swansea.
Moellering (1963): Howard A. Moellering, *Plutarch's De superstitione, its Place in the Changing Meaning of Deisidaimonia and in the Context of his Theological Writings*, Boston.
Moggi (1992): Mauro Moggi, "Straniero due volte: il barbaro e il mondo greco", in: Maurizio Bettini (ed.), *Lo straniero ovvero l'identità culturale a confronto*, Rome and Bari, 51–76.
Moles (1988): John L. Moles, *Plutarch: The Life of Cicero*, with an introduction, translation and commentary, Warminster.
Moles (1994): John L. Moles, "Xenophon and Callicratidas", *JHS* 114, 70–84.
Momigliano (1993): Arnaldo Momigliano, *The Development of Greek Biography*, Cambridge, Mass.
Montanari (1981): Enrico Montanari, *Il mito dell'autoctonia: Linee di une dinamica mitico-politica ateniese*, Rome.
Montes Cala et al. (1999): José Guillermo Montes Cala, Manuel Sánchez Ortiz de Landaluce and Rafael Jesús Gallé Cejudo (eds.), *Plutarco, Dioniso y el vino. Actas del VI Simposio español sobre Plutarco, Cádiz, 14–16 de mayo de 1998*, Madrid.
Mossman (1988): Judith Mossman, "Tragedy and Epic in Plutarch's *Alexander*", *JHS* 108, 83–93.
Mossman (1991): Judith Mossman, "Plutarch's use of statues", in: Michael A. Flower and Mark Toher (eds.), *Georgica: Greek studies in honour of George Cawkwell* (*BICS* Supplement 58), London, 98–119.
Mossman (1992): Judith Mossman, "Plutarch, Pyrrhus, and Alexander", in: Philip Stadter (ed.), *Plutarch and the Historical Tradition*, London and New York, 90–108.
Mossman (1997): Judith Mossman, "Plutarch's *Dinner of the Seven Wise Men* and its place in *symposion* literature", in: Judith Mossman (ed.), *Plutarch and his Intellectual World*, London, 119–140.
Mossman (1999): Judith Mossman, "Is the Pen Mightier than the Sword? The Failure of Rhetoric in Plutarch's *Demosthenes*", *Histos* 3, 77–101.
Mossman (2005): Judith Mossman, "*Taxis ou Barbaros:* Greek and Roman in Plutarch's *Pyrrhus*", *CQ* 55, 498–517.
Mossman (2006): Judith Mossman, "Travel writing, history, and biography", in: Brian McGing and Judith Mossman (eds.), *The Limits of Ancient Biography*, Swansea and Oakville, CT, 281–303.
Most (1989): Glenn W. Most, "Zur Archäologie der Archaik", *A&A* 35, 1–23.
Muccioli (2012): Federicomaria Muccioli, *La storia attraverso gli esempi: protagonisti e interpretazioni del mondo greco in Plutarco*, Milan and Udine.
Müller (2014): Alexander Müller, "Das Orakel und das Dialogische. Zu Plutarchs Schriften *De defectu oraculorum* und *De Pythiae oraculis*", in: Sabine Föllinger and Gernot M. Müller (eds.), *Der Dialog in der Antike: Formen und Funktionen einer literarischen Gattung zwischen Philosophie, Wissensvermittelung und dramatischer Inszenierung*, Berlin, 65–86.
Muñoz-Gallarte (2010): Israel Muñoz-Gallarte, "Ciencia y religión en conflicto: fantasmas y sucesos paranormales en Plutarco", in: Luc Van der Stockt, Frances Titchener, Heinz-Gerd Ingenkamp and Aurelio Pérez Jiménez (eds.), *Gods, Daimones, Rituals, Myths and History of Religions in Plutarch's Works*, Málaga and Logan, Utah, 295–314.
Murdoch (2006): Jon Murdoch, *Post-structuralist Geography: A Guide to Relational Space*, London.

Nafissi (2010): Massimo Nafissi "The Great Rhetra (Plut. *Lyc.* 6): a retrospective and intentional construct?", in: Lin Foxhall, Hans-Joachim Gehrke and Nino Luraghi (eds.), *Intentional History. Spinning time in ancient Greece*, Stuttgart, 89 – 113.

Nagy (1994): Gregory Nagy, *Le meilleur des Achéens: la fabrique du héros dans la poésie grecque archaïque*, Transl. J. Carlier and N. Loraux, Paris.

Nederlof (1940): Arie B. Nederlof, *Plutarchus' Leven van Pyrrhus: historische commentaar*, Amsterdam.

Nesselrath (2007): Heinz-Günther Nesselrath, "Later Greek Voices on the Predicament of Exile: From Teles to Plutarch and Favorinus", in: Jan F. Gaertner (ed.) *Writing Exile: The Discourse of Displacement in Greco-Roman Antiquity and Beyond*, Leiden and Boston, 87 – 108.

Neu (1971): Jerome Neu "Plato's Analogy of State and Individual: The *Republic* and the Organic Theory of the State", *Philosophy* 46, 238 – 254.

Nevin (2014): Sonya Nevin, "Negative Comparison: Agamemnon and Alexander in Plutarch's *Agesilaus–Pompey*", *GRBS* 54, 45 – 68.

Newmyer (2006): Stephen T. Newmyer, *Animals, Rights and Reason in Plutarch and Modern Ethics*, New York.

Nicolai (1992): Roberto Nicolai, *La storiografia nell'educazione antica*, Pisa.

Nikolaidis (1986): Anastasios G. Nikolaidis, "Ἑλληνικός-βαρβαρικός. Plutarch on Greek and Barbarian Characteristics", *WS* 99, 229 – 244.

Nikolaidis (1997): Anastasios G. Nikolaidis, "Plutarch's Criteria for Judging his Historical Sources", in: Carlos Schrader, Vicente Ramón Palerm and José Vela (eds.), *Plutarco y la historia. Actas del V Simposio Español sobre Plutarco, Zaragoza, 20 – 22 de junio de 1996*, Zaragoza, 329 – 341.

Nikolaidis (2005): Anastasios G. Nikolaidis, "Plutarch's methods: his cross-references and the sequence of the *Parallel Lives*", in: Aurelio Pérez Jiménez and Frances Titchener (eds.), *Historical and Biographical Values of Plutarch's Works. Studies Devoted to Professor Philip A. Stadter by The International Plutarch Society*, Málaga and Logan, Utah, 283 – 324.

Nikolaidis (2012): Anastasios G. Nikolaidis, "Aspects of Plutarch's notion of Philotimia", in: Geert Roskam, Maarten de Pourcq and Luc van der Stockt (eds.), *The Lash of Ambition. Plutarch, Imperial Greek Literature and the Dynamics of Philotimia* (Collection d'Études Classiques 25), Leuven, Namur, Paris, Walpole, Mass., 31 – 53.

Nikolaidis (2013): Anastasios G. Nikolaidis, "Plutarch's Views on Art and Especially on Painting and Sculpture", in: Germán Santana Henríquez (ed.), *Plutarco y las artes. XI Simposio Internacional de la Sociedad Española de Plutarquistas*, Madrid, 169 – 181.

Nikolaidis (2014): Anastasios G. Nikolaidis, "Morality, Characterization, and Individuality", in: Mark Beck (ed.), *A Companion to Plutarch*, Malden, Mass. and Oxford, 350 – 372.

Nikolaidis (forthcoming): Anastasios G. Nikolaidis, "Compositional Techniques and Strategies in the *Comparisons* of Plutarch's *Lives*", in: Luc Van der Stockt and Philip A. Stadter (eds.), *Studia Hellenistica* (forthcoming).

Norden (1909): Eduard Norden, *Die antike Kunstprosa*, 2nd ed. 2 vols., Leipzig and Berlin.

North Fowler (1960): Harold North Fowler, *Plutarch's Moralia*, Vol. X, Loeb Classical Library, Cambridge, Mass. and London.

North (1966): Helen North, *Sophrosyne: Self-Knowledge and Self-Restraint in Greek Literature*, *Cornell Studies in Classical Philology* 35, Ithaca, NY.

Ober (1989): Josiah Ober, *Mass and Elite in Democratic Athens: Rhetoric, Ideology, and the Power of the People*, Princeton.

Obsieger (2013): Hendrik Obsieger, *Plutarch: 'De E apud Delphos'. Über das Epsilon am Apolltempel in Delphi: Einführung, Ausgabe und Kommentar* (*Palingenesia* Band 101), Stuttgart.

Ogilvie (1967): Robert M. Ogilvie, "The date of the *De Defectu Oraculorum*", *Phoenix* 21, 108 – 119.

Oikonomopoulou (2007): Katerina Oikonomopoulou, *The Symposia in the Greek and Roman World of the High Empire: Literary Forms and Intellectual Realities*, DPhil. Diss., University of Oxford.

Oikonomopoulou (2011): Katerina Oikonomopoulou, "Peripatetic Knowledge in Plutarch's *Table Talk*", in: Frieda Klotz and Katerina Oikonomopoulou (eds.), *The Philosopher's Banquet: Plutarch's Table Talk in the Intellectual Culture of the Roman Empire*, Oxford and New York, 105–130.

Oikonomopoulou (2013), Katerina Oikonomopoulou, "Ancient question-and-answer literature and its role in the tradition of dialogue", in: Sabine Föllinger and Gernot M. Müller (eds.), *Der Dialog in der Antike: Formen und Funktionen einer literarischen Gattung zwischen Philosophie, Wissensvermittlung und dramatischer Inszenierung* (Beiträge zur Altertumskunde Band 315), Berlin and Boston, 37–63.

Oikonomopoulou (2015): Katerina Oikonomopoulou, "Rhetoricians on Homer. The *Antinomikon Zētēma* of Plutarch's *Quaestiones Convivales* 9.13", in: Sandrine Dubel, Anne-Marie Favreau-Linder and Estelle Oudot (eds.), *À l'école d'Homère. La culture des orateurs et des sophistes*, Paris, 191–201.

Opsomer (1994): Jan Opsomer, "L'âme du monde et l'âme de l'homme chez Plutarque", in: Manuela García Valdés (ed.), *Estudios sobre Plutarco: ideas religiosas. Actas del III Simposio Internacional sobre Plutarco, Oviedo 30 de abril a 2 de mayo de 1992*, Madrid, 33–49.

Opsomer (1996): Jan Opsomer, "Ζητήματα: structure et argumentation dans les *Quaestiones Platonicae*", in: José A. Fernández Delgado and Francisca Pordomingo Pardo (eds.), *Estudios Sobre Plutarco: Aspectos Formales* (*IV Simposio Espanol Sobre Plutarco, Salamanca, 26 a 28 de Mayo de 1994*), Madrid, 71–83.

Opsomer (2007): Jan Opsomer, "Plutarch on the One and the Dyad", in: Robert Sharples and Richard Sorabji (eds.), *Greek and Roman Philosophy. 100 BC – 200 AD*, London, 379–395.

Opsomer (2009): Jan Opsomer, "M. Annius Ammonius. A philosophical profile", in: Mauro Bonazzi and Jan Opsomer (eds.), *The Origins of the Platonic System. Platonisms of the Early Empire and their Philosophical Context*, Leuven and Walpole, Mass., 123–186.

Osborne (2009), Robin Osborne, *Greece in the Making: 1200–479 BC*, London and New York.

Oudot (2006): Estelle Oudot, "Au commencement était Athènes. Le Panathénaïque d'Ælius Aristide ou l'histoire abolie", *Ktèma* 31, 227–238.

Oudot (2008): Estelle Oudot, "Aelius Aristides and Thucydides: Some Remarks About the *Panathenaic Oration*", in: William V. Harris and Brooke Holmes (eds.), *Aelius Aristides between Greece, Rome, and the Gods*, Leiden and Boston, 31–49.

Oudot (2011): Estelle Oudot, "'Marathon, l'Eurymédon, Platées, laissons-les aux écoles des sophistes!' Les Guerres Médiques au second siècle de notre ère", in: Pierre-Louis Malosse, Marie-Pierre Noël and Bernard Schouler (eds.), *Clio sous le regard d'Hermès. L'utilisation de l'histoire dans la rhétorique ancienne de l'époque hellénistique à l'Antiquité tardive*, Alessandria, 143–157.

Oudot (2012): Estelle Oudot, "'*Elle fut aussitôt antique*'. L'image de la ville d'Athènes sous le Haut-Empire et ses enjeux", in: Sandrine Dubel, Sophie Gotteland and Estelle Oudot (eds), *Éclats de littérature grecque d'Homère à Pascal Quignard, Mélanges offerts à Suzanne Saïd*, Paris, 108–122.

Pace and Volpe Cacciatore (2013): Giovanna Pace and Paola Volpe Cacciatore (eds.), *Gli scritti di Plutarco: tradizione, traduzione, ricezione, commento. Atti del IX Convegno internazionale della International Plutarch Society, Ravello-Auditorium Oscar Niemeyer, 29 settembre-1° ottobre 2011*, Naples.

Pade (2014): Marianne Pade, "The Reception of Plutarch from Antiquity to the Italian Renaissance", in: Mark Beck (ed.), *A Companion to Plutarch*, Malden, Mass. and Oxford, 531–543.

Palyvou (2007): Clairy Palyvou, "La mise en images d'un paysage historique; 'l'unification des sites archéologiques d'Athènes' reconsidérée", http://etudesbalkaniques.revues.org/249 (Accessed 11.11.2016).
Parker (1987): Robert Parker, "Myths of Early Athens", in: Jan Bremmer (ed.), *Interpretations of Greek Mythology*, London, 187–214.
Parker (1989): Robert Parker, "Spartan religion", in: Anton Powell (ed.), *Classical Sparta. Techniques Behind her Success*, London, 142–172.
Patillon (1997): Michel Patillon, *Aelius Theon: Progymnasmata, Texte établi et traduit, avec l'assistance, pour l'arménien, de G. Bolognesi*, Budé, Paris.
Payen (1998): Pascal Payen, "Rhétorique et Géographie dans les *Questions romaines* et *Questions grecques* de Plutarque", in: Pascal Payen (ed.), *Plutarque: Grecs et Romains en Questions* Entretiens d'archéologie et d'Histoire, Saint-Bertrand-de-Comminges, Musée archéologique départemental, 39–73.
Payen (2006): Pascal Payen, "Archaïsme et époque archaïque en Grèce ancienne. Remarques sur la constitution d'une origine", *Ktèma* 31, 17–31.
Payen (2011): Pascal Payen, "'Si l'on supprime les hommes d'action, on n'aura plus d'hommes de lettres'. Thucydide et Polybe dans le corpus des œuvres de Plutarque", in: Pierre-Louis Malosse, Marie-Pierre Noël and Bernard Schouler (eds.), *Clio sous le regard d'Hermès. L'utilisation de l'histoire dans la rhétorique ancienne de l'epoque hellénistique à l'Antiquité tardive*, Alessandria, 13–25.
Payen (2014): Pascal Payen, "Plutarch the antiquarian", in: Mark Beck (ed.), *A Companion to Plutarch*, Malden, Mass. and Oxford, 235–248.
Pearson and Sandbach (1965): Lionel Pearson and Francis H. Sandbach, *Plutarch, Moralia*, Vol. XI, Cambridge Mass. and London.
Pelling (1979): Christopher B.R. Pelling, "Plutarch's method of work in the Roman *Lives*", *JHS* 99, 74–96; repr. with a postscript in: Barbara Scardigli (ed.) (1995), *Essays on Plutarch's Lives*, Oxford, 265–318; repr. with revisions in: Christopher B.R. Pelling (2002), *Plutarch and History: Eighteen Studies*, Swansea, 1–44.
Pelling (1980): Christopher B.R. Pelling, "Plutarch's adaptation of his source material", *JHS* 100: 127–140; repr. in: Christopher B.R. Pelling (2002), *Plutarch and History: Eighteen Studies*, Swansea, 91–115.
Pelling (1985): Christopher B.R. Pelling, "Plutarch and Catiline", *Hermes* 113, 311–329; repr. with revisions in: Christopher B.R. Pelling (2002), *Plutarch and History: Eighteen Studies*, Swansea, 45–63.
Pelling (1986a): Christopher B.R. Pelling, "*Synkrisis* in Plutarch's *Lives*", in: Frederick E. Brenk and Italo Gallo (eds.), *Miscellanea Plutarchea. Atti del I convegno di studi su Plutarco, Roma, 23 novembre 1985, QGFF* 8, Ferrara, 83–96; repr. with postscript in: Christopher B.R. Pelling (2002), *Plutarch and History: Eighteen Studies*, Swansea, 349–363.
Pelling (1986b): Christopher B.R. Pelling, "Plutarch and Roman Politics", in: Ian S. Moxon, James D. Smart and Anthony J. Woodman (eds.), *Past Perspectives: Studies in Greek and Roman Historical Writing. Papers presented at a conference in Leeds, 6–8 April 1983*, Cambridge, 359–368; repr. in: Christopher B.R. Pelling (2002), *Plutarch and History: Eighteen Studies*, Swansea, 207–236.
Pelling (1988): Christopher B.R. Pelling, *Plutarch: Life of Antony*, Cambridge.
Pelling (1989): Christopher B.R. Pelling, "Plutarch: Roman heroes and Greek culture", in: Miriam T. Griffin and Jonathan Barnes (eds.), *Philosophia Togata: Essays on Roman Philosophy and Society*, Oxford, 199–232.
Pelling (1992): Christopher B.R. Pelling, "Plutarch and Thucydides", in: Philip A. Stadter (ed.), *Plutarch and the Historical Tradition*, London, 10–40; repr. with revisions in: Christopher B.R. Pelling (2002), *Plutarch and History: Eighteen Studies*, Swansea, 117–141.

Pelling (1995): Christopher B.R. Pelling, "The Moralism of Plutarch's *Lives*", in: Donald A. Russell, Doreen C. Innes, Harry M. Hine and Christopher B.R. Pelling (eds.), *Ethics and Rhetoric: Classical Essays for Donald Russell on his Seventy-Fifth Birthday*, Oxford, 205–220; repr. in: Christopher B.R. Pelling (2002), *Plutarch and History: Eighteen Studies*, Swansea, 237–251.

Pelling (1997a): Christopher B.R. Pelling, "Plutarch: Roman Heroes and Greek Culture", in: Miriam T. Griffin and Jonathan Barnes (eds.), *Philosophia Togata: Essays on Roman Philosophy and Society*, Oxford, 199–232.

Pelling (1997b): Christopher B.R. Pelling, "Tragical dreamer: some dreams in the Roman historians", *G&R* 44.2, 197–213.

Pelling (1999a): Christopher B.R. Pelling, "'Making myth look like history': Plato in Plutarch's *Theseus–Romulus*", in: Aurelio Pérez Jiménez, José García López and Rosa M. Aguilar (eds.), *Plutarco, Platón y Aristóteles. Actas del V Congreso Internacional de la I.P.S., Madrid-Cuenca, 4–7 de mayo de 1999*, Madrid, 431–443; expanded version in: Christopher B.R. Pelling (2002), *Plutarch and History: Eighteen Studies*, Swansea, 171–195.

Pelling (1999b): Christopher B.R. Pelling, "Epilogue", in: Christina S. Kraus (ed.), *The Limits of Historiography: Genre and Narrative in Ancient Historical Texts*, Leiden, Boston and Cologne, 325–357.

Pelling (2000): Christopher B.R. Pelling, "Rhetoric, *Paideia*, and Psychology in Plutarch's *Lives*", in: Luc Van der Stockt (ed.), *Rhetorical Theory and Praxis in Plutarch. Acta of the IVth International Congress of the International Plutarch Society, Leuven, July 3–6, 1996*, Leuven and Namur, 331–339.

Pelling (2002): Christopher B.R. Pelling, *Plutarch and History: Eighteen Studies*, Swansea.

Pelling (2007a): Christopher B.R. Pelling, "*De malignitate Plutarchi*. Plutarch, Herodotus, and the Persian Wars", in: Emma Bridges, Edith Hall and Peter J. Rhodes (eds.), *Cultural Responses to the Persian Wars. Antiquity to the Third Millennium*, Oxford, 145–164.

Pelling (2007b): Christopher B.R. Pelling, "Sophocles' Learning Curve", in: Patrick J. Finglass, Christopher Collard and Nicholas J. Richarson (eds.), *Hesperos. Studies in Ancient Greek Poetry Presented to M.L. West on his Seventieth Birthday*, Oxford, 204–227.

Pelling (2010): Christopher B.R. Pelling, "Plutarch's Tale of the Two Cities: Do the *Parallel Lives* combine as global histories?", in: Noreen Humble (ed.), *Plutarch's Lives: Parallelism and Purpose*, Swansea, 217–236.

Pelling (2011a): Christopher B.R. Pelling, "Putting the -viv- into 'convivial': The *Table Talk* and the *Lives*", in: Frieda Klotz and Katerina Oikonomopoulou (eds.), *The Philosopher's Banquet. Plutarch's Table Talk in the Intellectual Culture of the Roman Empire*, Oxford, 207–231.

Pelling (2011b): Christopher B.R. Pelling, *Plutarch: Caesar. Translated with an Introduction and Commentary*, Oxford.

Pelling (2011c): Christopher B.R. Pelling, "Herodotus and Samos", *BICS* 54.1, 1–19.

Pelling (2012): Christopher B.R. Pelling, "Plutarch on Roman *Philotimia*", in: Geert Roskam, Maarten De Pourcq and Luc Van der Stockt (eds.), *The Lash of Ambition: Plutarch, Imperial Greek Literature and the Dynamics of Philotimia* (Collection d'Études Classiques, 25), Leuven, Namur, Paris, Walpole, Mass., 55–67.

Pelling (2014): Christopher B.R. Pelling, "Political Philosophy", in: Mark Beck (ed.), *A Companion to Plutarch*, Malden, Mass. and Oxford, 149–162.

Pelling (2016): Christopher B.R. Pelling, "Plutarch the Multiculturalist: is West always Best?", *Ploutarchos n.s.* 13, 33–51.

Pérez Jiménez (1995): Aurelio Pérez Jiménez, "*Proairesis:* las formas de acceso a la vida pública y el pensamiento político de Plutarco", in: Italo Gallo and Barbara Scardigli (eds.), *Teoria e prassi politica nelle opere di Plutarco. Atti del V Convegno plutarcheo, Certosa di Pontignano, 7–9 giugno 1993*, Naples, 363–381.

Pérez Jiménez (1996): Aurelio Pérez Jiménez, “Δεισιδαιμονία: el Miedo a los Dioses en Plutarco”, in: Luc Van der Stockt (ed.), *Plutarchea Lovaniensia. A Miscellany of Essays on Plutarch*, Leuven, 195 – 225.

Pérez-Jiménez (2005): Aurelio Pérez Jiménez, “La estructura literaria de la *Vida de Teseo* de Plutarco”, in: Aurelio Pérez-Jiménez and Frances Titchener (eds.), *Historical and Biographical Values of Plutarch’s Works: Studies devoted to Professor Philip A. Stadter by the International Plutarch Society*, Málaga and Logan, Utah, 341 – 354.

Pérez-Jiménez (2013): Aurelio Pérez-Jiménez, “Interpretación moral de las obras de arte en Plutarco”, in: Germán Santana Henríquez (ed.), *Plutarco y las artes. XI Simposio Internacional de la Sociedad Española de Plutarquistas*, Madrid, 191 – 202.

Perlman (1964): Shalom Perlman, “The Causes and the Outbreak of the Corinthian War”, *CQ* 14, 64 – 81.

Perrin (1920): Bernadotte Perrin, *Plutarch’s Lives*, Vol. IX, Loeb Classical Library, Cambridge, Mass. and London.

Philipp (1984): Günther B. Philipp, “Herakles und die frühgriechische Dichtung. Zu Plut., *De Hdt. mal.* 14, 857E–F”, *Gymnasium* 91, 327 – 340.

Piccirilli (1977): Luigi Piccirilli, “Cronologia relativa e fonti della *Vita Solonis* di Plutarco”, *ASNSP* ser. 3, 7.3, 999 – 1016.

Piccirilli (1983): Luigi Piccirilli, “Introduzione. Bibliografia” and “Commento”, in: Carlo Carena, Mario Manfredini and Luigi Piccirilli, *Plutarco. Le vite di Temistocle e di Camillo*, Milan, IX-LIX, 219 – 350.

Piettre (2012): Renée K. Piettre, “Atome du providence? La *Vie de Timoléon* de Plutarque, ou comment faire de l’histoire avec des atomes”, in: Stella Georgoudi, Renée K. Piettre and Francis Schmidt (eds.), *La raison des signes: présages, rites, destin dans les sociétés de la Méditerranée ancienne*, Leiden, 177 – 219.

Pinnoy (1984): Maurits Pinnoy, “Plutarch’s Comment on Sophocles’ Style”, *QUCC* 45.1, 159 – 164.

Podlecki (1975): Anthony Podlecki, *The Life of Themistocles. A Critical Survey of the Literary and Archaeological Evidence*, Montreal and London.

Porter (1989): Stanley E. Porter, *Verbal Aspect in the Greek of the New Testament with Reference to Tense and Mood* (Studies in Biblical Greek 1), New York.

Preston (2001): Rebecca Preston, “Roman questions, Greek answers: Plutarch and the construction of identity”, in: Simon Goldhill (ed.), *Being Greek under Rome: Cultural Identity, the Second Sophistic and the Development of Empire*, Cambridge, 86 – 119.

Priestley (2014): Jessica Priestley, *Herodotus and Hellenistic Culture. Literary Studies in the Reception of the Histories*, Oxford.

Puech (1992): Bernadette Puech, “Prosopographie des amis de Plutarque”, *ANRW* II.33.6, 4831 – 4893.

Purves (2010): Alex C. Purves, *Space and Time in Ancient Greek Narrative*, Cambridge.

Raaflaub (2009): Kurt A. Raaflaub, “Early Greek Political Thought in Its Mediterranean Context”, in: Ryan K. Balot (ed.), *A Companion to Greek and Roman Political Thought*, Malden, Mass. and Oxford, 37 – 56.

Rademaker (2005): Adriaan Rademaker, *Sophrosyne and the Rhetoric of Self–Restraint. Polysemy and Persuasive Use of an Ancient Greek Term*, Leiden.

Radice (1969): Betty Radice, *The Letters of Pliny the Younger*, London.

Ragogna (2002): Gabriele Ragogna, “Alcune considerazioni sul *De Herodoti malignitate* di Plutarco”, *Patavium* 19, 23 – 41.

Ramelli (2005): Ilaria Ramelli, *I Sette Sapienti. Vite e opinioni nell’edizione di Bruno Snell*, Milan.

Ramon Palerm (1992): Vicente Ramón Palerm, *Plutarco y Nepote: fuentes e interpretacion del modelo biografico plutarqueo*, Zaragoza.

Redondo (1990): Jordi Redondo, "Precisione sobre la lengua de los Moralia", in: Aurelio Pérez Jiménez and Gonzalo Del Cerro Calderón (eds.), *Estudios sobre Plutarco: obra y tradicion. Actas del I Symposion español sobre Plutarco, Fuengirola 1988*, Málaga, 135–139.
Reichel (1909): Georg Reichel, *Quaestiones progymnasmaticae*, Diss. Leipzig, 1909.
Reisigl (2009): Martin Reisigl, "Das Konzept der 'lieux de mémoire' aus rhetorischer und linguistischer Perspektive", in: Benoît Majerus, Sonja Kmec, Michel Margue and Pit Péporté (eds.), *Dépasser le cadre national des "lieux de mémoire" / Nationale Erinnerungsorte Hinterfragt: Innovations méthodologiques, approches comparatives*, Brussels, 117–140.
Reynolds (1988): Joyce Reynolds, "Cities", in: David Braund (ed.), *The Administration of the Roman Empire: 241 BC–AD 193*, Exeter, 15–51.
Rhodes (1988): Peter J. Rhodes, *Thucydides: History II*, Warminster.
Richardson (1992): Nicholas J. Richardson, "Panhellenic cults and panhellenic poets", in: *The Cambridge Ancient History*, Vol. 5, Cambridge, 223–244.
Richer (2007): Nicholas Richer, "The religious system at Sparta", in: Daniel Odgen (ed.), *A Companion to Greek Religion*, Malden, Mass. and Oxford, 236–252.
Richter (2001): Daniel S. Richter, "Plutarch on Isis and Osiris: Text, Cult, and Cultural Appropriation", *TAPA* 131, 191–216.
Richter (2011): Daniel S. Richter, *Cosmopolis. Imagining Community in Late Classical Athens and the Early Roman Empire*, Cambridge.
Ricoeur (1984): Paul Ricoeur, *Time and Narrative*, Vol. 1, transl. K. McLaughlin and D. Pellauer, Chicago.
Rigotti (1989): Francesca Rigotti, *Metafore della politica*, Bologna.
Rijksbaron (1988): Albert Rijksbaron, "The discourse function of the imperfect", in: Albert Rijksbaron, Henk A.J. Mulder and Gerry C. Wakker (eds.), *In the footsteps of Raphael Kühner: Proceedings of the International Colloquium in commemoration of the 150th anniversary of the publication of Raphael Kühner's Ausführliche Grammatik der griechischen Sprache*, ii: *Syntaxe*, Amsterdam, 237–254.
Rijksbaron (2002): Albert Rijksbaron, *The Syntax and Semantics of the Verb in Classical Greek: An Introduction*, Chicago.
Roberts (1897): William Rhys Roberts, "Caecilius of Calacte", *AJPh* 18, 302–312.
Roisman (2003): Joseph Roisman, "The Rhetoric of Courage in the Athenian Orators", in: Ralph M. Rosen and Ineke Sluiter (eds.), *Andreia: Studies in Manliness and Courage in Classical Antiquity* (*Mnemosyne* Supplements 238), Leiden and Boston, 127–143.
Romeri (2002): Luciana Romeri, *Philosophes entre mots et mets. Plutarque, Lucien et Athénée autour de la table de Platon*, Grenoble.
Romm (1994): James Romm, *The Edges of the Earth in Ancient Thought*, Princeton.
Romm and Mensch (2012): James Romm and Pamela Mensch (eds.), *Plutarch: Lives that made Greek history*, Indianapolis and Cambridge.
Rosenmeyer (1992): Thomas G. Rosenmeyer, "Beginnings in Plutarch's *Lives*", *YCS* 29, 205–230.
Rosivach (1987): Vincent J. Rosivach, "Autochthony and the Athenians", *CQ* N.S. 37, 294–305.
Roskam (2002): Geert Roskam, "*A Paideia for the Ruler. Plutarch's Dream of Collaboration Between Philosopher and Ruler*", in: Philip A. Stadter and Luc Van Der Stockt (eds.), *Sage and Emperor. Plutarch, Greek Intellectuals, and Roman Power in the Time of Trajan, 98–117 A.D.*, Leuven, 175–189.
Roskam (2004): Geert Roskam, "Plutarch on Self and Others", *AncSoc* 34, 245–273.
Roskam (2005): Geert Roskam, *On the Path to Virtue. The Stoic Doctrine of Moral Progress and its Reception in (Middle-)Platonism*, Leuven.
Roskam (2009): Geert Roskam, *Plutarch's Maxime cum principibus philosopho esse disserendum. An Interpretation with Commentary*, Leuven.

Roskam (2011): Geert Roskam, "Ambition and Love of Fame in Plutarch's *Lives of Agis, Cleomenes and the Gracchi*", *CPh* 106, 208–225.
Roskam (2014): Geert Roskam, "Ἀποθεωρεῖν / ἀποθεώρησις: A Semasiological Study", *Glotta* 90, 180–191.
Roskam et al. (2012): Geert Roskam, Maarten de Pourcq and Luc Van der Stockt (eds.), *The Lash of Ambition. Plutarch, Imperial Greek Literature and the Dynamics of Philotimia* (Collection d'Études Classiques, 25), Leuven, Namur, Paris and Walpole, Mass.
Russell (1966a): Donald A. Russell, "On Reading Plutarch's *Lives*", *G&R* 13, 139–154; repr. in: Barbara Scardigli (ed.) (1995), *Essays on Plutarch's Lives*, Oxford, 75–94.
Russell (1966b): Donald A. Russell, "The Loeb *Moralia* again", *CR* 16, 180–182.
Russell (1972): Donald A. Russell, *Plutarch*, London.
Russell (1981): Donald A. Russell, *Criticism in Antiquity*, London.
Russell (1993): Donald A. Russell, *Plutarch: Selected Essays and Dialogues. A New Translation*, Oxford.
Saïd (2005): Suzanne Saïd, "Plutarch and the People in the *Parallel Lives*", in: Lukas de Blois, Jeroen Bons, Ton Kessels and Dirk M. Schenkeveld (eds.), *The Statesman in Plutarch's Works, Volume II: The Statesman in Plutarch's Greek and Roman Lives* (*Mnemosyne* Supplements 250.2), Leiden and Boston, 7–25.
Salmeri (2000): Giovanni Salmeri, "Dio, Rome, and the Civic Life of Asia Minor", in: Simon Swain (ed.), *Dio Chrysostom. Politics, Letters, and Philosophy*, Oxford, 53–92.
Santaniello (2000): Carlo Santaniello, "Rapporti fra generi letterari e pubblico nel *Corpus* plutarcheo", in: Italo Gallo and Claudio Moreschini (eds.), *I generi letterari in Plutarco. Atti del VIII Convegno plutarcheo, Pisa, 2–4 giugno 1999* (*Collectanea* 19), Naples, 271–286.
Scheid (2012a): John Scheid, *Plutarch. Römische Fragen. Ein virtueller Spaziergang im Herzen des alten Rom* (Texte zur Forschung 10), Darmstadt.
Scheid (2012b): John Scheid, *À Rome sur les pas de Plutarque*, Paris.
Schenkeveld (1996): Dirk Schenkeveld, "Plutarch's First Table Talk (612E–615C)", in: Luc Van der Stockt (ed.), *Plutarchea Lovaniensia. A Miscellany of Essays on Plutarch*, Leuven, 257–264.
Schenkeveld (1997): Dirk Schenkeveld, "Plutarch's *Table Talks* VIII.4 as a literary work of art", *Mnemosyne* 50, 597–602.
Schepens (2000): Guido Schepens, "Plutarch's View of Ancient Rome: Some Comments on the *Life of Pyrrhus*", in: Leon Mooren (ed.), *Politics, Administration and Society in the Hellenistic and Roman Worlds. Proceedings of the international colloquium, Bertorino, 19–24 July 1997*, Leuven, 349–364.
Schettino (2014): Maria Teresa Schettino, "The Use of Historical Sources", in: Mark Beck (ed.), *A Companion to Plutarch*, Malden, Mass. and Oxford, 417–436.
Schmidt and Fleury (2010): Thomas Schmidt and Pascale Fleury (eds.), *Perceptions of the Second Sophistic and its Times. Regards sur la Seconde Sophistique et son époque*, Toronto.
Schmidt (1999): Thomas S. Schmidt, *Plutarque et les Barbares. La rhétorique d'une image*, Leuven and Namur.
Schmidt (2008): Thomas S. Schmidt, "*Les Questions barbares* de Plutarque: un essai de reconstitution", in: Martine Chassignet (ed.), *L'étiologie dans la pensée antique*, Turnhout, 165–183.
Schmitz (2012): Thomas A. Schmitz, "Sophistic *Philotimia* in Plutarch", in: Geert Roskam, Maarten de Pourcq and Luc van der Stockt (eds.), *The Lash of Ambition. Plutarch, Imperial Greek Literature and the Dynamics of* philotimia (Collection d'Études Classiques, 25), Leuven, Namur, Paris and Walpole, Mass., 69–84.
Schmitz (2014): Thomas A. Schmitz, "Plutarch and the Second Sophistic", in: Mark Beck (ed.), *A Companion to Plutarch*, Malden, Mass. and Oxford, 32–42.

Schorn (2011): Stefan Schorn, "Aristoxenus' Biographical Method", in: Carl A. Huffman (ed.), *Aristoxenus. Text, Translation and Discussion*, New Brunswick and London, 177–221.
Schröder (1990): Stephan Schröder, *Plutarchs Schrift 'De Pythiae oraculis'. Text, Einleitung und Kommentar* (Beiträge zur Altertumskunde Band 8), Stuttgart.
Scott (2010): Michael Scott, *Delphi and Olympia. The Spatial Politics of Panhellenism in the Archaic and Classical Periods*, Cambridge.
Scott (2014): Michael Scott, *Delphi. A History of the Center of the Ancient World*, Princeton.
Scott-Kilvert and Pelling (2011): Ian Scott-Kilvert and Christopher B.R. Pelling, *Plutarch: Rome in Crisis*, London and New York.
Scuderi (1996): Rita Scuderi, "L'incontro fra Grecia e Roma nelle biografie plutarchee di Filopemene e Flaminino", in: Emilio Gabba, Paolo Desideri and Sergio Roda (eds.), *Italia sul Baetis: studi di storia romana in memoria di Fernando Gascó*, Turin, 65–89.
Seavey (1991): William Seavey, "Forensic Epistolography and Plutarch's *de Herodoti malignitate*", *Hellas* 2, 33–45.
Sfameni Gasparro (1996): Giulia Sfameni Gasparro, "Plutarco e la religione delfica: il Dio 'filosofo' e il suo esegeta", in: Italo Gallo (ed.), *Plutarco e la religione. Atti del VI convegno plutarcheo, Ravello, 29–31 maggio 1995*, Naples, 157–188.
Shapiro (1992): H. Alan Shapiro, "Theseus in Kimonian Athens: The Iconography of Empire", *Med. Hist. Rev.* 7, 29–49.
Shapiro (1995): H. Alan Shapiro, "Autochthony and the Visual Arts in Fifth-Century Athens", in: Deborah D. Boedeker and Kurt A. Raaflaub (eds.), *Democracy, Empire, and the Arts in Fifth-Century Athens*, Cambridge, Mass., 127–151.
Sheldrake (2001): Philip Sheldrake, *Spaces for the Sacred. Place, Memory and Identity*, Baltimore.
Sherwin-White (1969): Adrian N. Sherwin-White, "Pliny, the Man and his Letters", *G&R* 16.1, 76–90.
Shipley (1997): Donald R. Shipley, *A Commentary on Plutarch's Life of Agesilaos. Response to Sources in the Presentation of Character*, Oxford.
Sirinelli (2000): Jean Sirinelli, *Plutarque*, Paris.
Skinner (2002): Quentin Skinner, *Visions of Politics, Vol. 1: Regarding Method*, Cambridge.
Sluiter (2005): Ineke Sluiter, "Homer in the Dining Room: An Ancient Rhetorical Interpretation of the Duel between Paris and Menelaus (Plut. *Quaest. Conv.* 9.13)", *CW* 98.4, 379–396.
Smadja (1995): Elisabeth Smadja, "Héraclès, héros et dieu (mythe et histoire)", *Dialogues d'histoire ancienne* 21, 241–246.
Snell (1971): Bruno Snell, *Leben und Meinungen der Sieben Weisen*, Tübingen.
Spawforth (1992): Antony J. Spawforth, "Spartan cults under the Roman Empire: some notes", in: Jan Motyka Sanders (ed.), *Philolakon: Lakonian studies in honour of Hector Catling*, London, 227–238.
Stadter (1975): Philip A. Stadter, "Plutarch's Comparison of Pericles and Fabius Maximus", *GRBS* 16, 77–85.
Stadter (1988): Philip A. Stadter, "The proems of Plutarch's *Lives*", *ICS* 13.2, 275–295.
Stadter (1989): Philip A. Stadter, *A Commentary on Plutarch's Pericles*, Chapel Hill, NC and London.
Stadter (1992): Philip A. Stadter, "Paradoxical Paradigms: Lysander and Sulla", in: Philip A. Stadter (ed.), *Plutarch and the Historical Tradition*, London and New York, 41–55.
Stadter (1996): Philip A. Stadter, "Anecdotes and the Thematic Structure of Plutarchean Biography", in: José A. Fernández Delgado and Francisca Pordomingo Pardo (eds.), *Estudios Sobre Plutarco: Aspectos Formales. IV Simposio Espanol Sobre Plutarco, Salamanca, 26 a 28 de Mayo de 1994*, Madrid, 291–303.

Stadter (2002a): Philip A. Stadter, "Introduction: Setting Plutarch in His Context", in: Philip A. Stadter and Luc Van der Stockt (eds.), *Sage and Emperor. Plutarch, Greek Intellectuals, and Roman Power in the Time of Trajan, 98–117 A.D.*, Leuven, 1–26.

Stadter (2002b): Philip A. Stadter, "Plutarch's *Lives* and their Roman Readers", in: Erik N. Ostenfeld (ed.), *Greek Romans and Roman Greeks: Studies in Cultural Interaction*, Aarhus, 123–135.

Stadter (2004): Philip A. Stadter, "Plutarch: diplomat for Delphi?", in: Lukas de Blois, Jeroen Bons, Ton Kessels and Dirk M. Schenkeveld (eds.), *The Statesman in Plutarch's Works, Volume I: Plutarch's Statesman and His Aftermath – Political, Philosophical, and Literary Aspects* (*Mnemosyne* Supplements 250.1), Leiden, 19–32.

Stadter (2005): Philip A. Stadter, "Plutarch and Apollo of Delphi", in: Rainer Hirsch-Luipold (ed.), *Gott und die Götter bei Plutarch: Götterbilder – Gottesbilder – Weltbilder*, Berlin and New York, 197–214.

Stadter (2007): Philip A. Stadter, "Plutarco e la formazione dell'ideologia traianea", in: Paola Volpe Cacciatore and Franco Ferrari (eds.), *Plutarco e la cultura della sua età. Atti del X Convegno plutarcheo, Fisciano-Paestum, 27–29 ottobre 2005*, Naples, 189–204.

Stadter (2014a): Philip A. Stadter, "Plutarch and Rome", in: Mark Beck (ed.), *A Companion to Plutarch*, Malden, Mass. and Oxford, 11–31.

Stadter (2014b): Philip A. Stadter, "Plutarch's Compositional Technique: The Anecdote Collections and the *Parallel Lives*", *GRBS* 54, 665–686.

Stadter (2015): Philip A. Stadter, *Plutarch and his Roman Readers*, Oxford.

Stadter and Van Der Stockt (2002): Philip A. Stadter and Luc Van Der Stockt (eds.), *Sage and Emperor. Plutarch, Greek Intellectuals, and Roman Power in the Time of Trajan, 98–117 A.D.*, Leuven.

Stein-Hölkeskamp (1999): Elke Stein-Hölkeskamp, "Kimon und die athenische Demokratie", *Hermes* 127, 145–164.

Stephens (2003): Susan A. Stephens, *Seeing Double: Intercultural Poetics in Ptolemaic Alexandria*, Berkeley.

Stoneman (1995): Richard Stoneman, "Naked philosophers: the Brahmans in the Alexander historians and the Alexander Romance", *JHS* 115, 99–114.

Strauss Clay (1989): Jenny Strauss Clay, *The Politics of Olympus: Form and Meaning in the Major Homeric Hymns*, Princeton.

Strauss Clay (2011): Jenny Strauss Clay, *Homer's Trojan Theater: Space, Vision, and Memory in the Iliad*, Cambridge and New York.

Syme (1964): Ronald Syme, *Sallust*, Berkeley, Los Angeles and London.

Swain J.W. (1940): Joseph Ward Swain, "The Theory of the Four Monarchies: Opposition History Under the Roman Empire", *CPh* 35, 1–21.

Swain (1988): Simon C.R. Swain, "Plutarch's Philopoemen and Flamininus", *ICS* 13, 335–347.

Swain (1989a): Simon C.R. Swain, "Plutarch: Chance, Providence and History", *AJP* 110, 272–302.

Swain (1989b): Simon C.R. Swain, "Plutarch's Aemilius and Timoleon", *Historia* 38, 314–334.

Swain (1989c): Simon C.R. Swain, "Plutarch's *De Fortuna Romanorum*", *CQ* 39.2, 504–516.

Swain (1989d): Simon C.R. Swain, "Character change in Plutarch", *Phoenix* 43, 62–68.

Swain (1990a): Simon C.R. Swain, "Plutarch's *Lives* of Cicero, Cato and Brutus", *Hermes* 118.2, 192–203.

Swain (1990b): Simon C.R. Swain, "Hellenic Culture and the Roman Heroes of Plutarch", *JHS* 110, 126–145; repr. in: Barbara Scardigli (ed.) (1995), *Essays on Plutarch's Lives*, Oxford, 229–264.

Swain (1991): Simon C.R. Swain, "Plutarch, Hadrian, and Delphi", *Historia* 40, 318–330.

Swain (1992a): Simon C.R. Swain, "Plutarchan *Synkrisis*", *Eranos* 90, 101–111.

Swain (1992b): Simon C.R. Swain, "Plutarch's Characterization of Lucullus", *RhM* 135, 307–316.

Swain (1996): Simon C.R. Swain, *Hellenism and Empire. Language, Classicism and Power in the Greek World, AD 50–250*, Oxford.
Swain (2000): Simon C.R. Swain (ed.), *Dio Chrysostom. Politics, Letters and Philosophy*, Oxford.
Talamo (2007): Clara Talamo, "Plutarco, Delfi e il Panellenio", in: Paola Volpe Cacciatore and Franco Ferrari (eds.), *Plutarco e la cultura della sua età. Atti del X Convegno plutarcheo, Fisciano – Paestum, 27–29 ottobre 2005*, Naples, 205–220.
Talbert, Scott-Kilvert and Pelling (2005): Richard J.A. Talbert, Ian Scott-Kilvert and Christopher B.R. Pelling (eds.), *Plutarch on Sparta*, London.
Tatum (1996): Jeffrey Tatum, "The regal image in Plutarch's *Lives*", *JHS* 116, 135–151.
Tatum (2010): Jefferey Tatum, "Another Look at *Tyche* in Plutarch's *Aemilius Paulus–Timoleon*", *Historia* 59, 448–461.
Teodorsson (1989/1990/1996): Sven-Tage Teodorsson, *A Commentary on Plutarch's Table Talks*, 3 vols., Göteborg.
Teodorsson (1995): Sven-Tage Teodorsson, "La politica nelle *Questioni Conviviali*", in: Italo Gallo and Barbara Scardigli (eds.), *Teoria e prassi politica nelle opere di Plutarco. Atti del V Convegno plutarcheo (Certosa di Pontignano, 7–9 giugno 1993)*, Naples, 433–437.
Teodorsson (1996): Sven-Tage Teodorsson, "Principles of composition in the *Quaestiones Convivales*", in: José A. Fernández Delgado and Francisca Pordomingo Pardo (eds.), *Estudios Sobre Plutarco: Aspectos Formales (IV Simposio Espanol Sobre Plutarco, Salamanca, 26 a 28 de Mayo de 1994)*, Madrid, 39–47.
Teodorsson (1997): Sven-Tage Teodorsson, "Ethical Historiography. Plutarch's Attitude to Historical Criticism", in: Carlos Schrader, Vicente Ramón Palerm and José Vela; (eds.), *Plutarco y la historia. Actas del V Simposio Español sobre Plutarco. Zaragoza, 20–22 de junio de 1996*, Zaragoza, 439–447.
Teodorsson (2005): Sven-Tage Teodorsson, "Plutarco, innovatore del vocabolario greco", in: Aurelio Pérez Jiménez and Frances Titchener (eds.), *Valori letterari delle opere di Plutarco. Studi offerti al professore Italo Gallo dall'International Plutarch Society*, Málaga and Logan, Utah, 405–418.
Teodorsson (2009): Sven-Tage Teodorsson, "The place of Plutarch in the literary genre of *symposium*", in: José Ribeiro Ferreira, Delfim Leão, Manuel Tröster and Paula Barata Dias (eds.), *Symposion and Philanthropia in Plutarch*, Coimbra, 3–16.
Thalmann (2011): William G. Thalmann, *Apollonius of Rhodes and the Spaces of Hellenism*, Oxford.
Theander (1951): Carl Theander, *Plutarch und die Geschichte*, Lund.
Thiolier (1985): Jean-Claude Thiolier (ed.), *Plutarque. De Gloria Atheniensium*, Paris.
Thomas (1974): Carol G. Thomas, "On the role of the Spartan kings", *Historia* 23, 257–270.
Thonemann (2016): Peter Thonemann, "Croesus and the Oracles", *JHS* 136, 152–167.
Throm (1932): Hermann Throm, *Die Thesis, ein Beitrag zu ihrer Entstehung und Geschichte*, Paderborn.
Tirelli (2005): Aldo Tirelli, *Plutarco: Ad un governante incolto*, Naples.
Titchener (2002): Frances B. Titchener, "Plutarch and Roman(ized) Athens", in: Erik N. Ostenfeld (ed.), *Greek Romans and Roman Greeks: Studies in Cultural Interaction*, Aarhus, 136–141.
Titchener (2009): Frances B. Titchener, "The Role of Reality in Plutarch's *Quaestiones Convivales*", in: José Ribeiro Ferreira, Delfim Leão, Manuel Tröster and Paula Barata Dias (eds.), *Symposion and Philanthropia in Plutarch*, Coimbra, 395–401.
Titchener (2014): Frances B. Titchener, "Fate and Fortune", in: Mark Beck (ed.), *A Companion to Plutarch*, Malden, Mass. and Oxford, 479–487.
Torraca (1998): Luigi Torraca, "Problemi di lingua e stile nei 'Moralia' di Plutarco", *ANRW* II.34.4, 3487–3510.
Toye (2000): David L. Toye, "Plutarch on Poetry, Prose, and *Politeia*", *AncW* 31, 173–181.

Trapp (1990): Michael B. Trapp, "Plato's *Phaedrus* in Second-Century Greek Literature", in: Donald A. Russell (ed.), *Antonine Literature*, Oxford, 141–173.
Trapp (2007): Michael B. Trapp, *Philosophy in the Roman Empire: Ethics, Politics and Society*, Aldershot.
Tribulato (2014): Olga Tribulato, "'Not even Menander would use this word!' Perceptions of Menander's Language in Greek Lexicography", in: Alan H. Sommerstein (ed.), *Menander in Contexts*, New York and London, 199–214.
Tritle (1992): Lawrence A. Tritle, "Plutarch's *Life of Phocion:* An Analysis and Critical Report", *ANRW* II.33.6, 4258–4297.
Tröster (2004): Manuel Tröster, "Hellenism and *Tryphē* in Plutarch's *Life of Lucullus*", in: Lukas de Blois, Jeroen Bons, Ton Kessels and Dirk M. Schenkeveld (eds.), *The Statesman in Plutarch's Works, Volume II: The Statesman in Plutarch's Greek and Roman Lives* (*Mnemosyne* Supplements 250.2), Leiden and Boston, 303–313.
Tsagalis (2008): Christos Tsagalis, "'CEG 594' and Euripides' *Erechtheus*", *ZPE* 162, 9–13.
Tuan (1977): Yi-Fu Tuan, *Space and Place. The Perspective of Experience*, Minneapolis and London.
Tusa Massaro (2000): Laura Tusa Massaro, "Il Labirinto dialogico. Divagazione e scrittura della filosofia nei *Pythikoi logoi* di Plutarco", in: Italo Gallo and Claudio Moreschini (eds.), *I generi letterari in Plutarco. Atti del VIII Convegno plutarcheo, Pisa, 2–4 giugno 1999* (Collectanea 19), Naples, 117–132.
Usher (1993): Stephen Usher, *Demosthenes on the Crown (De Corona)*, Warminster.
Ustinova (2013): Yulia Ustinova, "Modes of Prophecy, or Modern Arguments in Support of the Ancient Approach", *Kernos* 26, 25–44.
Valgiglio (1992): Ernesto Valgiglio, *Plutarco. Gli oracoli della Pizia*, Naples.
Vamvouri Ruffy (2004): Maria Vamvouri Ruffy, *Les Hymnes de Callimaque à la lumière des Hymnes homériques et des hymnes épigraphiques*, Liège.
Vamvouri Ruffy (2011): Maria Vamvouri Ruffy, "Symposium, Physical and Social Health in Plutarch's *Table Talk*", in: Frieda Klotz and Katerina Oikonomopoulou (eds.), *The Philosopher's Banquet: Plutarch's Table Talk in the Intellectual Culture of the Roman Empire*, Oxford, 131–157.
Vamvouri Ruffy (2012): Maria Vamvouri Ruffy, *Les Vertus thérapeutiques du banquet. Médecine et idéologie dans les Propos de Table de Plutarque*, Paris.
Van der Stockt (1990): Luc Van der Stockt, "La peinture, l'histoire et la poésie dans *de gloria Atheniensium* (*Mor.* 346F–347C)", in: Aurelio Pérez Jiménez and Gonzalo Del Cerro Calderón (eds.), *Estudios sobre Plutarco: obra y tradicion. Actas del I Symposion español sobre Plutarco, Fuengirola 1988*, Málaga, 173–177.
Van der Stockt (1992): Luc Van der Stockt, *Twinkling and Twilight: Plutarch's Reflections on Literature*, Brussels.
Van der Stockt (1996): Luc Van der Stockt (ed.), *Plutarchea Lovaniensia. A Miscellany of Essays on Plutarch*, Leuven.
Van der Stockt (1999): Luc Van der Stockt, "A Plutarchan *Hypomnema* on Self-Love", *AJPh* 120, 575–599.
Van der Stockt (2000a): Luc Van der Stockt (ed.), *Rhetorical Theory and Praxis in Plutarch. Acta of the IVth International Congress of the International Plutarch Society, Leuven, July 3–6, 1996*, Leuven and Namur.
Van der Stockt (2000b): Luc Van der Stockt, "Aspects of the Ethics and Poetics of the Dialogue in the *Corpus Plutarcheum*", in: Italo Gallo and Claudio Moreschini (eds.), *I generi letterari in Plutarco. Atti del VIII Convegno plutarcheo, Pisa, 2–4 giugno 1999* (Collectanea 19), Naples, 93–116.
Van der Stockt (2002): Luc Van der Stockt, "Καρπὸς ἐκ φιλίας ἡγεμονικῆς (*Mor.* 814C): Plutarch's Observations on the "old-Boy Network", in: Philip A. Stadter and Luc Van Der Stockt (eds.),

Sage and Emperor: Plutarch, Greek Intellectuals, and Roman Power in the Time of Trajan, 98–117 A.D., Leuven, 115–140.
Van der Stockt (2004): Luc Van der Stockt, "Plutarch in Plutarch: the Problem of the *Hypomnemata*", in: Italo Gallo (ed.), *La biblioteca di Plutarco. Atti del IX Convegno plutarcheo, Pavia, 13–15 giugno 2002*, Naples, 331–340.
Van Hoof (2010): Lieve Van Hoof, *Plutarch's Practical Ethics. The Social Dynamics of Philosophy*, Oxford.
Van Nuffelen (2011): Peter Van Nuffelen, *Rethinking the Gods. Philosophical Readings of Religion in the Post-Hellenistic Period*, Oxford.
Vasunia (2002): Phiroze Vasunia, "Plutarch and the Return of the Archaic", in: Anthony J. Boyle and William J. Dominik (eds.), *Flavian Rome. Culture, Image, Text*, Leiden and Boston, 369–389.
Vegetti (1989): Mario Vegetti, *L'etica degli antichi*, Rome and Bari.
Verdegem (2010): Simon Verdegem, *Plutarch's Life of Alcibiades: Story, Text and Moralism*, Leuven.
Vernière (1990): Yvonne Vernière, "La théorie de l'inspiration prophétique dans les *Dialogues pythiques* de Plutarque", *Kernos* 3, 359–366.
Vetta (2000): Massimo Vetta, "Plutarco e il 'genere simposio'", in: Italo Gallo and Claudio Moreschini (eds.), *I generi letterari in Plutarco. Atti del VIII Convegno plutarcheo, Pisa, 2–4 giugno 1999* (*Collectanea* 19), Naples, 217–229.
Vicente Sánchez (2005): Ana Vicente Sánchez, "Plutarco, *Sobre si es más útil el agua o el fuego:* una tesis progimnasmática", in: Montserrat Jufresa, Francesca Mestre, Pilar Gómez and Pau Gilabert (eds.), *Plutarc a la seva època: Paideia i societat. Actas del VIII simposio Espanol sobre Plutarco, Barcelona, 6–8 de noviembre de 2003*, Barcelona and Málaga, 507–515.
Vicente Sánchez (2009): Ana Vicente Sánchez, "Influencia de los *progymnasmata* en la composición de los *symposia* de Plutarco: el caso de *El Banquete de los Siete Sabios*", in: José Ribeiro Ferreira, Delfim Leão, Manuel Tröster and Paula Barata Dias (eds.), *Symposion and Philanthropia in Plutarch*, Coimbra, 75–85.
Vischer (1965): Rüdiger Vischer, *Das einfache Leben. Wort- und motivgeschichtliche Untersuchungen zu einem Wortbegriff der antiken Literatur*, Göttingen.
Volpe Cacciatore (2007): Paola Volpe Cacciatore, "Il concetto di ἁρμονία nel *de tranquillitate animi*", in: Alberto Bernabé and Ignacio Rodriguez Alfageme (eds.), *Φίλου σκιά. Studia philologiae in honorem Rosae Aguilar ab amicis et sodalibus dicata*, Madrid, 97–102.
Wälchli (2003): Philipp Wälchli, *Studien zu den literarischen Beziehungen zwischen Plutarch und Lukian, ausgehend von Plutarch: De genio Socratis und Lukian: Philopseudeis*, Leipzig.
Von Arnim (1898): Hans Von Arnim, *Leben und Werke des Dio von Prusa*, Berlin.
Walker A.D. (1993): Andrew D. Walker, "*Enargeia* and the spectator in Greek historiography", *TAPA* 123, 353–377.
Walker (1995): Henry J. Walker, *Theseus and Athens*, New York and Oxford.
Walsh (1992): Joseph J. Walsh, "Syzygy, theme and history: a study in Plutarch's *Philopoemen* and *Flamininus*", *Philologus* 136, 208–233.
Wardman (1967): Alan Wardman, "Description of personal appearance in Plutarch and Suetonius. The use of statues as evidence", *CQ* 17.1, 414–420.
Wardman (1974): Alan Wardman, *Plutarch's Lives*, London.
Warf (2010): Barney Warf, "Relative/Relational Space", in: Barney Warf (ed.), *Encyclopedia of Geography*, Thousand Oaks, CA, 2403–2406.
Warf and Arias (2009): Barney Warf and Santa Arias (eds.), *The Spatial Turn. Interdisciplinary Perspectives*, London and New York.
Waterfield (1998): Robin Waterfield (transl.), *Plutarch. Greek Lives: A Selection of Nine Greek Lives*, with introduction and notes by Philip Stadter, Oxford and New York.

Waterfield (1999): Robin Waterfield (transl.), *Plutarch. Roman Lives: A Selection of Eight Roman Lives*, with introduction and notes by Philip Stadter, Oxford and New York.
Watkins (1984): Oscar D. Watkins, *A Commentary on Plutarch's Life of Pompey, chapters 1–46.4*, DPhil. Diss., University of Oxford.
Weissenberger (1895): Burkard Weissenberger, *Die Sprache Plutarchs von Chaeronea und die pseudoplutarchischen Schriften*, Straubing 1895; transl. in Italian by: Giovanni Indelli and Italo Gallo (1994), *La lingua di Plutarco di Cheronea e gli scritti pseudoplutarchei*, Naples.
Whitehead (1983): David Whitehead, "Competitive Outlay and Community Profit: φιλοτιμία in Democratic Athens", *C&M* 34, 55–74.
Whitehead (1993): David Whitehead, "Cardinal Virtues: The Language of Public Approbation in Democratic Athens", *C&M* 44, 37–75.
Whitmarsh (2001a): Tim Whitmarsh, *Greek Literature and the Roman Empire. The Politics of Imitation*, Oxford.
Whitmarsh (2001b): Tim Whitmarsh, "'Greece is the World': exile and identity in the Second Sophistic", in: Simon Goldhill (ed.), *Being Greek under Rome. Cultural Identity, the Second Sophistic and the Development of Empire*, Cambridge, 269–305.
Whitmarsh (2002): Tim Whitmarsh, "Alexander's Hellenism and Plutarch's textualism", *CQ* 52, 174–192.
Whitmarsh (2005): Tim Whitmarsh, "Quickening the Classics. The Politics of Prose in Roman Greece", in: James I. Porter (ed.), *Classical Pasts. The Classical Traditions of Greece and Rome*, Princeton, 353–374.
Whitmarsh (2010): Tim Whitmarsh, "Thinking Local", in: Tim Whitmarsh (ed.), *Local Knowledge and Microidentities in the Imperial Greek World*, Cambridge, 1–16.
Whitmarsh (2013): Tim Whitmarsh, *Beyond the Second Sophistic. Adventures in Greek Postclassicism*, Berkeley.
Wilamowitz-Möllendorf (1922): Ulrich von Wilamowitz-Möllendorf, "Plutarch als Biograph", *Reden und Vorträge* ii (5), 247–279.
Williams (1973): Bernard Williams, "The Analogy of City and Soul in Plato's *Republic*", in: Edward N. Lee, Alexander P.D. Mourelatos and Richard Rorty (eds.), *Exegesis and Argument. Studies in Greek Philosophy Presented to Gregory Vlastos*, Assen, 196–206.
Wright et al. (1974): George E. Wright, Charles L. Redman and Jean Perrot, "The Tell: Basic Unit for Reconstructing Complex Societies of the Near East", in: Charlotte B. Moore (ed.), *Reconstructing Complex Societies: An Archaeological Colloquium, Supplement to the Bulletin of the American Schools of Oriental Research* 20, 123–143.
Wycherley (1978): Richard E. Wycherley, *The Stones of Athens*, Princeton.
Xenophontos (2010): Sophia Xenophontos, "The complementarity of Plutarch's Πῶς δεῖ τὸν νέον ποιημάτων ἀκούειν and Περὶ τοῦ ἀκούειν: Two educational 'sessions' (re)visited", *Rosetta* 8.5, 164–185.
Xenophontos (2012): Sophia Xenophontos, "Περὶ ἀγαθοῦ στρατηγοῦ: Plutarch's *Life* of Fabius Maximus and the Ethics of Generalship", *Hermes* 140.2, 160–183.
Xenophontos (2016): Sophia Xenophontos, *Ethical Education in Plutarch: Moralising Agents and Contexts*, Berlin.
Zaccarini (2015): Matteo Zaccarini, "The Return of Theseus to Athens: A Case Study in Layered Tradition and Reception", *Histos* 9, 174–198.
Zadorojnyi (1997): Alexei V. Zadorojnyi, "The Roman Poets in Plutarch's Stories", in: Carlos Schrader, Vicente Ramón Palerm and José Vela (eds.), *Plutarco y la historia. Actas del V Simposio Español sobre Plutarco, Zaragoza, 20–22 de junio de 1996*, Zaragoza, 497–506.
Zadorojnyi (2005a): Alexei V. Zadorojnyi, "Stabbed with large pens: trajectories of literacy in Plutarch's *Lives*", in: Lukas de Blois, Jeroen Bons, Ton Kessels and Dirk M. Schenkeveld

(eds.), *The Statesman in Plutarch's Works, Volume II: The Statesman in Plutarch's Greek and Roman Lives* (*Mnemosyne* Supplements 250.2), Leiden and Boston, 113 – 137.
Zadorojnyi (2005b): Alexei V. Zadorojnyi, "Plutarch and the Forbidden City: *Demosthenes* 1 – 2", in: Aurelio Pérez-Jiménez and Frances Titchener (eds.), *Historical and Biographical Values of Plutarch's Works: Studies devoted to Professor Philip A. Stadter by the International Plutarch Society*, Málaga and Logan, Utah, 493 – 512.
Zadorojnyi (2006): Alexei V. Zadorojnyi, "King of his Castle: Plutarch, *Demosthenes* 1 – 2", *PCPhS* 52, 102 – 127.
Zadorojnyi (2012): Alexei V. Zadorojnyi, "*Mimesis* and the (Plu)past in Plutarch's *Lives*", in: Jonas Grethlein and Christopher B. Krebs (eds.), *Time and Narrative in Ancient Historiography: the 'Plupast' from Herodotus to Appian*, Cambridge, 175 – 198.
Zadorojnyi (2013): Alexei V. Zadorojnyi, "Shuffling surfaces: epigraphy, power, and integrity in the Greco-Roman narratives", in: Peter Liddel and Polly Low (eds.), *Inscriptions and their Uses in Greek and Latin* Literature, Oxford, 365 – 386.
Zanker (1981): Graham Zanker, "*Enargeia* in the ancient criticism of poetry", *RhM* 124, 297 – 311.
Zanker (1995): Paul Zanker, *Die Maske des Sokrates. Das Bild des Intellectuellen in der antiken Kunst*, Munich.
Ziegler (1949): Konrat Ziegler, *Plutarchos von Chaironeia*, Stuttgart, 2nd edn. 1964; repr. in *RE* 21.1 (1951), 636 – 962.
Ziegler (1965): Konrat Ziegler, *Plutarco*, ed. Bruno Zucchelli, transl. Maria Rosa Zancan Rinaldini, Brescia.
Ziolkowski (1981): John E. Ziolkowski, *Thucydides and the Tradition of Funeral Speeches at Athens*, New York.
Zuiderhoek (2009): Arjan Zuiderhoek, *The Politics of Munificence in the Roman Empire. Citizens, Elites and Benefactors in Asia Minor*, Cambridge.

Index of subjects

Index of ancient and modern authors

Index of passages

The section numbering for Plutarch's *Lives* follows that of the Loeb Classical Library. Where the Teubner edition has been used, the sections are marked with (T.)